MICROBIAL FERMENTATIONS IN NATURE AND AS DESIGNED PROCESSES

MICROBIAL FERMENTATIONS IN NATURE AND AS DESIGNED PROCESSES

Edited by

CHRISTON J. HURST
Cincinnati, OH, USA
Universidad del Valle, Santiago de Cali, Valle, Colombia

WILEY

This edition first published 2023

Registered Offices
John Wiley & Sons, Inc., 111 River Street, Hoboken, NJ 07030, USA
John Wiley & Sons Ltd, The Atrium, Southern Gate, Chichester, West Sussex, PO19 8SQ, UK

For details of our global editorial offices, customer services, and more information about Wiley products visit us at www.wiley.com.

Wiley also publishes its books in a variety of electronic formats and by print-on-demand. Some content that appears in standard print versions of this book may not be available in other formats.

Library of Congress Cataloging-in-Publication Data applied for

[ISBN: 9781119849971; ePDF: 9781119849988; epub: 9781119849995; oBook: 9781119850007]

Cover Design: Wiley
Cover Images: Courtesy of Keith Thomas; © Connor Ki, CC BY-SA 4.0,/Wikimedia Commons; Ibex73, CC BY-SA 4.0/ Wikimedia Commons; Aviv Hod, CC BY 3.0/Wikimedia Commons; Kagor at the Ukrainian language Wikipedia, CC BY-SA 3.0/Wikimedia Commons; Basotxerri/Wikimedia Commons

Set in 10/12pt Times by Straive, Pondicherry, India

DEDICATION

Daniel Yee-Chak Fung was born in Hong Kong on May 15, 1942 and loving support from his family enabled Daniel to emerge from the Pacific War with a tremendous sense of optimism about life. Daniel eventually began his university training with a scholarship at the International Christian University in Tokyo, Japan, where he received a Bachelor of Arts Degree during 1965. He next earned a Master of Science degree in Public Health at the University of North Carolina, Chapel Hill during 1967. Soon afterward, during 1969 he was awarded a Doctor of Philosophy degree from Iowa State University. Daniel then found a place that truly felt like home when he became a professor of food science at Kansas State University in Manhattan, Kansas. He also found humor in explaining to his relatives that yes, he was at a university in Manhattan, but it was in the middle of the Kansas prairie and not in New York City. Daniel's professional interest was to understand and explain the microbiology of food science. He earned tremendous international recognition for his efforts on the rapid detection of microbial contamination in food. Although, perhaps Daniel had the most fun when teaching a course on authentic Chinese cooking in what he described as small-town USA.

I had the extremely good fortune to meet Daniel around 1990 during the annual meetings of the American Society for Microbiology. From then onward, I hoped to see him each time that I attended a microbiology conference. Daniel was one of the most enthusiastic people who ever could be imagined. His positive attitude constantly was encouraging to everyone around him. His smile and kindness seemed to brighten any room.

Eventually, many years ago I came to guess that something was wrong with Daniel's health when he did not return a telephone call. I then spoke with the chairman of his department and learned that Daniel was in a group home for Alzheimer's patients. I sent to Daniel a greeting card and included a photograph of myself with my name on the back of the photograph. I guessed that Daniel might recognize my face as being familiar even if remembering my name and the circumstances of our acquaintance might be challenging. Unfortunately, my good friend Daniel passed away on December 1, 2019.

Knowing Daniel was a great honor. The memory of his presence remains with me as a source of inspiration and I do very proudly dedicate my efforts on this book to him.

Daniel Yee-Chak Fung 1942–2019

CONTENTS

LIST OF FIGURES

LIST OF CONTRIBUTORS

Lourdes Georgina Michelena Álvarez, ICIDCA-Instituto Cubano de Investigaciones de los Derivados de la caña de azúcar. Ciudad de La Habana, Cuba

Roberto Arredondo Valdés, Faculty of Chemical Sciences of the Autonomous University of Coahuila, Coahuila, México

Victoria Bell, Faculty of Pharmacy, University of Coimbra; Pólo das Ciências da Saúde, Azinhaga de Santa Comba, Coimbra, Portugal

HD Bhimani, Division of Microbial & Environmental Biotechnology, Navsari Agricultural University, Gujarat, India

Simana Bora, Department of Physics, Silapathar College, Silapathar, Department of Botany, Bahona College, Assam, India

François Bourdichon, Facoltà di Scienze Agrarie, Alimentarie Ambientali, Università Cattolica Del Sacro Cuore, Piacenza, Italy

Magdalena Calusinska, Environmental Research and Innovation Department, Luxembourg Institute of Science and Technology, Belvaux, Luxembourg

Nathaly Cancino-Padilla, Centro Regional de Investigación Carillanca; Instituto de Investigaciones Agropecuarias INIA, Vilcún, La Araucanía, Chile

Sagarika Ekanayake, Department of Biochemistry, Faculty of Medical Sciences, University of Sri Jayewardenepura, Nugegoda, Sri Lanka

Dania Alonso Estrada, Faculty of Chemical Sciences of the Autonomous University of Coahuila, Coahuila, México

Evelyn Faife Pérez, ICIDCA-Instituto Cubano de Investigaciones de los Derivados de la caña de azúcar. Ciudad de La Habana, Cuba

Tito H. Fernandes, CIISA - Centre for Interdisciplinary Research in Animal Health, Faculty of Veterinary Medicine, University of Lisbon, Lisboa, Portugal.
Former CEIL, Lúrio University, Nampula, Mozambique

Alessandra Fontana, Facoltà di Scienze Agrarie, Alimentarie Ambientali, Università Cattolica Del Sacro Cuore, Piacenza, Italy

Gomathy M, Department of Agricultural Microbiology, Tamil Nadu Agricultural University, Tamil Nadu, India

Xavier Goux, Environmental Research and Innovation Department, Luxembourg Institute of Science and Technology, Belvaux, Luxembourg

José Guina, Instituto Superior de Estudos Universitários, Polytechnic University, Nampula, Mozambique

Christon J. Hurst, Cincinnati, OH, USA; Universidad del Valle, Santiago de Cali, Valle del Cauca, Colombia

Anna Iliná, Faculty of Chemical Sciences of the Autonomous University of Coahuila, Coahuila, México.

Priya John, Department of Plant Pathology, Navsari Agricultural University, Gujarat, India

Marta Laranjo, MED-Mediterranean Institute for Agriculture, Environment and Development & CHANGE-Global Change and Sustainability Institute, IIFA-Institute for Advanced Studies and Research, Universidade de Évora, Évora, Portugal

Li Liu, Department of East Asian Languages and Cultures, Stanford University Stanford, CA, USA

Tong Liu, Department of Forest Ecology and Management, Swedish University of Agricultural Sciences, Umeå, Sweden

Uma Maheswari, Department of Agricultural Microbiology, Tamil Nadu Agricultural University, Tamil Nadu, India

Manoj M, Department of Microbiology, Navsari Agricultural University, Gujarat, India

José Luis Martínez-Hernández, Faculty of Chemical Sciences of the Autonomous University of Coahuila, Coahuila, México

Erin A. McKenney, Department of Applied Ecology, North Carolina State University, Raleigh, NC, USA

Lorenzo Morelli, Facoltà di Scienze Agrarie, Alimentarie Ambientali, Università Cattolica Del Sacro Cuore, Piacenza, Italy

Arianna Núñez Caraballo, Autonomous University of Coahuila, Coahuila, México

Nayra Ochoa Viñals, Faculty of Chemical Sciences of the Autonomous University of Coahuila, Coahuila, México

Ken Oda, National Research Institute of Brewing, Higashi-hiroshima, Hiroshima, Japan

Vania Patrone, Facoltà di Scienze Agrarie, Alimentarie Ambientali, Università Cattolica Del Sacro Cuore, Piacenza, Italy

Rajakumar D, Department of Agronomy, Tamil Nadu Agricultural University, Tamil Nadu, India

Anu S. Rajan, Department of Agricultural Microbiology, Kerala Agricultural University, Kerala, India

Ramya P, Department of Agricultural Microbiology, Tamil Nadu Agricultural University, Tamil Nadu, India

Sabarinathan KG, Department of Agricultural Microbiology, Tamil Nadu Agricultural University, Tamil Nadu, India

Dilip Saikia, Department of Physics, Silapathar College, Silapathar, Department of Botany, Bahona College, Assam, India

Cassandra Suther, Department of Food Science, University of Massachusetts Amherst, Amherst, MA, USA

Norio Takeshita, Microbiology Research Center for Sustainability (MiCS), Faculty of Life and Environmental Sciences, University of Tsukuba, Tsukuba, Japan

Dilip Tamang, Kakajan Govt. M.E. School, Assam, India

Theradimani M, Department of Plant Pathology, Tamil Nadu Agricultural University, Tamil Nadu, India

Keith Thomas, Brewlab Ltd., University of Sunderland, England

Emilio M. Ungerfeld, Centro Regional de Investigación Carillanca; Instituto de Investigaciones Agropecuarias INIA, Vilcún, La Araucanía, Chile

Nelson Vera-Aguilera, Centro Regional de Investigación Carillanca; Instituto de Investigaciones Agropecuarias INIA, Vilcún, La Araucanía, Chile

Zwi G. Weinberg, Forage Preservation and By-Products Research Unit, Agricultural Research Organization, The Volcani Institute, Rishon Le Zion, Israel (Retired)

Maria Westerholm, Department of Molecular Sciences, Swedish University of Agricultural Sciences, Uppsala, Sweden

Tharanee Wijayaratne, Department of Biochemistry, Faculty of Medical Sciences, University of Sri Jayewardenepura, Nugegoda, Sri Lanka

PREFACE

Microbial Fermentations in Nature and as Designed Processes
Christon J. Hurst, Editor

Fermentation is one of the most important metabolic tools that biology has developed and microorganisms do in many ways seem to have become the true masters of fermentative metabolism. There are times when microbial fermentation processes seem to occupy nature's center stage, although microbial fermentation processes often function less obviously by serving as part of biology's stabilizing foundation and by providing the backstage support for more noticeable activities.

Animals and plants very importantly depend upon their own fermentative capabilities at times and in places where neither aerobic nor anaerobic respiration are energetically suitable. Those capabilites are overshadowed by the extensive and reliably successful symbiotic interactions which the same animals and plants have developed with their fermenting microbial partners.

Each of the fermentative microbial functions evolved to fit an energetic opportunity and each function has ecological value. It is important for us to acknowledge those values and understand the balance that exists between beneficial versus detrimental perspectives of microbial fermentation. For example, fermentative recycling processes which are so crucial in soil and aquatic sediments, and which also have a necessary role in the gastrointestinal tract of animals, can seem detrimental when they contribute to disease in animals and plants.

This book provides a summary of our understanding with regard to the commonalities and distinctions of aerobic versus anaerobic fermentations as performed by microorganisms. The authors explain how animals, including ruminants and termites, have evolved to use their symbiont fermentative microbes as tools for deconstruction of complex polysaccharides, and how we have adopted those natural technologies into our designed processes for anaerobic degradation of lignocellulosic materials. The important role of rhizosphere microbes that facilitate availability of inorganic and organic phosphates for plants also is explained. Those phosphates get stored in the plant's seeds. And, after animals ingest the seeds, enzymes produced by symbiotic microbial fermentation allow the animals to utilize their dietary phosphates.

The authors provide understanding of the ways that fermentation has been harnessed from prehistoric times to the present for food processing and food preservation, and that technology valuably helps to sustain our species by allowing seasonal harvests to provision us throughout the year. The authors also explain how natural microbial activities have been similarly harnessed for creation of fermented fodder that sustains many of our domesticated agricultural animals.

During my efforts to understand the microbiology of food fermentations, I have asked myself several times if using microbial fermentation to produce and preserve food items may represent a collaborative genetic selection process. The microbes which are prevalent in spontaneous fermentations of food would have selected for a human population which could survive ingesting those microbes. Similarly, those microbes which are spontaneously present in the silage materials that we produce as fodder for our ruminant livestock would have exerted a selection process on our livestock. In cooperative response, we have learned to design food and silage fermentation processes in ways that would favor growth of the most easily survivable microbial populations. Those microbial populations have become our symbionts. That collaborative selection process has not been perfect as evidenced by the opportunistic pathogenicity associated with some of the microbes that exist in fermented food products and fermented fodder.

This book also offers insight into the ways that we have learned to use microbial fermentations as an engineering tool for producing chemicals, including enzymes and pharmaceuticals, which improve the health of ourselves and our domesticated animals. Additionally, the book offers some insight into possible future research directions for the field of applied microbial fermentation that will help to advance agriculture and industry.

I wish to thank Scott Robbins and Andrea Siefring-Robbins who own Urban Stead Cheese, which is an artisanal cheese company in Cincinnati, Ohio, for very kindly having allowed me to watch their crew perform the almost magical process by which microbial fermentation produces cheddar cheese.

This book is the twenty-second and final microbiology volume that I will have edited during a period of thirty four and a half years, beginning in February of 1989 when I started organizing the first volume. Those volumes have contained in total five hundred and thirty eight chapters of which I wrote fifty two. I have enjoyed the honor of representing the six hundred and eighty seven colleagues who joined me on this long interesting adventure as authors and coauthors, especially those who also joined me as coeditors, and I offer my thanks to each of them. Edited books are collaborative efforts similar to good fermentations, intended to share collective knowledge with an appreciative audience. Many of those authors and coauthors kindly have contributed to more than a single book project. In particular, I thank Professor Gideon Wolfaardt who graciously always agreed to participate each of the five times he was asked to write a chapter that would help us present the unfolding story of science. Gideon, collaborative projects have success because of the helpful and reliable efforts made by people such as you. A dream of mine would be to invite all of those colleagues to meet together in a large pub, I would buy a pint of nicely fermented beer for each, and then I would raise my glass to them and say "Together, we have accomplished something very good". That could not happen, and many contributors to the earlier books have since passed away. Instead, I will think of them and say these lines from an old song,

> "So fill to me the parting glass ... Good night and joy be to you all".

CHRISTON J. HURST IN 2022

CHRISTON J. HURST
Cincinnati, Ohio

SECTION I

AN INTRODUCTION TO MICROBIAL FERMENTATION

CHAPTER 1

MICROBIAL FERMENTATION: UNDERSTANDING AND USING ONE OF NATURE'S TOOLS

CHRISTON J. HURST[1,2]
[1]Cincinnati, OH, USA
[2]Universidad del Valle, Santiago de Cali, Valle del Cauca, Colombia

CONTENTS

Microbial Fermentations in Nature and as Designed Processes, First Edition. Edited by Christon J. Hurst.

1.1 INTRODUCTION TO FERMENTATION PROCESSES

When we consider the ways in which living beings gain energy by degrading organic compounds, most of us first think of the aerobic respiration process which relies upon oxygen as its terminal electron acceptor and generates water molecules. Nearly all of the eukaryotic species that possess mitochondria, including many groups of microorganisms, use aerobic respiration with oxygen as their electron acceptor. Humans gain almost all of our energy by that process.

There are groups of microbes that survive and thrive under conditions where the presence of oxygen is limited. Those microbes often will degrade organic compounds by using anaerobic respiration processes which rely upon either nitrate, sulfate, or elemental sulfur as terminal electron acceptors (Lecomte et al. 2018). Microbes additionally may use metals and metalloids as electron acceptors, and that can occur under aerobic as well as anaerobic conditions (Hurst 2021).

Metabolic fermentation processes degrade organic compounds in ways that use either the same organic molecule, or a different organic molecule, as the terminal electron acceptor. Metabolic fermentation generates a wide range of products including organic acids and their conjugate bases, alcohols, aldehydes, ketones, carbon dioxide, and hydrogen. The amount of energy that can be gained by using metabolic fermentation to catabolize carbohydrates, lipids, and proteins is low in comparison with the amount of energy that could be gained if the organism were able to process those same compounds by either aerobic or anaerobic respiration. The energetic inefficiency of using fermentation to degrade organic materials is associated with the fact that potential energy remains in the molecular bonds of the organic products. Among the particularly well-known examples of metabolic fermentation processes are those which rely upon pyruvic acid as final electron acceptor to produce lactic acid, and the use of acetaldehyde as final electron acceptor which produces ethanol.

Macroorganisms do at times use metabolic fermentation processes as a supplemental means for obtaining energy. Animals, including humans, rely upon fermentation to produce lactic acid within our muscle tissues when those tissues are stressed by a shortage of intracellular oxygen. Plants growing in soil use fermentation processes in their root tissues when the surrounding soil is saturated with water and the level of oxygen supplied from photosynthesis performed in aerial parts of the plants is insufficient to supply the metabolic needs of those root tissues. Rice plants very notably can produce acetaldehyde and ethanol as a short-term emergency survival mechanism during darkness and oxygen deprivation, as might occur when the soil is completely flooded with turbid water (Boamfa et al. 2003).

Microorganisms living freely in the environment will use metabolic fermentation processes to satisfy their energy needs when the conditions of their surroundings are unfavorable for supporting either aerobic or anaerobic respiration. There additionally are symbiotic partnerships by which macroorganisms serve as physical hosts for many of the microbial fermenters and these are interdependencies that enable the macroorganisms to gain benefits from their fermenting microbial partners. Examples of such beneficial partnerships are the microbial fermentations that naturally occur in the gastrointestinal tracts of animals (Mississippi State University 2022) and humans (Oliphant and Allen-Vercoe 2019). Coevolution has developed those gastrointestinal processes as a way of helping both the animals and microbes to optimally benefit from the ingested food. Our developments in food technology include using *Propionibacterium* and *Anaerotignum propionicum* (previously *Clostridium propionicum*) for the propionic acid fermentations of cheese, and we understand that the natural ecology of those microbes includes residence in the rumen and intestinal tract, as well as on the skin of mammals (Ciani et al. 2010). Plants also have many types of partnerships with fermenting microbes, with an example being the fact that bacteria in the rhizosphere use metabolic fermentation processes to generate organic acids which assist with the availability of phosphates, as will be described a bit later in this chapter.

Many of the natural microbial fermentation processes that are so important to putrefaction are proteolytic and they serve by liberating amino acids from proteins. Our designed microbial fermentation processes that create miso and soy sauce from soybeans, and those processes that produce meat products like ham and salami, are examples of similarly using proteolytic microbial fermentation to liberate amino acids. And yet, there are unfortunate times when gastrointestinal microbes that naturally represent those same nutrient recycling processes cause disease (Diether and Willing 2019) including associations with intestinal infections, peritoneal infections, and gangrene (Ciani et al. 2010). Fermentations which produce amino acids are termed amino acid fermentations.

We additionally have applied our skills at technological design processes to harness some of the other microbial metabolic fermentation mechanisms. The numerous applications of these processes prominently include the preservation of food that is intended for us to consume (Tables 1.1 and 1.2).

The microbial ecology of food preservation greatly depends upon fermentations which accomplish anaerobic conversions of sugars into acids and ethanol. As you already may have guessed, fermentations, and the microbes which perform those metabolic activities, often are named descriptively based on the desired product. Acetic acid, butyric acid, lactic acid, and propionic acid, respectively, represent products from acetic acid fermentations, butyric acid fermentations, lactic acid fermentations, and propionic acid fermentations. The fermentations which typically produce ethanol are termed either alcoholic or ethanolic fermentations. All of these acids and the ethanol are chemical weapons that fermenting microbes use in an effort to control the ecology of their environment and these chemical products are important to the microbial successions that occur during fermentation processes.

Anaerobic lactic acid fermentations are performed by lactic acid bacteria and also by some yeasts. These fermentation processes are described as either homolactic or heterolactic. Homolactic fermentation converts one molecule of glucose initially into two molecules of pyruvate, then converts those into two molecules of lactic acid. Heterolactic fermentation performs that same conversion of glucose into lactic acid, but then further converts one of the lactic acid molecules into a

TABLE 1.1 Examples of bacteria that are present in food fermentations.

Genus	Species	Primary Material Being Fermented	Product	Geographical Origin
Acetobacter	*aceti*	Milk	Fermented beverage (Kefir)	North Caucasus
Acetobacter	*aceti*	Palm wine	Palm wine vinegar	Philippines
Acetobacter	*aceti*	Previously alcoholic fermented coconut water (liquid from coconuts plus sugar)	Coconut water vinegar	Philippines
Acetobacter	*aceti*	Previously alcoholic fermented fruit juices	Vinegar	Global
Acetobacter	*aceti*	Previously alcoholic fermented molasses (Molasses beer)	Molasses vinegar	Global
Acetobacter	*cerevisiae*	Previously alcoholic fermented fruit juices	Vinegar	Global
Acetobacter	*fabarum*	Cacao seeds	Cocoa beans	Ghana
Acetobacter	*fabarum*	Milk	Fermented beverage (Kefir)	North Caucasus
Acetobacter	*lovaniensis*	Milk	Fermented beverage (Kefir)	North Caucasus
Acetobacter	*lovaniensis*	Sugar	Fermented beverage (Tibicos, also called Water Kefir)	Global
Acetobacter	*malorum*	Previously alcoholic fermented fruit juices	Vinegar	Global
Acetobacter	*oeni*	Previously alcoholic fermented fruit juices	Vinegar	Global
Acetobacter	*orientalis*	Milk	Fermented beverage (Kefir)	North Caucasus
Acetobacter	*orleanensis*	*Vanilla* pods	Vanilla	Mexico
Acetobacter	*pasteurianus*	Cacao seeds	Cocoa beans	Equatorial South America
Acetobacter	*pasteurianus*	Cassava	Beer (Cassava beer, Chicha)	Africa (Cassava beer), Brazil (Chicha)
Acetobacter	*pasteurianus*	Chañar fruit	Beer (Chicha)	Chile
Acetobacter	*pasteurianus*	Kañiwa	Beer (Chicha)	Andean South America
Acetobacter	*pasteurianus*	Maize	Beer (Chicha)	Andean South America and Costa Rica

Acetobacter	*pasteurianus*	Oca	Beer (Chicha)	Brazil
Acetobacter	*pasteurianus*	Palm fruit	Beer (Chicha, Palm beer)	Central and South America (Chicha), Fiji (Palm beer)
Acetobacter	*pasteurianus*	Peanut	Beer (Chicha)	Peru
Acetobacter	*pasteurianus*	Potato	Beer (Chicha, Potato beer)	Germany and Netherlands (Potato beer), Northern South America (Chicha)
Acetobacter	*pasteurianus*	Previously alcoholic fermented fruit juices	Vinegar	Global
Acetobacter	*pasteurianus*	Previously alcoholic fermented pineapple peel and sugar	Pineapple peel vinegar	Asia, Latin America
Acetobacter	*pasteurianus*	Quinoa	Beer (Cusqueña)	Peru
Acetobacter	*pasteurianus*	Rice	Beer (Chicha, Rice beer)	Japan (Rice beer), South America (Chicha)
Acetobacter	*pomorum*	Previously alcoholic fermented fruit juices	Vinegar	Global
Acetobacter	*tropicalis*	*Vanilla* pods	Vanilla	Mexico
Acetobacterium	*tundrae*	Barley	Beer (Historical ale)	United Kingdom
Acidipropionibacterium	*acidipropionici*	Milk	Cheese (Swiss type)	Europe
Acidipropionibacterium	*jensenii*	Milk	Cheese (Swiss type)	Europe
Acidipropionibacterium	*thoenii*	Milk	Cheese (Swiss type)	Europe
Actinotignum	*urinale*	Barley	Beer (Historical ale)	United Kingdom
Aerococcus	*viridans*	Fish with rice	Fermented seasoning (Pla ra)	Thailand
Anaerotignum	*propionicum*	Milk	Cheese	Global
Bacillus	*amyloliquefaciens*	Barley	Beer (Historical ale)	United Kingdom
Bacillus	*amyloliquefaciens*	Lentils	Griddle cake (Dosa)	India
Bacillus	*amyloliquefaciens*	Rice	Steamed cake (Idli)	India and Sri Lanka
Bacillus	*amyloliquefaciens*	Soybean	Fermented soybean (Hawaijar, Peruyaan)	India

(*Continued*)

TABLE 1.1 (Continued)

Genus	Species	Primary Material Being Fermented	Product	Geographical Origin
Bacillus	*atrophaeus*	*Vanilla* pods	Vanilla	Mexico
Bacillus	*cereus*	Rice	Steamed cake (Idli)	India and Sri Lanka
Bacillus	*cereus*	Soybean	Fermented soybean (Hawaijar), Curry (Kinema)	India
Bacillus	*licheniformis*	Barley	Beer (Historical ale)	United Kingdom
Bacillus	*licheniformis*	Locust bean	Dry condiment (Eware, Irú)	Africa
Bacillus	*licheniformis*	Soybean	Fermented paste or curry (Bekang), Curry (Kinema)	India
Bacillus	*pumilus*	Locust bean	Dry condiment (Eware, Irú)	Africa
Bacillus	*pumilus*	Soybean	Fermented paste (Tungrymba), Paste or curry (Bekang)	India
Bacillus	*pumilus*	*Vanilla* pods	Vanilla	Mexico
Bacillus	*subtilis*	Barley	Beer (Historical ale)	United Kingdom
Bacillus	*subtilis*	Cassava	Fermented cassava root subsequently granulated into a flour (Garri)	West Africa
Bacillus	*subtilis*	Locust bean	Dry condiment (Eware, Irú)	Africa
Bacillus	*subtilis*	Maize	Porridge (Ogi)	Nigeria
Bacillus	*subtilis*	Millet	Porridge (Ogi)	Nigeria
Bacillus	*subtilis*	Rice	Steamed cake (Idli)	India and Sri Lanka
Bacillus	*subtilis*	*Sorghum* grain	Porridge (Ogi)	Nigeria
Bacillus	*subtilis*	Soybean	Coagulated soy milk (Tofu), Fermented bean served whole often as an alternative to meat (Hawaijar, Natto), Fermented paste (Miso, Tungrymbai)	China (Tofu), India (Hawaijar, Tungrymbai), Japan (Miso, Nattō)
Bacillus	*subtilis*	Tea leaves	Fermented tea leaves [To be brewed for beverage] (Qingzhuan brick tea)	China
Bacillus	*subtilis* subsp. *subtilis*	*Vanilla* pods	Vanilla	Mexico

Bacillus	*tequilensis*	Rice	Steamed cake (Idli)	India and Sri Lanka
Bacillus	*thuringiensis*	Soybean	Curry (Kinema)	India
Bacillus	*thuringiensis*	*Vanilla* pods	Vanilla	Mexico
Bifidobacterium	*animalis subsp. lactis*	Milk	Fermented beverage	Europe
Bifidobacterium	*breve*	Milk	Fermented beverage	Europe
Brachybacterium	*alimentarium*	Milk	Cheese (Gruyère type, including Beaufort)	France (Beaufort), Switzerland (Gruyère)
Brevibacillus	*brevis*	Soybean	Fermented paste or curry (Bekang)	India
Brevibacterium	*jeotgali*	Bivalve mollusks (intact)	Jeotgal	Korea
Brevibacterium	*jeotgali*	Cephalopods (intact)	Jeotgal	Korea
Brevibacterium	*jeotgali*	Crustaceans (intact)	Jeotgal	Korea
Brevibacterium	*jeotgali*	Fish (intact)	Jeotgal	Korea
Brevibacterium	*jeotgali*	Fish (internal organs)	Jeotgal	Korea
Brevibacterium	*jeotgali*	Gastropod mollusks (intact)	Jeotgal	Korea
Brevibacterium	*jeotgali*	Skate (intact)	Jeotgal	Korea
Brevibacterium	*linens*	Milk	Cheese (Washed rind cheese, also called "stinky cheese," examples are Limburger and Munster)	Belgium (Limburger), France (Munster)
Brochothrix	*thermosphacta*	Beef with Pork	Fermented cased sausage (Salami)	Northern Europe
Brochothrix	*thermosphacta*	Pork	Fermented cased sausage (Salami)	Italy
Citrobacter	*farmeri*	Tomato (in water with sugar)	Fermented tomato	Global
Citrobacter	*freundii*	Soybean	Fermented soybean (Tempeh)	Indonesia
Clostridium	*butyricum*	Milk	Cheese (Swiss type)	Europe
Clostridium	*sporogenes*	Milk	Cheese	Global
Clostridium	*tyrobutyricum*	Milk	Cheese	Global
Companilactobacillus	*alimentarius*	Cream	Buttermilk (Mattha)	Bangladesh
Companilactobacillus	*alimentarius*	Pork	Fermented cased sausage (Salami)	Italy
Companilactobacillus	*alimentarius*	Walleye pollock meat	Fermented fish meat (Sikhae)	Korea

(*Continued*)

TABLE 1.1 (Continued)

Genus	Species	Primary Material Being Fermented	Product	Geographical Origin
Companilactobacillus	*kimchii*	Cabbage	Fermented vegetable (Kimchi)	Korea
Companilactobacillus	*paralimentarius*	Pork	Fermented cased sausage (Salami)	Italy
Companilactobacillus	*versmoldensis*	Beef with Pork	European raw fermented sausage	Germany
Companilactobacillus	*versmoldensis*	Poultry with Pork	European raw fermented sausage	Germany
Corynebacterium	*afermentans*	Maize	Porridge (Ogi)	Nigeria
Corynebacterium	*afermentans*	Millet	Porridge (Ogi)	Nigeria
Corynebacterium	*afermentans*	*Sorghum* grain	Porridge (Ogi)	Nigeria
Corynebacterium	*ammoniagenes*	Milk	Cheese	Global
Corynebacterium	*kroppenstedtii*	Barley	Beer (Historical ale)	United Kingdom
Corynebacterium	*kutscheri*	Maize	Porridge (Ogi)	Nigeria
Corynebacterium	*kutscheri*	Millet	Porridge (Ogi)	Nigeria
Corynebacterium	*kutscheri*	*Sorghum* grain	Porridge (Ogi)	Nigeria
Corynebacterium	*striatum*	Maize	Porridge (Ogi)	Nigeria
Corynebacterium	*striatum*	Millet	Porridge (Ogi)	Nigeria
Corynebacterium	*striatum*	*Sorghum* grain	Porridge (Ogi)	Nigeria
Cronobacter	*sakazakii*	*Sorghum* grain	Bread	Saudi Arabia
Cytobacillus	*firmus*	Maize	Porridge (Ogi)	Nigeria
Cytobacillus	*firmus*	Millet	Porridge (Ogi)	Nigeria
Cytobacillus	*firmus*	*Sorghum* grain	Porridge (Ogi)	Nigeria
Desulfobacter	*psychrotolerans*	Barley	Beer (Historical ale)	United Kingdom
Enterobacter	*cloacae*	Rice	Steamed cake (Idli)	India and Sri Lanka
Enterobacter	*ludwigii*	Tomato (in water with sugar)	Fermented tomato	Global
Enterococcus	*avium*	Fish with rice	Fermented seasoning (Pla ra)	Thailand
Enterococcus	*camelliae*	Tea leaves	Fermented tea [Pickled leaves which are to be chewed instead of brewed] (Miang)	Thailand

Enterococcus	*casseliflavus*	Tea leaves	Fermented tea [Pickled leaves which are to be chewed instead of brewed] (Miang)	Thailand
Enterococcus	*durans*	Milk	Cheese (Including Semi-soft cheese, example is Mozzarella)	Global, Italy (Mozzarella)
Enterococcus	*durans*	Wheat	Bread (Sourdough)	Asia minor and Europe
Enterococcus	*faecalis*	Fish with rice	Fermented seasoning (Pla ra)	Thailand
Enterococcus	*faecalis*	Lentils	Griddle cake (Dosa)	India
Enterococcus	*faecalis*	Rice	Steamed cake (Idli)	India and Sri Lanka
Enterococcus	*faecalis*	Wheat	Deep-fried batter (Jalebi)	Western Asia
Enterococcus	*faecium*	Fish with rice	Fermented seasoning (Pla ra)	Thailand
Enterococcus	*faecium*	Milk	Cheese (Semi-soft cheese, example is Mozzarella)	Italy (Mozzarella)
Enterococcus	*faecium*	Soybean	Curry (Kinema)	India
Enterococcus	*hermanniensis*	Chautsi bass (intact fish)	Fermented fish (Chouguiyu)	China
Enterococcus	*hirae*	Fish with rice	Fermented seasoning (Pla ra)	Thailand
Enterococcus	*thailandicus*	Fish with rice	Fermented seasoning (Pla ra)	Thailand
Finegoldia	*magna*	Barley	Beer (Historical ale)	United Kingdom
Fructilactobacillus	*fructivorans*	Walleye pollock meat	Fermented fish meat (Sikhae)	Korea
Furfurilactobacillus	*rossiae*	Maize	Fermented pap beverage (Mageu)	South Africa
Fructilactobacillus	*sanfranciscensis*	Barley	Fermented pap beverage (Boza)	Possibly Turkey
Fructilactobacillus	*sanfranciscensis*	Maize	Fermented pap beverage (Boza)	Possibly Turkey
Fructilactobacillus	*sanfranciscensis*	Millet	Fermented pap beverage (Boza)	Possibly Turkey
Fructilactobacillus	*sanfranciscensis*	Wheat	Bread (Sourdough), Fermented pap beverage (Boza)	Asia Minor and Europe (Sourdough), Possibly Turkey (Boza)
Gluconacetobacter	*entanii*	Previously alcoholic fermented fruit juices	Vinegar	Global
Gluconacetobacter	*liquefaciens*	Barley	Beer (Historical ale)	United Kingdom

(*Continued*)

TABLE 1.1 (Continued)

Genus	Species	Primary Material Being Fermented	Product	Geographical Origin
Gluconacetobacter	*liquefaciens*	Previously alcoholic fermented fruit juices	Vinegar	Global
Gluconobacter	*oxydans*	Previously alcoholic fermented fruit juices	Vinegar	Global
Halomonas	*cibimaris*	Bivalve mollusks (intact)	Jeotgal	Korea
Halomonas	*cibimaris*	Cephalopods (intact)	Jeotgal	Korea
Halomonas	*cibimaris*	Crustaceans (intact)	Jeotgal	Korea
Halomonas	*cibimaris*	Fish (intact)	Jeotgal	Korea
Halomonas	*cibimaris*	Fish (internal organs)	Jeotgal	Korea
Halomonas	*cibimaris*	Gastropod mollusks (intact)	Jeotgal	Korea
Halomonas	*cibimaris*	Skate (intact)	Jeotgal	Korea
Halomonas	*xinjiangensis*	Barley	Beer (Historical ale)	United Kingdom
Klebsiella	*aerogenes*	*Sorghum* grain	Bread	Saudi Arabia
Klebsiella	*oxytoca*	Maize	Porridge (Ogi)	Nigeria
Klebsiella	*oxytoca*	Millet	Porridge (Ogi)	Nigeria
Klebsiella	*oxytoca*	*Sorghum* grain	Bread, Porridge (Ogi)	Nigeria (Ogi), Saudi Arabia (Bread)
Klebsiella	*pneumoniae*	*Sorghum* grain	Bread	Saudi Arabia
Klebsiella	*pneumoniae*	Soybean	Fermented soybean (Tempeh)	Indonesia
Kocuria	*atrinae*	Bivalve mollusks (intact)	Jeotgal	Korea
Kocuria	*atrinae*	Cephalopods (intact)	Jeotgal	Korea
Kocuria	*atrinae*	Crustaceans (intact)	Jeotgal	Korea
Kocuria	*atrinae*	Fish (intact)	Jeotgal	Korea
Kocuria	*atrinae*	Fish (internal organs)	Jeotgal	Korea
Kocuria	*atrinae*	Gastropod mollusks (intact)	Jeotgal	Korea
Kocuria	*atrinae*	Skate (intact)	Jeotgal	Korea

Kocuria	*rhizophila*	Barley	Beer (Historical ale)	United Kingdom
Kocuria	*rosea*	Barley	Beer (Historical ale)	United Kingdom
Kocuria	*salsicia*	Beef with Pork	Fermented cased sausage (Salami)	Northern Europe
Kocuria	*salsicia*	Fish and grain mixture (typically Righteye flounders and millet, cooked together then fermented)	Fermented fish (Sikhae)	Korea
Kocuria	*salsicia*	Pork	Fermented cased sausage (Salami)	Italy
Kocuria	*varians*	Pork	Fermented cased sausage (Salami)	Italy
Komagataeibacter	*europaeus*	Previously alcoholic fermented fruit juices	Vinegar	Global
Komagataeibacter	*intermedius*	Previously alcoholic fermented fruit juices	Vinegar	Global
Komagataeibacter	*intermedius*	Tea (aqueous extract of tea leaves) plus sugar	Fermented tea (Kombucha)	China
Komagataeibacter	*medellinensis*	Previously alcoholic fermented fruit juices	Vinegar	Global
Komagataeibacter	*oboediens*	Previously alcoholic fermented fruit juices	Vinegar	Global
Komagataeibacter	*rhaeticus*	Tea (aqueous extract of tea leaves) plus sugar	Fermented tea (Kombucha)	China
Komagataeibacter	*saccharivorans*	Tea (aqueous extract of tea leaves) plus sugar	Fermented tea (Kombucha)	China
Komagataeibacter	*xylinus*	Previously alcoholic fermented fruit juices	Vinegar	Global
Komagataeibacter	*xylinus*	Tea (aqueous extract of tea leaves) plus sugar	Fermented tea (Kombucha)	China
Lacticaseibacillus	*camelliae*	Barley	Beer (Historical ale)	United Kingdom
Lacticaseibacillus	*camelliae*	Tea leaves	Fermented tea [Pickled leaves which are to be chewed instead of brewed] (Miang)	Thailand

(*Continued*)

TABLE 1.1 (Continued)

Genus	Species	Primary Material Being Fermented	Product	Geographical Origin
Lacticaseibacillus	*casei*	Apple juice	Malolactic fermented apple juice (Non-alcoholic Cider)	Europe
Lacticaseibacillus	*casei*	Maize	Porridge (Ogi)	Nigeria
Lacticaseibacillus	*casei*	Milk	Cheese (Including Semi-soft cheese, example is Mozzarella)	Global, Italy (Mozzarella)
Lacticaseibacillus	*casei*	Millet	Porridge (Ogi)	Nigeria
Lacticaseibacillus	*casei*	Olive	Fermented olive	Mediterranean
Lacticaseibacillus	*casei*	*Sorghum* grain	Porridge (Ogi)	Nigeria
Lacticaseibacillus	*casei*	Vegetable juice	Fermented beverage	Global
Lacticaseibacillus	*casei*	Wheat	Bread (Sourdough)	Asia Minor and Europe
Lacticaseibacillus	*paracasei*	Barley	Fermented pap beverage (Boza), Distilled beverage (Malt whisky)	Possibly Turkey (Boza), Ireland and Scotland (Malt whisky)
Lacticaseibacillus	*paracasei*	Fish with rice	Fermented seasoning (Pla ra)	Thailand
Lacticaseibacillus	*paracasei*	Grape juice	Wine	Global
Lacticaseibacillus	*paracasei*	Maize	Fermented pap beverage (Boza)	Possibly Turkey
Lacticaseibacillus	*paracasei*	Milk	Cheese	Global
Lacticaseibacillus	*paracasei*	Millet	Fermented pap beverage (Boza)	Possibly Turkey
Lacticaseibacillus	*paracasei*	Sugar	Fermented beverage (Tibicos, also called Water Kefir)	Global
Lacticaseibacillus	*paracasei*	Wheat	Fermented pap beverage (Boza)	Possibly Turkey
Lacticaseibacillus	*paracasei* subsp. *paracasei*	Milk	Fermented beverage (Kefir)	North Caucasus
Lacticaseibacillus	*paracasei* subsp. *tolerans*	Pork	Fermented cased sausage (Salami)	Italy
Lacticaseibacillus	*rhamnosus*	Apple juice	Malolactic fermented apple juice (Non-alcoholic Cider)	Europe
Lacticaseibacillus	*rhamnosus*	Barley	Fermented pap beverage (Boza)	Possibly Turkey

Lacticaseibacillus	*rhamnosus*	Maize	Fermented pap beverage (Boza)	Possibly Turkey
Lacticaseibacillus	*rhamnosus*	Milk	Cheese	Global
Lacticaseibacillus	*rhamnosus*	Millet	Fermented pap beverage (Boza)	Possibly Turkey
Lacticaseibacillus	*rhamnosus*	Wheat	Fermented pap beverage (Boza)	Possibly Turkey
Lacticaseibacillus	*thailandensis*	Tea leaves	Fermented tea [Pickled leaves which are to be chewed instead of brewed] (Miang)	Thailand
Lactiplantibacillus	*argentoratensis*	Rice	Steamed cake (Idli)	India and Sri Lanka
Lactiplantibacillus	*paraplantarum*	Bivalve mollusks (intact)	Jeotgal	Korea
Lactiplantibacillus	*paraplantarum*	Cephalopods (intact)	Jeotgal	Korea
Lactiplantibacillus	*paraplantarum*	Crustaceans (intact)	Jeotgal	Korea
Lactiplantibacillus	*paraplantarum*	Fish (intact)	Jeotgal	Korea
Lactiplantibacillus	*paraplantarum*	Fish (internal organs)	Jeotgal	Korea
Lactiplantibacillus	*paraplantarum*	Gastropod mollusks (intact)	Jeotgal	Korea
Lactiplantibacillus	*paraplantarum*	Skate (intact)	Jeotgal	Korea
Lactiplantibacillus	*pentosus*	Barley	Fermented pap beverage (Boza), Distilled beverage (Malt whisky)	Ireland and Scotland (Malt whisky), Possibly Turkey (Boza)
Lactiplantibacillus	*pentosus*	Bivalve mollusks (intact)	Jeotgal	Korea
Lactiplantibacillus	*pentosus*	Cephalopods (intact)	Jeotgal	Korea
Lactiplantibacillus	*pentosus*	Crustaceans (intact)	Jeotgal	Korea
Lactiplantibacillus	*pentosus*	Fish (intact)	Jeotgal	Korea
Lactiplantibacillus	*pentosus*	Fish (internal organs)	Jeotgal	Korea
Lactiplantibacillus	*pentosus*	Gastropod mollusks (intact)	Jeotgal	Korea
Lactiplantibacillus	*pentosus*	Grape juice	Wine	Global
Lactiplantibacillus	*pentosus*	Maize	Fermented pap beverage (Boza), Porridge (Ogi),	Nigeria (Ogi), Possibly Turkey (Boza)
Lactiplantibacillus	*pentosus*	Millet	Fermented pap beverage (Boza), Porridge (Ogi)	Nigeria (Ogi), Possibly Turkey (Boza)

(*Continued*)

TABLE 1.1 (Continued)

Genus	Species	Primary Material Being Fermented	Product	Geographical Origin
Lactiplantibacillus	*pentosus*	Olive	Fermented olive	Mediterranean
Lactiplantibacillus	*pentosus*	Rice	Steamed cake (Idli)	India and Sri Lanka
Lactiplantibacillus	*pentosus*	Skate (intact)	Jeotgal	Korea
Lactiplantibacillus	*pentosus*	*Sorghum* grain	Porridge (Ogi)	Nigeria
Lactiplantibacillus	*pentosus*	Tea leaves	Fermented tea [Pickled leaves which are to be chewed instead of brewed] (Miang)	Thailand
Lactiplantibacillus	*pentosus*	Wheat	Fermented pap beverage (Boza)	Possibly Turkey
Lactiplantibacillus	*plantarum*	Apple juice	Malolactic fermented apple juice (Non-alcoholic Cider)	Europe
Lactiplantibacillus	*plantarum*	Barley	Fermented pap beverage (Boza)	Possibly Turkey
Lactiplantibacillus	*plantarum*	Bivalve mollusks (intact)	Jeotgal	Korea
Lactiplantibacillus	*plantarum*	Cabbage	Fermented vegetable (Kimchi, Gundruk, Sauerkraut)	Germany (Sauerkraut), Korea (Kimchi), Nepal (Gundruk)
Lactiplantibacillus	*plantarum*	Cassava	Beer (Cassava beer, Chicha), Fermented cassava root subsequently granulated into a flour (Garri)	Africa (Cassava beer), Brazil (Chicha), West Africa (Garri)
Lactiplantibacillus	*plantarum*	Cauliflower	Fermented vegetable (Gundruk)	Nepal
Lactiplantibacillus	*plantarum*	Cephalopods (intact)	Jeotgal	Korea
Lactiplantibacillus	*plantarum*	Chañar fruit	Beer (Chicha)	Chile
Lactiplantibacillus	*plantarum*	*Coffea* seeds	Coffee	Yemen
Lactiplantibacillus	*plantarum*	Crustaceans (intact)	Jeotgal	Korea
Lactiplantibacillus	*plantarum*	Cucumber	Fermented vegetable (Khalpi)	Himalayas
Lactiplantibacillus	*plantarum*	Fish (intact)	Jeotgal	Korea
Lactiplantibacillus	*plantarum*	Fish (internal organs)	Jeotgal	Korea
Lactiplantibacillus	*plantarum*	Fish with rice	Fermented seasoning (Pla ra)	Thailand
Lactiplantibacillus	*plantarum*	Gastropod mollusks (intact)	Jeotgal	Korea

Lactiplantibacillus	*plantarum*	Grape juice	Wine	Global
Lactiplantibacillus	*plantarum*	Kañiwa	Beer (Chicha)	Andean South America
Lactiplantibacillus	*plantarum*	Maize	Beer (Chicha), Fermented dough used to make a pap beverage (Pozol), Fermented pap beverage (Boza, Mageu)	Andean South America and Costa Rica (Chicha), Mexico (Pozol), Possibly Turkey (Boza), South Africa (Mageu)
Lactiplantibacillus	*plantarum*	Mango	Pickled fruit (Burong mangga)	Philippines
Lactiplantibacillus	*plantarum*	Milk	Cheese (Semi-hard cheese, example is Cheddar), Yogurt	England (Cheddar), Western Asia (Yogurt)
Lactiplantibacillus	*plantarum*	Millet	Fermented pap beverage (Boza)	Possibly Turkey
Lactiplantibacillus	*plantarum*	Mixed vegetables	Pickled vegetables	Global
Lactiplantibacillus	*plantarum*	Mung bean	Mung bean milk (Douzhi)	China
Lactiplantibacillus	*plantarum*	Mustard	Fermented vegetable (Gundruk)	Nepal
Lactiplantibacillus	*plantarum*	Oca	Beer (Chicha)	Brazil
Lactiplantibacillus	*plantarum*	Olive	Fermented olive	Mediterranean area
Lactiplantibacillus	*plantarum*	Palm fruit	Beer (Chicha, Palm beer)	Central America and South America (Chicha), Fiji (Palm beer)
Lactiplantibacillus	*plantarum*	Palm sap (from palm flowers and tapped trees)	Palm wine (Taberna, Tari)	Africa, Caribbean, Mexico (Taberna), Micronesia, Philippines, South America, South Asia (Tari), Southeast Asia
Lactiplantibacillus	*plantarum*	Peanut	Beer (Chicha)	Peru
Lactiplantibacillus	*plantarum*	Pork	Fermented cased sausage (Salami)	Italy
Lactiplantibacillus	*plantarum*	Potato	Beer (Chicha, Potato beer)	Germany and Netherlands (Potato beer), Northern South America (Chicha)
Lactiplantibacillus	*plantarum*	Quinoa	Beer (Cusqueña)	Peru
Lactiplantibacillus	*plantarum*	Radish roots	Fermented vegetable (Sinki)	Nepal
Lactiplantibacillus	*plantarum*	Rice	Steamed cake (Idli)	India and Sri Lanka
Lactiplantibacillus	*plantarum*	Skate (intact)	Jeotgal	Korea
Lactiplantibacillus	*plantarum*	Tomato (in water with sugar)	Fermented tomato	Global

(*Continued*)

TABLE 1.1 (Continued)

Genus	Species	Primary Material Being Fermented	Product	Geographical Origin
Lactiplantibacillus	*plantarum*	Vegetable juices	Fermented beverage	Global
Lactiplantibacillus	*plantarum*	Wheat	Fermented pap beverage (Boza)	Possibly Turkey
Lactiplantibacillus	*plantarum*	Wheat germ	Fermented wheat germ	Hungary
Lactiplantibacillus	*plantarum*	Yogurt with wheat flour	Yellow curd (a type of Kashk)	Iran
Lactobacillus	*acidophilus*	Apple juice	Malolactic fermented apple juice (Non-alcoholic Cider)	Europe
Lactobacillus	*acidophilus*	Milk	Fermented beverage (Acidophiline milk, Acidophilus milk)	Ukraine
Lactobacillus	*acidophilus*	Taro	Fermented paste (Poi)	Hawaii
Lactobacillus	*acidophilus*	Vegetable juices	Fermented beverage	Global
Lactobacillus	*amylolyticus*	Barley	Fermented beverage also used in soup (Borş)	Romania
Lactobacillus	*amylolyticus*	Wheat	Fermented beverage also used in soup (Borş)	Romania
Lactobacillus	*crispatus*	Milk	Cheese (Hard cheese, example is Parmesan)	Italy (Parmesan)
Lactobacillus	*delbrueckii*	Maize	Fermented pap beverage (Mageu)	South Africa
Lactobacillus	*delbrueckii*	Milk	Cheese (Semi-soft cheese, example is Mozzarella), (Hard cheese, example is Parmesan), Yogurt	Italy (Mozzarella, Parmesan), Western Asia (Yogurt)
Lactobacillus	*delbrueckii*	Taro	Fermented paste (Poi)	Hawaii
Lactobacillus	*delbrueckii*	Vegetable juices	Fermented beverage	Global
Lactobacillus	*delbrueckii*	Yogurt with wheat flour	Yellow curd (a type of Kashk)	Iran
Lactobacillus	*delbrueckii* subsp. *bulgaricus*	Cream	Buttermilk (Mattha)	Bangladesh
Lactobacillus	*delbrueckii* subsp. *bulgaricus*	Maize	Fermented pap beverage (Mageu)	South Africa

Lactobacillus	*delbrueckii* subsp. *bulgaricus*	Milk	Cheese (Kashkaval), Fermented beverage (Kefir), Yogurt	Balkans and Southeast Europe (Kashkaval), North Caucasus (Kefir), Western Asia (Yogurt)
Lactobacillus	*delbrueckii* subsp. *bulgaricus*	Wheat	Deep-fried batter (Jalebi)	Western Asia
Lactobacillus	*delbrueckii* subsp. *lactis*	Fish with rice	Fermented seasoning (Pla ra)	Thailand
Lactobacillus	*delbrueckii* subsp. *lactis*	Milk	Cheese (Dutch type, example is Gouda), (Semi-hard cheese, example is Cheddar)	England (Cheddar), Netherlands (Gouda)
Lactobacillus	*helveticus*	Barley	Distilled beverage (Malt whisky)	Ireland and Scotland
Lactobacillus	*helveticus*	Milk	Cheese (Semi-soft cheese, example is Mozzarella), (Semi-hard cheese, example is Cheddar), (Hard cheese, example is Parmesan), Fermented beverage (Kefir)	England (Cheddar), Italy (Mozzarella, Parmesan), North Caucasus (Kefir)
Lactobacillus	*helveticus*	Vegetable juices	Fermented beverage	Global
Lactobacillus	*kefiranofaciens*	Milk	Fermented beverage (Kefir)	North Caucasus
Lactobacillus	*paracasei subsp. paracasei*	Pork	Fermented cased sausage (Salami)	Italy
Lactobacillus	*paralimentarius*	Pork	Fermented cased sausage (Salami)	Italy
Lactococcus	*chungangensis*	Milk with Cream	Cream Cheese	United States
Lactococcus	*cremoris*	Butter (Ghee)	Fermented ghee	Bangladesh
Lactococcus	*cremoris*	Cream	Buttermilk (Mattha)	Bangladesh
Lactococcus	*cremoris*	Milk	Cheese (Acid-set cheese, also called Sour milk cheese, example is Cottage cheese)	United States (Cottage cheese)
Lactococcus	*lactis*	Barley	Beer (Historical ale)	United Kingdom
Lactococcus	*lactis*	Butter (Ghee)	Fermented ghee	Bangladesh
Lactococcus	*lactis*	Cashew	Vegan cheese	Canada
Lactococcus	*lactis*	Chautsi bass (intact fish)	Fermented fish (Chouguiyu)	China

(Continued)

TABLE 1.1 (Continued)

Genus	Species	Primary Material Being Fermented	Product	Geographical Origin
Lactococcus	*lactis*	*Coffea* seeds	Coffee	Yemen
Lactococcus	*lactis*	Cream	Buttermilk (Mattha)	Bangladesh
Lactococcus	*lactis*	Milk	Cheese (Acid-set cheese, also called Sour milk cheese, example is Cottage cheese), Fermented beverage (Kefir)	North Caucasus (Kefir), United States (Cottage cheese)
Lactococcus	*lactis*	Quinoa	Starter culture for making vegan cheese	Canada
Lactococcus	*lactis*	Rice	Steamed cake (Idli)	India and Sri Lanka
Lactococcus	*lactis*	Sugar	Fermented beverage (Tibicos, also called Water Kefir)	Global
Lactococcus	*lactis*	Taro	Fermented paste (Poi)	Hawaii
Lactococcus	*lactis*	Wheat	Deep-fried batter (Jalebi)	Western Asia
Lactococcus	*lactis* subsp. *cremoris*	Cream	Sour cream (Crème Fraiche)	Central America, Eastern Europe, France (Crème Fraiche)
Lactococcus	*lactis* subsp. *lactis*	Cream	Sour cream (Crème Fraiche)	Central America, Eastern Europe, France (Crème Fraiche)
Lactococcus	*lactis* subsp. *lactis*	*Sorghum* grain	Bread	Saudi Arabia
Lactococcus	*lactis* subsp. *lactis*	Vegetable juices	Fermented beverage	Global
Lactococcus	*lactis* subsp. *lactis* bv. *diacetylactis*	Cream	Sour cream (Crème Fraiche)	Central America, Eastern Europe, France (Crème Fraiche)
Lactococcus	*laudensis*	Milk	Cheese (Dutch type, example is Gouda)	Netherlands (Gouda)
Lactococcus	*raffinolactis*	Maize	Fermented dough used to make a pap beverage (Pozol)	Mexico
Lapidilactobacillus	*dextrinicus*	Rice	Steamed cake (Idli)	India and Sri Lanka
Latilactobacillus	*curvatus*	Beef with Pork	Fermented cased sausage (Salami)	Northern Europe
Latilactobacillus	*curvatus*	Pork	Fermented cased sausage (Salami)	Italy
Latilactobacillus	*graminis*	Pork	Fermented cased sausage (Salami)	Italy

Latilactobacillus	*graminis*	Walleye pollock meat	Fermented fish meat (Sikhae)	Korea
Latilactobacillus	*sakei*	Beef with Pork	Fermented cased sausage (Salami)	Northern Europe
Latilactobacillus	*sakei*	Pork	Fermented cased sausage (Salami)	Italy
Latilactobacillus	*sakei*	Rice	Rice wine (Saké)	Japan
Latilactobacillus	*sakei*	Walleye pollock meat	Fermented fish meat (Sikhae)	Korea
Lelliottia	*amnigena*	Maize	Porridge (Ogi)	Nigeria
Lelliottia	*amnigena*	Millet	Porridge (Ogi)	Nigeria
Lelliottia	*amnigena*	*Sorghum* grain	Porridge (Ogi)	Nigeria
Lentilactobacillus	*buchneri*	Maize	Porridge (Ogi)	Nigeria
Lentilactobacillus	*buchneri*	Millet	Porridge (Ogi)	Nigeria
Lentilactobacillus	*buchneri*	*Sorghum* grain	Porridge (Ogi)	Nigeria
Lentilactobacillus	*buchneri*	Wheat	Deep-fried batter (Jalebi)	Western Asia
Lentilactobacillus	*buchneri*	Wheat germ	Fermented wheat germ	Hungary
Lentilactobacillus	*diolivorans*	Wheat	Bread (Sourdough)	Asia Minor and Europe
Lentilactobacillus	*hilgardii*	Ginger plus sugar	Alcoholic Ginger beer	United Kingdom
Lentilactobacillus	*hilgardii*	Grape juice	Wine	Global
Lentilactobacillus	*hilgardii*	Sugar	Fermented beverage (Tibicos, also called Water Kefir)	Global
Lentilactobacillus	*kefiri*	Milk	Cheese (Hard cheese, example is Parmesan), Fermented beverage (Kefir)	Italy (Parmesan), North Caucasus (Kefir)
Lentilactobacillus	*kefiri*	Sugar	Fermented beverage (Tibicos, also called Water Kefir)	Global
Lentilactobacillus	*parabuchneri*	Barley	Distilled beverage (Malt whisky)	Ireland and Scotland
Lentilactobacillus	*parabuchneri*	Sugar	Fermented beverage (Tibicos, also called Water Kefir)	Global
Lentilactobacillus	*parakefiri*	Milk	Fermented beverage (Kefir)	North Caucasus
Leuconostoc	*citreum*	Barley	Beer (Historical ale)	United Kingdom
Leuconostoc	*citreum*	Cabbage	Fermented vegetable (Kimchi)	Korea

(*Continued*)

TABLE 1.1 (Continued)

Genus	Species	Primary Material Being Fermented	Product	Geographical Origin
Leuconostoc	*citreum*	Cashew	Vegan cheese	Canada
Leuconostoc	*citreum*	Quinoa	Starter culture for making vegan cheese	Canada
Leuconostoc	*fallax*	Cucumber	Fermented vegetable (Khalpi)	Himalayas
Leuconostoc	*gelidum*	Cabbage	Fermented vegetable (Kimchi)	Korea
Leuconostoc	*gelidum* subsp. *gasicomitatum*	Cabbage	Fermented vegetable (Kimchi)	Korea
Leuconostoc	*lactis*	Maize	Fermented dough used to make a pap beverage (Pozol), Fermented pap beverage (Mageu)	Mexico (Pozol), South Africa (Mageu)
Leuconostoc	*mesenteroides*	Bivalve mollusks (intact)	Jeotgal	Korea
Leuconostoc	*mesenteroides*	Cabbage	Fermented vegetable (Kimchi, Sauerkraut)	Germany (Sauerkraut), Korea (Kimchi)
Leuconostoc	*mesenteroides*	Cassava	Fermented cassava root subsequently granulated into a flour (Garri)	West Africa
Leuconostoc	*mesenteroides*	Cephalopods (intact)	Jeotgal	Korea
Leuconostoc	*mesenteroides*	Coconut	Pickled fruit (Aachaar, Chutney)	Indian subcontinent
Leuconostoc	*mesenteroides*	*Coffea* seeds	Coffee	Yemen
Leuconostoc	*mesenteroides*	Crustaceans (intact)	Jeotgal	Korea
Leuconostoc	*mesenteroides*	Cucumber	Fermented vegetable (Aachaar, Chutney)	Indian subcontinent
Leuconostoc	*mesenteroides*	Fish (intact)	Jeotgal	Korea
Leuconostoc	*mesenteroides*	Fish (internal organs)	Jeotgal	Korea
Leuconostoc	*mesenteroides*	Gastropod mollusks (intact)	Jeotgal	Korea
Leuconostoc	*mesenteroides*	Lentils	Griddle cake (Dosa)	India
Leuconostoc	*mesenteroides*	Maize	Fermented dough used to make a pap beverage (Pozol), Fermented pap beverage (Mageu)	Mexico (Pozol), South Africa (Mageu)

Leuconostoc	*mesenteroides*	Mango	Pickled fruit (Burong mangga)	Philippines
Leuconostoc	*mesenteroides*	Milk	Cheese (Semi-soft cheese, example is Mozzarella), Fermented beverage (Kefir), Yogurt	Italy (Mozzarella), North Caucasus (Kefir), Western Asia (Yogurt)
Leuconostoc	*mesenteroides*	Milk (curd or yogurt)	Fermented vegetable (Aachaar, Chutney)	Indian subcontinent
Leuconostoc	*mesenteroides*	Mint	Fermented vegetable (Aachaar, Chutney)	Indian subcontinent
Leuconostoc	*mesenteroides*	Mixed vegetables	Pickled vegetables, (Aachaar, Chutney)	Global, Indian subcontinent (Aachaar, Chutney)
Leuconostoc	*mesenteroides*	Onion	Pickled vegetables, (Aachaar, Chutney)	Global, Indian subcontinent (Aachaar, Chutney)
Leuconostoc	*mesenteroides*	Palm sap (from palm flowers and tapped trees)	Palm wine (Taberna, Tari)	Africa, Caribbean, Mexico (Taberna), Micronesia, Philippines, South America, South Asia (Tari), Southeast Asia
Leuconostoc	*mesenteroides*	Peanut	Pickled vegetables, (Aachaar, Chutney)	Global, Indian subcontinent (Aachaar, Chutney)
Leuconostoc	*mesenteroides*	Rice	Steamed cake (Idli)	India and Sri Lanka
Leuconostoc	*mesenteroides*	Skate (intact)	Jeotgal	Korea
Leuconostoc	*mesenteroides*	Sugar	Fermented beverage (Tibicos, also called Water Kefir)	Global
Leuconostoc	*mesenteroides*	Tomato (in water with sugar)	Fermented tomato	Global
Leuconostoc	*mesenteroides* subsp. *cremoris*	Cream	Sour cream (Crème Fraiche)	Central America, Eastern Europe, France (Crème Fraiche)
Leuconostoc	*mesenteroides* subsp. *cremoris*	Olive	Fermented olive	Mediterranean area
Leuconostoc	*mesenteroides* subsp. *dextranicum*	Barley	Fermented pap beverage (Boza)	Possibly Turkey
Leuconostoc	*mesenteroides* subsp. *dextranicum*	Maize	Fermented pap beverage (Boza)	Possibly Turkey

(*Continued*)

TABLE 1.1 (Continued)

Genus	Species	Primary Material Being Fermented	Product	Geographical Origin
Leuconostoc	*mesenteroides* subsp. *dextranicum*	Millet	Fermented pap beverage (Boza)	Possibly Turkey
Leuconostoc	*mesenteroides* subsp. *dextranicum*	Wheat	Fermented pap beverage (Boza)	Possibly Turkey
Leuconostoc	*mesenteroides* subsp. *mesenteroides*	Barley	Fermented pap beverage (Boza)	Possibly Turkey
Leuconostoc	*mesenteroides* subsp. *mesenteroides*	Maize	Fermented pap beverage (Boza)	Possibly Turkey
Leuconostoc	*mesenteroides* subsp. *mesenteroides*	Millet	Fermented pap beverage (Boza)	Possibly Turkey
Leuconostoc	*mesenteroides* subsp. *mesenteroides*	Wheat	Fermented pap beverage (Boza)	Possibly Turkey
Leuconostoc	*pseudomesenteroides*	Milk	Cheese (Dutch type, example is Gouda)	Netherlands (Gouda)
Levilactobacillus	*brevis*	Barley	Distilled beverage (Malt whisky)	Ireland, Scotland
Levilactobacillus	*brevis*	Cabbage	Fermented vegetable (Kimchi, Gundruk, Sauerkraut)	Germany (Sauerkraut), Korea (Kimchi), Nepal (Gundruk)
Levilactobacillus	*brevis*	Cauliflower	Fermented vegetable (Gundruk)	Nepal
Levilactobacillus	*brevis*	*Coffea* seeds	Coffee	Yemen
Levilactobacillus	*brevis*	Cucumber	Fermented vegetable (Khalpi)	Himalayas
Levilactobacillus	*brevis*	Grape juice	Wine	Global
Levilactobacillus	*brevis*	Maize	Fermented pap beverage (Mageu)	South Africa
Levilactobacillus	*brevis*	Mango	Pickled fruit (Burong mangga)	Philippines
Levilactobacillus	*brevis*	Milk	Cheese (Semi-hard cheese, example is Cheddar), Fermented beverage (Kefir)	England (Cheddar), North Caucasus (Kefir)
Levilactobacillus	*brevis*	Mustard	Fermented vegetable (Gundruk)	Nepal
Levilactobacillus	*brevis*	Pork	Fermented cased sausage (Salami)	Italy

Levilactobacillus	*brevis*	Radish roots	Fermented vegetable (Sinki)	Nepal
Levilactobacillus	*brevis*	*Sorghum* grain	Bread	Saudi Arabia
Levilactobacillus	*brevis*	Sugar	Fermented beverage (Tibicos, also called Water Kefir)	Global
Levilactobacillus	*brevis*	Vegetable juices	Fermented beverage	Global
Ligilactobacillus	*salivarius*	Milk	Cheese	Global
Ligilactobacillus	*salivarius*	Pork	Fermented cased sausage (Salami)	Italy
Ligilactobacillus	*salivarius*	Vegetable juices	Fermented beverage	Global
Loigolactobacillus	*coryniformis*	Barley	Fermented pap beverage (Boza)	Possibly Turkey
Loigolactobacillus	*coryniformis*	Maize	Fermented pap beverage (Boza)	Possibly Turkey
Loigolactobacillus	*coryniformis*	Millet	Fermented pap beverage (Boza)	Possibly Turkey
Loigolactobacillus	*coryniformis*	Wheat	Fermented pap beverage (Boza)	Possibly Turkey
Loigolactobacillus	*coryniformis subsp. torquens*	Pork	Fermented cased sausage (Salami)	Italy
Limosilactobacillus	*fermentum*	*Agave* sap	Distilled beverage (Tequila)	Mexico
Limosilactobacillus	*fermentum*	Bamboo shoot	Fermented sprout	Taiwan
Limosilactobacillus	*fermentum*	Barley	Distilled beverage (Malt whisky), Fermented pap beverage (Boza)	Ireland and Scotland (Malt whisky), Possibly Turkey (Boza)
Limosilactobacillus	*fermentum*	Cacao seeds	Cocoa beans	Equatorial South America
Limosilactobacillus	*fermentum*	Cassava	Beer (Chicha)	South America
Limosilactobacillus	*fermentum*	Lentils	Griddle cake (Dosa)	India
Limosilactobacillus	*fermentum*	Maize	Fermented pap beverage (Boza, Mageu)	Possibly Turkey (Boza), South Africa (Mageu)
Limosilactobacillus	*fermentum*	Milk	Cheese (Semi-soft cheese, example is Mozzarella), (Hard cheese, example is Parmesan),	Italy (Mozzarella, Parmesan)
Limosilactobacillus	*fermentum*	Millet	Fermented pap beverage (Boza)	Possibly Turkey
Limosilactobacillus	*fermentum*	Pork	Fermented cased sausage (Salami)	Italy

(*Continued*)

TABLE 1.1 (Continued)

Genus	Species	Primary Material Being Fermented	Product	Geographical Origin
Limosilactobacillus	*fermentum*	Radish roots	Fermented vegetable (Sinki)	Nepal
Limosilactobacillus	*fermentum*	*Sorghum* grain	Bread	Saudi Arabia
Limosilactobacillus	*fermentum*	Wheat	Bread (Sourdough), Deep-fried batter (Jalebi), Fermented pap beverage (Boza)	Asia Minor and Europe (Sourdough), Possibly Turkey (Boza), Western Asia (Jalebi)
Limosilactobacillus	*fermentum*	Yogurt with wheat flour	Yellow curd (a type of Kashk)	Iran
Limosilactobacillus	*pontis*	Yogurt with wheat flour	Yellow curd (a type of Kashk)	Iran
Liquorilactobacillus	*ghanensis*	Cacao seeds	Cocoa beans	Ghana
Liquorilactobacillus	*mali*	Grape juice	Wine	Global
Liquorilactobacillus	*oeni*	Grape juice	Wine	Global
Lysinibacillus	*fusiformis*	Soybean	Fermented paste or curry (Bekang)	India
Lysinibacillus	*sphaericus*	Soybean	Fermented paste or curry (Bekang, Kinema)	India
Macrococcus	*caseolyticus*	Chautsi bass (intact fish)	Fermented fish (Chouguiyu)	China
Malaciobacter	*marinus*	Beef with Pork	Fermented cased sausage (Salami)	Northern Europe
Malaciobacter	*marinus*	Pork	Fermented cased sausage (Salami)	Italy
Mammaliicoccus	*sciuri*	Soybean	Fermented soybean (Hawaijar)	India
Microbacterium	*arabinogalactanolyticum*	Tomato (in water with sugar)	Fermented tomato	Global
Micrococcus	*luteus*	Rice	Steamed cake (Idli)	India and Sri Lanka
Mogibacterium	*pumilum*	Barley	Beer (Historical ale)	United Kingdom
Niallia	*circulans*	Soybean	Fermented paste or curry (Bekang), Curry Kinema)	India
Novacetimonas	*hansenii*	Previously alcoholic fermented fruit juices	Vinegar	Global
Novacetimonas	*hansenii*	Tea (aqueous extract of tea leaves) plus sugar	Fermented tea (Kombucha)	China

Oenococcus	*oeni*	Apple juice	Malolactic fermented apple juice (Non-alcoholic Cider)	Europe
Oenococcus	*oeni*	Barley	Fermented pap beverage (Boza)	Possibly Turkey
Oenococcus	*oeni*	Grape juice	Wine	Global
Oenococcus	*oeni*	Maize	Fermented pap beverage (Boza)	Possibly Turkey
Oenococcus	*oeni*	Millet	Fermented pap beverage (Boza)	Possibly Turkey
Oenococcus	*oeni*	Wheat	Fermented pap beverage (Boza)	Possibly Turkey
Parabacteroides	*distasonis*	Barley	Beer (Historical ale)	United Kingdom
Paucilactobacillus	*suebicus*	Barley	Beer (Historical ale)	United Kingdom
Paucilactobacillus	*vaccinostercus*	Barley	Beer (Historical ale)	United Kingdom
Paucilactobacillus	*vaccinostercus*	Tea leaves	Fermented tea [Pickled leaves which are to be chewed instead of brewed] (Miang)	Thailand
Pediococcus	*acidilactici*	Barley	Beer (Historical ale)	United Kingdom
Pediococcus	*acidilactici*	Fish with rice	Fermented seasoning (Pla ra)	Thailand
Pediococcus	*acidilactici*	Milk	Cheese (European types)	Europe
Pediococcus	*acidilactici*	Mixed vegetables	Pickled vegetables	Global
Pediococcus	*acidilactici*	Soybean	Fermented soybean (Peruyaan)	India
Pediococcus	*acidilactici*	Yogurt with wheat flour	Yellow curd (a type of Kashk)	Iran
Pediococcus	*damnosus*	Barley	Beer (Historical ale)	United Kingdom
Pediococcus	*damnosus*	Olive	Fermented olive	Mediterranean area
Pediococcus	*damnosus*	Rice	Steamed cake (Idli)	India and Sri Lanka
Pediococcus	*inopinatus*	Barley	Beer (Historical ale)	United Kingdom
Pediococcus	*parvulus*	Wheat	Bread (Sourdough)	Asia Minor and Europe
Pediococcus	*pentosaceus*	Cabbage	Fermented vegetable (Gundruk, Sauerkraut)	Germany, (Sauerkraut), Nepal (Gundruk)
Pediococcus	*pentosaceus*	Cauliflower	Fermented vegetable (Gundruk)	Nepal
Pediococcus	*pentosaceus*	Cucumber	Fermented vegetable (Khalpi)	Himalayas

(Continued)

TABLE 1.1 (Continued)

Genus	Species	Primary Material Being Fermented	Product	Geographical Origin
Pediococcus	*pentosaceus*	Fish with rice	Fermented seasoning (Pla ra)	Thailand
Pediococcus	*pentosaceus*	Maize	Porridge (Ogi), Fermented pap beverage (Mageu)	Nigeria (Ogi), South Africa (Mageu)
Pediococcus	*pentosaceus*	Millet	Porridge (Ogi)	Nigeria
Pediococcus	*pentosaceus*	Milk	Cheese (European types)	Europe
Pediococcus	*pentosaceus*	Mustard	Fermented vegetable (Gundruk)	Nepal
Pediococcus	*pentosaceus*	Pork	Fermented cased sausage (Salami)	Italy
Pediococcus	*pentosaceus*	Rice	Pancake (Appam), Steamed cake (Idli)	India and Sri Lanka (Idli), South India (Appam)
Pediococcus	*pentosaceus*	*Sorghum* grain	Bread, Porridge (Ogi)	Nigeria (Ogi), Saudi Arabia (Bread)
Pediococcus	*siamensis*	Tea leaves	Fermented tea [Pickled leaves which are to be chewed instead of brewed] (Miang)	Thailand
Periweissella	*ghanensis*	Cacao seeds	Cocoa beans	Ghana
Propionibacterium	*freudenreichii*	Milk	Cheese (Swiss type)	Europe
Proteus	*mirabilis*	Soybean	Fermented soybean paste (Aakhone), Fermented soybean (Hawaijar)	India
Proteus	*terrae* subsp. *cibarius*	Bivalve mollusks (intact)	Jeotgal	Korea
Proteus	*terrae* subsp. *cibarius*	Cephalopods (intact)	Jeotgal	Korea
Proteus	*terrae* subsp. *cibarius*	Crustaceans (intact)	Jeotgal	Korea
Proteus	*terrae* subsp. *cibarius*	Fish (intact)	Jeotgal	Korea
Proteus	*terrae* subsp. *cibarius*	Fish (internal organs)	Jeotgal	Korea
Proteus	*terrae* subsp. *cibarius*	Gastropod mollusks (intact)	Jeotgal	Korea
Proteus	*terrae* subsp. *cibarius*	Skate (intact)	Jeotgal	Korea
Providencia	*rettgeri*	Soybean	Fermented soybean (Hawaijar)	India
Salinicoccus	*jeotgali*	Bivalve mollusks (intact)	Jeotgal	Korea

Salinicoccus	*jeotgali*	Cephalopods (intact)	Jeotgal	Korea
Salinicoccus	*jeotgali*	Crustaceans (intact)	Jeotgal	Korea
Salinicoccus	*jeotgali*	Fish (intact)	Jeotgal	Korea
Salinicoccus	*jeotgali*	Fish (internal organs)	Jeotgal	Korea
Salinicoccus	*jeotgali*	Gastropod mollusks (intact)	Jeotgal	Korea
Salinicoccus	*jeotgali*	Skate (intact)	Jeotgal	Korea
Schleiferilactobacillus	*harbinensis*	Milk	Cheese (Hard cheese, example is Parmesan)	Italy (Parmesan)
Schleiferilactobacillus	*harbinensis*	Sugar	Fermented beverage (Tibicos, also called Water Kefir)	Global
Secundilactobacillus	*collinoides*	Grape juice	Wine	Global
Secundilactobacillus	*malefermentans*	Barley	Beer (Historical ale)	United Kingdom
Serratia	*marcescens*	*Sorghum* grain	Bread	Saudi Arabia
Serratia	*odorifera*	*Sorghum* grain	Bread	Saudi Arabia
Shigella	*sonnei*	Barley	Beer (Historical ale)	United Kingdom
Staphylococcus	*aureus*	Milk	Yogurt	Western Asia
Staphylococcus	*aureus*	Soybean	Fermented soybean (Hawaijar)	India
Staphylococcus	*carnosus*	Beef with Pork	Fermented cased sausage (Salami)	Northern Europe
Staphylococcus	*carnosus*	Pork	Fermented cased sausage (Salami)	Italy
Staphylococcus	*chromogenes*	Pork	Fermented cased sausage (Salami)	Italy
Staphylococcus	*epidermidis*	Barley	Beer (Historical ale)	United Kingdom
Staphylococcus	*epidermidis*	Pork	Fermented cased sausage (Salami)	Italy
Staphylococcus	*equorum*	Pork	Fermented cased sausage (Salami)	Italy
Staphylococcus	*haemolyticus*	Tomato (in water with sugar)	Fermented tomato	Global
Staphylococcus	*hominis*	Pork	Fermented cased sausage (Salami)	Italy
Staphylococcus	*kloosii*	Pork	Fermented cased sausage (Salami)	Italy
Staphylococcus	*lugdunensis*	Pork	Fermented cased sausage (Salami)	Italy
Staphylococcus	*pasteuri*	Pork	Fermented cased sausage (Salami)	Italy

(*Continued*)

TABLE 1.1 (Continued)

Genus	Species	Primary Material Being Fermented	Product	Geographical Origin
Staphylococcus	*saprophyticus*	Beef with Pork	Fermented cased sausage (Salami)	Northern Europe
Staphylococcus	*saprophyticus*	Pork	Fermented cased sausage (Salami)	Italy
Staphylococcus	*saprophyticus*	Tomato (in water with sugar)	Fermented tomato	Global
Staphylococcus	*simulans*	Pork	Fermented cased sausage (Salami)	Italy
Staphylococcus	*warneri*	Pork	Fermented cased sausage (Salami)	Italy
Staphylococcus	*xylosus*	Beef with Pork	Fermented cased sausage (Salami)	Northern Europe
Staphylococcus	*xylosus*	Maize	Porridge (Ogi)	Nigeria
Staphylococcus	*xylosus*	Millet	Porridge (Ogi)	Nigeria
Staphylococcus	*xylosus*	Pork	Fermented cased sausage (Salami)	Italy
Staphylococcus	*xylosus*	*Sorghum* grain	Porridge (Ogi)	Nigeria
Stenotrophomonas	*maltophilia*	Barley	Beer (Historical ale)	United Kingdom
Streptococcus	*parauberis*	Chautsi bass (intact fish)	Fermented fish (Chouguiyu)	China
Streptococcus	*salivarius*	Milk	Cheese	Global
Streptococcus	*thermophilus*	Milk	(Semi-soft cheese, example is Mozzarella), (Hard cheese, example is Parmesan), (Kashkaval cheese), Fermented beverage (Kefir), Yogurt	Balkans and Southeast Europe (Kashkaval), Italy (Mozzarella, Parmesan), North Caucasus (Kefir), Western Asia (Yogurt)
Streptococcus	*thermophilus*	Wheat	Deep-fried batter (Jalebi)	Western Asia
Streptococcus	*thermophilus*	Yogurt with wheat flour	Yellow curd (a type of Kashk)	Iran
Tetragenococcus	*halophilus*	Fish with rice	Fermented seasoning (Pla ra)	Thailand
Vagococcus	*lutrae*	Soybean	Fermented soybean (Peruyaan)	India
Varibaculum	*cambriense*	Barley	Beer (Historical ale)	United Kingdom
Weissella	*cibaria*	Cabbage	Fermented vegetable (Kimchi)	Korea
Weissella	*cibaria*	Cashew	Vegan cheese	Canada
Weissella	*cibaria*	Chautsi bass (intact fish)	Fermented fish (Chouguiyu)	China
Weissella	*cibaria*	Fish with rice	Fermented seasoning (Pla ra)	Thailand

Weissella	*cibaria*	Quinoa	Starter culture for making vegan cheese	Canada
Weissella	*cibaria*	Rice	Steamed cake (Idli)	India and Sri Lanka
Weissella	*confusa*	Barley	Fermented pap beverage (Boza)	Possibly Turkey
Weissella	*confusa*	Cabbage	Fermented vegetable (Kimchi)	Korea
Weissella	*confusa*	Fish with rice	Fermented seasoning (Pla ra)	Thailand
Weissella	*confusa*	Maize	Fermented dough used to make a pap beverage (Pozol), Fermented pap beverage (Boza, Mageu)	Mexico (Pozol), Possibly Turkey (Boza), South Africa (Mageu)
Weissella	*confusa*	Milk	Cheese	Global
Weissella	*confusa*	Millet	Fermented pap beverage (Boza)	Possibly Turkey
Weissella	*confusa*	Rice	Gruel (Panta bhat), Steamed cake (Idli)	Bangladesh (Panta bhat), India and Sri Lanka, (Idli)
Weissella	*confusa*	Wheat	Fermented pap beverage (Boza)	Possibly Turkey
Weissella	*jogaejeotgali*	Bivalve mollusks (intact)	Jeotgal	Korea
Weissella	*jogaejeotgali*	Cephalopods (intact)	Jeotgal	Korea
Weissella	*jogaejeotgali*	Crustaceans (intact)	Jeotgal	Korea
Weissella	*jogaejeotgali*	Fish (intact)	Jeotgal	Korea
Weissella	*jogaejeotgali*	Fish (internal organs)	Jeotgal	Korea
Weissella	*jogaejeotgali*	Gastropod molluscs (intact)	Jeotgal	Korea
Weissella	*jogaejeotgali*	Skate (intact)	Jeotgal	Korea
Weissella	*koreensis*	Cabbage	Fermented vegetable (Kimchi)	Korea
Weissella	*paramesenteroides*	Barley	Fermented pap beverage (Boza)	Possibly Turkey
Weissella	*paramesenteroides*	Fish with rice	Fermented seasoning (Pla ra)	Thailand
Weissella	*paramesenteroides*	Maize	Fermented pap beverage (Boza)	Possibly Turkey
Weissella	*paramesenteroides*	Millet	Fermented pap beverage (Boza)	Possibly Turkey
Weissella	*paramesenteroides*	Olive	Fermented olive	Mediterranean area
Weissella	*paramesenteroides*	Wheat	Fermented pap beverage (Boza)	Possibly Turkey
Weissella	*thailandensis*	Walleye pollock meat	Fermented fish meat (Sikhae)	Korea

(*Continued*)

TABLE 1.1 (Continued)

Genus	Species	Primary Material Being Fermented	Product	Geographical Origin
Weissella	*viridescens*	Cassava	Beer (Cassava beer, Chicha)	Africa (Cassava beer), Brazil (Chicha)
Weissella	*viridescens*	Chañar fruit	Beer (Chicha)	Chile
Weissella	*viridescens*	Fish with rice	Fermented seasoning (Pla ra)	Thailand
Weissella	*viridescens*	Kañiwa	Beer (Chicha)	Andean South America
Weissella	*viridescens*	Oca	Beer (Chicha)	Brazil
Weissella	*viridescens*	Palm fruit	Beer (Chicha, Palm beer)	Central America and South America (Chicha), Fiji (Palm beer)
Weissella	*viridescens*	Peanut	Beer (Chicha)	Peru
Weissella	*viridescens*	Pork	Fermented cased sausage (Salami)	Italy
Weissella	*viridescens*	Potato	Beer (Chicha, Potato beer)	Germany and Netherlands (Potato beer), Northern South America (Chicha)
Weissella	*viridescens*	Quinoa	Beer (Cusqueña)	Peru
Weizmannia	*coagulans*	Soybean	Fermented paste or curry, (Bekang), Curry Kinema)	India

Many of these processes are wild fermentations that involve a mixture of microorganisms and often they are dominated in succession by different microbial species whose relative presence changes as the chemistry of the food becomes altered. *Enterococcus faecalis* contributes to the fermentation but presumably represents fecal contamination. The juices used as a base material for preparing vinegar typically are either fruit juices, potato juice, or rice juice that previously had been anaerobically fermented to produce ethanol. Taberna and Tari are examples of palm wine which is made using sap either from the flowers or tapped from the trees of *Acrocomia aculeata*, *Borassus flabellifer*, *Caryota urens*, *Cocos nucifera*, *Elaeis guineensis*, *Hyphaene coriacea*, *Phoenix dactylifera*, and *Phoenix sylvestris*.

TABLE 1.2 Examples of fungi that are present in food fermentations.

Genus	Species	Primary Material Being Fermented	Product	Geographical Origin
Acremonium	*charticola*	Barley	Beer (Historical ale)	United Kingdom
Actinomucor	*elegans*	Soybean	Coagulated soy milk (Tofu)	China
Actinomucor	*elegans*	Soybean curd	Fermented bean curd (Fermented Tofu)	China
Aspergillus	*clavatus*	Milk	Cheese (Washed rind cheese, also called "stinky cheese," examples are Limburger and Munster)	Belgium (Limburger), France (Munster)
Aspergillus	*flavus*	Maize	Porridge (Ogi)	Nigeria
Aspergillus	*flavus*	Milk	Cheese	Global
Aspergillus	*flavus*	Millet	Porridge (Ogi)	Nigeria
Aspergillus	*flavus*	*Sorghum* grain	Porridge (Ogi)	Nigeria
Aspergillus	*fumigatus*	Tea leaves	Fermented tea leaves [To be brewed for beverage] (Qingzhuan brick tea)	China
Aspergillus	*fumigatus*	Tomato (in water with sugar)	Fermented tomato	Global
Aspergillus	*luchuensis*	Barley	Distilled beverage (Shōchū)	Japan
Aspergillus	*luchuensis*	Brown sugar	Distilled beverage (Shōchū)	Japan
Aspergillus	*luchuensis*	Rice	Distilled beverage (Shōchū)	Japan
Aspergillus	*luchuensis*	Sweet potato	Distilled beverage (Shōchū)	Japan
Aspergillus	*luchuensis*	Tea leaves	Fermented tea [Dry fermented leaves subsequently brewed to produce a beverage] (Pu'er tea)	China
Aspergillus	*niger*	Maize	Porridge (Ogi)	Nigeria
Aspergillus	*niger*	Millet	Porridge (Ogi)	Nigeria
Aspergillus	*niger*	Olive press cake	Fermented paste	Morocco
Aspergillus	*niger*	*Sorghum* grain	Bread, Porridge (Ogi)	Nigeria (Ogi), Saudi Arabia (Bread)
Aspergillus	*oryzae*	Barley	Distilled beverage (Shōchū)	Japan
Aspergillus	*oryzae*	Brown sugar	Distilled beverage (Shōchū)	Japan
Aspergillus	*oryzae*	Rice	Distilled beverage (Shōchū), Pap (Amazake), Wine (Sake)	Japan

(*Continued*)

TABLE 1.2 (Continued)

Genus	Species	Primary Material Being Fermented	Product	Geographical Origin
Aspergillus	*oryzae*	Soybean	Coagulated soy milk (Tofu), Fermented paste (Miso), Liquid condiment (Soy sauce)	China (Soy sauce, Tofu), Japan (Miso)
Aspergillus	*oryzae*	Sweet potato	Distilled beverage (Shōchū)	Japan
Aspergillus	*sojae*	Soybean	Fermented paste (Miso), Liquid condiment (Soy sauce)	China (Soy sauce), Japan (Miso)
Blastobotrys	*adeninivorans*	*Coffea* seeds	Coffee	Yemen
Brettanomyces	*anomalus*	Milk	Fermented beverage (Kefir)	North Caucasus
Brettanomyces	*anomalus*	Wheat	Bread (Sourdough)	Asia Minor and Europe
Brettanomyces	*bruxellensis*	Barley	Beer, (Historical ale)	Belgium (Beer), United Kingdom (Historical ale)
Brettanomyces	*bruxellensis*	Grape juice	Wine	Global
Brettanomyces	*bruxellensis*	Sugar	Fermented beverage (Tibicos, also called Water Kefir)	Global
[*Candida*]	*boidinii*	Barley	Beer (Historical ale)	United Kingdom
[*Candida*]	*ethanolica*	Tea leaves	Fermented tea [Pickled leaves which are to be chewed instead of brewed] (Miang)	Thailand
[*Candida*]	*glabrata*	*Coffea* seeds	Coffee	Yemen
[*Candida*]	*glabrata*	Maize	Fermented pap beverage (Mageu)	South Africa
[*Candida*]	*glabrata*	*Saccharum* sap	Distilled beverage (Rum)	Global
[*Candida*]	*glabrata* (closest genomic match)	Sleeper shark meat	Fermented shark meat (Hákarl)	Iceland
[*Candida*]	*glaebosa*	Milk	Cheese	Global
[*Candida*]	*haemuloni*	Maize	Fermented pap beverage (Mageu)	South Africa
[*Candida*]	*inconspicua*	Milk	Cheese	Global
[*Candida*]	*intermedia*	Milk	Cheese	Global

[*Candida*]	*intermedia*	Palm sap (from palm flowers and tapped trees)	Palm wine (Taberna, Tari)	Africa, Caribbean, Mexico (Taberna), Micronesia, Philippines, South America, South Asia (Tari), Southeast Asia
[*Candida*]	*membranifaciens*	*Coffea* seeds	Coffee	Yemen
[*Candida*]	*mogii*	Milk	Cheese	Global
[*Candida*]	*saitoana*	*Coffea* seeds	Coffee	Yemen
[*Candida*]	*zeylanoides*	Maize	Beer (Chicha)	Argentina
[*Candida*]	*zeylanoides*	Milk	Cheese	Global
[*Candida*]	*zeylanoides*	Sleeper shark meat	Fermented shark meat (Hákarl)	Iceland
Candida	*parapsilosis*	*Coffea* seeds	Coffee	Yemen
Candida	*parapsilosis*	Maize	Beer (Chicha)	Argentina
Candida	*parapsilosis*	Milk	Cheese	Global
Candida	*parapsilosis*	Sleeper shark meat	Fermented shark meat (Hákarl)	Iceland
Candida	*parapsilosis*	*Sorghum* grain	Bread	Saudi Arabia
Candida	*parapsilosis*	Soybean	Curry (Kinema)	India
Candida	*tropicalis*	Cassava	Fermented cassava root subsequently granulated into a flour (Garri)	West Africa
Candida	*tropicalis*	Maize	Porridge (Ogi)	Nigeria
Candida	*tropicalis*	Millet	Porridge (Ogi)	Nigeria
Candida	*tropicalis*	Palm sap (from palm flowers and tapped trees)	Palm wine (Taberna, Tari)	Africa, Caribbean, Mexico (Taberna), Micronesia, Philippines, South America, South Asia (Tari), Southeast Asia
Candida	*tropicalis* (closest genomic match)	Sleeper shark meat	Fermented shark meat (Hákarl)	Iceland
Candida	*tropicalis*	*Sorghum* grain	Porridge (Ogi)	Nigeria
Cladosporium	*cladosporioides*	Milk	Cheese	Global
Cladosporium	*tenuissimum* (closest genomic match)	Sleeper shark meat	Fermented shark meat (Hákarl)	Iceland
Cutaneotrichosporon	*cutaneum*	Milk	Cheese	Global

(*Continued*)

TABLE 1.2 (Continued)

Genus	Species	Primary Material Being Fermented	Product	Geographical Origin
Cyberlindnera	*jadinii*	Milk	Cheese (Brick)	United States
Cystobasidium	*slooffiae*	Maize	Beer (Chicha)	Argentina
Debaryomyces	*coudertii*	Barley	Beer (Historical ale)	United Kingdom
Debaryomyces	*hansenii*	Barley	Beer (Historical ale) Bread (Sourdough)	Asia Minor and Europe (Sourdough), United Kingdom (Historical ale)
Debaryomyces	*hansenii*	*Coffea* seeds	Coffee	Yemen
Debaryomyces	*hansenii*	Grape juice	Wine	Global
Debaryomyces	*hansenii*	Lentils	Griddle cake (Dosa)	India
Debaryomyces	*hansenii*	Maize	Fermented pap beverage (Mageu)	South Africa
Debaryomyces	*hansenii*	Milk	Cheese (Brick)	Global, United States (Brick)
Debaryomyces	*hansenii*	Pork	Fermented cased sausage (Salami)	Italy
Debaryomyces	*hansenii*	Rice	Bread (Sourdough)	Asia Minor and Europe
Debaryomyces	*hanseni*	Teff (may contain as replacement flour Barley, Maize, Rice, or Wheat)	Flatbread (Injera)	East Africa
Debaryomyces	*hansenii*	Wheat	Bread (Sourdough)	Asia Minor and Europe
Diutina	*catenulata*	Milk	Cheese	Global
Filobasidium	*magnum*	Maize	Beer (Chicha)	Argentina
Geotrichum	*bryndzae*	Milk	Cheese	Global
Geotrichum	*candidum*	Milk	Cheese (Acid-set cheese also called Sourmilk cheese, example is Cottage cheese), (Surface mold ripened cheese, examples are Brie and Camembert), (Washed rind cheese, also called 'stinky cheese', examples are Limburger and Munster)	Africa, Belgium (Limburger), France (Brie, Camembert, Munster), United States (Cottage cheese)
Geotrichum	*candidum*	Soybean	Curry (Kinema)	India

Geotrichum	*candidum*	Taro	Fermented paste (Poi)	Hawaii
Hanseniaspora	*guilliermondii*	Grape juice	Wine	Global
Hanseniaspora	*guilliermondii*	Palm sap (from palm flowers and tapped trees)	Palm wine (Taberna, Tari)	Africa, Caribbean, Mexico (Taberna), Micronesia, Philippines, South America, South Asia (Tari), Southeast Asia
Hanseniaspora	*occidentalis*	Palm sap (from palm flowers and tapped trees)	Palm wine (Taberna, Tari)	Africa, Caribbean, Mexico (Taberna), Micronesia, Philippines, South America, South Asia (Tari), Southeast Asia
Hanseniaspora	*opuntiae*	*Coffea* seeds	Coffee	Yemen
Hanseniaspora	*uvarum*	Apple juice (Unfermented cider)	Alcoholic fermented apple juice (Variously identified as Cider or Hard Cider), Distilled beverage (Apple brandy, Calvados), Fractional freezing beverage (Applejack)	Central Asia and Europe (Cider), North America (Hard Cider), France (Apple brandy, Calvados), United States (Applejack)
Hanseniaspora	*uvarum*	*Coffea* seeds	Coffee	Yemen
Hanseniaspora	*uvarum*	Grape juice	Wine	Global
Hanseniaspora	*uvarum*	Maize	Beer (Chicha)	Argentina
Hanseniaspora	*vineae*	Grape juice	Wine	Global
Hyphodontia	*sambuci*	Barley	Beer (Historical ale)	United Kingdom
Hyphopichia	*burtonii*	*Coffea* seeds	Coffee	Yemen
Kazachstania	*exigua*	Milk	Fermented beverage (Kefir)	North Caucasus
Kazachstania	*exigua*	Palm sap (from palm flowers and tapped trees)	Palm wine (Taberna, Tari)	Africa, Caribbean, Mexico (Taberna), Micronesia, Philippines, South America, South Asia (Tari), Southeast Asia
Kazachstania	*humilis*	Barley	Bread (Sourdough)	Asia Minor and Europe
Kazachstania	*humilis*	Teff (may contain as replacement flour Barley, Maize, Rice, or Wheat)	Flatbread (Injera)	East Africa
Kazachstania	*humilis*	Wheat	Bread (Sourdough)	Asia Minor and Europe
Kazachstania	*servazzii*	*Sorghum* grain	Distilled beverage (Baijiu)	China
Kazachstania	*unispora*	Milk	Cheese, Fermented beverage (Kefir)	Global (Cheese), North Caucasus (Kefir)

(*Continued*)

TABLE 1.2 (Continued)

Genus	Species	Primary Material Being Fermented	Product	Geographical Origin
Kazachstania	*unispora*	Palm sap (from palm flowers and tapped trees)	Palm wine (Taberna, Tari)	Africa, Caribbean, Mexico (Taberna), Micronesia, Philippines, South America, South Asia (Tari), Southeast Asia
Kluyveromyces	*lactis*	Milk	Cheese, Fermented beverage (Kefir)	Global (Cheese), North Caucasus (Kefir)
Kluyveromyces	*lactis*	Sugar	Fermented beverage (Tibicos, also called Water Kefir)	Global
Kluyveromyces	*marxianus*	Apple juice (Unfermented cider)	Alcoholic fermented apple juice (Variously identified as Cider or Hard Cider), Distilled beverage (Apple brandy, Calvados), Fractional freezing beverage (Applejack)	Central asia and Europe (Cider), North America (Hard Cider), France (Apple brandy, Calvados), United States (Applejack)
Kluyveromyces	*marxianus*	Cassava	Beer (Chicha)	Africa, Brazil (Chicha)
Kluyveromyces	*marxianus*	Chañar fruit	Beer (Chicha)	Chile
Kluyveromyces	*marxianus*	Kañiwa	Beer (Chicha)	Andean South America
Kluyveromyces	*marxianus*	Maize	Beer (Chicha), Porridge (Ogi)	Andean South America and Costa Rica (Chicha), Nigeria (Ogi)
Kluyveromyces	*marxianus*	Milk	Cheese, Fermented beverage (Kefir)	Global (Cheese), North Caucasus (Kefir)
Kluyveromyces	*marxianus*	Millet	Porridge (Ogi)	Nigeria
Kluyveromyces	*marxianus*	Oca	Beer (Chicha)	Brazil
Kluyveromyces	*marxianus*	Palm fruit	Beer (Chicha)	Central America (Chicha), South America (Chicha), Fiji
Kluyveromyces	*marxianus*	Peanut	Beer (Chicha)	Peru
Kluyveromyces	*marxianus*	Potato	Beer (Chicha, Potato beer)	Germany and Netherlands (Potato beer), Northern South America (Chicha)
Kluyveromyces	*marxianus*	Quinoa	Beer (Cusqueña)	Peru
Kluyveromyces	*marxianus*	Rice	Beer (Chicha)	Japan, South America (Chicha),
Kluyveromyces	*marxianus*	*Sorghum* grain	Porridge (Ogi)	Nigeria

Kluyveromyces	*marxianus*	Teff (may contain as replacement flour Barley, Maize, Rice, or Wheat)	Flatbread (Injera)	East Africa
Kodamaea	*ohmeri*	Milk	Cheese	Global
Kodamaea	*ohmeri*	Palm sap (from palm flowers and tapped trees)	Palm wine (Taberna, Tari)	Africa, Caribbean, Mexico (Taberna), Micronesia, Philippines, South America, South Asia (Tari), Southeast Asia
Kregervanrija	*fluxuum*	Grape juice	Wine	Global
Kurtzmaniella	*quercitrusa*	*Coffea* seeds	Coffee	Yemen
Kurtzmaniella	*quercitrusa*	Grape juice	Wine	Global
Lachancea	*thermotolerans*	Grape juice	Wine	Global
Medusomyces	*gisevii*	Tea (aqueous extract of tea leaves) plus sugar	Fermented tea (Kombucha)	China
Metschnikowia	*pulcherrima*	Barley	Beer (Historical ale)	United Kingdom
Metschnikowia	*pulcherrima*	Grape juice	Wine	Global
Metschnikowia	*viticola*	Grape juice	Wine	Global
Meyerozyma	*caribbica*	Cassava	Beer (Chicha)	South America
Meyerozyma	*caribbica*	*Coffea* seeds	Coffee	Yemen
Meyerozyma	*caribbica*	Tomato (in water with sugar)	Fermented tomato	Global
Meyerozyma	*carpophila*	*Coffea* seeds	Coffee	Yemen
Meyerozyma	*carpophila*	Tomato (in water with sugar)	Fermented tomato	Global
Meyerozyma	*guilliermondii*	*Coffea* seeds	Coffee	Yemen
Meyerozyma	*guilliermondii*	Maize	Beer (Chicha)	Argentina
Meyerozyma	*guilliermondii*	Milk	Cheese	Global
Meyerozyma	*guilliermondii*	Palm sap (from palm flowers and tapped trees)	Palm wine (Taberna, Tari)	Africa, Caribbean, Mexico (Taberna), Micronesia, Philippines, South America, South Asia (Tari), Southeast Asia
Millerozyma	*farinosa*	Milk	Cheese	Global

(*Continued*)

TABLE 1.2 (Continued)

Genus	Species	Primary Material Being Fermented	Product	Geographical Origin
Moristroma	*quercinum* (closest genomic match)	Sleeper shark meat	Fermented shark meat (Hákarl)	Iceland
Mucor	*circinelloides*	Milk	Cheese (Washed rind cheese, also called 'stinky cheese', examples are Limburger and Munster), Yogurt	Belgium (Limburger), France (Munster), Wesern Asia (Yogurt)
Mucor	*genevensis*	Milk	Yogurt	Wesern Asia
Mucor	*hiemalis*	Milk	Cheese, Yogurt	Global (Cheese), Wesern Asia (Yogurt)
Mucor	*indicus*	Rice	Rice wine	China
Mucor	*plumbeus*	Milk	Cheese	Global
Mucor	*racemosus*	Maize	Porridge (Ogi)	Nigeria
Mucor	*racemosus*	Milk	Cheese, Yogurt	Global (Cheese), Wesern Asia (Yogurt)
Mucor	*racemosus*	Millet	Porridge (Ogi)	Nigeria
Mucor	*racemosus*	*Sorghum* grain	Porridge (Ogi)	Nigeria
Mucor	*spinosus*	Milk	Cheese	Global
Nakazawaea	*ernobii*	*Coffea* seeds	Coffee	Yemen
Ogataea	*minuta*	*Saccharum* sap	Distilled beverage (Rum)	Global
Ogataea	*naganishii*	Teff (may contain as replacement flour Barley, Maize, Rice, or Wheat)	Flatbread (Injera)	East Africa
Papiliotrema	*flavescens*	Maize	Beer (Chicha)	Argentina
Papiliotrema	*nemorosus*	Maize	Beer (Chicha)	Argentina
Penicillium	*adametzioides*	Cream	Sour cream (Crème Fraiche)	Central America, Eastern Europe, France (Crème Fraiche)
Penicillium	*aurantiogriseum*	Milk	Cheese (Semi-hard cheese, example is Cheddar)	England (Cheddar)
Penicillium	*bialowiezense*	Meat (presumably was pork)	Dry fermented sausage	Europe

Penicillium	*bialowiezense*	Milk	Cheese, Yogurt	Global (cheese), Western Asia (Yogurt)
Penicillium	*brevicompactum*	Milk	Cheese	Global
Penicillium	*camemberti*	Beef with Pork	Dry fermented cased sausage (Salami)	Europe
Penicillium	*camemberti*	Milk	Cheese (Surface mold ripened cheese, examples are Brie and Camembert), (Washed rind cheese, also called 'stinky cheese', examples are Limburger and Munster)	Belgium (Limburger), France (Brie, Camembert, Munster)
Penicillium	*camemberti*	Pork	Fermented cased sausage (Salami)	Italy
Penicillium	*caseifulvum*	Milk	Cheese (Washed rind cheese, also called 'stinky cheese', examples are Limburger and Munster)	Belgium (Limburger), France (Munster)
Penicillium	*citrinum*	Milk	Yogurt	Wesern Asia
Penicillium	*commune*	Milk	Cheese, Yogurt	Global (cheese), Western Asia (Yogurt)
Penicillium	*commune*	Olive	Fermented olive	Mediterranean
Penicillium	*corylophilum*	Milk	Cheese (Semi-hard cheese, example is Cheddar)	England (Cheddar)
Penicillium	*crustosum*	Milk	Yogurt	Western Asia
Penicillium	*crustosum*	Olive	Fermented olive	Mediterranean
Penicillium	*discolor*	Milk	Cheese	Global
Penicillium	*expansum*	Milk	Cheese (Fresh cheese)	Global
Penicillium	*expansum*	Olive	Fermented olive	Mediterranean
Penicillium	*fellutanum*	Milk with Cream	Cream Cheese	United States
Penicillium	*implicatum*	Milk	Cheese (Semi-hard cheese, example is Cheddar)	England (Cheddar)
Penicillium	*nalgiovense*	Pork	Dry fermented cased sausage	Europe
Penicillium	*nordicum*	Pork	Dry fermented cased sausage	Europe
Penicillium	*oxalicum*	Maize	Porridge (Ogi)	Nigeria
Penicillium	*oxalicum*	Millet	Porridge (Ogi)	Nigeria
Penicillium	*oxalicum*	*Sorghum* grain	Porridge (Ogi)	Nigeria
Penicillium	*palitans*	Milk with Cream	Cream Cheese	United States
Penicillium	*paneum*	Olive	Fermented olive	Mediterranean
Penicillium	*paxilli*	Milk	Cheese (Semi-hard cheese, example is Cheddar)	England (Cheddar)

(*Continued*)

TABLE 1.2 (Continued)

Genus	Species	Primary Material Being Fermented	Product	Geographical Origin
Penicillium	*polonicum*	Olive	Fermented olive	Mediterranean
Penicillium	*roqueforti*	Cream	Sour cream (Crème Fraiche)	Central America, Eastern Europe, France (Crème Fraiche)
Penicillium	*roqueforti*	Milk	Cheese (Surface mold ripened cheese, examples are Brie and Camembert), (Blue veined cheese, examples are Roquefort and Stilton)	England (Stilton), France (Brie, Camembert, Roquefort)
Penicillium	*roqueforti*	Olive	Fermented olive	Mediterranean
Penicillium	*salamii*	Pork	Dry fermented cased sausage	Europe
Penicillium	*solitum*	Milk	Cheese (Semi-hard cheese, example is Cheddar), Yogurt	England (Cheddar), Western Asia (Yogurt)
Penicillium	*verrucosum*	Milk (specifically from sheep)	Cheese (Pecorino)	Italy
Phialotubus	*microsporus*	Barley	Beer (Historical ale)	United Kingdom
Pichia	*fermentans*	*Coffea* seeds	Coffee	Yemen
Pichia	*fermentans*	Milk	Cheese, Fermented beverage (Kefir)	Global (Cheese), North Caucasus (Kefir)
Pichia	*fermentans*	Wheat	Bread (Sourdough)	Asia Minor and Europe
Pichia	*kluyveri*	Grape juice	Wine	Global
Pichia	*kluyveri*	Palm sap (from palm flowers and tapped trees)	Palm wine (Taberna, Tari)	Africa, Caribbean, Mexico (Taberna), Micronesia, Philippines, South America, South Asia (Tari), Southeast Asia
Pichia	*kudriavzevii*	Cacao seeds	Cocoa beans	Equatorial South America
Pichia	*kudriavzevii*	Cassava	Fermented cassava root subsequently granulated into a flour (Garri)	West Africa
Pichia	*kudriavzevii*	Grape juice	Wine	Global
Pichia	*kudriavzevii*	Maize	Porridge (Ogi)	Nigeria
Pichia	*kudriavzevii*	Milk	Cheese, Fermented beverage (Kefir)	Global (Cheese), North Caucasus (Kefir)

Pichia	*kudriavzevii*	Millet	Porridge (Ogi)	Nigeria
Pichia	*kudriavzevii*	Palm sap (from palm flowers and tapped trees)	Palm wine (Taberna, Tari)	Africa, Caribbean, Mexico (Taberna), Micronesia, Philippines, South America, South Asia (Tari), Southeast Asia
Pichia	*kudriavzevii*	*Saccharum* sap	Distilled beverage (Rum)	Global
Pichia	*kudriavzevii*	*Sorghum* grain	Distilled beverage (Baijiu), Porridge (Ogi)	China (Baijiu), Nigeria (Ogi)
Pichia	*manshurica*	*Sorghum* grain	Distilled beverage (Baijiu)	China
Pichia	*membranifaciens*	Grape juice	Wine	Global
Pichia	*membranifaciens*	Maize	Beer (Chicha)	Argentina
Pichia	*membranifaciens*	Milk	Cheese	Global
Pichia	*membranifaciens*	*Saccharum* sap	Distilled beverage (Rum)	Global
Pichia	*membranifaciens*	Wheat	Bread (Sourdough)	Asia Minor and Europe
Pichia	*norvegensis*	*Sorghum* grain	Bread	Saudi Arabia
Pichia	*occidentalis*	Grape juice	Wine	Global
Pichia	*scaptomyzae*	*Sorghum* grain	Distilled beverage (Baijiu)	China
Pichia	*terricola*	Grape juice	Wine	Global
Priceomyces	*carsonii*	Grape juice	Wine	Global
Rhizopus	*arrhizus*	Rice	Rice wine	China
Rhizopus	*arrhizus*	Soybean	Liquid condiment (Soy sauce)	China
Rhizopus	*arrhizus*	Soybean curd	Fermented bean curd (Fermented Tofu)	China
Rhizopus	*microsporus*	Soybean curd	Fermented bean curd (Fermented Tofu)	China
Rhizopus	*microsporus var. oligosporus*	Soybean	Fermented soybean (Tempeh)	Indonesia
Rhizopus	*stolonifer*	Maize	Porridge (Ogi)	Nigeria
Rhizopus	*stolonifer*	Millet	Porridge (Ogi)	Nigeria
Rhizopus	*stolonifer*	*Sorghum* grain	Porridge (Ogi)	Nigeria
Rhodotorula	*glutinis*	*Sorghum* grain	Bread	Saudi Arabia
Rhodotorula	*mucilaginosa*	*Coffea* seeds	Coffee	Yemen
Rhodotorula	*mucilaginosa*	Maize	Beer (Chicha)	Argentina

(*Continued*)

TABLE 1.2 (Continued)

Genus	Species	Primary Material Being Fermented	Product	Geographical Origin
Rhodotorula	*mucilaginosa*	Pork	Fermented cased sausage (Salami)	Italy
Rhodotorula	*mucilaginosa*	Teff (may contain as replacement flour Barley, Maize, Rice, or Wheat)	Flatbread (Injera)	East Africa
Saccharomyces	*bayanus*	Apple juice (Unfermented cider)	Alcoholic fermented apple juice (Variously identified as Cider or Hard Cider), Distilled beverage (Apple brandy, Calvados), Fractional freezing beverage (Applejack)	Central Asia and Europe (Cider), North America (Hard Cider), France (Apple brandy, Calvados), United States (Applejack)
Saccharomyces	*bayanus*	Barley	Beer (Historical ale)	United Kingdom
Saccharomyces	*bayanus*	*Coffea* seeds	Coffee	Yemen
Saccharomyces	*bayanus*	Grape juice	Wine	Global
Saccharomyces	*bayanus*	Honey	Wine (Mead)	Africa, Asia, and Europe
Saccharomyces	*bayanus*	Wheat	Deep-fried batter (Jalebi)	Western Asia
Saccharomyces	*cerevisiae*	*Agave* sap	Distilled beverage (Tequila)	Mexico
Saccharomyces	*cerevisiae*	Anchovy (intact) plus other ingredients	Worcestershire sauce	England
Saccharomyces	*cerevisiae*	Apple juice (Unfermented cider)	Alcoholic fermented apple juice (Variously identified as Cider or Hard Cider), Distilled beverage (Apple brandy, Calvados), Fractional freezing beverage (Applejack)	Central Asia and Europe (Cider), North America (Hard Cider), France (Apple brandy, Calvados), United States (Applejack)
Saccharomyces	*cerevisiae*	Barley	Beer (Ale, Historical ale), Distilled beverage (Malt whisky), Fermented pap beverage (Boza)	Europe (Ale), Ireland and Scotland (Malt whisky), Possibly Turkey (Boza), United Kingdom (Historical ale)
Saccharomyces	*cerevisiae*	Brown sugar	Distilled beverage (Shōchū)	Japan
Saccharomyces	*cerevisiae*	Cacao seeds	Cocoa beans	Equatorial South America
Saccharomyces	*cerevisiae*	Cassava	Beer (Chicha)	Africa, Brazil (Chicha)
Saccharomyces	*cerevisiae*	Chañar fruit	Beer (Chicha)	Chile

Saccharomyces	*cerevisiae*	Coconut water (liquid from coconuts) plus sugar	Fermented beverage	Asia
Saccharomyces	*cerevisiae*	*Coffea* seeds	Coffee	Yemen
Saccharomyces	*cerevisiae*	Fruit	Distilled beverage (Vodka)	Eastern Europe
Saccharomyces	*cerevisiae*	Grape juice	Wine	Global
Saccharomyces	*cerevisiae*	Honey	Wine (Mead), Distilled beverage (Vodka)	Africa, Asia, Europe (Mead), Eastern Europe (Vodka)
Saccharomyces	*cerevisiae*	Kañiwa	Beer (Chicha)	Andean South America
Saccharomyces	*cerevisiae*	Lentils	Griddle cake (Dosa)	India
Saccharomyces	*cerevisiae*	Maize	Beer (Chicha), Distilled beverage (Grain whisky, Vodka), Fermented pap beverage (Boza, Mageu), Porridge (Ogi)	Andean South America and Costa Rica (Chicha), Canada and United States (Grain whisky), Eastern Europe (Vodka), Nigeria (Ogi), Possibly Turkey (Boza), South Africa (Mageu)
Saccharomyces	*cerevisiae*	Maple sap	Distilled beverage (Vodka)	Eastern Europe
Saccharomyces	*cerevisiae*	Milk	Cheese, Fermented beverage (Kefir)	Global (Cheese), North Caucasus (Kefir)
Saccharomyces	*cerevisiae*	Millet	Fermented pap beverage (Boza), Porridge (Ogi)	Nigeria (Ogi), Possibly Turkey (Boza)
Saccharomyces	*cerevisiae*	Molasses	Beer (Molasses beer), Distilled beverage (Rum)	Global
Saccharomyces	*cerevisiae*	Oca	Beer (Chicha)	Brazil
Saccharomyces	*cerevisiae*	Palm fruit	Beer (Chicha)	Central America and South America (Chicha), Fiji
Saccharomyces	*cerevisiae*	Palm sap (from palm flowers and tapped trees)	Palm wine (Taberna, Tari)	Africa, Caribbean, Mexico (Taberna), Micronesia, Philippines, South America, South Asia (Tari), Southeast Asia
Saccharomyces	*cerevisiae*	Peanut	Beer (Chicha)	Peru
Saccharomyces	*cerevisiae*	Pineapple peel plus sugar	Fermented beverage (Tepache)	Asia, Mexico (Tepache)
Saccharomyces	*cerevisiae*	Potato	Beer (Chicha, Potato beer), Distilled beverage (Vodka)	Eastern Europe (Vodka), Germany and Netherlands (Potato beer), Northern South America (Chicha)
Saccharomyces	*cerevisiae*	Quinoa	Beer (Cusqueña)	Peru

(*Continued*)

TABLE 1.2 (Continued)

Genus	Species	Primary Material Being Fermented	Product	Geographical Origin
Saccharomyces	*cerevisiae*	Rice	Beer (Chicha), Distilled beverage (Shōchū, Vodka), Pap (Amazake), Wine (Sake), Steamed cake (Idli)	Eastern Europe (Vodka), India and Sri Lanka (Idli), Japan (Amazake, Sake, Shōchū), South America (Chicha)
Saccharomyces	*cerevisiae*	Rye	Distilled beverage (Grain whisky, Vodka)	Canada and United States (Grain whisky), Eastern Europe (Vodka)
Saccharomyces	*cerevisiae*	*Saccharum* sap	Distilled beverage (Rum)	Global
Saccharomyces	*cerevisiae* (closest genomic match)	Sleeper shark meat	Fermented shark meat (Hákarl)	Iceland
Saccharomyces	*cerevisiae*	*Sorghum* grain	Distilled beverage (Baijiu, Vodka), Porridge (Ogi)	China (Baijiu), Eastern Europe (Vodka), Nigeria (Ogi)
Saccharomyces	*cerevisiae*	*Sorghum* sap	Distilled beverage (Sorghum Whiskey)	United States (Sorghum Whiskey)
Saccharomyces	*cerevisiae*	Sugar	Distilled beverage (Vodka), Fermented beverage (Tibicos, also called Water Kefir)	Eastern Europe (Vodka), Global (Tibicos)
Saccharomyces	*cerevisiae*	Sweet potato	Distilled beverage (Shōchū)	Japan
Saccharomyces	*cerevisiae*	Tea (aqueous extract of tea leaves) plus sugar	Fermented tea (Kombucha)	China
Saccharomyces	*cerevisiae*	Wheat	Bread (Sourdough), Deep-fried batter (Jalebi), Distilled beverage (Grain whisky, Vodka), Fermented pap beverage (Boza)	Asia Minor and Europe (Sourdough), Canada and United States (Grain whisky), Eastern Europe (Vodka), Possibly Turkey (Boza), Western Asia (Jalebi)
Saccharomyces	*chevalieri*	Palm sap (from palm flowers and tapped trees)	Palm wine (Taberna, Tari)	Africa, Caribbean, Mexico (Taberna), Micronesia, Philippines, South America, South Asia (Tari), Southeast Asia
Saccharomyces	*chevalieri*	*Saccharum* sap	Distilled beverage (Rum)	Global
Saccharomyces	*eubayanus*	Honey	Wine (Mead)	Africa, Asia, Europe (Mead)
Saccharomyces	*pastorianus*	Barley	Beer (Lager)	Bohemia

Saccharomyces	*pastorianus*	*Saccharum* sap	Distilled beverage (Rum)	Global
Saccharomyces	*uvarum*	Barley	Fermented pap beverage (Boza)	Possibly Turkey
Saccharomyces	*uvarum*	Grape juice	Wine	Global
Saccharomyces	*uvarum*	Maize	Fermented pap beverage (Boza)	Possibly Turkey
Saccharomyces	*uvarum*	Millet	Fermented pap beverage (Boza)	Possibly Turkey
Saccharomyces	*uvarum*	Wheat	Fermented pap beverage (Boza)	Possibly Turkey
Saccharomycodes	*ludwigii*	Barley	Beer (low alcohol)	Europe
Saccharomycodes	*ludwigii*	Palm sap (from palm flowers and tapped trees)	Palm wine (Taberna, Tari)	Africa, Caribbean, Mexico (Taberna), Micronesia, Philippines, South America, South Asia (Tari), Southeast Asia
Saccharomycopsis	*capsularis*	Maize	Fermented pap beverage (Mageu)	South Africa
Saturnispora	*mendoncae*	Milk	Cheese	Global
Schizosaccharomyces	*japonicus*	*Saccharum* sap	Distilled beverage (Rum)	Global
Schizosaccharomyces	*pombe*	Grape juice	Wine	Global
Schizosaccharomyces	*pombe*	Palm sap (from palm flowers and tapped trees)	Palm wine (Taberna, Tari)	Africa, Caribbean, Mexico (Taberna), Micronesia, Philippines, South America, South Asia (Tari), Southeast Asia
Schizosaccharomyces	*pombe*	*Saccharum* sap	Distilled beverage (Rum)	Global
Schwanniomyces	*polymorphus*	*Coffea* seeds	Coffee	Yemen
Starmerella	*apicola*	Honey	Wine (Mead)	Africa, Asia, Europe (Mead)
Starmerella	*bacillaris*	Grape juice	Wine	Global
Starmerella	*etchellsii*	Milk	Cheese	Global
Starmerella	*stellata*	Grape juice	Wine	Global
Stereum	*sanguinolentum*	Barley	Beer (Historical ale)	United Kingdom
Sugiyamaella	*smithiae*	*Coffea* seeds	Coffee	Yemen
Suhomyces	*xylopsoci*	Milk	Cheese	Global
Torulaspora	*delbrueckii*	Cassava	Beer (Chicha)	South America
Torulaspora	*delbrueckii*	*Coffea* seeds	Coffee	Yemen

(*Continued*)

TABLE 1.2 (Continued)

Genus	Species	Primary Material Being Fermented	Product	Geographical Origin
Torulaspora	*delbrueckii*	Grape juice	Wine	Global
Torulaspora	*delbrueckii*	Maize	Beer (Chicha)	Argentina
Torulaspora	*delbrueckii*	Milk	Cheese	Global
Torulaspora	*delbrueckii*	*Saccharum* sap	Distilled beverage (Rum)	Global
Torulaspora	*delbrueckii*	Wheat	Beer	Germany
Torulaspora	*microellipsoides*	*Saccharum* sap	Distilled beverage (Rum)	Global
Trichosporon	*beigelii*	Lentils	Griddle cake (Dosa)	India
Trichosporon	*coremiiforme*	Milk	Cheese	Global
Trichosporon	*lactis*	Milk	Cheese	Global
Vishniacozyma	*carnescens*	Maize	Beer (Chicha)	Argentina
Wickerhamiella	*pararugosa*	Milk	Cheese	Global
Wickerhamiella	*sorbophila*	Maize	Fermented pap beverage (Mageu)	South Africa
Wickerhamomyces	*anomalus*	*Coffea* seeds	Coffee	Yemen
Wickerhamomyces	*anomalus*	Maize	Beer (Chicha)	Argentina
Wickerhamomyces	*anomalus*	*Saccharum* sap	Distilled beverage (Rum)	Global
Wickerhamomyces	*anomalus*	*Sorghum* grain	Distilled beverage (Baijiu)	China
Wickerhamomyces	*anomalus*	Wheat	Deep-fried batter (Jalebi)	Western Asia
Wickerhamomyces	*onychis*	Tomato (in water with sugar)	Fermented tomato	Global
Wickerhamomyces	*subpelliculosus*	*Coffea* seeds	Coffee	Yemen
Wickerhamomyces	*sydowiorum*	*Coffea* seeds	Coffee	Yemen
Yarrowia	*lipolytica*	Milk	Cheese	Global
Yarrowia	*lipolytica*	Pork	Fermented cased sausage (Salami)	Italy
Zygoascus	*hellenicus*	Grape juice	Wine	Global
Zygosaccharomyces	*bailii*	Grape juice	Wine	Global

Zygosaccharomyces	*lentus*	Grape juice	Wine	Global
Zygosaccharomyces	*mellis*	*Sorghum* sap	Distilled beverage (Sorghum Whiskey)	United states (Sorghum Whiskey)
Zygosaccharomyces	*rouxii*	Honey	Wine (Mead)	Africa, Asia, Europe (Mead)
Zygosaccharomyces	*rouxii*	Milk	Cheese	Global
Zygosaccharomyces	*rouxii*	*Saccharum* sap	Distilled beverage (Rum)	Global
Zygosaccharomyces	*rouxii*	*Sorghum* sap	Distilled beverage (Sorghum Whiskey)	United states (Sorghum Whiskey)
Zygosaccharomyces	*rouxii*	Soybean	Liquid condiment (Soy sauce)	China
Zygotorulaspora	*florentina*	Ginger plus sugar	Alcoholic Ginger beer	United Kingdom
Zygotorulaspora	*florentina*	Grape juice	Wine	Global
Zygotorulaspora	*florentina*	Milk	Fermented beverage (Kefir)	North Caucasus
Zymomonas	*mobilis*	Palm sap (from palm flowers and tapped trees)	Palm wine (Taberna, Tari)	Africa, Caribbean, Mexico (Taberna), Micronesia, Philippines, South America, South Asia (Tari), Southeast Asia

Mead is wine made by fermenting honey that has been diluted with water. The term "Honey wine" often refers to wine made from fruit juices that have been sweetened with honey.

molecule of carbon dioxide plus a molecule of ethanol. The products of heterolactic fermentation thus are a combination of carbon dioxide, ethanol, and lactic acid. Generation of the three additional organic acids that have earned prominence in the creation and preservation of food, those being propionic, butyric, and acetic, all follow lactic acid production. Fermentation by propionic acid bacteria anaerobically converts lactic acid to acetic and propionic acids, carbon dioxide, and water. Fermentation by butyric acid bacteria anaerobically converts lactic acid to acetic and butyric acids, carbon dioxide, and hydrogen. Butyric acid production occurs when the pH of either food or silage fermentations is not sufficiently low to exclusively allow control by microbes that produce lactic acid fermentation.

The ethanol produced by microbial metabolic fermentations can be used to preserve food but mostly is used as a recreational drug. Humans do at times include beverages containing ethanol as part of religious ceremonies. And, interestingly, numerous cultures have assigned dieties both to beverages that contain ethanol and to the processes used for creating those beverages. Ethanol additionally is used as a biofuel. During ethanol fermentation yeasts anaerobically convert monosaccharide, disaccharide, and trisaccharide sugars first into pyruvate, then acetaldehyde, and finally ethanol. That ethanol fermentation process also generates carbon dioxide. Acetic acid bacteria aerobically then can convert either the ethanol or sugars to acetic acid.

We also use microbial metabolic fermentations to preserve as silage the food that we supply to our livestock animals (Tables 1.3 and 1.4). Fermentation additionally is used as a microbial process to modify the aroma and flavor of tobacco (Tables 1.5 and 1.6) and to compost organic wastes (Tables 1.7 and 1.8). I have applied standard usage for listing fermentation of *Zea mays*. Food fermentations tend to list that species as maize. Silage fermentations tend to list that species as corn. Processes that rely upon microorganisms to produce industrial chemicals and enzymes, including pharmaceuticals, broadly are termed to be microbial fermentations although the involved microbial activity may be either aerobic or anaerobic respiration and not metabolic fermentation (Tables 1.9–1.11).

1.2 EXAMPLES OF MICROBIAL FERMENTATIONS IN NATURE

Before we try to understand the role of microbial fermentations in designed processes, we need to understand that microbial fermentations are natural processes that have been optimized by evolution. We have learned to satisfy some of our own needs by harnessing and then adapting the microbial fermentation processes.

1.2.1 Fermentation Contributes to the Nutrient Recycling Capabilities Within Extreme Environments

Microbial fermentation has a role in nutrient recycling within some extreme environments including oxygen-minimum zones within the ocean (Srain et al. 2020). Microbial fermentation also is among those capabilities present within bottom sediments of the Dead Sea and the Great Salt Lake (Oren 1986).

The important ecological role of yeasts is represented by their responsibility to fermentatively decompose organic material even in the cryosphere. As one example, *Rhodotorula frigidialcoholis* survives the cold temperatures of permafrost by switching from respiration to the production of ethanol by fermentation (Touchette et al. 2022). Yeasts also will recycle the organic material in glacial ice and it may be reasonable to presume that fermentation similarly is involved there. de Garcia et al. (2012) isolated and identified more than 20 species of yeasts from glacial ice and

TABLE 1.3 Examples of bacteria that are present in silage fermentations.

Genus	Species	Material Being Ensiled
Acetobacter	*indonesiensis*	Corn
Acetobacter	*pasteurianus*	Corn
Achromobacter	*xylosoxidans*	Corn
Acidipropionibacterium	*acidipropionici*	Corn, *Sorghum*
Acinetobacter	*johnsonii*	Corn
Acinetobacter	*soli*	Corn
Agrobacterium	*larrymoorei*	Wheat
Alcaligenes	*faecalis*	Corn
Alkalihalobacillus	*clausii*	Corn, *Sorghum* kernel
Arthrobacter	*agilis*	Wheat
Bacillus	*atrophaeus*	Corn, Corn kernel, *Sorghum* kernel
Bacillus	*cereus*	Corn kernel, Mixed grass and corn, *Sorghum* kernel
Bacillus	*licheniformis*	Corn kernel, False wheatgrass, *Sorghum* kernel
Bacillus	*magaterium*	Oat
Bacillus	*pumilus*	Barley, Corn, Corn kernel, *Sorghum* kernel
Bacillus	*subtilis*	Corn, Wheat
Bacillus	*vallismortis*	Corn kernel, *Sorghum* kernel
Bacillus	*velezensis*	Corn
Brevibacillus	*brevis*	Corn, Corn kernel, *Sorghum* kernel
Brevibacillus	*invocatus*	Corn, Corn kernel, *Sorghum* kernel
Brevundimonas	*bullata*	Barley with molasses
Carnobacterium	*maltaromaticum*	Alfalfa
Citrobacter	*koseri*	Corn
Clostridium	*baratii*	Corn
Clostridium	*beijerinckii*	Corn
Clostridium	*botulinum*	Grass
Clostridium	*butyricum*	Corn
Clostridium	*ljungdahlii*	Grass
Clostridium	*luticellarii*	Oat
Clostridium	*pabulibutyricum*	Grass
Clostridium	*perfringens*	Corn
Clostridium	*sporogenes*	Alfalfa
Clostridium	*tyrobutyricum*	Corn, Oat
Companilactobacillus	*farciminis*	Paper mulberry
Companilactobacillus	*furfuricola*	Paper mulberry
Companilactobacillus	*insicii*	Paper mulberry
Cytobacillus	*horneckiae*	*Sorghum* kernel
Cytobacillus	*kochii*	Paper mulberry
Empedobacter	*brevis*	Barley

(*Continued*)

TABLE 1.3 (Continued)

Genus	Species	Material Being Ensiled
Enterobacter	*asburiae*	Corn kernel
Enterococcus	*casseliflavus*	Alfalfa with guinea grass, Corn kernel
Enterococcus	*durans*	Wheat
Enterococcus	*faecalis*	Alfalfa, Alfalfa with guinea grass, Corn
Enterococcus	*faecium*	Alfalfa, Barley, Corn, Corn kernel
Enterococcus	*gallinarum*	Banana stem and leaf, Fish (Gibel carp and Klunzinger's ponyfish), Pineapple peel
Enterococcus	*gilvus*	Corn kernel
Enterococcus	*hirae*	Alfalfa with guinea grass
Enterococcus	*mundtii*	Alfalfa with guinea grass
Escherichia	*coli*	Alfalfa, Wheat
Fructilactobacillus	*fructivorans*	Alfalfa, Corn
Fructilactobacillus	*lindneri*	Wheat
Gluconobacter	*oxydans*	Corn
Glutamicibacter	*bergerei*	Wheat
Heyndrickxia	*oleronia*	*Sorghum* kernel
Klebsiella	*oxytoca*	Corn
Klebsiella	*pneumoniae*	Corn, Corn kernel
Kluyvera	*ascorbata*	Corn
Komagataeibacter	*hansenii*	Oat
Kosakonia	*cowanii*	Corn kernel
Lacrimispora	*saccharolytica*	Corn
Lacticaseibacillus	*casei*	Alfalfa with Guinea grass, Banana stem and leaf, Corn, Papaya peel, Paper mulberry, Pineapple peel
Lacticaseibacillus	*paracasei*	Corn, Corn kernel
Lacticaseibacillus	*rhamnosus*	Corn, Corn kernel, Potato pulp
Lacticaseibacillus	*zeae*	Corn
Lactiplantibacillus	*paraplantarum*	Alfalfa with Guinea grass, Banana stem and leaf, Stylo
Lactiplantibacillus	*pentosus*	Corn
Lactiplantibacillus	*plajomi*	Paper mulberry
Lactiplantibacillus	*plantarum*	Alfalfa, Banana stem and leaf, Barley, Corn, Corn kernel, Fish (Gibel carp and Klunzinger's ponyfish), Oat, Papaya peel, Pineapple peel, Rice straw, *Sorghum*, *Sorghum* kernel, Wheat
Lactiplantibacillus	*plantarum subsp. plantarum*	Wheat
Lactobacillus	*acetotolerans*	Barley with molasses, Corn, Paper mulberry
Lactobacillus	*acidophilus*	Alfalfa, Corn
Lactobacillus	*delbrueckii*	Alfalfa
Lactobacillus	*delbrueckii subsp. bulgaricus*	Corn

TABLE 1.3 (Continued)

Genus	Species	Material Being Ensiled
Lactobacillus	*delbrueckii subsp. lactis*	Alfalfa
Lactobacillus	*gasseri*	Alfalfa
Lactobacillus	*helveticus*	Alfalfa
Lactobacillus	*intestinalis*	Alfalfa
Lactococcus	*lactis*	Alfalfa, Corn, Corn kernel, Grass
Latilactobacillus	*curvatus*	Alfalfa, Wheat
Latilactobacillus	*sakei*	Rice straw
Lelliottia	*amnigena*	Corn
Lentilactobacillus	*buchneri*	Alfalfa, Barley with molasses, Corn, Corn kernel, Paper mulberry, *Sorghum* kernel, Wheat
Lentilactobacillus	*diolivorans*	Corn
Lentilactobacillus	*hilgardii*	Alfalfa, Corn, Sugarcane
Lentilactobacillus	*parabuchneri*	Corn kernel, Oat
Lentilactobacillus	*parafarraginis*	Corn
Lentilactobacillus	*raoultii*	Paper mulberry
Leuconostoc	*fallax*	Wheat
Leuconostoc	*mesenteroides*	Corn kernel, Rice straw, Wheat
Leuconostoc	*mesenteroides subsp. mesenteroides*	Alfalfa, Corn
Leuconostoc	*pseudomesenteroides*	Corn kernel
Levilactobacillus	*brevis*	Alfalfa, Corn, Corn kernel, Fish (Gibel carp and Klunzinger's ponyfish), Oat, *Sorghum* kernel, Wheat
Levilactobacillus	*hammesii*	Paper mulberry
Levilactobacillus	*namurensis*	Paper mulberry
Ligilactobacillus	*acidipiscis*	Corn kernel, Oat
Limosilactobacillus	*fermentum*	Banana stem and leaf, Papaya peel, Stylo
Limosilactobacillus	*ingluviei*	Corn
Liquorilactobacillus	*nagelii*	Papaya peel
Listeria	*innocua*	Corn, Sugarcane
Listeria	*ivanovii*	Corn, Sugarcane
Listeria	*monocytogenes*	Corn, Sugarcane
Listeria	*seeligeri*	Corn, Sugarcane
Listeria	*welshimeri*	Corn, Sugarcane
Loigolactobacillus	*coryniformis*	Rice straw
Loigolactobacillus	*coryniformis subsp. coryniformis*	Alfalfa
Lysinibacillus	*boronitolerans*	Corn, Corn kernel, *Sorghum* kernel
Lysinibacillus	*fusiformis*	Corn
Lysinibacillus	*sphaericus*	Corn, Corn kernel, *Sorghum* kernel

(*Continued*)

TABLE 1.3 (Continued)

Genus	Species	Material Being Ensiled
Lysinibacillus	*telephonicus*	Paper mulberry
Myroides	*odoratimimus*	Corn
Niallia	*circulans*	Corn, Corn kernel, Oat, Paper mulberry, *Sorghum* kernel
Paenibacillus	*cineris*	Corn
Paenibacillus	*glucanolyticus*	Corn kernel
Paenibacillus	*lautus*	Corn kernel
Paenibacillus	*macerans*	Corn, *Sorghum* kernel
Paenibacillus	*polymyxa*	Corn
Paenibacillus	*rhizosphaerae*	*Sorghum* kernel
Pantoea	*agglomerans*	Wheat
Pediococcus	*acidilactici*	Alfalfa, Corn kernel, Fish (Gibel carp and Klunzinger's ponyfish), Ryegrass, *Sorghum* kernel, Wheat
Pediococcus	*pentosaceus*	Alfalfa, Corn, Corn kernel, *Sorghum* kernel, Stylo
Priestia	*megaterium*	*Sorghum* kernel
Providencia	*rettgeri*	Corn
Providencia	*vermicola*	Oat
Pseudomonas	*putida*	Corn
Pseudomonas	*stutzeri*	Corn
Raoultella	*planticola*	Barley
Rummeliibacillus	*stabekisii*	Paper mulberry
Schleiferilactobacillus	*harbinensis*	Corn
Schleiferilactobacillus	*perolens*	Papaya peel, Pineapple peel
Secundilactobacillus	*collinoides*	Alfalfa
Secundilactobacillus	*pentosiphilus*	Corn, Paper mulberry
Serratia	*ficaria*	Wheat
Serratia	*marcescens*	Corn
Serratia	*nematodiphila*	Corn
Sphingobacterium	*alimentarium*	Barley with molasses
Staphylococcus	*saprophyticus*	Corn kernel
Stenotrophomonas	*maltophilia*	Corn
Streptococcus	*equinus*	Corn
Thermoactinomyces	*vulgaris*	Corn, Hay
Ureibacillus	*massiliensis*	Paper mulberry
Weissella	*cibaria*	Corn kernel, Wheat
Weissella	*confusa*	Alfalfa, Corn
Weissella	*hellenica*	Mixed grass (Timothy and Orchard)
Weissella	*paramesenteroides*	Corn kernel
Weissella	*viridescens*	Alfalfa

The material being ensiled was the whole harvested plant unless only the kernel is mentioned as having been ensiled. Corn kernel was rehydrated before ensiling. Sorghum kernel was rehydrated before ensiling. The type of grass being ensiled was not always specified. False wheatgrass is *Leymus chinensis*. Guinea grass is *Megathyrsus maximus*. Paper mulberry is *Broussonetia papyrifera*. Stylo is *Stylosanthes guianensis*. Timothy grass is *Phleum pratense*. Orchard grass is *Dactylis glomerata*.

TABLE 1.4 Examples of fungi and oomycetes that are present in silage fermentations.

Genus	Species	Material Being Ensiled
Alternaria	*alternata*	Corn, Hay, Oat, Wheat
Alternaria	*tenuissima*	Corn
Amylomyces	*rouxii*	Corn
Aspergillus	*amstelodami*	Hay
Aspergillus	*brasiliensis*	Corn
Aspergillus	*candidus*	Corn, Oat
Aspergillus	*chevalieri*	Hay
Aspergillus	*clavatus*	Corn, Grass
Aspergillus	*flavus*	Barley, Corn, Grass
Aspergillus	*fumigatus*	Barley, Corn, Grass, Hay
Aspergillus	*glaucus*	Grass
Aspergillus	*intermedius*	Hay
Aspergillus	*lentulus*	Wheat
Aspergillus	*melleus*	Corn
Aspergillus	*niger*	Corn, Oat
Aspergillus	*ochraceus*	Corn
Aspergillus	*parasiticus*	Corn
Aspergillus	*pseudoglaucus*	Corn, Hay
Aspergillus	*terreus*	Corn
Aspergillus	*tubingensis*	Corn, Grass
Aspergillus	*versicolor*	Corn, Hay
Aureobasidium	*pullulans* (closest genomic match)	Corn
Brettanomyces	*bruxellensis*	Corn
Brettanomyces	*custersianus*	Corn
[*Candida*]	*ethanolica*	Corn
[*Candida*]	*glabrata*	Corn, Grass
[*Candida*]	*intermedia*	Corn
[*Candida*]	*mesenterica*	Corn
[*Candida*]	*sake*	Corn
Candida	*tropicalis*	Corn
Cladosporium	*cladosporioides*	corn
Cladosporium	*herbarum*	Hay, Wheat
Cladosporium	*tenuissimum*	Corn
Clavispora	*lusitaniae*	Corn
Debaryomyces	*hansenii*	Corn
Didymella	*pomorum*	Grass
Diutina	*rugosa*	Corn
Epicoccum	*nigrum*	Corn, Grass, Hay
Filobasidium	*globisporum*	Oat
Fusarium	*avenaceum*	Corn

(*Continued*)

TABLE 1.4 (Continued)

Genus	Species	Material Being Ensiled
Fusarium	*cerealis*	Corn
Fusarium	*chlamydosporum*	Corn
Fusarium	*commune*	Corn
Fusarium	*culmorum*	Corn
Fusarium	*equiseti*	Corn
Fusarium	*graminearum*	Corn, Hay, *Sorghum* kernel
Fusarium	*incarnatum*	*Sorghum* kernel
Fusarium	*oxysporum*	Corn
Fusarium	*poae*	Corn
Fusarium	*proliferatum*	Corn, *Sorghum* kernel
Fusarium	*pseudograminearum*	Corn, Wheat straw
Fusarium	*solani*	Corn
Fusarium	*sporotrichioides*	Corn
Fusarium	*subglutinans*	Corn
Fusarium	*thapsinum*	*Sorghum* kernel
Fusarium	*tricinctum*	Corn
Fusarium	*verticillioides*	Corn, Grass, *Sorghum* kernel
Geotrichum	*candidum*	Corn, Grass
Hanseniaspora	*osmophila*	Sugarcane
Hanseniaspora	*vineae*	Sugarcane
Hyphopichia	*burtonii*	Corn, Corn kernel
Kazachstania	*aerobia*	Corn
Kazachstania	*barnettii*	Corn
Kazachstania	*bulderi*	Corn, Corn kernel
Kazachstania	*exigua*	Barley, Corn
Kazachstania	*humilis*	Barley, Corn
Kazachstania	*turicensis*	Barley
Kazachstania	*unispora*	Corn, Corn kernel
Kluyveromyces	*lactis*	Corn, Sugarcane
Kluyveromyces	*marxianus*	Corn
Kurtzmaniella	*quercitrusa*	Corn
Lichtheimia	*ramosa*	Oat
Magnusiomyces	*capitatus*	Hay
Malassezia	*restricta*	Oat
Metschnikowia	*pulcherrima*	Corn
Meyerozyma	*caribbica*	Corn
Meyerozyma	*guilliermondii*	Corn, Corn kernel
Monascus	*fuliginosus*	Oat
Monascus	*purpureus*	Corn

TABLE 1.4 (Continued)

Genus	Species	Material Being Ensiled
Monascus	*ruber*	Barley, Corn, Grass
Mrakia	*blollopis*	Wheat
Mucor	*circinelloides*	Corn
Mucor	*hiemalis*	Hay
Mucor	*racemosus*	Corn, Oat, Wheat
Naumovozyma	*castellii*	Corn kernel
Naumovozyma	*dairenensis*	Corn kernel
Paecilomyces	*laguncularíae*	Corn, Grass
Paecilomyces	*niveus*	Barley, Corn, Grass
Paecilomyces	*variotii*	Corn
Penicillium	*aurantiogriseum*	Corn
Penicillium	*brevicompactum*	Barley, Corn
Penicillium	*carneum*	Corn
Penicillium	*chrysogenum*	Oat
Penicillium	*citrinum*	Corn
Penicillium	*commune*	Corn
Penicillium	*crustosum*	Corn
Penicillium	*cyclopium*	Hay
Penicillium	*expansum*	Corn, Hay
Penicillium	*farinosum*	Corn
Penicillium	*glabrum*	Corn
Penicillium	*griseofulvum*	Corn
Penicillium	*implicatum*	Hay
Penicillium	*oxalicum*	Corn
Penicillium	*paneum*	Corn, Grass
Penicillium	*polonicum*	Corn
Penicillium	*roqueforti*	Corn, Grass
Penicillium	*solitum*	Corn, Grass
Penicillium	*thomii*	Hay
Pichia	*deserticola*	Corn
Pichia	*fermentans*	Corn
Pichia	*kudriavzevii*	Barley, Corn, Corn kernel, Sugarcane
Pichia	*manshurica*	Corn, Corn kernel
Pichia	*membranifaciens*	Corn, Corn kernel
Pichia	*sporocuriosa*	Corn kernel
Protobremia	*sphaerosperma*	Hay
Pseudopyrenochaeta	*lycopersici*	Oat

(Continued)

TABLE 1.4 (Continued)

Genus	Species	Material Being Ensiled
Rhizomucor	*pusillus*	Corn
Rhizopus	*arrhizus*	Potato pulp
Rhizopus	*microsporus*	Oat
Rhizopus	*stolonifer*	Corn
Rhodotorula	*mucilaginosa*	Corn, Corn kernel
Saccharomonospora	*viridis*	Hay
Saccharomyces	*cariocanus*	Corn kernel
Saccharomyces	*cerevisiae*	Corn, Corn kernel
Saccharomyces	*paradoxus*	Corn, Corn kernel
Saccharopolyspora	*rectivirgula*	Hay
Saturnispora	*diversa*	Corn
Scedosporium	*apiospermum* (closest genomic match)	Corn
Scedosporium	*boydii*	Barley
Schizosaccharomyces	*pombe*	Sugarcane
Scopulariopsis	*brevicaulis*	Barley
Septoria	*dysentericae*	Oat
Sporidiobolus	*metaroseus*	Barley with molasses
Sporisorium	*reilianum*	Oat
Starmerella	*apicola*	Corn
Syncephalastrum	*monosporum*	Corn
Talaromyces	*amestolkiae*	Oat
Talaromyces	*duclauxii*	Grass
Talaromyces	*funiculosus*	Corn
Talaromyces	*islandicus*	Corn
Talaromyces	*purpureogenus*	Corn
Talaromyces	*variabilis*	Grass
Torulaspora	*delbrueckii*	Sugarcane
Trichoderma	*asperellum*	Corn
Trichoderma	*harzianum*	Corn
Trichosporon	*asahii*	Corn
Trichosporon	*japonicum*	Corn
Wickerhamiella	*pararugosa* (closest genomic match)	Corn
Wickerhamomyces	*anomalus* (closest genomic match in corn, confirmed in corn kernel)	Corn, Corn kernel
Yarrowia	*lipolytica*	Fish (hake)
Zygosaccharomyces	*bailii*	Corn, Sugarcane

The material being ensiled was the whole harvested plant unless only the grain or kernel is mentioned as having been ensiled. Corn kernel was rehydrated before ensiling. Sorghum kernel was rehydrated before ensiling. The type of grass being ensiled was not always specified. *Protobremia sphaerosperma* is a member of the Phylum Oomycota.

TABLE 1.5 Examples of bacteria that are present in tobacco fermentations.

Genus	Species	Primary Material Being Fermented	Product	Geographical Origin
Aerococcus	*viridans*	Tobacco leaves	Toscano cigar tobacco	Italy
Bacillus	*cereus*	Tobacco leaves	Toscano cigar tobacco	Italy
Bacillus	*licheniformis*	Tobacco leaves	Toscano cigar tobacco	Italy
Bacillus	*mojavensis*	Tobacco leaves	Toscano cigar tobacco	Italy
Bacillus	*pumilus*	Tobacco leaves	Toscano cigar tobacco	Italy
Bacillus	*subtilis*	Tobacco leaves	Cigarette tobacco, Toscano cigar tobacco	Global (Cigarette), Italy (Toscano)
Bacillus	*thuringiensis*	Tobacco leaves	Toscano cigar tobacco	Italy
Corynebacterium	*ammoniagenes*	Tobacco leaves	Toscano cigar tobacco	Italy
Priestia	*megaterium*	Tobacco leaves	Cigar tobacco	Global
Ruoffia	*tabacinasalis*	Tobacco leaves	Toscano cigar tobacco	Italy

TABLE 1.6 Examples of fungi that are present in tobacco fermentations.

Genus	Species	Primary Material Being Fermented	Product	Geographical Origin
Aspergillus	*flavus*	Tobacco leaves	Cigar tobacco	Global
Aspergillus	*niger*	Tobacco leaves	Cigar tobacco	Global
Aspergillus	*ochraceus*	Tobacco leaves	Cigar tobacco	Global
Aspergillus	*terreus*	Tobacco leaves	Cigar tobacco	Global
Candida	*parapsilosis*	Tobacco leaves	Toscano cigar tobacco	Italy
Debaryomyces	*hansenii*	Tobacco leaves	Toscano cigar tobacco	Italy
Debaryomyces	*maramus*	Tobacco leaves	Toscano cigar tobacco	Italy
Debaryomyces	*prosopidis*	Tobacco leaves	Toscano cigar tobacco	Italy
Diutina	*rugosa*	Tobacco leaves	Toscano cigar tobacco	Italy
Schwanniomyces	*vanrijiae*	Tobacco leaves	Toscano cigar tobacco	Italy
Scopulariopsis	*brevicaulis*	Tobacco leaves	Cigar tobacco	Global

Johnson (1934) mentioned a fungal species *Oospora nicotianae* in fermenting tobacco. *O. nicotianae* no longer is a valid taxonomic name and that species is not represented in this table of data because I was unable to determine its current taxonomic name.

meltwater in the Patagonia Andes area of Argentina. Among these *Aureobasidium pullulans* which they recovered from ice also occurs in vermicompost (Table 1.8); [*Candida*] *mesenterica* which they recovered from both ice and meltwater is also found in corn silage (Table 1.4); *Debaryomyces hansenii* which they recovered from ice frequently is a component of both food fermentations (Table 1.2) and tobacco fermentations (Table 1.6); and *Rhodotorula mucilaginosa* which they recovered from ice occurs in silage fermentations of whole corn plants (corn) and of corn kernel (Table 1.4) (de Garcia et al. 2012).

TABLE 1.7 Examples of bacteria that are present in compost fermentations.

Genus	Species	Material Being Composted
Acetivibrio	*thermocellus*	Food waste
Ammoniibacillus	*agariperforans*	Sewage sludge with corn straw
Anoxybacillus	*flavithermus*	Food waste
Arenimonas	*composti*	Farm waste
[*Bacillus*]	*thermocloacae*	Sewage sludge with corn straw
Bacillus	*amyloliquefaciens*	Sewage sludge with corn straw
Bacillus	*badius*	Food waste
Bacillus	*benzoevorans*	Food waste
Bacillus	*benzoevorans* (closest genomic match)	Mixed agricultural byproducts
Bacillus	*cereus*	Food waste
Bacillus	*licheniformis*	Food waste
Bacillus	*licheniformis* (closest genomic match)	Mixed agricultural byproducts
Bacillus	*licheniformis*	Sewage sludge with corn straw
Bacillus	*pumilus*	Food waste
Bacillus	*pumilus* (closest genomic match)	Mixed agricultural byproducts
Bacillus	*pumilus*	Vermicompost
Bacillus	*subtilis*	Food waste
Bacillus	*subtilis* (closest genomic match)	Mixed agricultural byproducts
Bacillus	*subtilis*	Sewage sludge with corn straw
Bacillus	*subtilis*	Vermicompost
Bradyrhizobium	*japonicum*	Vermicompost
Brevibacillus	*brevis*	Food waste
Brevibacillus	*brevis*	Urban compost
Brevibacillus	*thermoruber*	Sewage sludge with corn straw
Brevundimonas	*diminuta*	Urban compost
Caldibacillus	*hisashii*	Sewage sludge with corn straw
Caldibacillus	*thermoamylovorans*	Sewage sludge with corn straw
Caldibacillus	*thermolactis*	Sewage sludge with corn straw
Chelativorans	*composti*	Sewage sludge with corn straw
Chryseobacterium	*gleum*	Urban compost
Chryseobacterium	*indologenes*	Urban compost
Comamonas	*kerstersii*	Food waste
Comamonas	*kerstersii* (closest genomic match)	Mixed agricultural byproducts
Cronobacter	*sakazakii*	Food waste
Cronobacter	*sakazakii* (closest genomic match)	Mixed agricultural byproducts
Cupriavidus	*oxalaticus*	Vermicompost
Flavobacterium	*johnsoniae*	Urban compost
Gemmatimonas	*aurantiaca*	Food waste
Geobacillus	*stearothermophilus*	Food waste
Geobacillus	*thermodenitrificans*	Food waste

TABLE 1.7 (Continued)

Genus	Species	Material Being Composted
Gottschalkia	*acidurici*	Food waste
Hoyosella	*subflava*	Food waste
Klebsiella	*pneumoniae*	Food waste
Klebsiella	*pneumoniae* (closest genomic match)	Mixed agricultural byproducts
Klebsiella	*pneumoniae*	Urban compost
Kocuria	*flava*	Food waste
Kocuria	*flava* (closest genomic match)	Mixed agricultural byproducts
Kyrpidia	*tusciae*	Food waste
Leuconostoc	*mesenteroides*	Urban compost
Lysinibacillus	*fusiformis*	Food waste
Lysinibacillus	*sphaericus*	Food waste
Mahella	*australiensis*	Food waste
Mammaliicoccus	*sciuri*	Food waste
Mammaliicoccus	*sciuri* (closest genomic match)	Mixed agricultural byproducts
Methylorubrum	*extorquens*	Urban compost
Moraxella	*bovis*	Urban compost
Mycobacterium	*xenopi*	Food waste
Mycolicibacterium	*thermoresistibile*	Food waste
Niallia	*circulans*	Food waste
Niallia	*nealsonii*	Food waste
Paenibacillus	*amylolyticus*	Sewage sludge with corn straw
Paenibacillus	*mucilaginosus*	Food waste
Paenibacillus	*polymyxa*	Food waste
Priestia	*flexa*	Food waste
Priestia	*flexa* (closest genomic match)	Mixed agricultural byproducts
Priestia	*megaterium*	Food waste
Priestia	*megaterium*	Vermicompost
Priestia	*paraflexa*	Food waste
Pseudomonas	*corrugata*	Vermicompost
Pseudomonas	*fluorescens*	Urban compost
Pseudomonas	*fragi*	Urban compost
Pseudomonas	*marginalis*	Urban compost
Pseudomonas	*mendocina*	Food waste
Pseudomonas	*oryzihabitans*	Urban compost
Pseudomonas	*putida*	Urban compost
Pseudomonas	*putida*	Vermicompost
Pseudomonas	*stutzeri*	Urban compost
Pseudomonas	*taetrolens*	Urban compost
Psychrobacter	*immobilis*	Urban compost

(*Continued*)

TABLE 1.7 (Continued)

Genus	Species	Material Being Composted
Rhizobium	*leguminosarum*	Vermicompost
Rhodothermus	*marinus*	Food waste
Serratia	*entomophila*	Urban compost
Serratia	*marcescens*	Food waste
Serratia	*marcescens*	Urban compost
Siminovitchia	*thermophila*	Sewage sludge with corn straw
Sinibacillus	*soli*	Sewage sludge with corn straw
Sinorhizobium	*meliloti*	Vermicompost
Solibacillus	*silvestris*	Food waste
Sorangium	*cellulosum*	Food waste
Sphaerobacter	*thermophilus*	Food waste
Sphaerobacter	*thermophilus*	Sewage sludge with corn straw
Sphingobacterium	*mizutaii*	Urban compost
Sporosarcina	*pasteurii*	Vermicompost
Staphylococcus	*aureus*	Food waste
Staphylococcus	*capitis* subsp. *urealyticus*	Urban compost
Staphylococcus	*xylosus*	Food waste
Stenotrophomonas	*maltophilia*	Urban compost
Streptomyces	*antibioticus*	Food waste
Streptomyces	*cinnabarinus*	Food waste
Streptomyces	*griseus*	Food waste
Streptomyces	*roseus*	Food waste
Streptosporangium	*roseum*	Food waste
Symbiobacterium	*thermophilum*	Food waste
Terribacillus	*halophilus*	Food waste
Thermaerobacter	*marianensis*	Food waste
Thermoactinomyces	*vulgaris*	Food waste
Thermobacillus	*composti*	Food waste
Thermobifida	*fusca*	Food waste
Thermobispora	*bispora*	Food waste
Thermoflavimicrobium	*dichotomicum*	Food waste
Thermomonospora	*curvata*	Food waste
Thermosediminibacter	*oceani*	Food waste
Weizmannia	*coagulans*	Food waste
Xanthomonas	*campestris*	Food waste
Xanthomonas	*oryzae*	Urban compost

The material identified as sewage sludge was from a food factory and corn straw presumably was stover.
The material identified as urban compost was mostly leaves, grass, and food waste.
The material identified as Mixed agricultural byproducts was stated to be a mixture of rice bran (15 kg), wheat bran (10 kg), rice husk (10 kg), other additives like grass and leaves (5 kg) each, ash (2.5 kg), cow dung (25 kg), mustard oil cake (10 kg), cow urine (40 l), molasses (4 l), and approximately 0.6% by weight of calcium oxide.

TABLE 1.8 Examples of fungi that are present in compost fermentations.

Genus	Species	Material Being Composted
Acremonium	*charticola*	Composting plant debris
Acremonium	*chrysogenum*	Composting plant debris
Acremonium	*chrysogenum*	Vermicompost
Acremonium	*fusidioides*	Composting plant debris
Acremonium	*persicinum*	Composting plant debris
Acremonium	*persicinum*	Vermicompost
Acremonium	*sclerotigenum*	Composting plant debris
Acremonium	*sclerotigenum*	Vermicompost
Acrodontium	*griseum*	Composting plant debris
Acrophialophora	*fusispora*	Composting plant debris
Acrophialophora	*fusispora*	Vermicompost
Akanthomyces	*lecanii*	Composting plant debris
Akanthomyces	*lecanii*	Vermicompost
Alternaria	*alternariae*	Composting plant debris
Alternaria	*alternata*	Composting plant debris
Alternaria	*alternata*	Vermicompost
Amesia	*nigricolor*	Composting plant debris
Apiospora	*montagnei*	Vermicompost
Apiotrichum	*sporotrichoides*	Vermicompost
Arthrobotrys	*thaumasia*	Municipal solid waste with oil contaminated sawdust and wet industrial sewage sludge
Arthroderma	*tuberculatum*	Vermicompost
Arthroderma	*uncinatum*	Municipal solid waste
Ascodesmis	*microscopica*	Composting plant debris
Aspergillus	*amstelodami*	Composting plant debris
Aspergillus	*amstelodami*	Vermicompost
Aspergillus	*candidus*	Vermicompost
Aspergillus	*chevalieri*	Vermicompost
Aspergillus	*fischeri* var. *fischeri*	Vermicompost
Aspergillus	*flavus*	Food waste
Aspergillus	*flavus*	Municipal solid waste
Aspergillus	*flavus* var. *columnaris*	Composting plant debris
Aspergillus	*flavus* var. *columnaris*	Vermicompost
Aspergillus	*flavus* var. *flavus*	Composting plant debris
Aspergillus	*fumigatus*	Composting plant debris
Aspergillus	*fumigatus*	Food waste
Aspergillus	*fumigatus*	Municipal solid waste
Aspergillus	*fumigatus*	Vermicompost
Aspergillus	*fumigatus* var. *fumigatus*	Composting plant debris

(*Continued*)

TABLE 1.8 (Continued)

Genus	Species	Material Being Composted
Aspergillus	*fumigatus* var. *fumigatus*	Vermicompost
Aspergillus	*intermedius*	Vermicompost
Aspergillus	*montevidensis*	Composting plant debris
Aspergillus	*neoafricanus*	Composting plant debris
Aspergillus	*neoniveus*	Vermicompost
Aspergillus	*nidulans* var. *nidulans*	Vermicompost
Aspergillus	*niger*	Composting plant debris
Aspergillus	*niger*	Food waste
Aspergillus	*niger*	Municipal solid waste
Aspergillus	*niger*	Vermicompost
Aspergillus	*ochraceus*	Composting plant debris
Aspergillus	*ochraceus*	Vermicompost
Aspergillus	*oryzae*	Composting plant debris
Aspergillus	*puniceus*	Composting plant debris
Aspergillus	*ruber*	Vermicompost
Aspergillus	*spinosus*	Vermicompost
Aspergillus	*sulphureus*	Composting plant debris
Aspergillus	*terreus* var. *terreus*	Composting plant debris
Aspergillus	*terreus* var. *terreus*	Vermicompost
Aspergillus	*versicolor*	Composting plant debris
Aspergillus	*versicolor*	Vermicompost
Aspergillus	*wentii*	Composting plant debris
Aureobasidium	*pullulans*	Vermicompost
Beauveria	*bassiana*	Vermicompost
Beauveria	*brongniartii*	Vermicompost
Berkeleyomyces	*basicola*	Vermicompost
Boeremia	*exigua* var. *exigua*	Vermicompost
Botrytis	*cinerea*	Vermicompost
[*Candida*]	*sake*	Municipal solid waste with oil contaminated sawdust and wet industrial sewage sludge
[*Candida*]	*sake*	Municipal solid waste with oil contaminated sawdust and dewater sludge of mixed industrial and household origin
Cephalotrichum	*microsporum*	Composting plant debris
Cephalotrichum	*purpureofuscum*	Vermicompost
Chaetomium	*globosum*	Vermicompost
Chrysosporium	*merdarium*	Vermicompost
Chrysosporium	*queenslandicum*	Vermicompost
Chrysosporium	*tropicum*	Vermicompost

TABLE 1.8 (Continued)

Genus	Species	Material Being Composted
Cladosporium	*cladosporioides*	Composting plant debris
Cladosporium	*cladosporioides*	Vermicompost
Cladosporium	*herbarum*	Composting plant debris
Cladosporium	*herbarum*	Vermicompost
Cladosporium	*oxysporum*	Composting plant debris
Cladosporium	*oxysporum*	Vermicompost
Cladosporium	*sphaerospermum*	Composting plant debris
Cladosporium	*sphaerospermum*	Vermicompost
Collariella	*bostrychodes*	Composting plant debris
Coniochaeta	*hoffmannii*	Vermicompost
Coprinus	*cordisporus*	Municipal solid waste with oil contaminated sawdust and wet industrial sewage sludge
Coprinus	*cordisporus*	Municipal solid waste with oil contaminated sawdust and dewater sludge of mixed industrial and household origin
Cordyceps	*memorabilis*	Composting plant debris
Corynascus	*sepedonium*	Composting plant debris
Corynascus	*sepedonium*	Vermicompost
Cunninghamella	*elegans*	Vermicompost
Dichotomopilus	*funicola*	Vermicompost
Dipodascus	*australiensis*	Municipal solid waste with oil contaminated sawdust and wet industrial sewage sludge
Dipodascus	*australiensis*	Municipal solid waste with oil contaminated sawdust and dewater sludge of mixed industrial and household origin
Entomortierella	*chlamydospora*	Composting plant debris
Entomortierella	*echinosphaera*	Composting plant debris
Entomortierella	*echinosphaera*	Vermicompost
Epicoccum	*nigrum*	Composting plant debris
Epicoccum	*nigrum*	Vermicompost
Exophiala	*moniliae*	Composting plant debris
Exophiala	*pisciphila*	Vermicompost
Fusarium	*haematococcum*	Composting plant debris
Fusarium	*haematococcum*	Vermicompost
Fusarium	*oxysporum*	Composting plant debris
Fusarium	*oxysporum*	Food waste
Fusarium	*verticillioides*	Food waste
Geotrichum	*candidum*	Municipal solid waste with oil contaminated sawdust and wet industrial sewage sludge

(Continued)

TABLE 1.8 (Continued)

Genus	Species	Material Being Composted
Geotrichum	*candidum*	Municipal solid waste with oil contaminated sawdust and dewater sludge of mixed industrial and household origin
Gibellulopsis	*nigrescens*	Vermicompost
Graphiopsis	*chlorocephala*	Composting plant debris
[*Humicola*]	*grisea* var. *thermoidea*	Vermicompost
Humicola	*fuscoatra* var. *fuscoatra*	Composting plant debris
Humicola	*fuscoatra* var. *fuscoatra*	Vermicompost
Keratinophyton	*terreum*	Vermicompost
Leptosphaeria	*coniothyrium*	Composting plant debris
Leptosphaeria	*coniothyrium*	Vermicompost
Lichtheimia	*corymbifera*	Composting plant debris
Linnemannia	*hyalina*	Composting plant debris
Malbranchea	*cinnamomea*	Composting plant debris
Malbranchea	*cinnamomea*	Vermicompost
Metschnikowia	aff. *Fructicola*	Municipal solid waste with oil contaminated sawdust and dewater sludge of mixed industrial and household origin
Microascus	*cirrosus*	Vermicompost
Moniliella	*suaveolens* var. *nigra*	Vermicompost
Mortierella	*alliacea*	Composting plant debris
Mortierella	*alpina*	Composting plant debris
Mortierella	*alpina*	Vermicompost
Mortierella	*globalpina*	Composting plant debris
Mortierella	*indohii*	Composting plant debris
Mucor	*griseocyanus*	Vermicompost
Nannizzia	*fulva*	Municipal solid waste
Nannizzia	*gypsea*	Municipal solid waste
Paecilomyces	*variotii*	Composting plant debris
Paecilomyces	*variotii*	Vermicompost
Paraphyton	*cookei*	Municipal solid waste
Parascedosporium	*putredinis*	Vermicompost
Parengyodontium	*album*	Composting plant debris
Parengyodontium	*album*	Vermicompost
Penicillium	*aurantiogriseum var. aurantiogriseum*	Composting plant debris
Penicillium	*aurantiogriseum* var. *aurantiogriseum*	Vermicompost
Penicillium	*brevicompactum*	Composting plant debris
Penicillium	*brevicompactum*	Vermicompost

TABLE 1.8 (Continued)

Genus	Species	Material Being Composted
Penicillium	*canescens*	Composting plant debris
Penicillium	*chermesinum*	Vermicompost
Penicillium	*chrysogenum*	Composting plant debris
Penicillium	*chrysogenum*	Vermicompost
Penicillium	*citrinum*	Composting plant debris
Penicillium	*citrinum*	Food waste
Penicillium	*citrinum*	Vermicompost
Penicillium	*dierckxii*	Composting plant debris
Penicillium	*digitatum*	Composting plant debris
Penicillium	*digitatum*	Vermicompost
Penicillium	*expansum*	Composting plant debris
Penicillium	*glabrum*	Composting plant debris
Penicillium	*glabrum*	Vermicompost
Penicillium	*glandicola*	Vermicompost
Penicillium	*herquei*	Composting plant debris
Penicillium	*herquei*	Vermicompost
Penicillium	*implicatum*	Composting plant debris
Penicillium	*implicatum*	Vermicompost
Penicillium	*italicum*	Vermicompost
Penicillium	*janczewskii*	Composting plant debris
Penicillium	*janczewskii*	Vermicompost
Penicillium	*jensenii*	Composting plant debris
Penicillium	*jensenii*	Vermicompost
Penicillium	*ochrochloron*	Composting plant debris
Penicillium	*paxilli*	Composting plant debris
Penicillium	*purpurescens*	Vermicompost
Penicillium	*restrictum*	Composting plant debris
Penicillium	*restrictum*	Vermicompost
Penicillium	*rolfsii*	Composting plant debris
Penicillium	*roqueforti*	Composting plant debris
Penicillium	*roseopurpureum*	Composting plant debris
Penicillium	*roseopurpureum*	Vermicompost
Penicillium	*simplicissimum*	Composting plant debris
Penicillium	*simplicissimum*	Vermicompost
Penicillium	*spinulosum*	Vermicompost
Penicillium	*verrucosum*	Composting plant debris
Penicillium	*waksmanii*	Composting plant debris
Penicillium	*waksmanii*	Vermicompost

(*Continued*)

TABLE 1.8 (Continued)

Genus	Species	Material Being Composted
Peroneutypa	*scoparia*	Composting plant debris
Phialemonium	*obovatum*	Composting plant debris
Plectosphaerella	*cucumerina*	Vermicompost
Podila	*humilis*	Composting plant debris
Preussia	*fleischhakii*	Vermicompost
Pseudogymnoascus	*pannorum*	Composting plant debris
Pseudogymnoascus	*pannorum*	Vermicompost
Pseudogymnoascus	*roseus*	Composting plant debris
Rhizopus	*arrhizus*	Composting plant debris
Rhizopus	*stolonifer*	Food waste
Rollandina	*capitata*	Composting plant debris
Sagenomella	*humicola*	Composting plant debris
Sarocladium	*strictum*	Composting plant debris
Sarocladium	*strictum*	Vermicompost
Scedosporium	*boydii*	Composting plant debris
Scedosporium	*boydii*	Municipal solid waste with oil contaminated sawdust and wet industrial sewage sludge
Scedosporium	*boydii*	Municipal solid waste with oil contaminated sawdust and dewater sludge of mixed industrial and household origin
Scedosporium	*boydii*	Vermicompost
Scopulariopsis	*brevicaulis*	Vermicompost
Scopulariopsis	*brumptii*	Composting plant debris
Scopulariopsis	*brumptii*	Vermicompost
Scopulariopsis	*candida*	Composting plant debris
Scopulariopsis	*candida*	Vermicompost
Scopulariopsis	*koningii*	Composting plant debris
Scopulariopsis	*koningii*	Vermicompost
Scopulariopsis	*sphaerospora*	Vermicompost
Scytalidium	*lignicola*	Composting plant debris
Skoua	*fertilis*	Composting plant debris
Stachybotrys	*chartarum*	Vermicompost
Staphylotrichum	*coccosporum*	Composting plant debris
Staphylotrichum	*coccosporum*	Vermicompost
Syncephalastrum	*racemosum*	Vermicompost
Talaromyces	*diversus*	Composting plant debris
Talaromyces	*flavus* var. *flavus*	Vermicompost
Talaromyces	*helicus* var. *helicus*	Vermicompost
Talaromyces	*helicus* var. *major*	Composting plant debris

TABLE 1.8 (Continued)

Genus	Species	Material Being Composted
Talaromyces	*helicus* var. *major*	Vermicompost
Talaromyces	*islandicus*	Vermicompost
Talaromyces	*minioluteus*	Composting plant debris
Talaromyces	*piceae*	Vermicompost
Talaromyces	*purpureogenus*	Composting plant debris
Talaromyces	*purpureogenus*	Vermicompost
Talaromyces	*rugulosus*	Vermicompost
Thermomyces	*dupontii*	Food waste
Thermomyces	*lanuginosus*	Composting plant debris
Thermomyces	*lanuginosus*	Vermicompost
Thermothelomyces	*heterothallicus*	Vermicompost
Thysanophora	*penicillioides*	Vermicompost
Trichocladium	*gilmaniellae*	Vermicompost
Trichoderma	*hamatum*	Composting plant debris
Trichoderma	*harzianum*	Composting plant debris
Trichoderma	*viride*	Composting plant debris
Trichophyton	*mentagrophytes*	Municipal solid waste
Trichophyton	*quinckeanum*	Municipal solid waste
Trichophyton	*verrucosum*	Municipal solid waste
Westerdykella	*dispersa*	Vermicompost

Vermicompost contained 70% animal dung, 30% plant debris, and the worms were *L. rubellus* (Anastasi et al. 2005).

TABLE 1.9 Examples of Archaea used in the fermentation production of chemicals, enzymes, and pharmaceuticals.

Genus	Species	Product
Pyrococcus	*furiosus*	α-Glucan phosphorylase
Thermococcus	*kodakarensis*	Protease

There are fermentation-derived proteases which are not identified with great specificity.

1.2.2 Fermentative Microbes Support the Biological Availability of Phosphates in Soil

Phosphorus is a component of soil that is critically important for supporting plant growth (Hanyabui et al. 2020). The sources of phosphorus in soil are primary phosphate minerals, secondary phosphate minerals, and organic phosphates.

Primary and secondary phosphate minerals are released into the soil by weathering activities. Among the primary phosphate minerals are appatites, which are calcium phosphates whose general formula is written either as $Ca_5(F,Cl,OH)(PO_4)_3$ or $Ca_{10}(PO_4)_6(OH,F,Cl)_2$, and we commonly

TABLE 1.10 Examples of bacteria used in the fermentation production of chemicals, enzymes, and pharmaceuticals.

Genus	Species	Product
Acetivibrio	*thermocellus*	Ethanol, Hydrogen
Acetobacter	*aceti*	Acetic acid
Acetobacter	*pasteurianus*	Acetic acid
Actinobacillus	*succinogenes*	Succinic acid
Agrobacterium	*tumefaciens*	Catalase
Alkalihalobacillus	*alcalophilus*	Lipase
Anaerobiospirillum	*succiniciproducens*	Succinic acid
Anoxybacillus	*kamchatkensis*	Xylanase
Bacillus	*acidopullulyticus*	Pullulanase
Bacillus	*amyloliquefaciens*	Amylase, Subtilisin
Bacillus	*licheniformis*	Amylase, Lipase, Subtilisin
Bacillus	*pumilus*	Lipase
Bacillus	*subtilis*	Amylase, Catalase, Laccase, Lipase, Protease, Riboflavin, Subtilisin, Thiamine
Bacillus	*thuringiensis*	Insecticidal endotoxin
Basfia	*succiniciproducens*	Succinic acid
[*Brevibacterium*]	*flavum*	Glutamic acid, Lysine, Malic acid
Brevundimonas	*naejangsanensis*	Butyrate, Hydrogen
Burkholderia	*cepacia*	Lipase
Burkholderia	*multivorans*	Lipase
[*Butyribacterium*]	*methylotrophicum*	Acetic acid, Butyrate
Citrobacter	*freundii*	Cobalamin, Hydrogen
Citrobacter	*intermedius*	Hydrogen
Clostridium	*acetobutylicum*	Acetone, Butanol, Butyrate, Hydrogen
Clostridium	*beijerinckii*	Butanol, Butyrate, Hydrogen
Clostridium	*butyricum*	Butyric acid, Hydrogen
Clostridium	*diolis*	Butyrate, Hydrogen
Clostridium	*paraputrificum*	Hydrogen
Clostridium	*pasteurianum*	Hydrogen
Clostridium	*tyrobutyricum*	Acetic acid, Butyrate, Hydrogen
Corynebacterium	*ammoniagenes*	Riboflavin
Corynebacterium	*glutamicum*	Lysine
Enterobacter	*cloacae*	Hydrogen
Escherichia	*coli*	Acetic acid, Amidase, Chymosin, Ethanol, Hydrogen, Lactic acid, Pantothenic acid, Succinic acid
Geobacillus	*stearothermophilus*	Lipase
Gluconobacter	*oxydans*	Acetic acid, Glucuronic acid
Kitasatospora	*aureofaciens*	Tetracycline
Klebsiella	*aerogenes*	2,3-Butanediol, Hydrogen

TABLE 1.10 (Continued)

Genus	Species	Product
Klebsiella	*oxytoca*	Ethanol
Klebsiella	*pneumoniae*	2,3-Butanediol, Cobalamin
Komagataeibacter	*xylinus*	Cellulose (bacterial cellulose)
Laceyella	*sacchari*	Xylanase
Lachnoclostridium	*phytofermentans*	Ethanol
Lactococcus	*lactis* subsp. *lactis*	Nisin
Lysinibacillus	*sphaericus*	Insecticidal endotoxin
Macrococcus	*caseolyticus*	Lipase
Micrococcus	*luteus*	Catalase
Micromonospora	*echinospora*	Calicheamicin
Micromonospora	*marina*	Thiocoraline
Mitsuokella	*jalaludinii*	Phytase
Paenibacillus	*polymyxa*	2,3-Butanediol
Propionibacterium	*freudenreichii subsp. shermanii*	Cobalamin
Pseudomonas	*aeruginosa*	Laccase, Lipase
Pseudomonas	*chlororaphis*	Nitrile hydratase
Pseudomonas	*denitrificans*	Cobalamin
Pseudomonas	*fragi*	Lactic acid
Ruminococcus	*albus*	Hydrogen
Saccharopolyspora	*erythraea*	Erythromycin
Sorangium	*cellulosum*	Epothilone
Streptoalloteichus	*tenebrarius*	Apramycin, Tobramycin
Streptomyces	*achromogenes*	Streptozotocin
Streptomyces	*albus*	Xylose isomerase
Streptomyces	*antibioticus*	Actinomycin D, Pentostatin
Streptomyces	*argillaceus*	Plicamycin
Streptomyces	*atroolivaceus*	Plicamycin
Streptomyces	*caespitosus*	Mitomycin C
Streptomyces	*calvus*	α-Amylase inhibitors (TAI-A and TAI-B)
Streptomyces	*carzinostaticus*	Neocarzinostatin
Streptomyces	*clavuligerus*	Clavulanic acid
Streptomyces	*corchorusii*	α-Amylase inhibitor (Paim)
Streptomyces	*galilaeus*	Aclarubicin
Streptomyces	*griseus*	Chromomycin A_2, Chromomycin A_3, Tobramycin
Streptomyces	*hygroscopicus*	Sirolimus
Streptomyces	*macromomyceticus*	Neocarzinostatin
Streptomyces	*natalensis*	Natamycin
Streptomyces	*peucetius*	Daunorubicin, Doxorubicin

(Continued)

TABLE 1.10 (Continued)

Genus	Species	Product
Streptomyces	*plicatus*	Plicamycin
Streptomyces	*rubiginosus*	Xylose isomerase
Streptomyces	*toxytricini*	Lipstatin
Streptomyces	*tsukubaensis*	Tacrolimus
Streptomyces	*verticillus*	Bleomycin
Streptomyces	*viridosporus*	Lactic acid
Symbiobacterium	*toebii*	Tyrosine phenol-lyase
Thermoactinomyces	*vulgaris*	Lactic acid
Weizmannia	*coagulans*	Lipase, Xylose isomerase
Xanthomonas	*campestris*	Ascorbic acid
Zymomonas	*mobilis*	Ethanol

There are many fermentation-derived enzymes which are not identified with great specificity.

TABLE 1.11 Examples of fungi used in the fermentation production of chemicals, enzymes, and pharmaceuticals.

Genus	Species	Product
Aspergillus	*awamori*	Cellulase, Pectinase, Xylanase
Aspergillus	*ficuum*	Phytase
Aspergillus	*flavus*	Lipase
Aspergillus	*niger*	Amyloglucosidase, α-L-Arabinofuranosidase, Catalase, Cellulase, Chymosin, Citric acid, Gluconic acid, Glucose oxidase, Lipase, Oxalic acid, Pectinase, Phytase, Protease
Aspergillus	*oryzae*	Amylase, Lipase
Aureobasidium	*pullulans*	Gluconic acid
Blakeslea	*trispora*	β-Carotene
[*Candida*]	*glabrata*	Pyruvic acid
Candida	*tropicalis*	Citric acid
Choanephora	*cucurbitarum*	β-Carotene
Debaryomyces	*hansenii*	Fusel alcohols
Diutina	*catenulata*	Citric acid
Eremothecium	*ashbyi*	Riboflavin
Eremothecium	*gossypii*	Riboflavin
Fusarium	*oxysporum*	Lipase
Fusarium	*venenatum*	Mycoprotein
Kluyveromyces	*lactis*	Amylase, Catalase, Cellulase, Cephalosporin, Laccase, Lipase, Pectinase, Penicillin, Peroxidase, Phosphorylase (example is sucrose phosphorylase), Protease, Xylanase, Xylose isomerase
Kluyveromyces	*marxianus*	Fusel alcohols, Lactase

TABLE 1.11 (Continued)

Genus	Species	Product
Meyerozyma	*guilliermondii*	Citric acid
Moesziomyces	*antarcticus*	Lipase
Mucor	*circinelloides*	β-Carotene
Nodulisporium	*sylviforme*	Paclitaxel
Penicillium	*chrysogenum*	Glucose oxidase
Penicillium	*janczewskii*	α-L-Arabinofuranosidase, β-Xylosidase, Zylanase
Penicillium	*stoloniferum*	Mycophenolic acid
Phycomyces	*blakesleeanus*	β-Carotene
Pichia	*kudriavzevii*	Glycerol
Rhizomucor	*miehei*	Lipase
Rhizopus	*arrhizus*	Lactic acid
Saccharomyces	*cerevisiae*	Catalase, Ethanol, Invertase
Talaromyces	*brunneus*	Amylase
Talaromyces	*funiculosus*	Cellulase
Talaromyces	*purpureogenus*	Glucose oxidase
Talaromyces	*variabilis*	Catalase
Taxomyces	*andreanae*	Paclitaxel
Thermoascus	*aurantiacus*	Catalase
Thermomyces	*lanuginosus*	Lipase
Tolypocladium	*niveum*	Cyclosporine A
Trichoderma	*reesei*	Cellulase
Trichothecium	*roseum*	Protease
Yarrowia	*lipolytica*	Citric acid, Retinol

There are many fermentation-derived enzymes which are not identified with great specificity.

use appatites as a source of phosphate in soil fertilizers. The secondary phosphate minerals include aluminum and iron phosphates in addition to calcium phosphates. Secondary phosphates can occur naturally, for example, as berlinite which is an aluminum phosphate with the chemical formula $AlPO_4$, and vivianite which is a hydrated iron phosphate with the chemical formula $Fe_3(PO_4)_2{\cdot}8H_2O$. Vivianite can have small amounts of calcium, magnesium, or manganese substituted in place of iron atoms. Secondary phosphate compounds also can be created by engineered processes.

Organic phosphate compounds can be categorized into three general groups, and these are phosphate monoesters, phosphate diesters, and the inositol phosphates which include phytic acid. Phytic acid is chemically a sixfold dihydrogen phosphate ester of inositol. Phytic acid exists mostly as phytates, and largely is present in seeds where the phytic acid serves as a phosphate storage reservoir for seedling plants. Inositol esters with less than six phosphates occur as catabolites of phytic acid. All three groups of organic phosphate compounds are found in soil. The phosphate monoesters account for the greatest amount of organic phosphorus compounds found in soil (Cabugao et al. 2021). The inositol forms are considered stable as contrasted with the mono- and di-phosphate esters which are considered labile. The three groups of organic phosphate compounds are mineralized by different types of phosphatase enzymes which are designated phosphomonoesterases, phosphodiesterases, and phytases.

Phosphate solubilizing bacteria that are present in soil produce organic acids by fermentation processes and those acids help to mobilize inorganic phosphorus, which makes that phosphorus more available for use by plants (Menezes-Blackburn et al. 2016). Those organic acids which accomplish this phosphate solubilization include ascorbic acid, citric and isocitric acids, fumaric acid, gluconic acid, lactic acid, and malic acid. The phosphate solubilizing bacteria that are present in soil also produce phosphatases in addition to their generation of organic acids (Jarrett et al. 2014; Jatuwong et al. 2020; Kobae 2019).

Some fungi also can help plants to utilize phosphates that are in soil. An important aspect of this process is the conversion of phosphate source materials present in soil into monobasic and dibasic phosphates which are more easily absorbed by the plants (Kobae 2019). de Oliveira Mendes et al. (2014) found that *Aspergillus niger* was the most effective fungus at being able to solubilize phosphates of aluminum, calcium, and iron, when compared against *Penicillium canescens*, *Penicillium ludwigii*, and *Talaromyces islandicus* (previously named *Penicillium islandicum*). Fungi of the genus *Trichoderma* also can promote plant growth by solubilizing those phosphate compounds that are present in acidic soils (Bononi et al. 2020). The fungal activity results in benefits to plant growth that are agriculturally valuable (Bononi et al. 2020).

Plant roots secrete both organic acids (Wu et al. 2018) and phosphatases (Tadano and Sakai 1991) which help to release the phosphorus that is present in the soil. Together, the acids and enzymes released into the soil increase the bioavailability of phosphate near the roots (Cabugao et al. 2021). The relative concentrations of available phosphorus and aluminum in soil have other ecological consequences, including their effect on the extent to which viruses can survive by adsorption to soil particles (Hurst et al. 1980).

1.2.3 Fermentative Microbes Are Important for the Biological Availability of Dietary Phosphate

Many animal tissues produce phytases which allow bioavailability of the phosphorus which is bound in phytic acid. Ruminant animals contain fermentative microbes in their rumens which secrete phytases and those microbes enable their host animal to utilize the phosphorus supplied in dietary seed materials (Jarrett et al. 2014). Monogastric animals such as birds, dogs, fish, humans, pigs, and poultry lack ruminant microbes which could secrete phytase and therefore phytic acid in the diet of these animals tends to pass through the gastrointestinal tract and be excreted. Providing exogenous phytase as a dietary supplement helps monogastric animals to utilize the phosphorus which is present as phytic acid in their food. There are designed microbial fermentations that use either bacteria (Table 1.10) or fungi (Table 1.11) to produce phytases which can be incorporated into feed supplements. *Aspergillus ficuum* is perhaps the microbe most notably used to produce phytase on a commercial scale (Jatuwong et al. 2020). Some edible mushrooms similarly are capable of producing phytases, and that ability facilitates growth of those fungi in their natural habitats (Jatuwong et al. 2020).

1.2.4 Natural Ethanol Production in Sap, Its Ingestion by Animals, and Human Intestinal Production of Ethanol

Many yeast species produce ethanol when spontaneously fermenting sugars and the ethanol is secreted as a waste product. We have adapted this capability as a technology and ingest that microbial waste product as part of our dining habits, and yet we certainly are not the only species that ingests the results of ethanol fermentations. It should not be surprising that our close relative the chimpanzee *Pan troglodytes* ingests fermenting sap from tapped *Raphia hookeri* palm trees in Guinea (Hockings et al. 2015). Numerous other animals including the pen-tailed tree shrew *Ptilocercus lowii* consume ethanol when ingesting naturally fermented nectar of the palm *Eugeissona tristis*, and that alcohol may serve as an attractant compound which entices the shrew to participate in pollination of the plant (Wiens et al. 2008). Ingestion by bats of the raw and fermented date palm sap in collection pots has a possible association with the zoonotic transmission of Nipah henipavirus that occurs when humans subsequently ingest sap from those same containers (Islam et al. 2021). Islam and coauthors (2021) did not mention the species of tree, but I will guess that it may have been *Phoenix dactylifera*. Overgrowth of colonizing yeasts within the human intestine can result in gut fermentation syndrome, also known as Auto-Brewery Syndrome, which is characterized by ethanol production leading to intoxication (Akhavan et al. 2019; Saverimuttu et al. 2019).

1.3 DESIGNING FERMENTATION PROCESSES

Microbial fermentations generally are categorized as being either submerged or solid state. Additional design characteristics that must be considered include the concentrations and types of nutrients, the oxygen concentration, pH, and temperature, all of which affect the population balance of microbes that performs the fermentation.

Very often, the chemical products such as acids and carbon dioxide which we value as being generated during microbial fermentations are indeed metabolic wastes when considered from the perspective of the fermenting microbes and those wastes may contribute to successful ecological suppression of competing microbes. Produced antibiotic compounds also have importance for the competitive strategies which microbes use in an effort to control the availability of resource materials.

1.3.1 The Levels of Liquid and Oxygen Are Perhaps the Two Most Defining Characteristics of Fermentations

Submerged fermentations occur when the microbes are growing in liquid. These can be roughly divided into two categories depending upon whether the microbes are growing freely in suspension versus growing on the surface of a solid carrier material, although there is a continuum with those two as the extremes. Submerged fermentations can be designed for either aerobic or anaerobic performance. Examples of the anaerobic submerged fermentations used in food processing are modification of beverages by generation of ethanol, during which the microbes are freely suspended, and the preservation of vegetables by generation of lactic acid during which the vegetables being preserved are a solid carrier. Aerobic submerged fermentation is used for converting dissolved ethanol into acetic acid which creates vinegar from cider and wine. Submerged fermentations similarly are important to the production of many chemicals, enzymes, and pharmaceuticals.

Solid state fermentations occur when the microbes are growing either on or within a solid substrate that is not submerged in liquid. These fermentations helpfully mimic the conditions under which many filamentous fungi naturally grow in substrate materials such as agricultural residues. Additional advantages associated with solid state fermentation include the use of relatively simpler fermentation equipment, relatively higher concentrations of the products, and less generation of process effluents (Liu and Kokare 2017). Solid state fermentations are designed to have various levels of moisture and they can be either aerobic or anaerobic. Examples of the products made by anaerobic solid state fermentations are bread, cheese, cocoa beans, coffee beans, some cured tea leaves, tobacco leaves for use in manufacturing cigars and cigarettes, silage, and some types of compost. Aerobic solid state fermentations are used for processing some types of tea leaves and some types of compost. Additionally, solid state fermentations are important in the production of chemicals, enzymes, and pharmaceuticals.

1.3.2 Microbial Ecology Largely Determines the Chemical Profile of Designed Fermentation Products

We have learned several techniques that can be used to adjust the microbial ecology of fermentations. The important formation of lactic acid in anaerobic food fermentations can be regulated by changing the concentrations of salt and sugar. Without adequate initial generation of lactic acid, the microbial ecology can change to allow production of butyric acid. If that change in microbial ecology occurs during fermentations that create silage, the result can be a high level of butyric acid in the silage that cattle will find unpalatable. The temperature at which materials ferment is another factor that affects their microbial ecology and controls the rate at which fermentation progresses. Consequently, we often need to use either heating or cooling to adjust the temperature of fermentations. It may be necessary to pasteurize or boil liquid raw materials as a pretreatment to destroy unwanted competitive microbes before a fermentation process begins. Thermal pretreatment also helps to reduce the level of pathogens that might be present in the final products of our fermentations. Solid state fermentations are critically affected by the moisture content of those materials that are being processed. Maintaining the desired microbial ecology of a solid state fermentation often requires the materials to either be kept within an enclosed chamber with controlled humidity or covered with plastic to prevent loss of moisture.

1.3.3 Spontaneous Fermentations and the Use of Back Slopping

We can consider microbial fermentations as being either spontaneous or controlled. Spontaneous fermentation occurs when those microorganisms naturally associated with the material that is being processed are allowed to establish and subsequently maintain the fermentation. For thousands of years, human usage of microbial fermentation as a designed process would have represented ritualistic efforts to duplicate the conditions which had resulted in favorable spontaneous fermentations. Much of our current fermentation technology, particularly the household level preparation of food, the fermentation of tobacco, and fermentations that create silage, generally still represents spontaneous fermentations.

Controlling the results of spontaneous fermentation can be done by a technique that is termed back slopping, during which fermented products having desired characteristics are added to the next batch of raw material that will be fermented. Back slopping still is done in many noncommercial productions of food. Back slopping typically is done for composting where it can be accomplished by reusing bulking materials. I will again mention inclusion of bulking materials later in this chapter.

We typically try to avoid spontaneous fermentations and back slopping in commercial food production both for reasons of product quality control and product safety.

1.3.4 The Importance of Using Defined Starter Cultures and Monitoring Fermentations for Product Quality Control

During the past century and a half, we have added to our knowledge of fermentation technology by learning to genetically select microorganisms that will produce products with the most favorable chemical profiles. We then use those microbes as defined starter cultures.

Commercial food fermentations are one example of how important starter cultures are for achieving product standardization (Waché et al. 2018).

Breeding of yeast strains allows us to change the balance of chemicals and correspondingly change the flavor profile which those microbes will produce during fermentations of food (Chen et al. 2022). Research that is being done for fermentation of meat is a helpful example of using standardized microbial inocula to create food (Laranjo et al. 2019). When we cannot eliminate unwanted microbes, we often begin the food fermentation by adding an overwhelming level of starter culture microbes with an example being those fermentations that produce wine from fruit juices. The types of product failures that can result from inability to control the microbial ecology of food production include possibility of spontaneous fermentation occurring which would create an unsellable product.

Using defined starter cultures is particularly important to the fermentations that generate commercially used chemicals including enzymes and pharmaceuticals. Some of the microbes used in starter cultures have been genetically engineered as a means of aiding standardization of the product and extending the metabolic capabilities of a microbial species. When you look at the microbes listed in Table 1.10 you may wonder how it is that *Escherichia coli* produces ethanol. That interesting result represents genetic engineering. Engineered microbes also can help to assure product safety and product quality control.

Monitoring the progress of ongoing fermentations is important for quality control. Depending upon the type of product being produced, monitoring can include chemical analyses, flavor assessment, and aroma assessment. Quality control analysis often requires sampling and culturing those microbes which are present in the material that is being fermented. High-throughput sequencing can be helpful for detecting the less dominant microbial species that participate in fermentation processes (Mendoza et al. 2017).

It is important to avoid the hazard of viruses which, if present in the starting materials or starter cultures, could destroy the desired microbes resulting in a fermentation that possibly becomes dominated by the wrong microbes. Table 1.12 is a master list of viral families that affect prokaryotes. Table 1.13 is a master list of viral families and unassigned (floating) genera that affect fungi.

The bacteria that produce lactic acid also are noted for producing bacteriocins and for producing hydrogen peroxide, both of which help to competitively control other microbes that might be present in the fermentation.

1.4 THE DISPOSAL OF WASTE PRODUCTS FROM FERMENTATIONS

Fortunately, there are many ways of using fermentation waste products, and the marketing for that reuse sometimes is termed valorization.

One commonly recognized example is the liquid fraction, called liquid whey, which remains after milk has been curdled with rennet for producing cheese. Liquid whey can be used as a food ingredient for pigs. Spray drying of liquid whey produces whey powder, which has several applications in food manufacturing including its function as a component material for dry mixes that are used to make cheese sauce. Whey powder similarly is part of the base material for some ice creams. Whey powder additionally is marketed as a dietary supplement both for animal feed and human body building.

Brewery wastes from the production of beer can serve as another example. Those wastes include Brewer's spent grain, hot trub, and spent yeast. Brewer's spent grain is a lignocellulosic biomass that results from the mashing process. The spent grain consists mostly of husks and is removed before the wort boiling process. Spent grain largely is used in either wet or dry form as

TABLE 1.12 A master list of viral families that affect prokaryotes.

Viral Family	Genome Type	Virion Structure	Known Infectious Host Range
Ackermannviridae	Double-stranded DNA	Head-tail, Non-enveloped	Bacteria
Ampullaviridae	Double-stranded DNA	Bottle shaped, Enveloped	Archaea
Atkinsviridae	Positive sense single-stranded RNA	Icosahedral, Non-enveloped	Bacteria
Autographiviridae	Double-stranded DNA	Head-tail, Non-enveloped	Bacteria
Autolykiviridae	Double-stranded DNA	Icosahedral, Non-enveloped	Bacteria
Bicaudaviridae[a]	Double-stranded DNA	Lemon shaped, Non-enveloped (presumably)	Archaea
Blumeviridae	Positive sense single-stranded RNA	Icosahedral, Non-enveloped	Bacteria
Chaseviridae	Double-stranded DNA	Head-tail, Non-enveloped	Bacteria
Corticoviridae	Double-stranded DNA	Icosahedral, Non-enveloped	Bacteria
Cystoviridae	Double-stranded RNA	Double shelled Icosahedral, Enveloped	Bacteria
Demerecviridae	Double-stranded DNA	Head-tail, Non-enveloped	Bacteria
Drexlerviridae	Double-stranded DNA	Head-tail, Non-enveloped	Bacteria
Duinviridae	Positive sense single-stranded RNA	Icosahedral, Non-enveloped	Bacteria
Fiersviridae	Positive sense single-stranded RNA	Icosahedral, Non-enveloped	Bacteria
Finnlakeviridae	Single-stranded DNA	Icosahedral, Internal membrane, Non-enveloped	Bacteria
Fuselloviridae	Double-stranded DNA	Lemon-shaped to Pleomorphic, Enveloped, Short tail fibers extend from one polar end	Archaea
Globuloviridae	Double-stranded DNA	Helical, Enveloped, Spherical	Archaea
Guelinviridae	Double-stranded DNA	Head-tail, Non-enveloped	Bacteria
Guttaviridae	Double-stranded DNA	Helical (presumably), Enveloped, Egg-shaped, Fibers	Archaea
Halspiviridae	Double-stranded DNA	Lemon shaped, Enveloped, One tail	Archaea
Herelleviridae	Double-stranded DNA	Head-tail, Non-enveloped	Bacteria
Inoviridae	Positive sense single-stranded DNA	Helical, Non-enveloped	Bacteria
Leviviridae	Positive sense single-stranded RNA	Icosahedral, Non-enveloped	Bacteria
Lipothrixviridae	Double-stranded DNA	Helical, Enveloped, Rod-shaped	Archaea
Microviridae	Positive sense single-stranded DNA	Icosahedral, Non-enveloped	Bacteria
Myoviridae	Double-stranded DNA	Head-tail, Non-enveloped	Archaea, Bacteria, Cyanobacteria
Ovaliviridae	Double-stranded DNA	Ellipsoidal stacked disk structure, Enveloped	Archaea

TABLE 1.12 (Continued)

Viral Family	Genome Type	Virion Structure	Known Infectious Host Range
Plasmaviridae[b]	Double-stranded DNA	Helical (presumably), Enveloped, Approximately spherical	Bacteria
Plectroviridae	Positive sense single-stranded DNA	Helical, Rod-shaped, Non-enveloped	Bacteria
Pleolipoviridae	Double or single-stranded DNA	Pleomorphic, Enveloped	Archaea
Podoviridae	Double-stranded DNA	Head-tail, Non-enveloped	Bacteria, Cyanobacteria
Portogloboviridae	Double-stranded DNA	Icosahedral, Internal lipid layer	Archaea
Rountreeviridae	Double-stranded DNA	Head-tail, Non-enveloped	Bacteria
Rudiviridae	Double-stranded DNA	Helical, Non-enveloped, Rod shaped with tail fibers on one end	Archaea
Salasmaviridae	Double-stranded DNA	Head-tail, Non-enveloped	Bacteria
Schitoviridae	Double-stranded DNA	Head-tail, Non-enveloped	Bacteria
Siphoviridae	Double-stranded DNA	Head-tail, Non-enveloped	Archaea, Bacteria
Solspiviridae	Positive sense single-stranded RNA	Icosahedral, Non-enveloped	Bacteria
Sphaerolipoviridae	Double-stranded DNA	Icosahedral, Internal membrane, Non-enveloped	Archaea
Spiraviridae	Positive sense single-stranded DNA	Coil-shaped, Non-enveloped	Archaea
Steitzviridae	Positive sense single-stranded RNA	Icosahedral, Non-enveloped	Bacteria
Tectiviridae	Double-stranded DNA	Icosahedral, Non-enveloped	Bacteria
Thaspiviridae	Double-stranded DNA	Spindle shaped with short fibers on one end, Non-enveloped	Archaea
Tristromaviridae	Double-stranded DNA	Helical, Enveloped, Filamentous with fibers on one end	Archaea
Turriviridae	Double-stranded DNA	Icosahedral, Internal membrane, Non-enveloped	Archaea
Zobellviridae	Double-stranded DNA	Icosahedral, Non-enveloped	Bacteria

[a] Note for Bicaudaviridae: Virions can develop two tail like extensions on opposide polar ends after budding.

[b] Note for Plasmaviridae: The viral genome is integrated into host chromosome and transcribed by host polymerase.

supplemental animal feed for mammalian livestock, poultry, and fish. Spent grain also can be pulped and manufactured into paper products. Hot trub is the sediment formed during wort boiling and mostly consists of coagulated colloidal protein. Hot trub less frequently is used as animal feed because the presence of spent hops contributes to an unpleasant flavor. Brewers' spent yeast sediments during and after fermentation. The spent yeast subsequently can be collected and reused a limited number of times. Spent yeast often is sold wet as animal feed but either heat treatment or

TABLE 1.13 A master list of viral families and unassigned (floating) genera that affect fungi.

Viral Family	Genome Type	Virion Structure	Known Infectious Host Range
Alphaflexiviridae	Positive sense single-stranded RNA	Helical, Non-enveloped	Fungi, Plants
Barnaviridae	Positive sense single-stranded RNA	Polyhedral, Non-enveloped	Fungi
Benyviridae	Positive sense single-stranded RNA	Helical, Non-enveloped	Fungi, Plants, Invertebrates
Botourmiaviridae	Positive sense single-stranded RNA	Bacilliform polyhedral, Non-enveloped	Fungi, Plants (mitochondrial endosymbionts)
Chrysoviridae	Double-stranded RNA	Icosahedral, Non-enveloped	Fungi, Plants, (and possibly Invertebrates)
Curvulaviridae	Double-stranded RNA	Virion undefined	Fungi
Deltaflexiviridae	Positive sense single-stranded RNA	Uncertain	Fungi, Plants
Endornaviridae	Positive sense single-stranded RNA	No true capsid	Algae, Fungi, Oomycetes, Plants
Fusariviridae	Positive sense single-stranded RNA	Might not form a true virion	Fungi
Gammaflexiviridae	Positive sense single-stranded RNA	Helical, Non-enveloped	Fungi
Genomoviridae	Single-stranded DNA	Icosahedral, Non-enveloped	Fungi, Vertebrates
Hypoviridae	Double-stranded RNA	No true capsid	Fungi
Megabirnaviridae	Double-stranded RNA	Icosahedra, Non-enveloped	Fungi
Metaviridae	Positive sense single-stranded RNA	Icoshedral, Enveloped	Fungi, Invertebrates, Plants, Vertebrates
Mitoviridae	Positive sense single-stranded RNA	No true virion	Fungi, Plants
Mymonaviridae	Negative sense single-stranded RNA	Filamentous, Possibly enveloped	Fungi
Narnaviridae	Positive sense single-stranded RNA	No true virion	Fungi, Invertebrates, Plants
Nodaviridae	Positive sense single-stranded RNA	Icosahedral, Non-enveloped	Fungi (experimentally), Invertebrates, Vertebrates
Partitiviridae	Double-stranded RNA	Icosahedral, Non-enveloped	Fungi, Plants
Polymycoviridae	Double-stranded RNA	No true capsid (genomic segments have a protein coat, possibly also coated with cellular colloids)	Fungi
Pseudoviridae	Positive sense single-stranded RNA	Icosahedral, Non-enveloped	Fungi
Quadriviridae	Double-stranded RNA	Icosahedral, Non-enveloped	Fungi
Reoviridae	Double-stranded RNA	Icosahedral, Non-enveloped	Algae, Fungi, Invertebrates, Plants, Vertebrates
Totiviridae	Double-stranded RNA	Icoshedral, Non-enveloped	Algae, Fungi, Invertebrates, Protozoa

Floating Genus	Genome Type	Virion Structure	Known Infectious Host Range
Rhizidiovirus	Double-stranded DNA	Icosahedral, Non-enveloped	Fungi

addition of acids is used to inactive the yeast so as to avoid gastrointestinal difficulties if the yeast will be fed to pigs (Rachwał et al. 2020).

The residues from commercial fermentation industries can cause loading problems for municipal sewage systems with the list of concerns including the biological oxygen demand (BOD) associated with alcohol, hops, pulped plant materials, spent grains, sugars, trub, and yeast. The distillery process creates between 8 and 20 l of residual liquid, termed stillage, for every liter of ethanol produced (Gebreeyessus et al. 2019). Mikucka and Zielińska (2020) have provided a good review of efforts to use aerobic as well as anaerobic processes to resolve problems associated with treatment and valorization of distillery stillage.

There also can be pH problems associated with releasing acidic final products, acidic cleaning solutions, and caustic cleaning solutions, either into the environment or into wastewater collection systems. Releasing boiling liquids into the sewage system additionally can cause thermal problems (City of Calgary 2022).

1.5 FERMENTATIONS THAT PRODUCE AND PRESERVE FOOD FOR HUMANS

Fermentation technology valuably has helped us to produce such food materials as beer, bread, and cheese, which do not exist in nature. Fermentation also helps us to preserve harvested items like plant materials and meat in ways that do not depend upon merely drying and salting.

1.5.1 Food Fermentation Is Largely About Lactic Acid and Ethanol

The two microbial activities that are most used for producing and preserving food materials are those which generate lactic acid and ethanol. Both of these activities are anaerobic.

Lactic acid is produced by some groups of bacteria, fungi, protozoa, and oomycetes. The ability of those microorganisms to excrete lactic acid into their surrounding materials successfully inhibits the activity of many other microorganisms. Lactic acid fermentations additionally produce a range of organic compounds including acetic acid and ethanol. Lactic acid fermentation is mostly used in the souring of milk products including buttermilk, cheese, sour cream, and yogurt. It similarly is used in fermentation of brined fruits and vegetables, and for the maturation of fermented sausages (Ciani et al. 2010). These fermentations do in many ways mimic the ecosystem which exists in the intestinal tract of warm-blooded animals where lactic acid bacteria, bifidobacteria, and clostridia coexist with other enteric bacteria, all of which are involved in the transformation of ingested food substrates via different fermentation pathways. Within the intestine, lactic acid bacteria help to maintain a balanced microfloral composition that suppresses the replication of clostridia (Ciani et al. 2010).

Ethanol is produced by heterolactic fermentation, but the presence of ethanol in food largely results from yeasts converting mono-, di-, and trisaccharides into ethanol. We rely upon that process by yeasts to generate beer, cider, and wine. Importantly, the processing of any raw materials which largely contain their carbohydrates as starches may require action by additional microbes which produce amylases, and those amylases convert the starches into the mono-, di-, and trisaccharides that can be used to generate ethanol. Many microbial groups can produce and release low levels of ethanol into their surrounding materials. However, ability of the facultative anaerobic yeast *Saccharomyces cerevisiae* to tolerate higher ethanol levels when fermenting sugars results in *S. cerevisiae* eventually winning its contests against competing microbes.

Examples of combined processes that use bacteria to produce lactic acid and yeast to produce ethanol are the submerged fermentations of milk which create kefir and kumis, plus those solid state fermentations which produce cocoa, coffee, and some types of bread.

Fermentation also can increase the digestibility of some food materials by reducing the lectin content of beans and reducing the cyanogenic glycoside content of plant products among which are cassava and flaxseed.

1.5.2 Fermentations That Preserve Fruits and Vegetables

Anaerobic submerged fermentation is used to preserve in brine fresh fruit and vegetable materials such as apples, beets, cabbages, carrots, cucumbers, onions, parsnips, peppers, and tomatoes. Establishing appropriate conditions for the fermentation of fruits and vegetables includes the necessary submergence of those plant materials in the brine liquid to reduce exposure of the fermenting vegetables to oxygen. This type of fermentation often is called fermentation pickling and the liquid for that brine may be drawn from the materials being preserved by adding salt to those materials (Behera et al. 2020). The necessity of adding salt is critical. The salt can be added either in the form of sodium chloride, naturally saline water, or sea salt. The salinity changes water availability and is ecologically important to the fermentation process by controlling the ability of different microorganisms to grow during the fermentation process. The salinity critically helps to prevent the growth of pathogenic microorganisms and also reduces the growth of spoilage microorganisms. The fermentation process will occur only at a suitably warm temperature. These fermentations very often are designed to not be stirred and the materials being processed may be in stationary position during the entire fermentation period. The microbes will be growing not only in suspension, but also growing on the surface of the food material which serves as a solid carrier and microbes will be growing within the food material. Some early understanding of the microbial ecology of plant material fermentations was summarized by Daeschel et al. (1987).

Fermentation of fruits and vegetables is mostly done by bacteria which break down sugars into acids. The fermentations of fruits and vegetables primarily produce lactic acid and appropriately are described as lactic acid fermentations, although other acids will be produced including acetic and propionic. The microbes also generate additional compounds which contribute to the flavor profile of the food, and the fermentation can generate carbon dioxide gas.

The yeast *Nakaseomyces delphensis* is found on dried figs. Aouizerat et al. (2019) found yeast of the species *N. delphensis* in containers that presumably once held fermented figs thousands of years ago, which could suggest the yeast had performed spontaneous fermentation of the figs.

Fermentations of fruits and vegetables naturally occur in a stepwise manner that is associated with changes in the composition of their microbial community. The microbial community is dominated in sequence by different microbial species whose ability to grow depends upon the chemistry of the fermentation (Bah et al. 2019). The primary fermentation of fruits and vegetables often is dominated by *Leuconostoc mesenteroides* and *Lapidilactobacillus dextrinicus* (previously named *Pediococcus cerevisiae* subsp. *dextrinicus*) which initially begin the generation of acid and carbon dioxide. The produced acidity eventually inhibits and can even be lethal to many of the primary fermentation microbes. Secondary fermentation then takes over which further lowers the pH and adds complexity to the flavor profile. The secondary fermentation often is done by microbes that include *Levilactobacillus brevis* and *Lactiplantibacillus plantarum* (previously named *Lactobacillus plantarum*) Eventually, *L. plantarum* finishes the fermentation process owing to its greater acid tolerance. The fermentation is completely terminated by acidity which the microbes generate (Eifert et al. 2014). As a homemade process, these generally are spontaneous fermentations. Defined starter cultures often are used for commercial production of fermented fruit and vegetable materials.

There also are processes termed chemical pickling which involve storage of fruits and vegetables in either high-salt brine, ethanol, or vinegar. Chemical pickling does not involve fermentation of the fruit and vegetable materials being preserved although microbial fermentations may have been used to produce the ethanol or vinegar.

1.5.3 Fermentations of Grains and Other Seeds Including Beans

Bread typically is prepared by a solid state anaerobic fermentation of grain that previously had been ground to produce a dry flour, then rehydrated to form a cohesive mass. Starchy tubers or dried beans that have been ground into a flour can substitute for the grain. The rehydrated mass used for preparing bread often is elastic, which is described as being glutinous, and is defined as being a dough. Bread dough is sufficiently cohesive that it easily can be shaped by hand. Its volume may expand, which is termed rising of the dough, by the action of yeast which ferment sugars to generate carbon dioxide. Ethanol also is produced by the yeast during these fermentations. And, notably in the case of sourdough bread, the microbial fermentation additionally can include bacterially produced lactic acid. Fermentations of bread dough typically are performed by bacteria of the order Lactobacillales and fungi of the order Saccharomycetales. The most commonly used fungus is *S. cerevisiae*, which is called baker's yeast. The fermented material subsequently will be solidified by either baking, boiling, or steaming before it is ingested. Not all breads have the volume of their dough expanded by microbial fermentation. Expansion of the dough can instead be done by adding sodium bicarbonate which releases carbon dioxide. Some food materials that are described as bread do not use any rising of the dough and are not fermented, with an example being corn tortillas which are a flat bread prepared from a cohesive mass that is created by rehydrating maize flour. Bread has a history of at least 14 000 years (Arranz-Otaegui et al. 2018).

Pancakes are similar to bread but the rehydrated material is more dilute and described as a batter. The batter can be made to rise and simultaneously flavored by fermentation, although the addition of sodium bicarbonate often is used as a faster alternative means of rising the mixture. Pancake batter is transferred by either pouring or spooning into shallow pans for baking, hence the name pancake. The batter alternatively can be transferred onto a flat cooking surface described as a griddle where the batter is fried, producing a griddle cake. Pancake batters also can be poured into either heated oil or fat for deep frying, often done by pouring the batter through a funnel which produces a funnel cake. The microbial ecology of fermenting batter includes bacteriocins such as those produced by *Pediococcus pentosaceus* (Kumar et al. 2018). The consistency of cooked pancakes is similar to a soft cooked bread.

Colloidal suspensions that have been made by boiling either sliced, crushed, or coarsely ground grains are called porridges. There are times when dried, cracked grain is fermented in water and then cooked as a porridge. Porridges are sufficiently viscous that they typically are eaten with a spoon. If the suspension is made of more finely ground material such as a flour, then the result is a pap. Paps may either be consumed as a thick beverage or as a paste. Paps also can be made by boiling in water, milk, or a mixture of both, other starchy plants that have been either crushed or ground.

Paps can be fermented before they are consumed.

Dried intact grain often is submerged in water and fermented in that water before boiling the mixture and eating it as cooked whole grain.

Dry grain also can be fermented in water, then redried, before it is ground into a flour for use in making breads, pancakes, or paps.

Sorghum is a cereal grain commonly used for preparing fermented food in Africa (Adebo 2020). There additionally are many types of ethnic food made by naturally fermenting whole soybeans (Tamang 2015) with examples being kinema and nattō.

Microbiological concerns have been associated with some of the fungal species that are found in barley kernels (Rabie et al. 1997) and some of those fungal species produce toxins (Welling 1974). Those concerns are a reason for not performing solid state fermentations of rehydrated grain under conditions which might allow the toxic fungi to grow.

1.5.4 Fermentations That Produce Alcoholic Beverages

Ethanol fermentation, also called alcoholic fermentation, is a submerged fermentation process done under anaerobic conditions. The microbes creating these fermentations are either anaerobic or facultatively anaerobic. They act by converting mono-, di- and trisaccharides into ethanol with carbon dioxide produced as a byproduct. An example of the fermentation process is that one glucose molecule yields two molecules each of ethanol and carbon dioxide. The alcoholic beverages that we produce by microbial fermentations include cider and wine, beer, and distilled beverages. The best wild yeasts for producing beer and wine often are found on flowers and plants (Aouizerat et al. 2019) rather than in soil.

Humans have a long history of creating alcoholic beverages by using microbial fermentations to produce ethanol. The production of beer has a history of approximately 9000 years in China where consumption of an alcoholic beverage made from rice with other ingredients including Job's tears may have been associated with social activities such as funerary rituals (Wang et al. 2021a; Chapter 6, pages 209–224 of this book by Liu). A mixed fermentation of rice, honey, and hawthorn fruit as a beverage was done in China as early as 7000 BCE (McGovern et al. 2004). Archeological excavations have revealed that production of beer from grains and tubers was done as early as perhaps 5000 BCE in the Jordan valley and Galilee (Rosenberg et al. 2021).

It is important to note that not all yeast strains which can be isolated from environmental samples will produce beer. Many of those wild yeast strains that do produce beer create a product that lacks acceptable flavor and aroma, and some wild yeasts produce toxic compounds. Aouizerat et al. (2019) isolated yeast cells from clay containers which presumably had held fermented beverages several thousand years ago in areas that now are part of Israel. It was presumed that the isolated yeast cells found by Aouizerat et al. (2019) were descendants of yeast which had been present in fermented beverages once held in those containers, and those isolated yeast successfully were used to produce beer that was considered acceptably flavorful and aromatic. Thomas et al. (2021 and Chapter 7, pages 225–243 of this book) have used bioarchaeology and shipwrecked beer to understand the more recent history of brewing and fermenting beer.

Beer often is produced by fermenting an aqueous extract called wort, which is prepared from partially sprouted grain that is described as having been malted. The wort may contain other plant materials that have a high starch content with examples being the tubers of cassava and potato (Chacón Mayorga et al. 2021). Starches must be converted into sugars before the fermentation can produce ethanol, and the conversion is done by amylases which often are produced by allowing that sprouting of grain before the wort is prepared. *S. cerevisiae* typically is used for the fermentations that produce beer because that yeast species can ferment a broad range of sugars including the maltose and maltotriose which are found in wort prepared from malted grain. *Saccharomyces ludwigii* is instead used for creating low-alcohol beer from malted grain because it can ferment the sugar's glucose, fructose, and sucrose which are present in the wort, but it cannot ferment either the maltose or maltotriose contributed by the malted grain.

Cider and wine are made from fruit juice, sap, sugar solutions, or diluted honey. Autochthonous yeast populations can cause spontaneous fermentations (Lopandic et al. 2008) with an example being the spontaneous fermentation of *Eucalyptus gunnii* tree sap to produce a cider-like alcoholic beverage, which was a natural process used by aboriginal populations in Tasmania (Varela et al. 2020). Fermentations that produce beverages such as cider and wine may be designed by using a mixture of yeast species which will create a desired chemical profile in the product. Although, the weak ethanol tolerance of many yeast species, in combination with the acidification which occurs during fermentation of the liquids, means that the microbial ecology of these fermentations becomes dominated by *S. cerevisiae* (Ciani et al. 2013). Eventually, the ethanol which *S. cerevisiae* produces becomes inhibitory even to its role in the fermentation process at which point ethanol production ends. *S. cerevisiae* is considered to be a domesticated species because of its selection during designed food processes (Carlsen et al. 1991; Ciani et al. 2010).

Rice wine and whiskey are examples of beverages that are produced by fermenting a mash, which is a slurry of crushed grain in water that may contain added sugar. Fermentation of rice to produce the alcoholic beverage saki occurs with simultaneous saccharification by *Aspergillus oryzae* which produces amylases, and conversion of the sugars to ethanol by *S. cerevisiae* (Yasui et al. 2020; Takeshita and Oda Chapter 9, pages 255–260 of this book).

Research into ethanol production by microbial fermentation has had notable, and Nobel, results. Eduard Buchner won the 1907 Nobel Prize in Chemistry for demonstrating that a cell-free liquid yeast extract could convert sugar into ethanol and carbon dioxide. Arthur Harden published an excellent history of scientific discoveries regarding ethanol fermentation (Harden 1914). Those discoveries included developing an understanding that ethanol and carbon dioxide are waste products generated when yeast use sugar to sustain their growth process. Arthur Harden and Hans von Euler-Chelpin jointly won the 1929 Nobel Prize in Chemistry for understanding the molecular chemistry of alcoholic fermentation. Interestingly, one of Hans von Euler-Chelpin's sons; Ulf von Euler, shared the 1970 Nobel Prize in Physiology or Medicine for studying neurotransmitters.

1.5.5 Fermentations That Produce Vinegar

The conversion of ethanol to acetic acid is an aerobic process that is considered to be a submerged fermentation, and it is performed by obligately aerobic microorganisms. This process is termed both acetic acid fermentation and vinegar fermentation. A typical example is the production of Molasses beer and Molasses vinegar by successive fermentations of molasses in water using *S. cerevisiae* to produce ethanol followed by *Acetobacter aceti* (Patel and Pandya 2015) to convert the ethanol into acetic acid. Preparation of kombucha by fermentation of tea containing added sugar is another example (May et al. 2019). Kombucha contains a very low level of ethanol plus it contains acetic acid, gluconic acid, and glucuronic acid. The fermentation that produces kombucha contains the yeasts *Medusomyces gisevii* and *S. cerevisiae*. It also contains the bacteria *Komagataeibacter xylinus* which produces cellulose (Volova et al. 2022) that provides structure to the pellicle which forms on the fermenting liquid.

Acetic acid also can be produced by fermentation of lactic acid under conditions of limited oxygen.

1.5.6 Fermentations That Create Dairy Products

Cheese is a particularly interesting and very well studied product of microbial fermentation. The making of cheese presumably originated in West Asian countries about 8000 years ago (Zheng et al. 2021) and one type of cheese that may have been produced 4000 years ago still is being made (Levitt 2018). The main proteins in milk are casein and whey. The main carbohydrates in milk are citrate, lactate, and lactose. Creation of cheese begins as a submerged fermentation of the milk and occurs simultaneously with a coagulation of the casein. The coagulated casein solids are termed curd, and the curd separates from the liquid portion of the milk which is called whey. That liquid portion contains the whey protein. During this first stage of cheese making, a primary fermentation occurs when lactic acid bacteria anaerobically produce lactic acid by fermenting the lactose. The food creation process then changes to being a solid state fermentation in the collected curd. Subsequent addition of salt to the curd slows production of lactic acid. The collected curd is termed green cheese in contrast with the eventual final fermented product that is considered to be either matured or ripened cheese.

During the ripening process of cheese, which is its secondary fermentation, citrate is converted into acetate, acetoin, butanediol, and diacetyl, while some remains as citric acid. Lactic acid fermentation continues to convert lactose into lactic acid, and also converts lactose into lactate. Propionic acid fermentation can metabolize lactate to produce short chain fatty acids including propionic acid. Propionic acid similarly is one of the short chain fatty acids which are produced by gut microbiota from indigestible carbohydrates that are consumed as dietary fiber. Propionic acid fermentation of cheese also produces acetate, carbon dioxide, and some amino acids. Carbon dioxide that is generated during the production of propionic acid in cheese causes the characteristic bubbles that are found in Swiss-type cheeses. Bücher et al. (2021) have reviewed applications of propionic acid bacteria in the food industry, and those bacteria primarily are used in Swiss-type cheeses. Together, the organic acids in ripened cheese largely consist of citric, lactic, malic, propionic, pyruvic, and succinic. Butyric acid can be an unwanted side product in cheese, it is fermentatively produced by *Clostridium butyricum* and *Clostridium tyrobutyricum*, and its occurrence tends to happen when the level of lactic acid production is not adequate to maintain a sufficiently low pH. Cheese fermentation additionally converts triglycerides into hydroxy fatty acids among which are octanoic acid and decanoic acid, plus it generates esters and thioesters, n-alkane-2-ones, and n-alkane-2-ols. The casein in fermenting cheese is converted into aldehydes, ketones, ammonia, and sulfur compounds (Zheng et al. 2021).

Pasteurizing the milk that is to be used for producing cheese very beneficially reduces the potential of pathogenesis associated with ingesting the produced cheese. Lipases are present in milk, and fat hydrolysis notably contributes to the ripening of some types of cheese such as cheddar. Lipases can be inactivated by pasteurization of the cream and milk which are used to produce cheese. Proteases naturally are present in milk, and the heating that accomplishes pasteurization decreases protein decomposition within the ripening cheese in ways that can have an undesired impact upon the flavor profile and texture of the matured cheese (Lane and Hammer 1935a). Addition of proteolytic enzymes such as erepsin, pepsin, and trypsin after pasteurization can aid with the proteolysis that helps to characterize cheddar cheese (Lane and Hammer 1941). Using appropriately selected starter cultures of *Lacticaseibacillus casei* can resolve the problems of flavor and texture (Lane and Hammer 1935b). We also intentionally alter the microbiology of cheese by selectively including molds, notably penicillia, as part of the starter culture at the beginning of the cheese fermentation process (Bryant and Hammer 1940). *Penicillium camemberti* is used in the production of cheese types such as Camembert and Brie. *Penicillium roqueforti* produces the veining coloration that is important to the ripening of blue cheese including Roquefort (Lane and Hammer 1935c).

Not all of the microbes that naturally are present in milk should be considered unwanted. Members of the bacterial genus *Enterococcus* are commonly present in milk and correspondingly found in the facilities that process milk to produce dairy products. *Enterococcus* can survive pasteurization, grow during refrigeration, and fermentatively produce lactic acid. While some members of that genus are pathogenic, others are an important part of the dairy fermentations that produce cheese (Bhardwaj et al. 2008). *P. pentosaceus*, which also is naturally present in milk, produces both lactic acid and phenylactic acid during fermentations of cheese. Those two acids valuably help to reduce spoilage that would be caused by the growth of unwanted fungi (Shi and Maktabdar 2022). Gilbert (2017) has provided a basic explanation of cheese making.

Ethanol is produced anaerobically by fungi during direct fermentation of the whey fraction because the low lactose content of whey consequently allows for only limited production of lactic acid. Microbial proteolysis of whey creates a substrate from which fermentation can produce propionic acid.

The beverage known as kefir is another interesting product which is created by a fermentation of milk. During kefir production, a combination of bacteria and yeast growing together in the milk forms a matrix of exopolysaccharide and protein that is called a kefir grain (Bourrie et al. 2016). The kefir grains are retained and used for back slopping. People do as well eat the kefir grains.

1.5.7 Fermentations That Preserve Meat Products

Meat products such as ham, salami, fish, and shellfish are preserved by anaerobic solid state fermentations that primarily lower the pH by producing lactic acid.

Rapidly and adequately lowering the pH of fermenting salami is an important safety factor which suppresses growth of clostridia within the meat matrix. Addition of sugar such as dextrose to the meat matrix, along with carefully selected starter cultures, helps to quickly begin that production of lactic acid. Microbes also will transfer from natural casings into the fresh meat matrix (Pisacane et al. 2015). The relative presence of different microbes successively changes during the salami fermentation process, which is called ripening, and the microbial community very notably differs between types of salami.

The microbial community composition also changes during fermentation of crustaceans, fish, and mollusks. For example, Koo et al. (2016) found that during production of sikhae by fermentations of *Gadus chalcogrammus* (walleye pollock) which often include the salt tolerant organism *Latilactobacillus sakei* (previously named *Lactobacillus sakei*), *L. sakei* represented only 2% of the microbial population before the fermentation. But, when the fermentation had reached a level of 10 °C *L. sakei* dominated the microbial community to the level of 74%, and *L. sakei* was present at 90% when the temperature achieved 20 °C. The fact that *L. sakei* is a bacteriocin producer may help its competitive dominance. *L. sakei* largely serves by anaerobically fermenting sugars to lactic acid and acetic acid.

1.5.8 The Microbial Safety of Fermented Food

Iron age salt miners in Europe subsisted upon a diet presumed to include blue type cheeses and beer, as evidenced by the presence of *P. roqueforti* and *S. cerevisiae* in paleofeces (Maixner et al. 2021). Fermented food still constitutes a major portion of human diets in areas of the world where refrigeration is not available (Eifert et al. 2014) and Sandor Katz, who is an enthusiast of homemade fermentations, interestingly has stated that fermentation of food represents belief in "... the idea of peaceful coexistance with microbes rather than warfare." (Katz 2003).

Unfortunately, unsanitary handling of the raw materials and also inadequate protection of the fermented food products, particularly the use of unsanitary containers for preparing and vending fermented products, can result in the sale and consumption of fermented materials that contain dangerous microbial contamination (Skowron et al. 2022).

There needs to be a sense of caution when saying that the microbes in fermented food products are not only safe but beneficially healthy as "Probiotics." The reason for caution comes from understanding that many of the bacteria and fungi which are considered to be probiotic are opportunistically pathogenic for mammals (Hurst 2018) including pathogenicity for humans (Hurst 2019).

Food spoilage microorganisms are of tremendous economic concern and critically important for people who have a dietary dependence upon fermented foods. I personally consume the products of food fermentations each day.

Several bacterial species have been associated with food spoilage and food associated illness. Among these is *Bacillus cereus* which produces gastrointestinal toxins and is found in fermentations of both rice as well as soybean that are used for production of food. *Listeria monocytogenes* which causes systemic infections is a particularly concerning opportunistic pathogen that is found in raw cow, goat, and sheep milk. *L. monocytogenes* is a problem in cheese because of its resistance to lactic acid, it actually produces lactic acid, plus it is a psychrotroph (Ghosh 2016) which allows that microbe to grow in products that are being refrigerated. *Staphylococcus aureus* causes contagious mastitis in cattle and is frequently present in raw milk. *S. aureus* expectedly has been found in milk fermentations and interestingly also is found in soybean fermentations. *S. aureus* produces staphylococcal enterotoxin in food which causes staphylococcal food poisoning (Ghosh 2016).

A number of fungal species have been associated with food spoilage and food associated illnesses. *Aspergillus clavatus* is found in molded barley and that microbe causes a hypersensitivity pneumonitis which is called malt-worker's lung. *Geotrichum candidum* which can cause invasive cutaneous infections is found in fermentations of milk, soybean, and taro. Cultured *P. camemberti* is added to the milk at the start of fermentations that produce some varieties of cheese including Camembert and Brie. *P. camemberti* can cause intestinal invasion and disseminated disease including cerebral infection in humans (Hurst 2019). Inhaling *P. camemberti* contained in water-based inoculum which is sprayed onto cased sausages can result in hypersensitivity pneumonitis (Marchisio et al. 1999). Inhaling *Penicillium caseifulvum* that is present on the rind surface of blue cheeses, where it causes brown spots, can cause a hypersensitivity pneumonitis that is called cheese-washer's disease (UpToDate 2022). *Penicillium expansum*, which can produce toxin but more commonly causes spoilage, is found in both milk and olive fermentations. *Penicillium palitans* which has been found in cream cheese produces the fungal neurotoxin cyclopiazonic acid (Ciegler 1969). *P. roquefortii* is one of the microbes that produces blue veining in some types of cheese, including Roquefort. *P. roquefortii* also is found in fermentations of olives and fermentations that produce sour cream. *P. roquefortii* is a hazardous species because its presence on cheese rind causes the hypersensitivity pneumonitis identified as cheese-washer's disease and it also causes the infection known as fungus ball maxillary sinusitis disease in humans (Hurst 2019). *Rhodotorula glutinis*, which is an opportunistic pathogen that causes nail infections of humans, has been found in *Sorghum* grain fermentations used for production of bread. *R. mucilaginosa* which can cause fungemia is found in fermentations of coffee seeds, maize, pork, and teff that are used for production of food. Even some strains of *S. cerevisiae* are opportunistically pathogenic in immunocompromised humans (Pérez-Torrado and Querol 2016) and *S. cerevisiae* is found in a tremendous number of fermentations used for food production.

1.6 SILAGE FERMENTATIONS THAT PRODUCE AND PRESERVE FOOD FOR LIVESTOCK

Ensiling is a solid state anaerobic microbial fermentation process which preserves forage plant materials that are intended as food for ruminant livestock. Ensiling generally uses the entire above ground part of the crop plants and that material typically is chopped and compacted before the fermentation begins. Silage fermentation of plant material typically is a batch process undertaken by microorganisms which naturally are present in the material being ensiled. In order to better anticipate the outcome of plant material ensiling processes it is important to know the chemical composition of the crops that will be ensiled, particularly their water soluble carbohydrate profiles and buffering capacity (Lin et al. 1992). Animal health concerns require avoiding conditions that would allow aerobic growth of spoilage microorganisms during the creation and storage of ensiled materials. Artturi Ilmari Virtanen very importantly won the 1945 Nobel prize in chemistry in part for his studies on ensilage (Virtanen 1945).

Ensiling begins aerobically followed by a rapid and progressive loss of oxygen. The silage also very quickly becomes acidic as different bacterial groups variously convert sugars into acids that include acetic acid, lactic acid, and propionic acid (Ávila and Carvalho 2020). Epiphytic lactic acid bacteria are particularly important for the first day of the ensiling process.

Quickly lowering the pH of silage materials is critical to safely reduce the growth of toxin producing members of the clostridia and suppress the formation of butyric acid (Driehuis et al. 2018). Bacterial production of butyric acid during silage fermentation is quite common when the pH is not low enough to ensure a more exclusive activity of lactic acid producing bacteria. Butyric acid makes the silage unpalatable.

Some members of the clostridia that have been found in silage generally are considered safe because they are non-toxigenic, among which are *Clostridium beijerinckii* found in corn silage, *Clostridium sporogenes* found in alfalfa silage, and *C. tyrobutyricum* found in both corn silage and also oat silage. Both *C. butyricum*, which can be toxigenic, and *C. tyrobutyricum* unfortunately contribute to the production of butyric acid. Several others of the *Clostridium* species that can be found in silage are associated with disease and they are mentioned in Section 1.6.2.

The ensiling material becomes warmer during the fermentation process. Progressive changes in the oxygen level and temperature within the ensiling material are accompanied by a succession of microbial dominance. The competitive skills of bacteria which participate in silage fermentation processes include not only their contributions to reducing the material's pH, but also their production of bacteriocins and production of compounds which inhibit the growth of competing microbial groups. An example of chemical competitiveness is that acetic acid and propionic acid which are produced by bacteria during ensiling beneficially can have a synergistically suppressive effect against fungi (Ávila and Carvalho 2020).

It is important to understand that strains of the same bacterial species which have been collected from different plants will vary with respect to such intrinsic characteristics as thermotolerance and ability to acidify the material being ensiled. Sometimes those naturally present microbes are supplemented by the addition of microbial inoculants. Starter cultures for corn silage fermentation can include *Enterococcus faecium*, *L. plantarum*, and *Lentilactobacillus buchneri* which in particular increases the amount of acetic acid and thus decreases the yeast counts.

Fermented materials also are given to animals other than ruminants, with an example being that feeding chickens with products fermented using *Bacillus subtilis* as a supplement beneficially increases weight gain (Chen and Yu 2021). Silage also gets made from some unusual base materials! Yang et al. (2016) considered the possibility of using silage fermentation to convert fruit processing residue into a potential food resource for livestock. Levin and Witkowski (1991) examined the microbiology of silage created from fish.

1.6.1 The Effect of Nitrate in Fertilizers

Nitrate fertilizers that are applied during the cultivation of crop plants contain ammonium, calcium, magnesium, potassium, and sodium salts of nitrate (NO_3). Nitrates that subsequently are present in the crop plants become reduced during the ensiling process by bacteria such as *Lacticaseibacillus rhamnosus* and *Lactobacillus acidophilus*. The result is generation of nitric oxide (NO) gas in fermenting silage. Nitric oxide combines with molecular oxygen in the air to form nitrogen dioxide (NO_2) and nitrogen dioxide exists in equilibrium with dinitrogen tetroxide (N_2O_4). All three of these nitrogen oxide gases are toxic to animals and humans. Carbon dioxide also is produced during the ensiling process. These three nitrogen oxide gasses and carbon dioxide collectively are termed silo gas. When ensiling is performed in a silo, during the initial few weeks of the fermentation silo gas accumulates inside the silo above the ensiling material. The concentrations of these gases reach levels that are hazardous and can be fatal for workers who enter the silo without a self-contained breathing apparatus (Lallemand Animal Nutrition 2021).

1.6.2 The Microbial Safety of Fermented Fodder

Unfortunately, as with concerns about the microbial safety of food, there are concerns about the microbial safety of silage that is fed to livestock animals.

Using slurry, which is a mixture of cattle manure and water, for fertilization of forage crops results in fecal contamination of those crops. The result is that silage made from those crops can be a vehicle for foodborne transmission of infections to ruminant livestock. Inadequate fermentation and poor handling of the materials can make that bad situation worse by allowing some of those pathogens to proliferate in the silage (Queiroz et al. 2018).

The health concerns associated with bacteria in silage include *B. cereus* which produces gastrointestinal toxins. *B. cereus* is found in silages of corn kernel, corn, grass, and *Sorghum*. *Clostridium botulinum* and *Clostridium perfringens* are found in silage and hazardous because they also will produce toxins (Goldsztejn et al. 2020). *C. botulinum*, which is found in grass silage, produces the neurotoxin botulinum and causes the disease botulism. *C. perfringens* is found in corn silage and produces a multitude of toxins. *C. perfringens* also causes food poisoning as well as intestinal infections that can be lethal. *Clostridium baratii* and *C. butyricum* additionally can produce botulinum and both of those species are found in corn silage. *Pantoea agglomerans* which causes systemic infections is found in silage fermentations of wheat.

The health concerns associated with fungi in silage include *G. candidum* which can cause invasive cutaneous infections and is found in silages of corn and grass. Some of the *Aspergillus* species which can grow in ensiled plant materials represent a danger of infections and toxicosis (Hurst 2019). In this regard, *Aspergillus fumigatus* can represent a significant hazard. *A. fumigatus*, which is found in silages of barley, corn, grass, and hay, causes mycosis. *Aspergillus* also produce aflatoxins and ochratoxins that may be present in those ensiled forages. Additional mycotoxins that could be present in silages include fumonisins and zearalenone which are produced by *Fusarium*. *Fusarium* are found in silages of corn, grass, hay, *Sorghum* kernel, and wheat straw. Some of the *Penicillium* species which can grow in ensiled plant materials also represent a danger of toxicosis (Hurst 2019). *Penicillium cyclopium* which has been found in ensiled hay produces the fungal neurotoxin cyclopiazonic acid. *P. expansum*, which can produce the toxin patulin but more commonly causes spoilage, is present in silage fermentations of both corn and hay. *Penicillium* additionally produce the toxins mycophenolic acid and roquefortine C. *Trichoderma* are found in corn silage and *Trichoderma* produce the trichothecene toxins.

Considering microbial antagonisms as an applied science has the potential to improve the safety of silage (Wambacq et al. 2018).

1.7 TOBACCO FERMENTATIONS

Fermentation of tobacco leaves uses a spontaneous anaerobic solid state fermentation process during which there is a rapidly progressive loss of oxygen. Tobacco fermentation is done to change the aroma and flavor of cigars which will be produced from those leaves (Di Giacomo et al. 2007; Johnson 1934). Tobacco leaves that will be used in manufacturing cigarettes also are fermented (Li et al. 2020).

1.8 COMPOST FERMENTATIONS AND THE ANAEROBIC DIGESTION OF FOOD WASTES FOR GENERATING BIOGAS

Composting is done by solid state fermentations which can be aerobic, microaerobic, or anaerobic. The loss of oxygen in these fermentations may be progressive and, importantly, aerobic composting proceeds more rapidly than does anaerobic composting. Bulking materials such as wood chips often are added to give physical structure to the fermenting mass. Those bulking materials can be separated at the end of the fermentation process and reused, which represents a type of back slopping. The fermentations tend to be self-heating and may be classified as either mesophilic or thermophilic depending upon the temperature reached.

The types of materials that are processed by composting include municipal solid waste to which wastewater solids may be added (Galitskaya et al. 2017), sewage sludge mixed with corn straw (Niu and li 2022) for which I will presume the corn straw was stover, and mixed agricultural byproducts to which calcium oxide may be added as a means for eliminating pH variation (Chandna et al. 2013).

Studies regarding the microbiology of composting material have included those by Dehghani et al. (2012) who considered the fungi present in composting municipal solid waste and Palaniveloo et al. (2020) who reviewed the bacteria and fungi in composting food waste.

One variation of composting is called vermicomposting, which means that worms have been added and their presence changes the manner by which the organic materials are processed. Achieving success with vermicomposting requires meeting biological needs of the worms. The process of vermicomposting is mesophilic and it must be aerobic. There also is a necessity to control temperature, pH, and provide adequate but not excessive moisture. It is important that the worm species chosen prefers residence in the compost rather than residence in the soil beneath the compost. Anastasi et al. (2005) studied the fungi in composting plant debris to which the humus earthworm *Lumbricus rubellus* had been added. A positive feedback relationship exists between the presence of *L. rubellus*, actinomycetes, and other bacterial groups in compost, and the eventual success of using that compost for growing barley. Pathma and Sakthivel (2012) reviewed the microbiology of vermicomposting.

I have chosen to include in this chapter section the anaerobic digestion of food wastes for generating biogas. Fermentative bacteria perform hydrolysis of the complex organic compounds and also accomplish acetogenesis. The products resulting from this microbial acetogenesis include acetic acid, butyric acid, and propionic acid, and alcohols. Microbial methanogenesis processes also occur, which serve as the basis of biogas by converting the microbial acetogenesis compounds to methane and carbon dioxide (Náthia-Neves et al. 2018).

1.9 FERMENTATIONS THAT PRODUCE CHEMICALS, ENZYMES, AND PHARMACEUTICALS

The types of microbial fermentations that I have presented earlier in Sections 1.5 through 1.8 and Tables 1.1 through 1.8, of this chapter generate products that either are immediately ready for final use, or at least nearly ready for use, and close to being marketable. When considering the application of fermentation for producing food, there are some final technical procedures such as sedimentations and distillations of beverages that may be required to generate a usable product. Compost may need to be processed by screening to remove bulking materials and worms before the fermented product can be marketed. And yet, processes such as sedimentation, screening, and even distillation represent fairly easy procedures.

All of that relatively simple understanding about generating and using microbial fermentation products changes at this point in the chapter. There is far greater technical complexity associated with gaining a marketable product from the microbial fermentations that are described in this section of the chapter and mentioned in Tables 1.9 through 1.11. Most of the fermentation processes mentioned in this section of the chapter generate raw materials that must be collected and chemically purified before their marketing, and many of the materials must be processed by use of complex chemical engineering and biological engineering technologies.

Many of the fermentation processes named in this section of the chapter are not performed by metabolic fermentation. They are instead performed by either aerobic or anaerobic respiration, although by tradition these processes still are termed fermentations. One example is the cultivation of *Fusarium venenatum* to produce mycoprotein from glucose. The mycoprotein then is processed for sale as a meat substitute. That production is described as being a fermentation even though it is a submerged aerobic respiration.

Some of these fermentation techniques are used for processing sewage and their products will be disposed into the environment without marketing. Sewage treatment fermentation processes are intended to reduce the environmental impact of disposed materials.

1.9.1 Basic Reactor Designs

Fermentations which are done to produce chemicals, enzymes, and pharmaceuticals variously are done as either submerged fermentations or solid state fermentations. Similar design technology is used for some wastewater treatment processes.

Submerged Fermentations These submerged fermentations use microbes to process materials that are in a continually moving aqueous fluid. That fluid supplies fresh nutrients and either dilutes or washes away spent nutrients. Flow of the liquid may be designed either as laminar or turbulent. Some of these fermentations are designed to have the microbes growing individually in free suspension and the movement of fluid may be done by continuous stirring. If the fermentative microbes are filamentous fungi then continuously stirred fermentations, unfortunately, can get clogged by the microbial mass. Optionally, instead of stirring, many types of submerged fermentations are shaken which reduces the problem of clogging.

There exist alternative designs for submerged fermentations that depend upon the microbes growing on and possibly within solid substrates which serve as carrier particles instead of having the microbes growing freely in suspension. The carrier particles may either be kept in motion by a gradual upward flow of the fermenting fluid and that is termed a fluidized bed reactor, or the particles may be kept stationary in what is called a fixed bed reactor. Carrier particles that are used for fluidized bed reactors and fixed bed reactors may be either permeable or impermeable.

Fluidized bed reactors traditionally depend upon the carrier particles being kept in suspension by a balance of lifting motion, which is provided by upward flow of the fermentation liquid, and gravity which pulls the particles back down. Examples of the materials used as carrier particles for fluidized bed reactors are alginates inside which the microbes can be encapsulated (Pörtner and Faschian 2019). There also exists a thus-far small-scale experimental concept which uses a magnetically assisted fluidized bed bioreactor for which the carrier particles contain magnetite (Velichkova et al. 2017).

A different version of fluidized bed technology is represented by Upflow Anaerobic Sludge Blanket (UASB) reactors which are used for processing waste effluents and also for sewage treatment. There are no solid carrier particles in a UASB reactor. Instead, the microbes exist as self-aggregating clumps, termed granules. Fermentation and methanogenesis can occur simultaneously in a UASB reactor depending upon the feed material.

Fixed bed reactors are designed to direct the flow of fermenting liquid either axially or radially through a bed of stationary carrier particles. The types of carrier particles used in fixed bed reactors include beechwood shavings, bone char, ceramic, clay brick, coke, glass beads, glass wool, polypropylene composite chips, polypropylene tow line (rope), polyurethane foam (Velichkova et al. 2017), slag, sugarcane bagasse, and vermiculite.

Rotary disc bioreactors represent a different option which instead relies upon microbes that are growing as a biofilm on submerged discs. The discs are fastened along a turning horizontal shaft that rotates the discs at a right angle relative to the flowpath by which liquid passes through an otherwise environmentally sealed reactor. There is space for the liquid to be in motion between the discs and rotation stirs the fluid. A similar design concept is used for rotating biological contactors that serve to process wastewater effluents, however, the traditional design for wastewater treatment has the discs only partially submerged and the top of the wastewater contactor is open to the atmosphere.

Solid State Fermentations Solid state fermentations are done either in the absence or near absence of free liquid. Growth of the involved microbes can occur on and within solid substrate particles. Microbial growth also can occur interstitially, within the spaces between the substrate particles. It is important to maintain both the moisture content and gas balance during solid state fermentations (Manan and Webb 2020).

A rotary drum fermenter contains unconsolidated matrix materials within a cylinder which rests horizontally on its long axis. Slow rotation of the cylinder along its longitudinal axis continuously mixes those solid materials. These fermenters can be as small as a 250 ml glass roller bottle. There is not a free flow of liquid in rotary drum fermenters. The fermenter design can either have a headspace above the fermenting material or diffuse pressurized air upward through the fermenting material. Gas exchange thus occurs between the fermenter's internal atmosphere and its matrix of materials. The fermenter drum may be sealed which enables operational control over both the humidity and net flow of gases through the fermenter.

A tray fermenter contains unconsolidated matrix materials in trays that can be vertically stacked. Each tray has a mesh bottom upon which the fermenting material rests, and the trays have open tops. When the trays are stacked, their top and bottom rims vertically align with the trays above and below them to form a vertical design for which the tray edges seal against leakage of gas. The bottom of the lowest tray aligns with the upper rim of a sealed lower section that has a gas inlet. The top of the uppermost tray aligns with the bottom rim of a sealed top section that has a gas outlet. This design allows for a forced vertical flow of gases upward through the fermenting materials and also allows for control of the humidity. The matrix materials are stationary during a tray fermentation process.

1.9.2 The Types of Microbes That Are Used for These Fermentations

Production of chemicals, enzymes, and pharmaceuticals by fermentation processes is done using microorganisms that have been genetically selected for their ability to create the desired organic molecules. Often, those abilities are naturally occurring, and some require cooperation by microbial consortia. The most broadly capable bacteria for use in these fermentations seem to be *B. subtilis* and *E. coli*. The most broadly capable fungi for use in these fermentations seem to be *A. niger* and *Kluyveromyces lactis*. These microbes are chosen because they easily adapt to fermentation technology and easily can be genetically engineered. Genetic engineering provides the means by which a microbe such as *E. coli* can produce ethanol.

1.9.3 The Types of Products and Their Applications

Many of the chemicals, enzymes, and pharmaceuticals upon which we have come to depend are created by these designed fermentation processes. Some of the created materials, such as ethanol, acetic acid, lactic acid, and propionic acid, are secreted by the microbes as metabolic waste products and that secretion can simplify the product collection process. Challengingly, numerous of the other desired fermentation products described in this section are retained within the living cells which generated those products, and obtaining those products often requires collecting the fermenting microbial cells after which a cellular lysate must be prepared before any further purification processing can begin. Nevertheless, the final products from these fermentation processes are of sufficient value to make their purification worthwhile. Table 1.14 lists some of the ways in which purified chemicals, enzymes, and pharmaceutical products from microbial fermentations are utilized.

Ethanol and hydrogen are examples of using microbial fermentations to produce chemicals that serve as biofuels. Sugar beet molasses is a manufacturing waste material produced during the extraction of sugar from sugar beets. Both Vučurović et al. (2018) and Detman et al. (2021) have studied the usage of sugar beet molasses as a feedstock for microbial fermentation processes that produce ethanol (Vučurović et al. 2018) which is termed bioethanol, and hydrogen (Detman et al. 2021) which is termed biohydrogen. Detman et al. (2021) found that correctly operating the fermentation in hydrogen production systems requires the contained microbial communities to have a stable balance of groups that produce hydrogen versus lactic acid.

A number of photosynthetic microbes have been studied for their ability to accumulate high concentrations of carbohydrates during their growth and potentially these microbes can be used as feedstock for fermentations that will produce biofuels (Hossain et al. 2019; Wang and Yin 2018). The cyanobacteria considered for this purpose have including *Anabaena cylindrica* and *Limnospira maxima*. The chlorophyte algae considered for this purpose have included *Chlamydomonas reinhardtii*, *Chlorococcum humicola*, *Chlorella sorokiniana*, *Chlorella vulgaris*, *Dunaliella salina*, and *Tetradesmus obliquus*. The rhodophyte algae considered for this purpose include *Porphyridium cruentum*. The highest carbohydrate production levels seem to be achieved by *C. vulgaris* (Hossain et al. 2019) and *P. cruentum* (Wang and Yin 2018). *C. vulgaris* can be grown in open ponds using a continuous harvesting process and *C. vulgaris* can be cultivated in the wastewater that results from raising swine, which is called piggery waste. Usage of piggery waste as a cultivation medium is a way of valorizing that wastewater which otherwise has little market value. The carbohydrates produced by *C. vulgaris* are complex and thus not immediately fermentable. However, after the *C. vulgaris* cells are collected they can be hydrolyzed to generate fermentable sugars. Their hydrolysis can be done either acidically or enzymatically, with an example being enzymatic hydrolysis that uses amyloglucosidase produced by *A. niger* (Phwan et al. 2019). Hydrolytically treated *C. vulgaris* can be fermented to produce either ethanol (Hossain et al. 2019; Phwan et al. 2019) or hydrogen (Wang and Yin 2018).

Microbial fermentations additionally are used to produce food supplements (Wang et al. 2021b) and feed supplements (Petri et al. 2019). Glucuronic acid is an example of using designed fermentations to generate industrial chemicals (Gomes et al. 2018) and it is sold as a food supplement although the benefits of ingesting glucuronic acid seem doubtful. My mother's personal least favorite food supplement is called fermented cod liver oil and despite the name of that material its production techniques generally do not involve microbial fermentation.

Liu and Kokare (2017) have published a good review of using microbial fermentation to produce commercially used enzymes among which are amyloglucosidase produced by *A. niger*. And, among the pharmaceuticals produced by designed fermentations is a fibrinolytic enzyme that is a promising candidate for thrombolytic therapy (Pan et al. 2019).

TABLE 1.14 Examples of how purified products from microbial fermentations are utilized.

Industry	Application	Microbial Product Used
Consumer chemicals and enzymes		
	Biofuels	Butanol, Ethanol, Hydrogen
	Detergent ingredients	Amylase, Cellulase, Lipase, Pectinase, Protease, Subtilisin, Xylanase
Dietary supplements		
	Dietary supplements for animals	Ascorbic acid, Cellulase, Citric acid, Cobalamin, Lysine, Pantothenic acid, Phytase, Retinol, Riboflavin, Thiamine
	Dietary supplements for humans	Ascorbic acid, Butyrate, Butyric acid, β-Carotene, Citric acid, Cobalamin, Lactase, Lysine, Pantothenic acid, Retinol, Riboflavin, Thiamine
Food industry		
	Food processing	Amylase, Amyloglucosidase, Catalase, Cellulase, Chymosin, Invertase, Laccase, Pectinase, Pullulanase, Xylose isomerase
	Food additives	Acetic acid, Butyric acid, Cellulose (bacterial cellulose), Citric acid, Ethanol, Fusel alcohols, Gluconic acid, Glutamic acid (as monosodium glutamate), Glycerol, Lactic acid, Malic acid, Succinate
	Food preservatives	Glucose oxidase, Nisin
	Food packaging	Cellulose (bacterial cellulose)
Industrial biochemical and chemical (non-food)		
	Biochemical manufacturing	Amidase, α-Glucan phosphorylase, Laccase, Peroxidase, Phosphorylase (example is sucrose phosphorylase), Tyrosine phenol-lyase
	Chemical manufacturing	Acetone, 2,3-Butanediol, Glucuronic acid, Glycerol, Laccase, Lactic acid, Nitrile hydratase, Succinic acid
	Leather industry	Gluconic acid
	Metal industry	Gluconic acid
	Paint industry	Acetic acid (as acetate)
	Perfume industry	Succinic acid
	Plastic industry	Butyric acid
	Polymer production	Acetic acid (as acetate)
	Textile industry	Butyric acid, Cellulase, Laccase
	Woodworking industry	Laccase, Oxalic acid, Succinic acid
Pesticide		
	Insecticide	Insecticidal endotoxin
Pharmaceuticals		
	Angina	Succinic acid (as succinate)
	Anticancer	Aclarubicin, Actinomycin D, Bleomycin, Calicheamicin, Chromomycin A_2, Chromomycin A_3, Daunorubicin, Doxorubicin, Epothilone, Mithramycin, Mitomycin C, Neocarzinostatin, Paclitaxel, Pentostatin, Plicamycin, Streptozotocin, Thiocoraline

(Continued)

TABLE 1.14 (Continued)

Industry	Application	Microbial Product Used
	Antimicrobial	Apramycin, Cephalosporin, Clavulanic acid, Erythromycin, Natamycin, Penicillin, Sirolimus, Tetracycline, Tobramycin
	Biomedical polymers	Cellulose (bacterial cellulose)
	Cataract prevention	Pyruvic acid
	Diabetes	α-Amylase inhibitors
	Hangover	Succinic acid (as succinate)
	Hypercholesterolemia	Laccase
	Hypertension	Succinic acid (as succinate)
	Immunosuppressant	Cyclosporine A, Mycophenolic acid, Sirolimus, Tacrolimus
	Menopause	Succinic acid (as succinate)
	Obesity	α-Amylase inhibitors, Lipstatin
	Product stabilizer	Succinic acid (as succinate)
	Skin care	Glycerol, Pyruvic acid, Retinol
	Stool softener	Glycerol

There are many fermentation-derived enzymes which are not identified with great specificity.

1.10 SUMMARY

Understanding and learning to use microbial fermentation does figuratively give us a handle for one of nature's most valuable tools. As a natural process, microbial fermentation helps us to remain healthy in ways that include assisting with our battle against those microbes that would do us harm. Designed fermentation has helped to keep our civilizations and our livestock fed. During my study efforts to understand the microbiology of food fermentations, I have asked myself several times if using microbial fermentation to produce and preserve food items may have resulted in a cooperative genetic selection process. The microbes which are prevalent in spontaneous fermentations of food will have selected for a human population which could survive ingesting those microbes. Similarly, those microbes which are spontaneously present in the silage materials that we produce as fodder for our ruminant livestock would have exerted a selection process on our livestock. In cooperative response, we have learned to design food and silage fermentation processes in ways that would favor growth of the most easily survivable microbial populations. Oftentimes, we have gotten drunk on our success.

REFERENCES

Adebo, O.A. (2020). African sorghum-based fermented foods: past, current and future prospects. *Nutrients* 12: 1111. https://doi.org/10.3390/nu12041111.

Akhavan, B.J., Ostrosky-Zeichner, L., and Thomas, E.J. (2019). Drunk without drinking: a case of auto-brewery syndrome. *ACG Case Reports Journal* 6 (9): e00208. https://doi.org/10.14309/crj.0000000000000208.

Anastasi, A., Varese, G.C., and Marchisio, V.F. (2005). Isolation and identification of fungal communities in compost and vermicompost. *Mycologia* 97 (1): 33–44. https://doi.org/10.3852/mycologia.97.1.33.

Aouizerat, T., Gutman, I., Paz, Y. et al. (2019). Isolation and characterization of live yeast cells from ancient vessels as a tool in bio-archaeology. *MBio* 10 (2): e00388-19. https://doi.org/10.1128/mBio.00388-19.

Arranz-Otaegui, A., Gonzalez Carretero, L., Ramsey, M.N. et al. (2018). Archaeobotanical evidence reveals the origins of bread 14,400 years ago in northeastern Jordan. *Proceedings of the National Academy of Sciences of the United States of America* 115 (31): 7925–7930. https://doi.org/10.1073/pnas.1801071115.

Ávila, C.L.S. and Carvalho, B.F. (2020). Silage fermentation-updates focusing on the performance of microorganisms. *Journal of Applied Microbiology* 128: 966–984. https://doi.org/10.1111/jam.14450.

Bah, A., Ferjani, R., Fhoula, I. et al. (2019). Microbial community dynamic in tomato fruit during spontaneous fermentation and biotechnological characterization of indigenous lactic acid bacteria. *Annales de Microbiologie* 69: 41–49. https://doi.org/10.1007/s13213-018-1385-4.

Behera, S.S., El Sheikha, A.F., Hammami, R. et al. (2020). Traditionally fermented pickles: how the microbial diversity associated with their nutritional and health benefits? *Journal of Functional Foods* 70: 103971. https://doi.org/10.1016/j.jff.2020.103971.

Bhardwaj, A., Malik, R.K., and Chauhan, P. (2008). Functional and safety aspects of enterococci in dairy foods. *Indian Journal of Microbiology* 48 (3): 317–325. https://doi.org/10.1007/s12088-008-0041-2.

Boamfa, E.I., Ram, P.C., Jackson, M.B. et al. (2003). Dynamic aspects of alcoholic fermentation of rice seedlings in response to anaerobiosis and to complete submergence: relationship to submergence tolerance. *Annals of Botany* 91: 279–290. https://doi.org/10.1093/aob/mcf205.

Bononi, L., Chiaramonte, J.B., Pansa, C.C. et al. (2020). Phosphorus-solubilizing *Trichoderma* spp. from Amazon soils improve soybean plant growth. *Scientific Reports* 10 (1): 2858. https://doi.org/10.1038/s41598-020-59793-8.

Bourrie, B.C., Willing, B.P., and Cotter, P.D. (2016). The microbiota and health promoting characteristics of the fermented beverage kefir. *Frontiers in Microbiology* 7: 647. https://doi.org/10.3389/fmicb.2016.00647.

Bryant, H.W. and Hammer, B.W. (1940). Bacteriology of cheese V. Defects of blue (Roquefort-Type) cheese. *Research Bulletin 283*. Agricultural Experiment Station, Iowa State College of Agriculture and Mechanic Arts, Ames, Iowa.

Bücher, C., Burtscher, J., and Domig, K.J. (2021). Propionic acid bacteria in the food industry: an update on essential traits and detection methods. *Comprehensive Reviews in Food Science and Food Safety* 20: 4299–4323. https://doi.org/10.1111/1541-4337.12804.

Cabugao, K.G., Yaffar, D., Stenson, N. et al. (2021). Bringing function to structure: Root-soil interactions shaping phosphatase activity throughout a soil profile in Puerto Rico. *Ecology and Evolution* 11 (3): 1150–1164. https://doi.org/10.1002/ece3.7036; Erratum in: *Ecology and Evolution* (2021). 11 (15): 10731.

Carlsen, H.N., Degn, H., and Lloyd, D. (1991). Effects of alcohols on the respiration and fermentation of aerated suspensions of baker's yeast. *Journal of General Microbiology* 137 (12): 2879–2883. https://doi.org/10.1099/00221287-137-12-2879.

Chacón Mayorga, G.A., Arias Palma, G.B., Sandoval-Cañas, G.J. et al. (2021). Ancestral fermented indigenous beverages from South America made from cassava (*Manihot esculenta*). *Food Science and Technology* 41 (Suppl 1): 360–367. https://doi.org/10.1590/fst.15220.

Chandna, P., Nain, L., Singh, S. et al. (2013). Assessment of bacterial diversity during composting of agricultural byproducts. *BMC Microbiology* 13: 99. https://doi.org/10.1186/1471-2180-13-99.

Chen, J.-Y. and Yu, Y.-H. (2021). *Bacillus subtilis*–fermented products ameliorate the growth performance and alter cecal microbiota community in broilers under lipopolysaccharide challenge. *Poultry Science* 100 (2): 875–886. https://doi.org/10.1016/j.psj.2020.10.070.

Chen, S., Perez-Samper, G., Herrera-Malaver, B. et al. (2022). Breeding of new *Saccharomyces cerevisiae* hybrids with reduced higher alcohol production for light aroma type *Xiaoqu* baijiu production. *Journal of the American Society of Brewing Chemists* https://doi.org/10.1080/03610470.2022.2033608.

Ciani, M., Comitini, F., Mannazzu, I. et al. (2010). Controlled mixed culture fermentation: a new perspective on the use of non-*Saccharomyces* yeasts in winemaking. *FEMS Yeast Research* 10 (2): 123–133. https://doi.org/10.1111/j.1567-1364.2009.00579.x.

Ciani, M., Comitini, F., and Mannazzu, I. (2013). Fermentation. Reference module in earth systems and environmental sciences. *Encycolopedia of Ecology* 2: 310–321. https://doi.org/10.1016/B978-0-12-409548-9.00693-X.

Ciegler, A. (1969). Tremorgenic toxin from *Penicillium palitans*. *Applied Microbiology* 18: 128–129. https://doi.org/10.1128/am.18.1.128-129.1969.

City of Calgary (2022). *Alcohol and Fermentation Wastewater*. Alberta, Canada: City of Calgary https://www.calgary.ca/content/dam/www/uep/water/documents/water-documents/alcohol-and-fermentation-wastewater.pdf.

Daeschel, M.A., Andersson, R.E., and Fleming, H.P. (1987). Microbial ecology of fermenting plant materials. *FEMS Microbiology Reviews* 46: 357–367.

Dehghani, R., Asadi, M.A., Charkhloo, E. et al. (2012). Identification of fungal communities in producing compost by windrow method. *Journal of Environmental Protection* 3: 61–67. https://doi.org/10.4236/jep.2012.31008.

Detman, A., Laubitz, D., Chojnacka, A. et al. (2021). Dynamics and complexity of dark fermentation microbial communities producing hydrogen from sugar beet molasses in continuously operating packed bed reactors. *Frontiers in Microbiology* 11: 612344. https://doi.org/10.3389/fmicb.2020.612344.

Di Giacomo, M., Paolino, M., Silvestro, D. et al. (2007). Microbial community structure and dynamics of dark fire-cured tobacco fermentation. *Applied and Environmental Microbiology* 73: 825–837. https://doi.org/10.1128/AEM.02378-06.

Diether, N.E. and Willing, B.P. (2019). Microbial fermentation of dietary protein: an important factor in diet–microbe–host interaction. *Microorganisms* 7: 19. https://doi.org/10.3390/microorganisms7010019.

Driehuis, F., Wilkinson, J.M., Jiang, Y. et al. (2018). Silage review: animal and human health risks from silage. *Journal of Dairy Science* 101: 4093–4110. https://doi.org/10.3168/jds.2017-13836.

Eifert, J.A., Boyer, R.R., Williams, R.C. et al. (2014). *Vegetable Fermentation*. Virginia Tech http://hdl.handle.net/10919/51491.

Galitskaya, P., Biktasheva, L., Saveliev, A. et al. (2017). Fungal and bacterial successions in the process of co-composting of organic wastes as revealed by 454 pyrosequencing. *PLoS One* 12 (10): e0186051. https://doi.org/10.1371/journal.pone.0186051.

de Garcia, V., Brizzio, S., and van Broock, M.R. (2012). Yeasts from glacial ice of Patagonian Andes, Argentina. *FEMS Microbiology Ecology* 82 (2): 540–550. https://doi.org/10.1111/j.1574-6941.2012.01470.x.

Gebreeyessus, G.D., Mekonen, A., and Alemayehu, E. (2019). A review on progresses and performances in distillery stillage management. *Journal of Cleaner Production* 232: 295–307. https://doi.org/10.1016/j.jclepro.2019.05.383.

Ghosh, J.S. (2016). Solid state fermentation and food processing: a short review. *Journal of Nutrition & Food Sciences* 6 (1): 1000453. https://doi.org/10.4172/2155-9600.1000453.

Gilbert S.M. (2017). Cheesy science. *ChemMatters* (December 2017/January 2018), p. 7–9.

Goldsztejn, M., Grenda, T., Kozieł, N. et al. (2020). Potential determinants of *Clostridium* spp. Occurrence in Polish Silage. *Journal of Veterinary Research* 64 (4): 549–555. https://doi.org/10.2478/jvetres-2020-0075.

Gomes, R.J., Borges, M.F., Rosa, M.F. et al. (2018). Acetic acid bacteria in the food industry: systematics, characteristics and applications. *Food Technology and Biotechnology* 56 (2): 139–151. https://doi.org/10.17113/ftb.56.02.18.5593.

Hanyabui, E., Apori, S.O., Frimpong, K.A. et al. (2020). Phosphorus sorption in tropical soils. *AIMS Agriculture and Food* 5 (4): 599–616. https://doi.org/10.3934/agrfood.2020.4.599.

Harden, A. (1914). *Monographs in Biochemistry: Alcoholic Fermentation*, 2e. London: Longman, Green and Co.

Hockings, K.J., Bryson-Morrison, N., Carvalho, S. et al. (2015). Tools to tipple: ethanol ingestion by wild chimpanzees using leaf-sponges. *Royal Society Open Science* 2 (6): 150150. https://doi.org/10.1098/rsos.150150.

Hossain, N., Mahlia, T.M.I., Zaini, J. et al. (2019). Techno-economics and sensitivity analysis of microalgae as commercial feedstock for bioethanol production. *Environmental Progress & Sustainable Energy* https://doi.org/10.1002/ep.13157.

Hurst, C.J. (2018). Opportunistic bacteria associated with mammalian livestock disease. In: *The Connections Between Ecology and Infectious Disease. Advances in Environmental Microbiology*, vol. 5 (ed. C.J. Hurst), 185–238. Cham: Springer https://doi.org/10.1007/978-3-319-92373-4_6.

Hurst, C.J. (2019). Dirt and disease: the ecology of soil fungi and plant fungi that are infectious for vertebrates. In: *Understanding Terrestrial Microbial Communities. Advances in Environmental Microbiology*, vol. 6 (ed. C.J. Hurst), 289–405. Cham: Springer https://doi.org/10.1007/978-3-030-10777-2_9.

Hurst, C.J. (ed.) (2021). *Microbial Metabolism of Metals and Metalloids*, Advances in Environmental Microbiology, vol. 10. Cham: Springer https://link.springer.com/book/10.1007/978-3-030-97185-4.

Hurst, C.J., Gerba, C.P., and Cech, I. (1980). Effects of environmental variables and soil characteristics on virus survival in soil. *Applied and Environmental Microbiology* 40: 1067–1079. https://doi.org/10.1128/aem.40.6.1067-1079.1980.

Islam, A., McKee, C., Ghosh, P.K. et al. (2021). Seasonality of date palm sap feeding behavior by bats in Bangladesh. *EcoHealth* 18 (3): 359–371. https://doi.org/10.1007/s10393-021-01561-9.

Jarrett, J.P., Wilson, J.W., Ray, P.P. et al. (2014). The effects of forage particle length and exogenous phytase inclusion on phosphorus digestion and absorption in lactating cows. *Journal of Dairy Science* 97: 411–418. https://doi.org/10.3168/jds.2013-7124.

Jatuwong, K., Suwannarach, N., Kumla, J. et al. (2020). Bioprocess for production, characteristics, and biotechnological applications of fungal phytases. *Frontiers in Microbiology* 11: 188. https://doi.org/10.3389/fmicb.2020.00188.

Johnson, J. (1934). Studies on the fermentation of tobacco. *Journal of Agricultural Research* 49 (2): 137–160. https://handle.nal.usda.gov/10113/IND43968481.

Katz, S.E. (2003). *Wild Fermentation: The Flavor, Nutrition and Craft of Live-Culture Foods*. White River Junction: Chelsea Green https://archive.org/details/WildFermentationSandorEllixKatz.

Kobae, Y. (2019). Dynamic phosphate uptake in arbuscular mycorrhizal roots under field conditions. *Frontiers in Environmental Science* 6: 159. https://doi.org/10.3389/fenvs.2018.00159.

Koo, O.K., Lee, S.J., Chung, K.R. et al. (2016). Korean traditional fermented fish products: jeotgal. *Journal of Ethnic Foods* 3 (2): 107–116. https://doi.org/10.1016/j.jef.2016.06.004.

Kumar, K.N., Devadas, S.M., Murugan, S. et al. (2018). Production and characterization of bacteriocin by lactic acid bacterium- *Pediococcus pentosaceus* NKSM1 isolated from fermented 'appam' batter. *Journal of Pure and Applied Microbiology* 12 (3): 1315–1330. https://doi.org/10.22207/JPAM.12.3.34.

Lallemand Animal Nutrition (2021). *Silage Safety Handbook*. Milwaukee, WI: Lallemand.

Lane, C.B. and Hammer, B.W. (1935a) Bacteriology of cheese I. Effect of pasteurizing the milk on the nitrogenous decomposition in cheddar cheese. *Research Bulletin 183*. Agricultural Experiment Station, Iowa State College of Agriculture and Mechanic Arts, Ames, Iowa.

Lane, C.B. and Hammer, B.W. (1935b) Bacteriology of cheese II. Effect of *Lactobacillus casei* on the nitrogenous decomposition and flavor development in cheddar cheese made from pasteurized milk. *Research Bulletin 190*. Agricultural Experiment Station, Iowa State College of Agriculture and Mechanic Arts, Ames, Iowa.

Lane, C.B. and Hammer, B.W. (1935c) Bacteriology of cheese III. Some factors affecting the ripening of blue (Roquefort type) cheese. *Research Bulletin* 237. Agricultural Experiment Station, Iowa State College of Agriculture and Mechanic Arts, Ames, Iowa.

Lane, C.B. and Hammer, B.W. (1941) Bacteriology of cheese VI. Relationship of fat hydrolysis to the ripening of cheddar cheese. *Research Bulletin 291*. Agricultural Experiment Station, Iowa State College of Agriculture and Mechanic Arts, Ames, Iowa.

Laranjo, M., Potes, M.E., and Elias, M. (2019). Role of starter cultures on the safety of fermented meat products. *Frontiers in Microbiology* 10: 853. https://doi.org/10.3389/fmicb.2019.00853.

Lecomte, S.M., Achouak, W., Abrouk, D. et al. (2018). Diversifying anaerobic respiration strategies to compete in the rhizosphere. *Frontiers in Environmental Science* 6: 139. https://doi.org/10.3389/fenvs.2018.00139.

Levin, R.E. and Witkowski, R. (1991). Characteristics and identity of obligately aerobic spoilage yeasts from fish silage. *The Journal of Applied Bacteriology* 71 (4): 354–359. https://doi.org/10.1111/j.1365-2672.1991.tb03800.x.

Levitt A. (2018). The life-giving properties of one of the world's oldest cheeses in Xinjiang, the living process dairy the way 4,000-year-old mummies did. https://www.atlasobscura.com/articles/cheese-buried-with-mummies (Accessed June 18, 2022).

Li, J., Zhao, Y., Qin, Y. et al. (2020). Influence of microbiota and metabolites on the quality of tobacco during fermentation. *BMC Microbiology* 20: 356. https://doi.org/10.1186/s12866-020-02035-8.

Lin, C., Brent, B.E., Bolsen, K.K. et al. (1992). Epiphytic lactic acid bacteria succession during the pre-ensiling and ensiling periods of alfalfa and corn. *Kansas Agricultural Experiment Station Research Reports* 0 (1): 113–119. https://doi.org/10.4148/2378-5977.2196.

Liu, X. and Kokare, C. (2017). *Biotechnology of Microbial Enzymes: Production, Biocatalysis and Industrial Applications* (ed. G. Brahmachari, A.L. Demain, and J.L. Adrio), 267–298. Amsterdam: Academic Press https://doi.org/10.1016/b978-0-12-803725-6.00011-x.

Lopandic, K., Tiefenbrunner, W., Gangl, H. et al. (2008). Molecular profiling of yeasts isolated during spontaneous fermentations of Austrian wines. *FEMS Yeast Research* 8 (7): 1063–1075. https://doi.org/10.1111/j.1567-1364.2008.00385.x.

Maixner, F., Sarhan, M.S., Huang, K.D. et al. (2021). Hallstatt miners consumed blue cheese and beer during the iron age and retained a non-Westernized gut microbiome until the Baroque period. *Current Biology* 31: 5149–5162. https://doi.org/10.1016/j.cub.2021.09.031.

Manan, M.A. and Webb, C. (2020). Newly designed multi-stacked circular tray solid-state bioreactor: analysis of a distributed parameter gas balance during solid-state fermentation with influence of variable initial moisture content arrangements. *Bioresources and Bioprocessing* 7: 16. https://doi.org/10.1186/s40643-020-00307-9.

Marchisio, V.F., Sulotto, F., Botta, G.C. et al. (1999). Aerobiological analysis in a salami factory: a possible case of extrinsic allergic alveolitis by *Penicillium camembertii*. *Medical Mycology* 37 (4): 285–289.

May, A., Narayanan, S., Alcock, J. et al. (2019). Kombucha: a novel model system for cooperation and conflict in a complex multi-species microbial ecosystem. *PeerJ* 7: e7565. https://doi.org/10.7717/peerj.7565.

McGovern, P.E., Zhang, J., Tang, J. et al. (2004). Fermented beverages of pre- and proto-historic China. *Proceedings of the National Academy of Sciences of the United States of America* 101 (51): 17593–17598. https://doi.org/10.1073/pnas.0407921102.

Mendoza, L.M., Neef, A., Vignolo, G. et al. (2017). Yeast diversity during the fermentation of Andean chicha: a comparison of high-throughput sequencing and culture-dependent approaches. *Food Microbiology* 67: 1–10. https://doi.org/10.1016/j.fm.2017.05.007.

Menezes-Blackburn, D., Paredes, C., Zhang, H. et al. (2016). Organic acids regulation of chemical-microbial phosphorus transformations in soils. *Environmental Science & Technology* 50 (21): 11521–11531. https://doi.org/10.1021/acs.est.6b03017.

Mikucka, W. and Zielińska, M. (2020). Distillery stillage: characteristics, treatment, and valorization. *Applied Biochemistry and Biotechnology* 192 (3): 770–793. https://doi.org/10.1007/s12010-020-03343-5.

Mississippi State University (2022). Understanding the ruminant animal digestive system. Publication Number P2503. http://extension.msstate.edu/publications/publications/understanding-the-ruminant-animal-digestive-system. (Accessed October 1, 2022).

Náthia-Neves, G., Berni, M., Dragone, G. et al. (2018). Anaerobic digestion process: technological aspects and recent developments. *International journal of Environmental Science and Technology* 15: 2033–2046. https://doi.org/10.1007/s13762-018-1682-2.

Niu, J. and Li, X. (2022). Effects of microbial inoculation with different indigenous bacillus species on physicochemical characteristics and bacterial succession during short-term composting. *Fermentation* 8: 152. https://doi.org/10.3390/fermentation8040152.

Oliphant, K. and Allen-Vercoe, E. (2019). Macronutrient metabolism by the human gut microbiome: major fermentation by-products and their impact on host health. *Microbiome* 7: 91. https://doi.org/10.1186/s40168-019-0704-8.

de Oliveira Mendes, G., Moreira de Freitas, A.L., Liparini Pereira, O. et al. (2014). Mechanisms of phosphate solubilization by fungal isolates when exposed to different P sources. *Annales de Microbiologie* 64: 239–249. https://doi.org/10.1007/s13213-013-0656-3.

Oren, A. (1986). The ecology and taxonomy of anaerobic halophilic eubacteria. *FEMS Microbiology Reviews* 2 (1–2): 23–29. https://doi.org/10.1111/j.1574-6968.1986.tb01838.x.

Palaniveloo, K., Amran, M.A., Norhashim, N.A. et al. (2020). Food waste composting and microbial community structure profiling. *Processes* 8: 723. https://doi.org/10.3390/pr8060723.

Pan, S., Chen, G., Wu, R. et al. (2019). Non-sterile submerged fermentation of fibrinolytic enzyme by marine *Bacillus subtilis* harboring antibacterial activity with starvation strategy. *Frontiers in Microbiology* 10: 1025. https://doi.org/10.3389/fmicb.2019.01025.

Patel, R. and Pandya, H.N. (2015). Production of acetic acid from molasses by fermentation process. *International Journal of Advance Research and Innovative Ideas in Education* 1 (2): 58–60.

Pathma, J. and Sakthivel, N. (2012). Microbial diversity of vermicompost bacteria that exhibit useful agricultural traits and waste management potential. *Springerplus* 1: 26. https://doi.org/10.1186/2193-1801-1-26.

Pérez-Torrado, R. and Querol, A. (2016). Opportunistic strains of *Saccharomyces cerevisiae*: a potential risk sold in food products. *Frontiers in Microbiology* 6: 1522. https://doi.org/10.3389/fmicb.2015.01522.

Petri, R.M., Mickdam, E., Klevenhusen, F. et al. (2019). Effects of the supplementation of plant-based formulations on microbial fermentation and predicted metabolic function in vitro. *Anaerobe* 57: 19–27. https://doi.org/10.1016/j.anaerobe.2019.03.001.

Phwan, C.K., Chew, K.W., Sebayang, A.H. et al. (2019). Effects of acids pre-treatment on the microbial fermentation process for bioethanol production from microalgae. *Biotechnology for Biofuels* 12: 191. https://doi.org/10.1186/s13068-019-1533-5.

Pisacane, V., Callegari, M.L., Puglisi, E. et al. (2015). Microbial analyses of traditional Italian salami reveal microorganisms transfer from the natural casing to the meat matrix. *International Journal of Food Microbiology* 207: 57–65. https://doi.org/10.1016/j.ijfoodmicro.2015.04.029.

Pörtner, R. and Faschian, R. (2019). Design and operation of fixed-bed bioreactors for immobilized bacterial culture. In: *Growing and Handling of Bacterial Cultures* (ed. M. Mishra), 81–92. London: Intechopen. https://www.intechopen.com/chapters/67945.

Queiroz, O.C.M., Ogunade, I.M., Weinberg, Z. et al. (2018). Silage review: foodborne pathogens in silage and their mitigation by silage additives. *Journal of Dairy Science* 101: 4132–4142. https://doi.org/10.3168/jds.2017-13901.

Rabie, C.J., Lübben, A., Marais, G.J. et al. (1997). Enumeration of fungi in barley. *International Journal of Food Microbiology* 35 (2): 117–127. https://doi.org/10.1016/s0168-1605(96)01210-x.

Rachwał, K., Waśko, A., Gustaw, K. et al. (2020). Utilization of brewery wastes in food industry. *PeerJ* 8: e9427. https://doi.org/10.7717/peerj.9427.

Rosenberg, D., Liu, L., Levin, M.J. et al. (2021). From hangovers to hierarchies: beer production and use during the Chalcolithic period of the southern Levant – new evidence from Tel Tsaf and Peqi'in cave. *Journal of Anthropological Archaeology* 64: 101361.

Saverimuttu, J., Malik, F., Arulthasan, M. et al. (2019). A case of auto-brewery syndrome treated with Micafungin. *Cureus* 11 (10): e5904. https://doi.org/10.7759/cureus.5904.

Shi, C. and Maktabdar, M. (2022). Lactic acid bacteria as biopreservation against spoilage molds in dairy products – a review. *Frontiers in Microbiology* 12: 819684. https://doi.org/10.3389/fmicb.2021.819684.

Skowron, K., Budzyńska, A., Grudlewska-Buda, K. et al. (2022). Two faces of fermented foods – the benefits and threats of its consumption. *Frontiers in Microbiology* 13: 845166. https://doi.org/10.3389/fmicb.2022.845166.

Srain, B.M., Sobarzo, M., Daneri, G. et al. (2020). Fermentation and anaerobic oxidation of organic carbon in the oxygen minimum zone of the upwelling ecosystem off Concepción, in Central Chile. *Frontiers in Marine Science* 7: 533. https://doi.org/10.3389/fmars.2020.00533.

Tadano, T. and Sakai, H. (1991). Secretion of acid phosphatase by the roots of several crop species under phosphorus-deficient conditions. *Soil Science & Plant Nutrition* 37 (1): 129–140. https://doi.org/10.1080/00380768.1991.10415018.

Tamang, J.P. (2015). Naturally fermented ethnic soybean foods of India. *Journal of Ethnic Foods* 2: 8–17. https://doi.org/10.1016/j.jef.2015.02.003.

Thomas, K., Ironside, K., Clark, L. et al. (2021). Preliminary microbiological and chemical analysis of two historical stock ales from Victorian and Edwardian brewing. *Journal of the Institute of Brewing* 127 (2): 167–175.

Touchette, D., Altshuler, I., Gostinčar, C. et al. (2022). Novel Antarctic yeast adapts to cold by switching energy metabolism and increasing small RNA synthesis. *The ISME Journal* 16: 221–232. https://doi.org/10.1038/s41396-021-01030-9.

UpToDate (2022). Hypersensitivity pneumonitis associated with farming and agriculture. https://www.medilib.ir/uptodate/show/69245 (Accessed June 30, 2022).

Varela, C., Sundstrom, J., Cuijvers, K. et al. (2020). Discovering the indigenous microbial communities associated with the natural fermentation of sap from the cider gum *Eucalyptus gunnii*. *Scientific Reports* 10: 14716. https://doi.org/10.1038/s41598-020-71663-x.

Velichkova, P.G., Ivanov, T.V., and Lalov, I.G. (2017). Magnetically assisted fluidized bed bioreactor for bioethanol production. *Bulgarian Chemical Communications* 49 (Special Issue L): 105–109.

Virtanen, A. I. (1945). The biological fixation of nitrogen and the preservation of fodder in agriculture, and their importance to human nutrition. Nobel lecture (12 December 1945).

Volova, T.G., Prudnikova, S.V., Kiselev, E.G. et al. (2022). Bacterial cellulose (BC) and BC composites: production and properties. *Nanomaterials (Basel)* 12 (2): 192. https://doi.org/10.3390/nano12020192.

Vučurović, V.M., Puškaš, V.S., and Miljić, U.D. (2018). Bioethanol production from sugar beet molasses and thick juice by free and immobilised *Saccharomyces cerevisiae*. *Journal of the Institute of Brewing* 125: 134–142.

Waché, Y., Do, T.L., Do, T.B. et al. (2018). Prospects for food fermentation in South-East Asia, topics from the tropical fermentation and biotechnology network at the end of the AsiFood Erasmus+Project. *Frontiers in Microbiology* 9: 2278. https://doi.org/10.3389/fmicb.2018.02278.

Wambacq, E., Audenaert, K., Höfte, M. et al. (2018). *Bacillus velezensis* as antagonist towards *Penicillium roqueforti s.l.* in silage: in vitro and in vivo evaluation. *Journal of Applied Microbiology* 125 (4): 986–996. https://doi.org/10.1111/jam.13944.

Wang, J. and Yin, Y. (2018). Fermentative hydrogen production using pretreated microalgal biomass as feedstock. *Microbial Cell Factories* 17: 22. https://doi.org/10.1186/s12934-018-0871-5.

Wang, J., Jiang, L., and Sun, H. (2021a). Early evidence for beer drinking in a 9000-year-old platform mound in southern China. *PLoS One* 16 (8): e0255833. https://doi.org/10.1371/journal.pone.0255833.

Wang, Y., Liu, L., Jin, Z. et al. (2021b). Microbial cell factories for green production of vitamins. *Frontiers in Bioengineering and Biotechnology* 9: 661562. https://doi.org/10.3389/fbioe.2021.661562.

Welling, B. (1974). Aspergillus – and Penicillium species on stored Danish Barley grain identification of species of fungi isolated from 40 barley samples in the harvest year 1970. *Tidsskrift for planteavl* 78: 1–8.

Wiens, F., Zitzmann, A., Lachance, M.A. et al. (2008). Chronic intake of fermented floral nectar by wild treeshrews. *Proceedings of the National Academy of Sciences of the United States of America* 105 (30): 10426–10431. https://doi.org/10.1073/pnas.0801628105.

Wu, L., Kobayashi, Y., Wasaki, J. et al. (2018). Organic acid excretion from roots: a plant mechanism for enhancing phosphorus acquisition, enhancing aluminum tolerance, and recruiting beneficial rhizobacteria. *Soil Science & Plant Nutrition* 64 (6): 697–704. https://doi.org/10.1080/00380768.2018.1537093.

Yang, J., Tan, H., and Cai, Y. (2016). Characteristics of lactic acid bacteria isolates and their effect on silage fermentation of fruit residues. *Journal of Dairy Science* 99: 5325–5334. https://doi.org/10.3168/jds.2016-10952.

Yasui, M., Oda, K., Masuo, S. et al. (2020). Invasive growth of *Aspergillus oryzae* in rice koji and increase of nuclear number. *Fungal Biology and Biotechnology* 7: 8. https://doi.org/10.1186/s40694-020-00099-9.

Zheng, X., Shi, X., and Wang, B. (2021). A review on the general cheese processing technology, flavor biochemical pathways and the influence of yeasts in cheese. *Frontiers in Microbiology* 12: 703284. https://doi.org/10.3389/fmicb.2021.703284.

CHAPTER 2

RHODOTORULA TORULOIDES AS A BIOFACTORY OF CAROTENOIDS, LIPIDS, AND ENZYMES

NAYRA OCHOA VIÑALS[1], DANIA ALONSO ESTRADA[1], EVELYN FAIFE PÉREZ[2], LOURDES GEORGINA MICHELENA ÁLVAREZ[2], JOSÉ LUIS MARTÍNEZ-HERNÁNDEZ[1], ROBERTO ARREDONDO VALDÉS[1], ARIANNA NÚÑEZ CARABALLO[3], AND ANNA ILINÁ[1]

[1]Nanobioscience Sciences Research Groups. Postgraduate Program in Food Science and Technology. Faculty of Chemical Sciences of the Autonomous University of Coahuila. Blvd. V. Carranza e Ing. José Cárdenas V., Col. República, Saltillo, CP 25280, Coahuila, México

[2]Instituto Cubano de Investigaciones de los Derivados de la Caña de Azúcar (ICIDCA), Vía Blanca #804 y Carretera Central, Zona postal 10, código 11 000, San Miguel del Padrón, Ciudad de La Habana, Cuba

[3]Faculty of Management Sciences of the Autonomous University of Coahuila, Blvd. Vito Alessio Robles 25015, Asturias, Coahuila, México

CONTENTS

Microbial Fermentations in Nature and as Designed Processes, First Edition. Edited by Christon J. Hurst.

2.1 INTRODUCTION

One of the most promising industrial oily yeasts is *Rhodotorula toruloides* due to its ability to consume multiple carbon and nitrogen sources and produce various chemicals with a broad spectrum of applications (Lopes et al. 2020). The advantage of its use lies in the possibility of propagating it and exerting metabolic routes in substrates that are considered industrial waste, which significantly increases the economic profitability of biotechnological processes.

R. toruloides is a natural producer of lipids (single-cell oil, SCO), high-value compounds such as carotenoids, terpenes, fatty acid esters, fatty alcohols, and medicinal and chemical enzymes, such as phenylalanine ammonia-lyase (PAL) and D-amino acid oxidase (Zhao et al. 2021). SCO is composed of triacylglycerides (TAGs) and could be used to produce oleochemicals such as biodiesel, cosmetics, and coatings, as well as a substitute for vegetable oil in fish feed (Szczepańska et al. 2021). Carotenoids are essential in foods, chemicals, medicines, and cosmetics. They can be converted to vitamin A and have antioxidant properties that have been proven to help prevent cancer and immune diseases. They also can help protect the skin against radiation, among other beneficial health effects. The global market for carotenoids is projected to reach US$2 billion by 2026 (Kot et al. 2019).

The microbial lipid and carotenoids production is influenced by culture parameters such as inoculum size, carbon and nitrogen sources in the growth medium, C/N concentration ratio, pH, temperature, culture period, agitation, and aeration rate. Oily yeast produces lipids with a carbon/nitrogen (C/N) ratio greater than 20. Batch, fed-batch, continuous, and repeated-batch culture techniques are developed for the propagation of oily yeasts (Jiru et al. 2017).

This review aims to describe the *R. toruloides* properties as a source of carotenoids, lipids, and enzymes, provides a better understanding of the methods applied to obtain these metabolites, and explains the important mechanisms for their biotechnological and industrial applications.

2.2 *RHODOTORULA TORULOIDES*

R. toruloides also has been known as *Rhodosporidium toruloides*, *Rhodotorula rubescens*, or *Rhodotorula gracilis*. It is a unicellular pigmented yeast of the *Phylum Basidiomycota*, *Sporidiobolaceae* family, order *Sporidiobolales*, class *Microbotryomycetes*, and subfamily *Rhodotoruloda*.

Due to the polyphyletic nature of the yeast species of *Pucciniomycotina phylum* of *Basidiomycota phylum*, a new phylogenetic classification proposed to rename *Rhodosporidium* as *Rhodotorula* (Wang et al. 2015). The cells are polyphyletic with a subglobose, ovoid, ellipsoidal, and long appearance. Reproduction most commonly occurs from polar and multilateral boots, although sexual reproduction occurs rarely in certain *Rhodotorula* strains in the form of pseudohyphae.

Rhodotorula species are widely distributed, nonpathogenic, aerophilic parasites that can be found in various habitats, including air, land, and ice, as well as in the bodies of animals, plants, and certain lower organisms (Wen et al. 2020). Yeast can utilize various carbon sources such as glucose, fructose, xylose, glycerol, starch, inulin, sucrose, and organic acids. They produced lipids in proportions greater than 70% of their biomass when using carbon sources such as xylose or glycerol (Park et al. 2018). The most significant biotechnological application is associated with its ability to convert industrial glycerol and lignocellulosic residues into lipids (Lopes et al. 2020). The recent discovery of their genetic accessibility paved the way for developing an economically viable lipid production process in these yeasts (Ram et al. 2020). The composition of the fatty acid profile produced by the *R. toruloides* species is mainly C16:0, C18:0, C18:1, and C18:2. The strains can produce a small amount of C18:3 and fatty acids with carbon length between C:20 and

C:24, the content of which is usually less than 6%. The fatty acid profile depends on the composition of the medium, the carbon/nitrogen (C/N) ratio, and the culture conditions (Osorio-González et al. 2019).

Patel et al. (2015) cultivated *R. toruloides* HIMPA1 in different carbon sources, finding that the content of monounsaturated fatty acids was higher in a medium with glucose and fructose. Unlike the results obtained with a sucrose-based medium, this increased the concentration of monounsaturated fatty acids and the polyunsaturated fatty acid content. Lipids produced in media containing rapeseed meal hydrolysate included higher C18:2 content and lower C16:0 content compared to those obtained using yeast extract as a nitrogen source.

The *R. toruloides* is resistant to biomass-derived inhibitors. Ammonium, nitrate, amino acids, and short peptides are good nitrogen sources for this yeast (Li et al. 2020). Oleaginous yeasts are frequently grown under nitrogen-limited conditions to accumulate large lipids (Evans and Ratledge 1984). Limiting nutrients such as inorganic phosphate, sulfate, and iron similarly aids lipid production (Wang et al. 2018).

2.3 MECHANISM OF LIPID BIOSYNTHESIS

The yeast *R. toruloides* belongs to the oleaginous microorganisms, described as organisms capable of generating and accumulating more than 20% lipids in the dry cellular matter. It is vital to understand cell metabolism in oleaginous yeasts. Methodologies based on the evaluation of enzymes that can be activated in the production of lipids have been used. Intracellular fluxes can be quantified by assessing anabolic and catabolic activities or by designing culture media. We can change the medium components, change feeding techniques during the fermentation process, and even have a direct influence on a vital metabolic pathway or an intermediate (Pinheiro et al. 2020).

One of the most notable differences between oily and non-oil yeasts is the amount of citrate in their mitochondria. Oily yeasts are grown on a carbon-rich medium with low nitrogen levels which can boost their ability to accumulate lipids (Figure 2.1). Under these conditions, a sequence of molecular steps leading to lipogenesis is activated: adenosine triphosphate (ATP) accumulation and adenosine monophosphate (AMP) depletion, inactivation of mitochondrial NAD+ isocitrate dehydrogenase, citrate transport from mitochondria to cytosol, cytosolic expression of ATP-citrate lyase, inhibition of ATP return, citrate lyase (ACL) activity, and citrate flux for production of long-chain acids and acetyl-CoA esters (Sitepu et al. 2014).

The activation of AMP deaminase in oleaginous yeasts causes ammonia to be internalized in nitrogen-deficient cells, reducing mitochondrial AMP content and stopping the tricarboxylic acid cycle (TCA). As a result, citrate accumulates in the mitochondria and is ejected into the cytosol, where it acts as a substrate for ATP citrate lyase in the production of acetyl-CoA in fatty acid biosynthesis (Zhu et al. 2012). Oxaloacetate in the cytosol is converted to malate, which is then used in the reflux system to remove citrate from the mitochondria (Ratledge and Wynn 2002).

Several authors agree that there is a relationship between lipid synthesis and the enzymatic activity of ATP-citrate lyase in oleaginous yeasts. They also report that citric acid is a donor of acetyl-CoA and that the malic enzyme controls lipid accumulation since it is a source of nicotinamide adenine dinucleotide phosphate (NADPH) for the synthesis of fatty acids (Tiukova et al. 2019a). Proteomic analyses have been performed to better understand changes in gene product expression under nitrogen-limited culture conditions with insufficient aeration. They revealed a link between protein upregulation in lipolysis pathways and reduced fatty acid synthesis during the early stationary phase. Glycolytic pathway enzymes showed that glucose was rapidly converted to pyruvate and acetyl-CoA. When the cells reached the stationary phase, the flux of

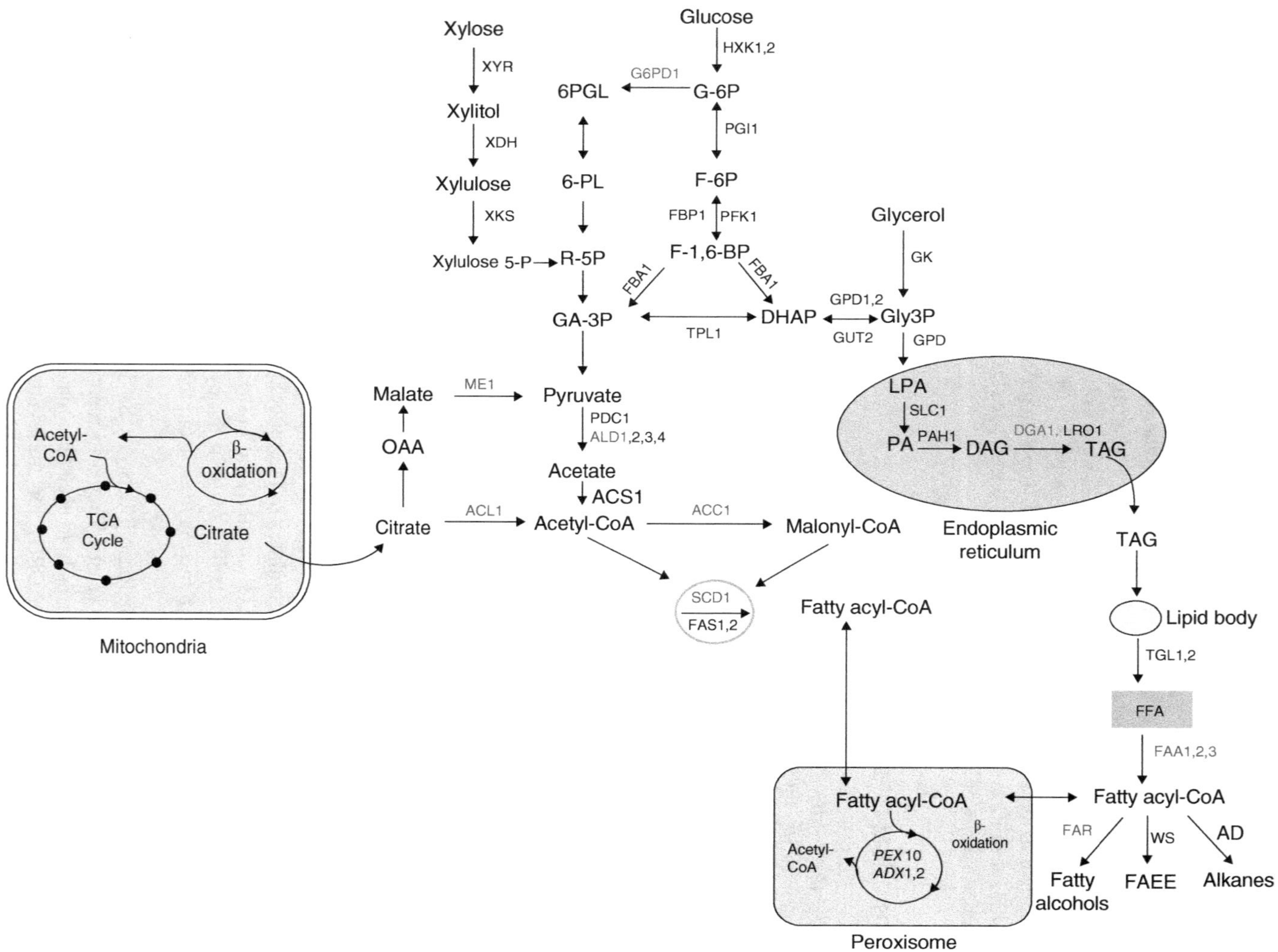

FIGURE 2.1 Metabolic pathways of *Rhodosporidium toruloides*. The metabolites connected by arrows (a single arrow represents a step) and involved enzymes (named in the center of the arrow) are represented. Abbreviations are shown for the commonly used names of metabolites and enzymes. The endoplasmic reticulum, mitochondria, lipid body, and peroxisome are also present as cellular compartments. The enzymes with names in green coincide with genes that have been previously defined. FA is an abbreviation for fatty acid; FAME is an abbreviation for fatty acid methyl ester; FFA is an abbreviation for free fatty acid; FAEE is an abbreviation for fatty acid ethyl ester; TAG is an abbreviation for triacylglycerol, and VLFA is an abbreviation for very long chain fatty acid. (*See colour plate section for colour representation of this figure.*)

acetyl-CoA promoted carotenoid formation over fatty acid synthesis. Fakankun et al. (2021) reported how culture conditions influence gene product expression levels, pathway usage, and end-product synthesis patterns.

2.4 FACTORS AFFECTING INTRACELLULAR LIPID BIOSYNTHESIS

Intracellular lipid production is influenced by several factors such as strain, carbon source, C/N ratio, medium components, and cultivation conditions. Complete genomic sequences are known for *R. toruloides* strains (Kumar et al. 2012; Morin et al. 2014; Hu and Ji 2016; Sambles et al. 2017; Coradetti et al. 2018). Proteomic analysis has been used to analyze the lipid production process throughout various phases of culture development in glucose medium and compare it with cells cultured in xylose medium.

Multiomics investigations have revealed metabolic changes in response to nitrogen and phosphate restriction (Wang et al. 2018). A greater lipid accumulation was demonstrated under nitrogen or phosphate limitation, which has been related to greater activation in nitrogen recycling, as well as lipid degradation and autophagy. During batch culture, microorganisms constantly change their physiological state in response to changes in external conditions (Shen et al. 2013), such as nutrient depletion, toxic compound accumulation, and ionic balance changes due to pH changes, all of which can lead to growth-limiting conditions. *R. toruloides* CCT 0783 grown in batches with glycerol medium found that an increase in extracellular pH (6.0–7.3) during cultivation had a limiting impact on yeast development before the carbon supply was exhausted. For *R. toruloides* CCT 0783 grown in batches with glycerol, an increase in extracellular pH (6.0–7.3) during culture was found to have a limiting impact on yeast development before the supply of carbon will be depleted (Azambuja et al. 2018).

The *R. toruloides* lipid production process is also influenced by the type of carbon source present in the culture medium. Glucose is the most suitable carbon source for lipid formation. Papanikolaou and Aggelis (2011) have studied in detail the conversion of glucose to lipids. The maximum theoretical conversion was 33 g of lipid per 100 g of glucose. Provided no other cellular material is produced, the maximum conversion of glucose to TAG is 33% (w/w). Therefore, a yield of 22% is the best that can be achieved in practice (Papanikolaou and Aggelis 2011).

Most bioprocesses use glucose as a carbon and energy source. However, other sugars such as xylose and other organic compounds are more profitable than glucose. Xylose is the most critical hemicellulose component. *R. toruloides*, in particular, has an advantage over other oily yeasts in that it can metabolize xylose without catabolic inhibition and at rates comparable to glucose. *R. toruloides* has metabolic flexibility to convert this and other substrates to lipids using different biotechnological processes (Table 2.1).

Kraisintu et al. (2010) used glucose as a carbon source for *R. toruloides* DMKU3-TK16 to produce lipids. The C/N ratio in the medium was between 115 and 140, with C/N at 140 being the maximum lipid content (approx. 62.3%).

Crude glycerol, a byproduct of biodiesel production, could be used as a carbon source (Table 2.1). Gao et al. (2016) found that with crude glycerol a higher lipid production was obtained than with pure glycerol, without the contaminants present in crude glycerol having a negative effect.

Singh et al. (2018) used *R. toruloides* ATCC 204091 in medium-containing glucose, xylose, and glycerol along with other carbon sources. A triauxic growth pattern was obtained with preferential use of glucose, xylose, and glycerol, resulting in increased growth, lipid accumulation, and carotenoid production.

Ye et al. (2021) studied the effects of sucrose, glucose, xylose, nitrogen sources (ammonium sulfate, ammonium nitrate), and C/N ratio on lipid synthesis capacity using the mutant strain *R. toruloides* ZL2. Sucrose and ammonium nitrate were the best combinations for lipid accumulation.

TABLE 2.1 Lipid synthesis and growth of various *Rhodotorula toruloides* strains in different processes.

Strain	Carbon Source	Biomass (g/L)	Lipids Content (%)	Culture Method	References
AS2.1389	Glycerol	26.5	38.1	Flask in two stages	Xu and Liu (2017)
AS2.1389	Crude glycerol	26.7	69.0	Batch bioreactor	Xu et al. (2012)
AS2.1389	Xylose	13.4	42.0	Flask in two stages	Lin et al. (2014)
AS 2.1389	Acetic acid	4.3	48.3	Batch bioreactor	Huang et al. (2016)
AS 2.1389	Bagasse hydrolysate	23.5	52.5	Batch bioreactor	Zhao et al. (2012)
AS2.1389	Glucose	8.7	61.8	Single-stage continuous bioreactor	Shen et al. (2013)
ATCC 10788	Crude glycerol/ Methanol/soap	21.2	53.3	Shake flask	Uprety et al. (2018)
DSM 4444	Crude glycerol	41.0	60.0	Fed-Batch bioreactor	Signori et al. (2016)
DSMZ 4444	Glucose/glycerol	17.4	63.0	Batch bioreactor	Bommareddy et al. (2017)
DSMZ 4444	Hydrolysate corn straw	54.0	59.0	Fed-Batch bioreactor	Fei et al. (2016)
CCT 0783	Molasses	34.8	44.0	Fed-Batch bioreactor	Vieira et al. (2014)
CCT 0783	Sorghum Hydrolysate	41.6	33.1	Batch bioreactor	Matsakas et al. (2015)
NCYC 921	Brewery effluent/ sugarcane molasses	42.5	29.9	Fed-batch	Dias et al. (2020)
NRRLY27012	Crude glycerol	23.8	47.1	Batch bioreactor	Tchakouteu et al. (2015)
Y-27012	Crude hydrolysate	32.9	36.4	Batch flasks	Carmona-Cabello et al. (2021)
CBS 14	Lignocellulose hydrolysate	29.8	39.3	Batch flasks	Brandenburg et al. (2021)
21167	Cassava starch	20.1	64.9	Co-culture bioreactor	Gen et al. (2014)
2F5	Inulin	15.6	70.4	Batch bioreactor	Wang et al. (2014)
32489	Crude glycerol	14.9	41.8	Conical flasks	Gao et al. (2016)
32489	Banana straw hydrolysate/ ethanol stillage	13.6	36.0	Batch flasks	Ma et al. (2021)
RM18	Tea waste hydrolysate	10.8	44.6	Batch flasks	Qi et al. (2020)

The *R. toruloides* was applied for the valorization of molasses in the generation of lipids and carotenoids. At 96 hours, the volumetric productivities in the molasses medium were 3.41, compared to 1.75 $gl^{-1}d^{-1}$ in the control medium. The lipid content of *R. toruloides* was 53% and the volumetric productivity of carotenoids (0.027 $mgl^{-1}h^{-1}$) (Lakshmidevi et al. 2021).

The *R. toruloides* NRRL Y-1091 cells were grown vigorously in two-stage flask culture using switchgrass hydrolyzate pretreated with dilute acid and saccharified with enzymes: 61% lipid per cell was produced and a lipid titer of 26.2 g/l. Despite the presence of inhibitors, this strain outperforms many other strains, including *Yarrowia lipolytica*, in terms of biomass, lipid titer, and total substrate consumption (Slininger et al. 2016).

The acidity of the culture medium also has a considerable effect on lipid biosynthesis. Dias et al. (2016) showed its effect on the culture of *R. toruloides* NCYC 921 oily strain. The highest productivity was achieved in a medium at pH 5.0.

Robles-Iglesias et al. (2021) evaluated the ability of *R. toruloides* to grow and accumulate lipids with acetic acid at concentrations up to 19 g/l. At concentrations of acetic acid higher than 15 g/l, the inhibition of its bioconversion process into lipids was observed, particularly in the latency phase. The lag phase was 50 hours in the presence of 19 g/l acetic acid compared to less than 10 hours with concentrations below 15 g/l.

Several strategies have been tested to increase the yield of lipid and carotenoid synthesis in media with lignocellulosic biomass. Colony selection and heterologous adaptive laboratory enhancement (ALE) techniques were applied to develop robust strains. Sugarcane hemicellulose culture medium was used to adapt ALE techniques for *R. toruloides*. The selected strain produced more biomass and more lipid content (60%). Acetic acid was shown to have a considerable effect in repressing xylose consumption during the *R. toruloides* culture (Liu et al. 2021a; Lopes et al. 2021).

The *R. toruloides* mutant RM18 with a high tolerance for the inhibitory effects of lignocellulosic hydrolysate was obtained using atmospheric room temperature plasma mutagenesis (ARTP) (Qi et al. 2014, 2017). UV-mutagenesis combined with ethanol, hydrogen peroxide, or cerulenin allowed the selection of a strain of *R. toruloides* with a lipid synthesis titer of 1.25 g/l compared to 0.94 g/l in control and a rate of 51.2 mg/lh versus 9.8 mg/lh. Transcriptional analysis revealed overexpression of isocitrate dehydrogenase and malate dehydrogenase, suggesting that lipid formation is driven by the stimulation of these enzymes and the production of NADPH, a crucial reducing agent (Yamada et al. 2017).

Liu et al. (2021b) explored the integration of sugar-containing wastewater treatment for biodiesel generation using the oily yeast *R. toruloides*. It was found that this microorganism achieved a total carbon decrease of 87.0±2.4%, with a lipid accumulation of 73.5±0.7%. The flow of NADPH from *R. toruloides* cultured in acetic acid was related to the malic enzyme pathway and not to the pentose phosphate pathway. This, according to in silico studies, allowed the reduction of CO_2 emissions and an increase of 15.6% in the accumulation of lipids.

Studies have been carried out to improve lipid synthesis, metabolism, and the genome of *R. toruloides*. Endogenous gene expression has been altered to enhance lipid production (Liu et al. 2016). Overexpression of DGAT1 and SCD1 genes under the control of the native xylose reductase promoter (XYL1), together with adaptive evolution, resulted in increased xylose consumption and tolerance to growth inhibitors (Díaz et al. 2018). Overexpression of ACC1 under the GAPDH promoter and DGA1 under the ACL promoter resulted in substantial lipid synthesis from 70 g/l glucose and 70 g/l xylose, with lipid yields of 61.1% and 43.4%, respectively. Overexpression of malic enzyme (ME), pyruvate carboxylase (PYC1), glycerol-3-P dehydrogenase (GPD), and stearoyl-CoA desaturase (SCD), as well as PEX10 removal, caused an accumulation of lipids upwards from 75.6% and altered oxidation (Zhang et al. 2016).

2.5 CAROTENOID BIOSYNTHESIS BY *RHODOTORULA TORULOIDES*

Carotenoids are natural colorants produced by some microorganisms. In particular, *R. toruloides* is characterized by a natural way of producing carotenoids, which are precursors of vitamin A (ß-carotene). Torularhodin, torulene, ß-carotene, and α-carotene account for 90% of the carotenoids produced by *Rhodotorula*. Depending on the pigment concentration, these can be reddish pink, orange, or yellow (Zoz et al. 2014). Table 2.2 summarizes the information regarding the production of carotenoids by different *R. toruloides* strains.

β-carotene ($C_{40}H_{56}$) is one of the best-known pigments, having been widely used in pharmaceutical formulations, food additives, cosmetics, eye care products, and other products in other industries. It is synthesized from carotenoids by the action of the enzyme β-carotene 15,15′-monooxygenase. The ß-ionone rings found at both ends of the molecule are its distinguishing feature. Carotene has three isomers: α-, β-, and γ- (Liu et al. 2018). It is soluble in ethanol, ethers, chloroform, benzene, and oil, and its melting point is 176–180 °C, with β-carotene being the most stable and effective. The pigment of *Rhodotorula* strains, β-carotene, represents almost 70% of the total carotenoids (Sharma and Ghoshal 2020).

Torulene ($C_{40}H_{54}$) and torularhodine ($C_{40}H_{52}O_{23}$) have one β-ionone ring connected to a chain of polyenes made up of 10 double links. This characteristic gives it powerful antioxidant properties (Kot et al. 2018). In petroleum ether, the absorption maxima of torulene are 460, 484, and 518 nm, and those of torularhodine are 465, 492, and 523 nm (Maldonade et al. 2008). In addition to antioxidant activity in both compounds and anticancer properties have been demonstrated *in vivo*. In anticancer experiments, torulene and torularhodine suppressed prostate cancer growth by triggering apoptosis in tumor cells, comparable to lycopene (Du et al. 2016). The antimicrobial

TABLE 2.2 Carotenoid biosynthesis by different *Rhodotorula toruloides* strains.

Rhodotorula toruloides Strain	Substrate	Cultivation Method	Carotenoids mg/L	References
DBVPG6739	Glucose	Batch	1.3	Buzzini et al. (2007)
NCYC 921	Glucose	Fed-batch	2.8	Dias et al. (2016)
NRRL Y-1091	Wheat straw hydrolysate	Batch	14.1	Liu et al. (2021c)
ATCC 204091	Waste extract (composed of peels of orange, sweet lemon, banana, mango, pea pods, and stubs of cauliflower/cabbage)	Batch	62.0	Singh et al. (2018)
ATCC 204091	Lignocellulosic waste hydrolysate	Batch	19.0	Sinha et al. (2021)
CBS 5490	Glycerol	Batch	28.5	Lee et al. (2014)
DSM 4444	*Camelina sativa* meal hydrolysates	Batch	16.0	Bertacchi et al. (2020)
CCT 7815	Xylose	Batch	44.0	Pinheiro et al. (2020)

activity of these two carotenoids has led to their use in infection prevention, particularly in implants, products, and medical preparations that require natural antimicrobials (Ungureanu and Ferdes 2012). Currently, neither torulene nor torularodine is produced on an industrial level. However, due to their valuable properties, carotenoids are required for different applications. This is why *Rhodotorula* strains are sought and yeast mutation methods that provide high yield and reduce production costs are applied (Kot et al. 2018).

2.6 MECHANISM OF CAROTENOID BIOSYNTHESIS

The metabolic processes of carotenoid synthesis in different *Rhodotorula* species were analyzed showing that there is an identical or conserved biosynthetic pathway for carotenoids, consisting of only three steps (Figure 2.2) (Kot et al. 2018).

The conversion of acetyl-CoA to 3-hydroxy-3-methylglutaryl-CoA (HMG-CoA) is catalyzed by hydroxymethylglutaryl-CoA synthase (HMG-CoA synthase). HMG-CoA is converted to C6, mevalonic acid (MVA), which is the first specific precursor in the terpenoid biosynthetic pathway. Then, MVA is converted to isopentenyl pyrophosphate (IPP) by a series of reactions classified by specific kinases and diphosphomevalonate decarboxylase. Eventually, IPP is isomerized to dimethylallyl pyrophosphate (DMAPP) by the sequential addition of three IPP molecules to DMAPP. These reactions are classified as prenyltransferase reactions and produce the C20 geranylgeranyl pyrophosphate compound (GGPP). The condensation of two molecules in this compound leads to the formation of phytoene through the enzyme phytoene synthase (crtYB). The following reactions are classified as phytoene desaturase (crtI).

Phytoene desaturase participates in the formation of neurospores and can catalyze up to three stages of desaturation. Neurosporene can be converted to lycopene or β-zeacarotene. The formation of γ-carotene is favored by the lycopene cycle or the dehydrogenation of β-zecarotene. The γ-carotene molecule serves as a precursor for the biosynthesis of ß-carotene and torulene, while torularodine is produced by a further torulene transformation including hydroxylation and oxidation (Figure 2.2) (Tang et al. 2019).

2.7 STRATEGIES TO INCREASE CAROTENOID SYNTHESIS

Understanding the effects of culture parameters, such as culture temperature, aeration, pH, lighting, and substrate composition, is essential for the industrial development of biotechnological processes for carotenoid synthesis. Production costs must be reduced, carotenoid yields must be improved, and ways to improve the recovery (extraction and separation) and preservation of these compounds must be optimized.

Oxidative stress, such as with hydrogen peroxide (H_2O_2) and UV radiation, has been shown to increase carotenoid synthesis. Culture conditions optimized under these factors can increase the production yields of these metabolites, including *R. toruloides*. Although the mechanism of cellular response to oxidative stress is unknown.

The existence of reactive oxygen species (ROS), such as H_2O_2, superoxide (O_2^-), and hydroxyl radicals (HO), is linked to this effect. ROS are powerful oxidants that can damage DNA, lipids, and proteins, among other biological components. There are enzymatic and non-enzymatic defense strategies in microbial cells against oxidative damage. The enzymes superoxide dismutase and catalase constitute the former, while the latter participates in the direct elimination of ROS or the recycling of oxidized substances such as ascorbate and glutathione, α-tocopherol, and carotenoids.

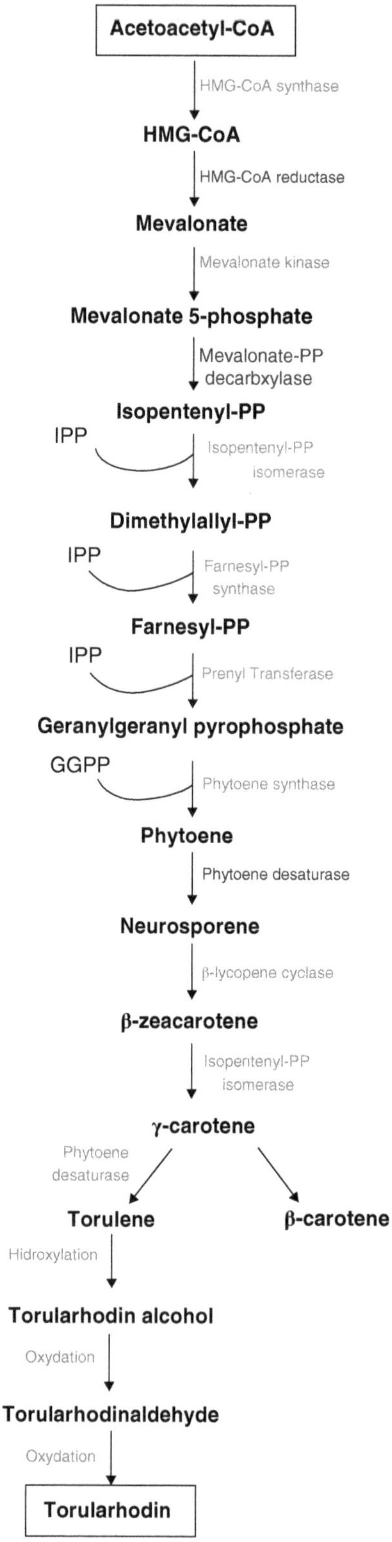

FIGURE 2.2 *Rhodotorula toruloides* carotenoid biosynthetic pathway.

Pinheiro et al. (Pinheiro et al. 2020) studied the accumulation of carotenes in *R. toruloides* CCT 7815 under oxidative stress (H_2O_2 and light irradiation) and adaptive evolution in the laboratory, discovering that light irradiation resulted in 70% more carotenoids than under ideal growth circumstances. However, hydrogen peroxide does not influence the synthesis of carotenoids. The modified strain contained 2.3 times the carotenoid content of the original strain and was in better physical condition.

R. toruloides MH023518 was cultured in the presence of ZnO nanoparticles, which cause cellular stress and the generation of ROS. As a result, carotene production was increased to combat ROS, up to 124 ± 4.5 mg/l (Ibrahim and Mahmoud 2021). The use of nanoparticles in the production of carotenoids by *R. toruloides* could be a viable solution to increase the economic viability of oleochemical synthesis.

Flow cytometry has been used to investigate the influence of culture medium pH on carotenoid and lipid synthesis by *R. toruloides* NCYC 921 (Martins et al. 2021a).

A dual pH method was used in conjunction with two-stage cultivation to optimize both lipid and carotenoid synthesis of *R. toruloides* in a batch time of 20 hours. Glucose and nutrients were administered at pH 4, promoting growth and lipid production, followed by a glucose-only stage at pH 5 for increased lipid and carotenoid production [0.29 g/lh]. This technique provided better biomass yields than the constant pH studies, demonstrating that a flexible process design suited to the requirements of lipogenesis could further optimize bioproduction. Variations in process parameters, such as pH and aeration rate, are easy to implement and are used on various scales (Dias et al. 2015).

Dias et al. (2016) used response surface methodology to investigate the role of culture medium components in lipid/carotene synthesis and found a 10% (w/v) increase in carotenoid production under optimum osmotic stress.

Martins et al. (2021b) studied sugar beet pulp hydrolysates to propagate *R. toruloides* PYCC 5615. This substrate mainly contains the neutral sugars D-glucose, L-arabinose, and D-galactose, as well as the acidic sugar D-galacturonic acid. A total carotenoid content of 255 g/g dry weight was achieved. Carob pulp syrup was used as a carbon source during fed-batch culture yielding 0.42 mg/dry cell weight (DCW) of carotenoids (Martins et al. 2018).

The generation of carotenoids from glycerol was studied using global metabolomics, which revealed a reduction of metabolites involved in TCA and amino acid biosynthesis (Lee et al. 2014).

When used *R. toruloides* mutant (RM18) obtained by ARTP mutagenesis and continuous adaptation in tea waste hydrolysate generated higher biomass, carotenoids, and α-linolenic acid than the control strain. Torularhodin (481.92 g/g DCW) and torulene (501 g/g DCW) yields were 12.86 and 1.5-fold greater, respectively than that of the control strain. Transcriptomic profiling has revealed that central metabolism and terpene biosynthesis resulted in increased carotenoid production, while aromatic amino acid synthesis and DNA damage detection and checkpoint likely involved tolerance to terpene from tea waste hydrolysate (Qi et al. 2020). This strain was also studied in sugarcane molasses at 35 and 70 g/l concentrations. The production of α-linolenate and torularhodin was 1.22 and 14.68 times higher than that of the wild-type strain. Transcriptome and Weighted correlation network analysis (WGCNA) revealed that genes involved in lipid and carotenoid synthesis, such as MYB, bHLH, Δ-4 desaturase, Δ-12 desaturase, and FA2H, were significantly overexpressed (Zheng et al. 2021).

The *R. toruloides* CCT 0783 received the Golden Gate DNA assembly system (RtGGA). Sequences were designed to characterize promoters and overexpress the carotenoid synthesis pathway. RtGGA was used to make three different variants of the carotenoid overexpression cassette, which resulted in a 41% increase in overall carotenoid concentration (Bonturi et al. 2022).

2.8 BIOSYNTHESIS OF PHENYLALANINE AMMONIA-LYASE BY *R. TORULOIDES*

The *R. toruloides,* depending on growth conditions, can synthesize numerous types of enzymes that can be used in a variety of industrial applications. It was found that the biomass of these yeasts could be a source of lipases, Δ12-desaturase, D-amino oxidase, and phenylalanine ammonia-lyase (PAL) (Tiukova et al. 2019b; Wu et al. 2021).

PAL is a member of the lyase family with EC 4.3.1.5. It was recently expanded to three enzymes: EC 4.3.1.24 (PALs), EC 4.3.1.25 (tyrosine ammonia lyases), and EC 4.3.1.26 (phenylalanine/tyrosine ammonia lyases). PAL catalyzes the conversion of L-phenylalanine to t-cinnamic acid (t-CA) and ammonia (Jun et al. 2018). PAL was initially discovered in higher plants as part of a defense mechanism. Later its expression was discovered in *Rhodotorula* strains grown with L-Phe as the only source of carbon and nitrogen (Sun et al. 2017). L-Phe was assigned the role of inducer.

The enzyme has great prospects for medicinal and industrial applications, including the treatment of phenylketonuria (PKU), a metabolic disease caused by decreased hepatic phenylalanine hydroxylase activity (Al Hafid and Christodoulou 2015). Furthermore, its substantial anticancer effect was revealed on cell lines *in vitro* and on mouse tumors *in vivo*. Babich et al. (2020) showed that PAL derived from *R. toruloides* had strong anticancer efficacy *in vitro* and *in vivo*, which was equivalent to PAL from *Rhodotorula glutinis*. No direct correlations were found between enzyme K_m and primary structure on the one hand and cytotoxicity on the other. The mechanism of the anticancer effect of PAL can be linked not only to the induction of amino acid deficiency but also to other biological characteristics.

The PAL gene from *R. toruloides* has been used as tac, pPGK: REP2, PLPR, to construct recombinant *Escherichia coli* expression vectors (Al Hafid and Christodoulou 2015; Sun et al. 2017; Babich et al. 2020). The high-resolution enzyme crystal structure has been determined, allowing structure-based molecular engineering approaches to be used to increase its therapeutic powers (Jun et al. 2018).

R. toruloides was grown in a two-phase culture with used cooking oil as the carbon source. The ability to simultaneously produce PAL enzyme (0.12 mol/mg), ß-carotene (4.5 g/ml), torularhodin (4.2 g/ml), and fatty acids (1 mg/ml) was demonstrated (Lee et al. 2021). The search for techniques to increase enzyme production is now ongoing. Conditions influence enzyme activity. For example, the acidity of the medium is one of these factors. For the deamination of L-phenylalanine to trans-cinnamic acid, the reaction pH should not be lower than 9.0 (Kawatra et al. 2020). Temperature is another critical factor influencing enzyme production. A temperature of 30 °C was considered the most suitable (Takaç et al. 1995).

Despite its promising properties, PAL from *R. toruloides* performed poorly in repeat-dose assays, even after extensive mutagenesis to minimize immunogenicity or increase stability. As a result, concerns such as immunogenicity, enzyme synthesis, and mode of administration remain an obstacle for industrial and medicinal applications (Bell et al. 2017).

2.9 CONCLUSIONS AND FUTURE TRENDS

R. toruloides has emerged as one of the most promising bioproduction yeasts. It can be metabolically modified to raise production levels and broaden the spectrum of produced chemicals. Particularly in the last decade, progress has been made in the development of biotechnological processes with *R. toruloides*. However, the biggest obstacle to their commercialization as a sustainable alternative to base oils or derivatives such as biodiesel is the cost of manufacturing lipids. Increased lipid productivity can be achieved with a variety of carbon sources during yeast proliferation. In addition, finding the appropriate microorganism for this is essential. Ideally, the strain of industrial interest should be an excellent natural producer that uses low-cost substrates and is amenable to process scaling effectively. Such constraints frequently necessitate the use of microbiology, biochemistry, and engineering concepts.

With the development of genetic engineering tools, the value of *R. toruloides* for industrial biotechnology has grown even further. Mutagenesis, laboratory adaptive evolution, and metabolic engineering have been used to improve tolerance to inhibitors and increase productivity. Mainly through metabolic engineering, it has been possible to improve the production of carotenoids and lipids by *R. toruloides*. Carotenoids and lipids are two industrially essential chemicals. Lipids are an excellent source of fuels and chemicals that could help replace oil. Carotenoids and lipids are two industrially essential chemicals. Lipids are an excellent source of fuels and chemicals that could help replace oil. Carotenoids are used in the food, feed, pharmaceutical, and cosmetic sectors as additives. *R. toruloides* produces significant levels of these compounds and enzymes, making it a suitable option for their production, which is currently under increasing research and development.

REFERENCES

Al Hafid, N. and Christodoulou, J. (2015). Phenylketonuria: a review of current and future treatments. *Translational Pediatrics* 4 (4): 304–317. https://doi.org/10.3978/J.ISSN.2224-4336.2015.10.07.

Azambuja, S.P.H., Bonturi, N., Miranda, E.A. et al. (2018). Physiology and lipid accumulation capacity of different *Yarrowia lipolytica* and *Rhodosporidium toruloides* strains on glycerol. *bioRxiv* 278523. https://doi.org/10.1101/278523.

Babich, O., Dyshlyuk, L., Prosekov, A. et al. (2020). Study of the potential of the capsule shell based on natural polysaccharides in targeted delivery of the L-phenylalanine ammonia-lyase enzyme preparation. *Pharmaceuticals* 13 (4): 63. https://doi.org/10.3390/PH13040063.

Bell, S.M., Wendt, D.J., Zhang, Y. et al. (2017). Formulation and PEGylation optimization of the therapeutic PEGylated phenylalanine ammonia lyase for the treatment of phenylketonuria. *PLoS One* 12 (3): https://doi.org/10.1371/JOURNAL.PONE.0173269.

Bertacchi, S., Bettiga, M., Porro, D. et al. (2020). *Camelina sativa* meal hydrolysate as sustainable biomass for the production of carotenoids by *Rhodosporidium toruloides*. *Biotechnology for Biofuels* 13 (1): 1–10. https://doi.org/10.1186/S13068-020-01682-3/TABLES/2.

Bommareddy, R.R., Sabra, W., and Zeng, A.P. (2017). Glucose-mediated regulation of glycerol uptake in *Rhodosporidium toruloides*: insights through transcriptomic analysis on dual substrate fermentation. *Engineering in Life Sciences* 17 (3): 282. https://doi.org/10.1002/ELSC.201600010.

Bonturi, N., Pinheiro, M.J., Oliveira, P.M. et al. (2022). Development of a dedicated Golden Gate Assembly platform (RtGGA) for *Rhodotorula toruloides*. *bioRxiv* 482697. https://doi.org/10.1101/2022.03.02.482697.

Brandenburg, J., Blomqvist, J., Shapaval, V. et al. (2021). Oleaginous yeasts respond differently to carbon sources present in lignocellulose hydrolysate. *Biotechnology for Biofuels* 14 (1): https://doi.org/10.1186/S13068-021-01974-2.

Buzzini, P., Innocenti, M., Turchetti, B. et al. (2007). Carotenoid profiles of yeasts belonging to the genera *Rhodotorula, Rhodosporidium, Sporobolomyces*, and *Sporidiobolus*. *Canadian Journal of Microbiology* 53 (8): 1024–1031. https://doi.org/10.1139/W07-068.

Carmona-Cabello, M., García, I.L., Papadaki, A. et al. (2021). Biodiesel production using microbial lipids derived from food waste discarded by catering services. *Bioresource Technology* 323: 124597. https://doi.org/10.1016/J.BIORTECH.2020.124597.

Coradetti, S.T., Pinel, D., Geiselman, G.M. et al. (2018). Functional genomics of lipid metabolism in the oleaginous yeast *Rhodosporidium toruloides*. *eLife* 7: https://doi.org/10.7554/ELIFE.32110.

Dias, C., Sousa, S., Caldeira, J. et al. (2015). New dual-stage pH control fed-batch cultivation strategy for the improvement of lipids and carotenoids production by the red yeast *Rhodosporidium toruloides* NCYC 921. *Bioresource Technology* 189: 309–318. https://doi.org/10.1016/J.BIORTECH.2015.04.009.

Dias, C., Silva, C., Freitas, C. et al. (2016). Effect of medium pH on *Rhodosporidium toruloides* NCYC 921 carotenoid and lipid production evaluated by flow cytometry. *Applied Biochemistry and Biotechnology* 179 (5): 776–787. https://doi.org/10.1007/S12010-016-2030-Y.

Dias, C., Reis, A., Santos, J.A.L. et al. (2020). Concomitant wastewater treatment with lipid and carotenoid production by the oleaginous yeast *Rhodosporidium toruloides* grown on brewery effluent enriched with sugarcane molasses and urea. *Process Biochemistry* 94: 1–14. https://doi.org/10.1016/J.PROCBIO.2020.03.015.

Díaz, T., Fillet, S., Campoy, S. et al. (2018). Combining evolutionary and metabolic engineering in *Rhodosporidium toruloides* for lipid production with non-detoxified wheat straw hydrolysates. *Applied Microbiology and Biotechnology* 102 (7): 3287–3300. https://doi.org/10.1007/S00253-018-8810-2.

Du, C., Li, Y., Guo, Y. et al. (2016). The suppression of torulene and torularhodin treatment on the growth of PC-3 xenograft prostate tumors. *Biochemical and Biophysical Research Communications* 469 (4): 1146–1152. https://doi.org/10.1016/J.BBRC.2015.12.112.

Evans, C.T. and Ratledge, C. (1984). Influence of nitrogen metabolism on lipid accumulation by *Rhodosporidium toruloides* CBS 14. *Journal of General Microbiology* 130 (7): 1705–1710. https://doi.org/10.1099/00221287-130-7-1705/CITE/REFWORKS.

Fakankun, I., Spicer, V., and Levin, D.B. (2021). Proteomic analyses of the oleaginous and carotenogenic yeast *Rhodotorula diobovata* across growth phases under nitrogen- and oxygen-limited conditions. *Journal of Biotechnology* 332: 11–19. https://doi.org/10.1016/J.JBIOTEC.2021.03.016.

Fei, Q., O'Brien, M., Nelson, R. et al. (2016). Enhanced lipid production by *Rhodosporidium toruloides* using different fed-batch feeding strategies with lignocellulosic hydrolysate as the sole carbon source. *Biotechnology for Biofuels* 9 (1): 1–12. https://doi.org/10.1186/S13068-016-0542-X/TABLES/3.

Gao, Z., Ma, Y., Wang, Q. et al. (2016). Effect of crude glycerol impurities on lipid preparation by *Rhodosporidium toruloides* yeast 32489. *Bioresource Technology* 218: 373–379. https://doi.org/10.1016/J.BIORTECH.2016.06.088.

Gen, Q., Wang, Q., and Chi, Z.M. (2014). Direct conversion of cassava starch into single cell oil by co-cultures of the oleaginous yeast *Rhodosporidium toruloides* and immobilized amylases-producing yeast *Saccharomycopsis fibuligera*. *Renewable Energy* 62: 522–526. https://doi.org/10.1016/J.RENENE.2013.08.016.

Hu, J. and Ji, L. (2016). Draft genome sequences of *Rhodosporidium toruloides* strains ATCC 10788 and ATCC 10657 with compatible mating types. *Genome Announcements* 4 (2): https://doi.org/10.1128/GENOMEA.00098-16.

Huang, X.F., Liu, J.N., Lu, L.J. et al. (2016). Culture strategies for lipid production using acetic acid as sole carbon source by *Rhodosporidium toruloides*. *Bioresource Technology* 206: 141–149. https://doi.org/10.1016/J.BIORTECH.2016.01.073.

Ibrahim, A.B.M. and Mahmoud, G.A.E. (2021). Chemical- vs sonochemical-assisted synthesis of ZnO nanoparticles from a new zinc complex for improvement of carotene biosynthesis from *Rhodotorula toruloides* MH023518. *Applied Organometallic Chemistry* 35 (2): e6086. https://doi.org/10.1002/AOC.6086.

Jiru, T.M., Groenewald, M., Pohl, C. et al. (2017). Optimization of cultivation conditions for biotechnological production of lipid by *Rhodotorula kratochvilovae* (syn, *Rhodosporidium kratochvilovae*) SY89 for biodiesel preparation. *3 Biotech* 7 (2): 1–11. https://doi.org/10.1007/S13205-017-0769-7/TABLES/2.

Jun, S.Y., Sattler, S.A., Cortez, G.S. et al. (2018). Biochemical and structural analysis of substrate specificity of a phenylalanine ammonia-lyase. *Plant Physiology* 176 (2): 1452–1468. https://doi.org/10.1104/PP.17.01608.

Kawatra, A., Dhankhar, R., Mohanty, A. et al. (2020). Biomedical applications of microbial phenylalanine ammonia lyase: current status and future prospects. *Biochimie* 177: 142–152. https://doi.org/10.1016/J.BIOCHI.2020.08.009.

Kot, A.M., Błażejak, S., Gientka, I. et al. (2018). Torulene and torularhodin: "New" fungal carotenoids for industry? *Microbial Cell Factories* 17 (1): 1–14. https://doi.org/10.1186/S12934-018-0893-Z/FIGURES/4.

Kot, A.M., Błażejak, S., Kieliszek, M. et al. (2019). Simultaneous production of lipids and carotenoids by the red yeast *Rhodotorula* from waste glycerol fraction and potato wastewater. *Applied Biochemistry and Biotechnology* 189 (2): 589–607. https://doi.org/10.1007/S12010-019-03023-Z/FIGURES/6.

Kraisintu, P., Yongmanitchai, W., and Limtong, S. (2010). Selection and optimization for lipid production of a newly isolated oleaginous yeast, *Rhodosporidium toruloides* DMKU3-TK16. *Agriculture and Natural Resources* 44 (3): 436–445. https://li01.tci-thaijo.org/index.php/anres/article/view/244948 (accessed 28 January 2022).

Kumar, S., Kushwaha, H., Bachhawat, A.K. et al. (2012). Genome sequence of the oleaginous red yeast *Rhodosporidium toruloides* MTCC 457. *Eukaryotic Cell* 11 (8): 1083–1084. https://doi.org/10.1128/EC.00156-12.

Lakshmidevi, R., Ramakrishnan, B., Ratha, S.K. et al. (2021). Valorisation of molasses by oleaginous yeasts for single cell oil (SCO) and carotenoids production. *Environmental Technology and Innovation* 21: https://doi.org/10.1016/J.ETI.2020.101281.

Lee, J.J.L., Chen, L., Shi, J. et al. (2014). Metabolomic profiling of *Rhodosporidium toruloides* grown on glycerol for carotenoid production during different growth phases. *Journal of Agricultural and Food Chemistry* 62 (41): 10203–10209. https://doi.org/10.1021/JF502987Q.

Lee, J.J.L., Li, A., Lyu, X. et al. (2021). Free fatty acids reduction in waste cooking oil by *Rhodosporidium toruloides* and simultaneous carotenoids, lipids, and PAL enzyme production in a two-phase culture system. *European Journal of Lipid Science and Technology* 123 (8): 2000354. https://doi.org/10.1002/EJLT.202000354.

Li, Q., Kamal, R., Wang, Q. et al. (2020). Lipid production from amino acid wastes by the oleaginous yeast *Rhodosporidium toruloides*. *Energies* 13 (7): 1576. https://doi.org/10.3390/EN13071576.

Lin, J., Li, S., Sun, M. et al. (2014). Microbial lipid production by oleaginous yeast in D-xylose solution using a two-stage culture mode. *RSC Advances* 4 (66): 34944–34949. https://doi.org/10.1039/C4RA01453G.

Liu, Y., Yap, S.A., Kohl, C.M.J. et al. (2016). Developing a set of strong intronic promoters for robust metabolic engineering in oleaginous *Rhodotorula* (*Rhodosporidium*) yeast species. *Microbial Cell Factories* 15 (1): 1–9. https://doi.org/10.1186/S12934-016-0600-X/TABLES/2.

Liu, C., Liu, Z., Sun, X. et al. (2018). Fabrication and characterization of β-lactoglobulin-based nanocomplexes composed of chitosan oligosaccharides as vehicles for delivery of astaxanthin. *Journal of Agricultural and Food Chemistry* 66 (26): 6717–6726. https://doi.org/10.1021/ACS.JAFC.8B00834.

Liu, Z., Radi, M., Mohamed, E.T.T. et al. (2021a). Adaptive laboratory evolution of *Rhodosporidium toruloides* to inhibitors derived from lignocellulosic biomass and genetic variations behind evolution. *Bioresource Technology* 333: https://doi.org/10.1016/J.BIORTECH.2021.125171.

Liu, X., Wang, D., and Li, A. (2021b). Biodiesel production of *Rhodosporidium toruloides* using different carbon sources of sugar-containing wastewater: experimental analysis and model verification. *Journal of Cleaner Production* 323: 129112. https://doi.org/10.1016/J.JCLEPRO.2021.129112.

Liu, Z., Natalizio, F., Dragone, G. et al. (2021c). Maximizing the simultaneous production of lipids and carotenoids by *Rhodosporidium toruloides* from wheat straw hydrolysate and perspectives for large-scale implementation. *Bioresource Technology* 340: 125598. https://doi.org/10.1016/J.BIORTECH.2021.125598.

Lopes, H.J.S., Bonturi, N., Kerkhoven, E.J. et al. (2020). C/N ratio and carbon source-dependent lipid production profiling in *Rhodotorula toruloides*. *Applied Microbiology and Biotechnology* 104 (6): 2639–2649. https://doi.org/10.1007/S00253-020-10386-5/FIGURES/5.

Lopes, H.J.S., Bonturi, N., and Miranda, E.A. (2021). Induction of resistance mechanisms in *Rhodotorula toruloides* for growth in sugarcane hydrolysate with high inhibitor content. *Applied Microbiology and Biotechnology* 105 (24): 9261–9272. https://doi.org/10.1007/S00253-021-11687-Z.

Ma, X., Zhang, M., Gao, Z. et al. (2021). Microbial lipid production from banana straw hydrolysate and ethanol stillage. *Environmental Science and Pollution Research* 28 (23): 29357–29368. https://doi.org/10.1007/S11356-021-12644-Z.

Maldonade, I.R., Rodriguez-Amaya, D.B., and Scamparini, A.R.P. (2008). Carotenoids of yeasts isolated from the Brazilian ecosystem. *Food Chemistry* 107 (1): 145–150. https://doi.org/10.1016/J.FOODCHEM.2007.07.075.

Martins, V., Dias, C., Caldeira, J. et al. (2018). Carob pulp syrup: a potential Mediterranean carbon source for carotenoids production by *Rhodosporidium toruloides* NCYC 921. *Bioresource Technology Reports* 3: 177–184. https://doi.org/10.1016/J.BITEB.2018.07.008.

Martins, J.A., Lopes da Silva, T., Marques, S. et al. (2021a). The use of flow cytometry to assess *Rhodosporidium toruloides* NCYC 921 performance for lipid production using Miscanthus sp. Hydrolysates. *Biotechnology Reports* 30: e00639. https://doi.org/10.1016/J.BTRE.2021.E00639.

Martins, L.C., Palma, M., Angelov, A. et al. (2021b). Complete utilization of the major carbon sources present in sugar beet pulp hydrolysates by the oleaginous red yeasts *Rhodotorula toruloides* and *R. mucilaginosa. Journal of Fungi (Basel, Switzerland)* 7 (3): https://doi.org/10.3390/JOF7030215.

Matsakas, L., Bonturi, N., Miranda, E.A. et al. (2015). High concentrations of dried sorghum stalks as a biomass feedstock for single cell oil production by *Rhodosporidium toruloides*. *Biotechnology for Biofuels* 8 (1): 1–6. https://doi.org/10.1186/S13068-014-0190-Y/TABLES/2.

Morin, N., Calcas, X., Devillers, H. et al. (2014). Draft genome sequence of *Rhodosporidium toruloides* CECT1137, an oleaginous yeast of biotechnological interest. *Genome Announcements* 2 (4): https://doi.org/10.1128/GENOMEA.00641-14.

Osorio-González, C.S., Hegde, K., Ferreira, P. et al. (2019). Lipid production in *Rhodosporidium toruloides* using C-6 and C-5 wood hydrolysate: a comparative study. *Biomass and Bioenergy* 130: 105355. https://doi.org/10.1016/J.BIOMBIOE.2019.105355.

Papanikolaou, S. and Aggelis, G. (2011). Lipids of oleaginous yeasts. Part I: biochemistry of single cell oil production. *European Journal of Lipid Science and Technology* 113 (8): 1031–1051. https://doi.org/10.1002/EJLT.201100014.

Park, Y.K., Nicaud, J.M., and Ledesma-Amaro, R. (2018). The engineering potential of *Rhodosporidium toruloides* as a workhorse for biotechnological applications. *Trends in Biotechnology* 36 (3): 304–317. https://doi.org/10.1016/J.TIBTECH.2017.10.013.

Patel, A., Pruthi, V., Singh, R.P. et al. (2015). Synergistic effect of fermentable and non-fermentable carbon sources enhances TAG accumulation in oleaginous yeast Rhodosporidium kratochvilovae HIMPA1. *Bioresource Technology* 188: 136–144. https://doi.org/10.1016/J.BIORTECH.2015.02.062.

Pinheiro, M.J., Bonturi, N., Belouah, I. et al. (2020). Xylose metabolism and the effect of oxidative stress on lipid and carotenoid production in *Rhodotorula toruloides*: insights for future biorefinery. *Frontiers in Bioengineering and Biotechnology* 8: https://doi.org/10.3389/FBIOE.2020.01008.

Qi, F., Kitahara, Y., Wang, Z. et al. (2014). Novel mutant strains of *Rhodosporidium toruloides* by plasma mutagenesis approach and their tolerance for inhibitors in lignocellulosic hydrolyzate. *Journal of Chemical Technology & Biotechnology* 89 (5): 735–742. https://doi.org/10.1002/JCTB.4180.

Qi, F., Zhao, X., Kitahara, Y. et al. (2017). Integrative transcriptomic and proteomic analysis of the mutant lignocellulosic hydrolyzate-tolerant *Rhodosporidium toruloides*. *Engineering in Life Sciences* 17 (3): 249–261. https://doi.org/10.1002/ELSC.201500143.

Qi, F., Shen, P., Hu, R. et al. (2020). Carotenoids and lipid production from *Rhodosporidium toruloides* cultured in tea waste hydrolysate. *Biotechnology for Biofuels* 13 (1): https://doi.org/10.1186/S13068-020-01712-0.

Ram, S., Mitra, M., Shah, F. et al. (2020). Bacteria as an alternate biofactory for carotenoid production: a review of its applications, opportunities and challenges. *Journal of Functional Foods* 67: 103867. https://doi.org/10.1016/J.JFF.2020.103867.

Ratledge, C. and Wynn, J.P. (2002). The biochemistry and molecular biology of lipid accumulation in oleaginous microorganisms. *Advances in Applied Microbiology* 51: 1–52. https://doi.org/10.1016/S0065-2164(02)51000-5.

Robles-Iglesias, R., Veiga, M.C., and Kennes, C. (2021). Carbon dioxide bioconversion into single cell oils (lipids) in two reactors inoculated with Acetobacterium woodii and *Rhodosporidium toruloides*. *Journal of CO2 Utilization* 52: 101668. https://doi.org/10.1016/J.JCOU.2021.101668.

Sambles, C., Middelhaufe, S., Soanes, D. et al. (2017). Genome sequence of the oleaginous yeast *Rhodotorula toruloides* strain CGMCC 2.1609. *Genomics Data* 13: 1–2. https://doi.org/10.1016/J.GDATA.2017.05.009.

Sharma, R. and Ghoshal, G. (2020). Characterization and cytotoxic activity of pigment extracted from Rhodotorula mucilaginosa to assess its potential as bio-functional additive in confectionary products. *Journal of Food Science and Technology* 58 (7): 2688–2698. https://doi.org/10.1007/S13197-020-04775-X.

Shen, H., Gong, Z., Yang, X. et al. (2013). Kinetics of continuous cultivation of the oleaginous yeast *Rhodosporidium toruloides*. *Journal of Biotechnology* 168 (1): 85–89. https://doi.org/10.1016/J.JBIOTEC.2013.08.010.

Signori, L., Ami, D., Posteri, R. et al. (2016). Assessing an effective feeding strategy to optimize crude glycerol utilization as sustainable carbon source for lipid accumulation in oleaginous yeasts. *Microbial Cell Factories* 15 (1): 1–19. https://doi.org/10.1186/S12934-016-0467-X/TABLES/3.

Singh, G., Sinha, S., Bandyopadhyay, K.K. et al. (2018). Triauxic growth of an oleaginous red yeast *Rhodosporidium toruloides* on waste "extract" for enhanced and concomitant lipid and β-carotene production. *Microbial Cell Factories* 17 (1): 182. https://doi.org/10.1186/S12934-018-1026-4.

Sinha, S., Singh, G., and Paul, D. (2021). Lipid and carotenoid production by *Rhodosporodium toruloides* ATCC 204091 using C5 and C6 sugars obtained from lignocellulosic hydrolysate y. *Journal of Environmental Biology* 42 (4): 938–944. https://doi.org/10.22438/JEB/42/4/MRN-1583.

Sitepu, I.R., Garay, L.A., Sestric, R. et al. (2014). Oleaginous yeasts for biodiesel: current and future trends in biology and production. *Biotechnology Advances* 32 (7): 1336–1360. https://doi.org/10.1016/J.BIOTECHADV.2014.08.003.

Slininger, P.J., Dien, B.S., Kurtzman, C.P. et al. (2016). Comparative lipid production by oleaginous yeasts in hydrolyzates of lignocellulosic biomass and process strategy for high titers. *Biotechnology and Bioengineering* 113 (8): 1676–1690. https://doi.org/10.1002/BIT.25928.

Sun, W., Yang, X., Wang, X. et al. (2017). Homologous gene targeting of a carotenoids biosynthetic gene in *Rhodosporidium toruloides* by Agrobacterium-mediated transformation. *Biotechnology Letters* 39 (7): 1001–1007. https://doi.org/10.1007/S10529-017-2324-3.

Szczepańska, P., Hapeta, P., and Lazar, Z. (2021). Advances in production of high-value lipids by oleaginous yeasts. *Critical Reviews in Biotechnology* 42 (1): 1–22. https://doi.org/10.1080/07388551.2021.1922353.

Takaç, S., Akay, B., and Özdamar, T.H. (1995). Bioconversion of trans-cinnamic acid to l-phenylalanine by l-phenylalanine ammonia-lyase of rhodotorula-glutinis – parameters and kinetics. *Enzyme and Microbial Technology* 17 (5): 445–452. https://doi.org/10.1016/0141-0229(94)00072-Y.

Tang, W., Wang, Y., Zhang, J. et al. (2019). Biosynthetic pathway of carotenoids in *Rhodotorula* and strategies for enhanced their production. *Journal of Microbiology and Biotechnology* 29 (4): 507–517. https://doi.org/10.4014/JMB.1801.01022.

Tchakouteu, S.S., Kalantzi, O., Gardeli, C. et al. (2015). Lipid production by yeasts growing on biodiesel-derived crude glycerol: strain selection and impact of substrate concentration on the fermentation efficiency. *Journal of Applied Microbiology* 118 (4): 911–927. https://doi.org/10.1111/JAM.12736.

Tiukova, I.A., Prigent, S., Nielsen, J. et al. (2019a). Genome-scale model of *Rhodotorula toruloides* metabolism. *Biotechnology and Bioengineering* 116 (12): 3396–3408. https://doi.org/10.1002/BIT.27162.

Tiukova, I.A., Brandenburg, J., Blomqvist, J. et al. (2019b). Proteome analysis of xylose metabolism in *Rhodotorula toruloides* during lipid production. *Biotechnology for Biofuels* 12 (1): 1–17. https://doi.org/10.1186/S13068-019-1478-8/FIGURES/4.

Ungureanu, C. and Ferdes, M. (2012). Evaluation of antioxidant and antimicrobial activities of torularhodin. *Advanced Science Letters* 18 (1): 50–53. https://doi.org/10.1166/ASL.2012.4403.

Uprety, B.K., Samavi, M., and Rakshit, S.K. (2018). Contribution of specific impurities in crude glycerol towards improved lipid production by *Rhodosporidium toruloides* ATCC 10788. *Bioresource Technology Reports* 3: 27–34. https://doi.org/10.1016/J.BITEB.2018.05.011.

Vieira, J.P.F., Ienczak, J.L., Rossell, C.E.V. et al. (2014). Microbial lipid production: screening with yeasts grown on Brazilian molasses. *Biotechnology Letters* 36 (12): 2433–2442. https://doi.org/10.1007/S10529-014-1624-0.

Wang, Z.P., Fu, W.J., Xu, H.M. et al. (2014). Direct conversion of inulin into cell lipid by an inulinase-producing yeast *Rhodosporidium toruloides* 2F5. *Bioresource Technology* 161: 131–136. https://doi.org/10.1016/J.BIORTECH.2014.03.038.

Wang, Q.M., Begerow, D., Groenewald, M. et al. (2015). Multigene phylogeny and taxonomic revision of yeasts and related fungi in the *Ustilaginomycotina*. *Studies in Mycology* 81: 55–83. https://doi.org/10.1016/J.SIMYCO.2015.10.004.

Wang, Y., Zhang, S., Zhu, Z. et al. (2018). Systems analysis of phosphate-limitation-induced lipid accumulation by the oleaginous yeast *Rhodosporidium toruloides*. *Biotechnology for Biofuels* 11 (1): 1–15. https://doi.org/10.1186/S13068-018-1134-8/FIGURES/4.

Wen, Z., Zhang, S., Odoh, C.K. et al. (2020). *Rhodosporidium toruloides* – a potential red yeast chassis for lipids and beyond. *FEMS Yeast Research* 20 (5): https://doi.org/10.1093/FEMSYR/FOAA038.

Wu, C.C., Ohashi, T., Kajiura, H. et al. (2021). Functional characterization and overexpression of Δ12-desaturase in the oleaginous yeast *Rhodotorula toruloides* for production of linoleic acid-rich lipids. *Journal of Bioscience and Bioengineering* 131 (6): 631–639. https://doi.org/10.1016/J.JBIOSC.2021.02.002.

Xu, J. and Liu, D. (2017). Exploitation of genus *Rhodosporidium* for microbial lipid production. *World Journal of Microbiology and Biotechnology* 33 (3): 1–13. https://doi.org/10.1007/S11274-017-2225-6.

Xu, J., Zhao, X., Wang, W. et al. (2012). Microbial conversion of biodiesel byproduct glycerol to triacylglycerols by oleaginous yeast *Rhodosporidium toruloides* and the individual effect of some impurities on lipid production. *Biochemical Engineering Journal* 65: 30–36. https://doi.org/10.1016/J.BEJ.2012.04.003.

Yamada, R., Kashihara, T., and Ogino, H. (2017). Improvement of lipid production by the oleaginous yeast *Rhodosporidium toruloides* through UV mutagenesis. *World Journal of Microbiology and Biotechnology* 33 (5): 1–9. https://doi.org/10.1007/S11274-017-2269-7.

Ye, Z., Sun, T., Hao, H. et al. (2021). Optimising nutrients in the culture medium of *Rhodosporidium toruloides* enhances lipids production. *AMB Express* 11 (1): 1–11. https://doi.org/10.1186/S13568-021-01313-6/TABLES/4.

Zhang, S., Ito, M., Skerker, J.M. et al. (2016). Metabolic engineering of the oleaginous yeast *Rhodosporidium toruloides* IFO0880 for lipid overproduction during high-density fermentation. *Applied Microbiology and Biotechnology* 100 (21): 9393–9405. https://doi.org/10.1007/S00253-016-7815-Y.

Zhao, X., Peng, F., Du, W. et al. (2012). Effects of some inhibitors on the growth and lipid accumulation of oleaginous yeast *Rhodosporidium toruloides* and preparation of biodiesel by enzymatic transesterification of the lipid. *Bioprocess and Biosystems Engineering* 35 (6): 993–1004. https://doi.org/10.1007/S00449-012-0684-6.

Zhao, Y., Song, B., Li, J. et al. (2021). *Rhodotorula toruloides*: an ideal microbial cell factory to produce oleochemicals, carotenoids, and other products. *World Journal of Microbiology & Biotechnology* 38 (1): https://doi.org/10.1007/S11274-021-03201-4.

Zheng, X., Hu, R., Chen, D. et al. (2021). Lipid and carotenoid production by the *Rhodosporidium toruloides* mutant in cane molasses. *Bioresource Technology* 326: https://doi.org/10.1016/J.BIORTECH.2021.124816.

Zhu, Z., Zhang, S., Liu, H. et al. (2012). A multi-omic map of the lipid-producing yeast *Rhodosporidium toruloides*. *Nature Communications* 3: https://doi.org/10.1038/NCOMMS2112.

Zoz, L., Carvalho, J.C., Soccol, V.T. et al. (2014). Torularhodin and Torulene: bioproduction, properties and prospective applications in food and cosmetics – a review. *Brazilian Archives of Biology and Technology* 58 (2): 278–288. https://doi.org/10.1590/S1516-8913201400152.

SECTION II

THE ROLE OF MICROBIAL FERMENTATION IN DIGESTION PROCESSES

CHAPTER 3

GUT MICROBIAL ECOLOGY

ERIN A. MCKENNEY
Department of Applied Ecology, North Carolina State University, 127 David Clark Labs, Campus Box 7617, 100 Eugene Brooks Avenue, Raleigh, NC 27695, USA

CONTENTS

3.1 INTRODUCTION

Microbes are ancient and highly adaptive. Bacteria have been around for perhaps 4 billion years, twice as long as multicellular organisms. This means that animals, including humans, evolved in a microbial world (McFall-Ngai et al. 2013). Microbes have always lived on and inside us, and they are inextricably linked to everything we do. In fact, across our evolutionary history, many animal species from carnivores to herbivores and omnivores, including humans, have separately learned to consume naturally fermented foods to maximize the energy harvested per feeding effort.

Of all the parts of our body, inside and out, microbes have made themselves most at home in our guts, specifically in our large intestine (or colon). Tens and even hundreds of trillions of microbes, belonging to at least 1000 different species, are living inside your gut. That complex community of microbes, known as the gut microbiome, is your own personal galaxy—it helps digest your food and educates and supports your immune system, every day of your life. The gut is literally a river of nutrients kept within a narrow temperature range and protected, for the most part, from the harsh extremes of the outside world. All that stability makes the gut a fantastic place to live and, because we, the host, digest different nutrients at different sites along the gut, different microbes can adapt to the conditions specific to each site.

Microbial Fermentations in Nature and as Designed Processes, First Edition. Edited by Christon J. Hurst.

3.2 VARIATION ALONG THE GUT

Microbes in the mouth can tolerate oxygen, and glue themselves to our teeth so that they're not swept away every time we eat, drink, cough, or sneeze (Martinez-Guryn et al. 2019). Oral microbes digest sugar and starch to produce acids that can cause dental caries. Microbes in the stomach must withstand peristaltic churning, digestive enzymes (e.g. the protease pepsin and gastric lipase), and varying levels of acidity, which carnivores and other animal species have evolved to minimize the risk of foodborne pathogens (Beasley et al. 2015; Martinez-Guryn et al. 2019; Stevens and Hume 2004). In the small intestine, microbes must survive bile and compete with host peptidases to break proteins down into amino acids (Martinez-Guryn et al. 2019). Host maltase, lactase, and sucrase in the small intestine convert disaccharides into glucose and fructose subunits, which microbes then ferment to produce short-chain fatty acids (SCFAs), alcohols, and gas (Kastl et al. 2020). By the time digesta reach the hindgut (i.e. the cecum and large intestine), most of the easily digested nutrients (i.e. sugars, starches, proteins, and lipids) have been absorbed into the bloodstream. This means that the bulk of gut content contains nutrients that the host cannot digest, including complex carbohydrate molecules collectively referred to as fiber. Therefore, the cecum and colon favor microbes that can digest recalcitrant starches and fibers in anaerobic conditions through a process called fermentation (Martinez-Guryn et al. 2019; Stevens and Hume 1998).

In addition to providing a high-fiber feast, the structure of the hindgut can also promote microbial diversity and digestive efficiency. The cecum and colon feature increased volume and surface area compared to the stomach and small intestine, and their structural complexity prolongs retention time and, thus, succession toward a more stable climax community with ample niche space for diverse taxa adapted to specific selective pressures (McKenney et al. 2015, 2018b; Stevens and Hume 1998). Indeed, species with a cecum host distinctly and significantly different gut microbial communities in the small versus large intestine (Greene et al. 2022), whereas species lacking a cecum host similar communities across the length of the gut (Gillman et al. 2020). In addition to the cecum, many species have also convergently evolved a narrow extension at the distal end of the cecum known as the appendix. Once thought to be a vestigial organ, the appendix is now appreciated as a "safe harbor" for beneficial bacteria that digest fiber and/or produce anti-inflammatory compounds (Greene and McKenney 2018). The appendix is particularly important for repopulating the gut after disruption associated with diarrhea/enteric illness (Randal Bollinger et al. 2007; Smith et al. 2009). Compared to the appendix, microbial communities in the cecum are statistically similar to those found in the large intestine and distal colon (Greene and McKenney 2018). In fact, feces are often collected as a non-invasive proxy for the gut microbiome. This method is often employed in studies of humans and wildlife, particularly endangered or elusive host species.

3.3 PHYLOGENY, GUT MORPHOLOGY, AND FEEDING STRATEGY SHAPE MICROBIAL COMMUNITIES

Across animals' evolutionary timeline, microbes moved into our guts and enabled us to digest different types of foods including high-fiber plant matter, essentially unlocking different feeding strategies. By opening the evolutionary doors to so many different lifestyles, gut microbes removed competition between different animal populations and ultimately drove speciation. There are approximately 4000 terrestrial herbivore species, compared to only 293 extant carnivore species across terrestrial and aquatic habitats (Ripple et al. 2015). This comparative preponderance of herbivorous species was likely facilitated, in part, by gut microbial fermentation (Greene et al. 2020). But animals must also evolve and adapt within the genetic constraints of each species.

Just as birds have wings, ungulates have hooves, and primates have large brains and opposable thumbs, we observe phylogenetic patterns of gut morphology across species that are consistent with our evolutionary relationships.

Different host species evolve gut morphologies that are specifically adapted to digest different diets depending on the relative proportions of protein, lipids, sugar, starch, and fiber in their preferred food items. These patterns in similar gut morphology provide an anatomical record of our ancestral feeding strategies (Stevens and Hume 1998). In turn, those distinct gastrointestinal landscapes and nutritional profiles select specific microbial communities associated with each unique animal species, population, and individual. Protein and fat are easy to digest, so carnivores tend to have short and simple guts with proportionally larger stomachs and long small intestines (Stevens and Hume 1998), and their simple guts favor relatively simple gut microbiomes with low alpha diversity compared to omnivores or herbivores (Reese and Dunn 2018). This simplicity also precludes differentiation among gut sites: microbial communities in the jejunum and colon are statistically indistinguishable in the American black bear (*Ursus americanus*) (Gillman et al. 2020). Yet, while carnivores generally harbor decreased alpha diversity, their simple guts and fast transit times also prevent host regulation of community membership, resulting in high inter-individual variation that is greatly influenced by the surrounding environment (Lafferty et al. 2022).

By comparison, the plant matter consumed by omnivores and herbivores is more varied and more structurally complex and thus selects for more complex gut morphologies and microbiomes (Greene et al. 2022; Reese and Dunn 2018; Stevens and Hume 1998). Plant roots, stems, leaves, and fruits are mostly made up of carbohydrates, which come in three general types: sugars (simple molecules of one or two units), starches (a series of glucose monomers linked together with alpha glycosidic bonds, in straight or branched chains), and fibers (the most complex molecules, still made up of glucose but linked with beta glycosidic bonds). No animal on the planet produces an enzyme that can digest the beta glycosidic bonds in cellulose, which is the main component of plant cell walls that gives plants their structure and makes celery tough to chew, or other fibers such as hemicellulose and lignin. We depend entirely on our gut microbes to digest those plant fibers and to release their energy in compounds called SCFAs, which we *can* use for energy. That means no herbivore should exist without microbes to digest their food for them!

Hosts with omnivorous feeding strategies tend to evolve guts of intermediate complexity, however, their diets are still relatively easy to digest, being relatively low in components such as plant fiber, and so, like carnivores, these hosts actually compete with gut microbial taxa (who also compete with each other) for nutrients. By contrast, herbivorous host species depend on their gut microbes to digest their high-fiber diets and therefore invest in a longer, more complex gut to facilitate microbial fermentation (Greene et al. 2022). This investment serves three purposes: (i) a complex gut provides more niche space to house a more functionally complex microbiome that can digest fiber; (ii) increased gut length and complexity provide more time for microbial fermentation and enhanced production of SCFAs; and (iii) complex guts provide more surface area for the host to absorb those SCFAs for energy. All SCFAs increase satiety, decreasing food intake, but different SCFAs are used in different parts of the body for different purposes. Acetate is transported to peripheral muscle tissues for adenosine triphosphate (ATP) production and increases adipogenesis. Propionate decreases liver lipogenesis as well as adipogenesis and also prevents colon cancer through antiproliferation and apoptosis. Butyrate is used on-site for ATP production by intestinal cells, and also prevents colon cancer through antiproliferation, apoptosis, and by triggering intestinal cell development. In addition, SCFAs provide communication signals that modulate host metabolism, physiology, moods, and behaviors (McKenney et al. 2018a).

Feeding strategy, gut passage rate, and gut morphology all promote or constrain microbial diversity, community composition, and variation among individuals (Table 3.1). Across species,

TABLE 3.1 Predicted responses to major drivers of gut microbial diversity.

		Predicted Response	
	Host Trait	Complexity (Alpha Diversity)	Variation (Beta Diversity)
Gut morphology	Short/simple gut, transit time	Low	High
	Long/complex gut, transit time	High	Low
Diet	Low fiber	Varies	High
	High fiber	High	Low
Feeding strategy	Generalist/omnivore	High	High
	Specialist	Varies	Low

feeding strategy dictates the nutritional composition of the diet as well as the variety of food items consumed. Gut morphology determines passage rate, which correlates to disturbance levels: longer guts have slower-moving content and more stable microbiomes, whereas short and simple guts with rapid passage rates are characterized by greater disruption and higher turnover. Gut morphology also determines oxygen levels, with oxygen decreasing along the length of the gut. Within a particular population or species, the greater variety of foods an animal consumes, whether compared to other individuals or in a particular season or habitat, the more niches are available to support diverse microbial taxa. Within more "strict" feeding strategies, in which animals consume a narrow range of food items or nutrients, nutritional composition dictates a more narrowly limited gut microbial diversity based on digestive function (Greene et al. 2020). Proteins and lipids are easily digested by many different microbes, so a carnivorous diet is not particularly selective and favors opportunistic taxa (Gillman et al. 2020). Sugars and starches are relatively simple carbohydrate structures containing only alpha glycosidic bonds, so frugivores (i.e. fruit-eating animals) tend to host similarly less specific and more variable microbiomes compared to folivores (i.e. leaf-eaters) or grazers (i.e. grass-eaters), whose high-fiber diets select for a relatively narrow range of fermentation specialists to digest their food (Greene et al. 2020; McKenney et al. 2015, 2018b).

Just as different species with different feeding strategies tend to harbor different gut microbiomes, humans who consume different diets and practice different lifestyles tend to harbor distinct gut microbiomes. For example, the Hadza of northern Tanzania and other hunter-gatherer communities harbor increased alpha diversity and distinct microbial membership compared to Western populations (Schnorr et al. 2014; Smits et al. 2017). Even short-term dietary changes incur rapid and predictable changes to the microbiome within just a few days. For example, individuals consuming animal-based diets favored the growth of bile-tolerant *Alistipes, Bilophila*, and *Bacteroides*, while plant-based diets favored carbohydrate-fermenting Firmicutes such as *Roseburia, Eubacterium rectale,* and *Ruminococcus bromii* (David et al. 2014). These shifts in community composition were accompanied by significant changes in microbial functional traits, indicating that different diets favor distinct microbial taxa based on their digestive abilities (David et al. 2014).

3.4 A PREDICTIVE ECOLOGICAL FRAMEWORK FOR THE GUT

In 1977, Grime published an ecological model describing three primary strategies in plants in response to environmental conditions and pressures (Grime 1977). Grime proposed that environments characterized by higher rates of disturbance would favor ruderal species associated with early stages of succession, with high nutritional requirements but low competitive ability, whereas

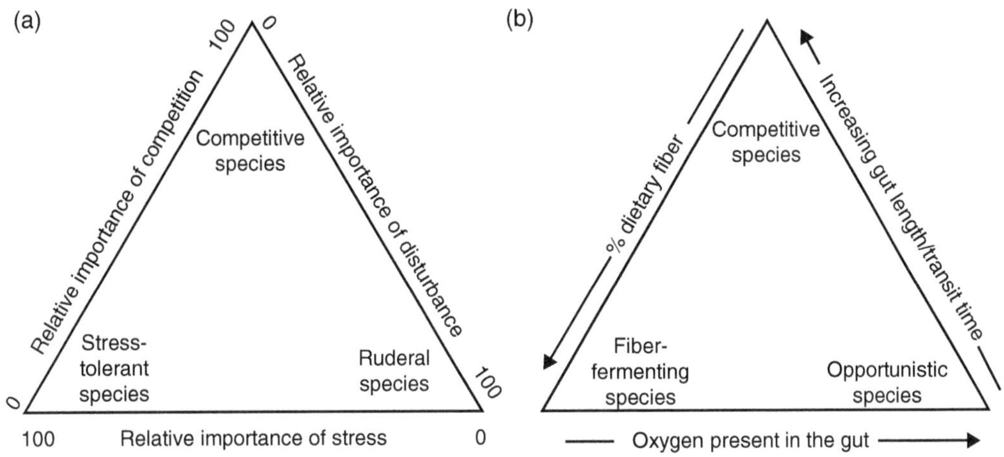

FIGURE 3.1 (a) Grime's ecological model describes three primary strategies in plants (ruderal, competitive, and stress-tolerant) in response to the environment. Grime originally defined "competitive species" as competing for resources and/or overlapping niche space. (b) Here, I adapt that framework to predict which types of microbial taxa are favored in species with different gut lengths and associated transit time; feeding strategies, specifically with regard to dietary fiber; and oxygen levels present in the gut. In my microbial model, "opportunistic species" include potential pathogens, and "competitive species" include microbial taxa that compete amongst themselves and with the host for easily digestible nutrients (e.g. sugar, starch, proteins, and lipids). Source: Figure adapted from Grime (1977).

more competitive environments would select for species able to compete for resources or overlapping niche space, and stable environments characterized by specific selective pressures would favor species able to tolerate those specific stressors (Figure 3.1a). Grime's ecological model provided a framework for predicting not only the general characteristics of species that would thrive under particular environmental conditions but also the changing presence of species types as communities progressed through successional stages. Here, I adapt that framework to predict which types of microbial taxa are favored in species with different gut lengths and associated transit time; feeding strategies, specifically with regards to dietary fiber; and oxygen levels present in the gut (Figure 3.1b).

Gut length and complexity correlate to transit time (i.e. the time between ingestion and defecation) and determine the amount of disturbance in the gut. Host species with short, simple guts have rapid transit times (i.e. high rates of disturbance) and little habitat variation across gut sites, and thus select for opportunistic taxa that thrive in oxygenated environments, on easily digested nutrients. Many opportunistic taxa are also associated with "dysbiosis" (e.g. imbalance compared to a reference state) and disease states (e.g. inflammatory bowel disease). However, pathogenicity (i.e. a microbe's ability to cause disease) is likely context-specific: many microbial taxa that are known pathogens in humans have also been isolated from healthy individuals across other species with short, simple guts and easily digestible diets, suggesting that these *potential* pathogens may traverse and exit shorter guts before they have time to induce a disease state.

Competitive taxa often compete among themselves and with their hosts for proteins, lipids, sugars, and starches. By contrast, omnivores' and herbivores' diets contain more fiber, so these species have evolved longer, more complex guts to digest their diets. The combination of gut length (which results in more stable, anoxic conditions) and fiber selects for a specific subset of stress-tolerant microbial taxa that can survive without oxygen and can digest fiber. In healthy animals under stable conditions, the mature microbial community can resist colonization by pathogenic taxa (McKenney et al. 2018a). However, disruption of the community, whether via antibiotic treatment, disease, or lifestyle changes, may upset the balance (i.e. induce dysbiosis) and favor potentially pathogenic opportunists. In some cases, the increased transit time associated with increased gut length may even *enable* or *increase* pathogenicity, by providing an infectious microbial taxon the time needed to interact with the host and possibly produce sufficient toxins to induce a disease state. This phenomenon has been documented extensively at the Duke Lemur Center, where folivorous *Propithecus coquereli* (with guts 16 times the length of their body) are uniquely susceptible to infection with *Cryptosporidium*, a protozoan that causes debilitating diarrhea in that species (McKenney et al. 2017). Cryptosporidiosis can be fatal in *P. coquereli*, yet other lemur species with shorter, simpler gut morphologies such as *Varecia variegata* and *Lemur catta* remain asymptomatic during annual outbreaks. Fortunately, *P. coquereli's* complex gut morphology and high-fiber diet provide enough selective pressure for their gut microbial communities to return to the healthy baseline after recovering from infection with *Cryptosporidium* (McKenney et al. 2017).

3.5 APPLICATIONS TO HOST MANAGEMENT

Beyond human studies and captive conditions, fecal microbiomes provide critical tools for monitoring and managing wildlife populations. Differences in gut microbial membership reflect local variations in feeding strategy among folivorous sifaka (Greene et al. 2019) and Alaskan brown bear (Trujillo et al. 2022) populations. Gut taxa and functions also shift predictably in Mexican howler monkeys (Amato et al. 2015) and Ethiopian geladas (Baniel et al. 2021), correlating with seasonal changes in food availability. Furthermore, gut microbiomes in mantled howler monkeys, which occur from Mexico to Ecuador, exhibit less variation across habitats or in response to disturbance, suggesting their communities are more plastic or flexible compared to the gut microbiomes of black howler monkeys, which occupy a more restricted habitat range (Amato et al. 2016).

In addition to facilitating specific host adaptations over evolutionary time scales, gut microbiomes may either help or hinder host flexibility in the face of climate change, urbanization, and other anthropogenic pressures. While the gut microbiome may compensate for seasonal (Mallott et al. 2022) or geographic (Trujillo et al. 2022) variations in diet, black howler monkeys living in suboptimal habitats have access to less diverse diets, resulting in decreased gut microbial diversity (Amato et al. 2013). Consumption of anthropogenic foods is also associated with decreased gut microbial diversity in American black bears (*U. americanus*) (Gillman et al. 2022). Do human-provisioned foods, then, decrease black bears' ability to adapt to a changing nutritional landscape?

In order to maximize the benefits of microbial fermentation, many species, particularly herbivores, practice coprophagy: they consume their feces, effectively doubling their digestion time to increase nutritional harvest (Stevens and Hume 1998). Humans may not find coprophagy palatable, but we do have a long history of fermenting foods (Amato et al. 2021b). In fact, many species consume fermented foods to maximize energy intake: carnivore species and human populations consume fermented meat (Campbell et al. 2022; Speth 2017), and omnivorous or herbivorous species may either cache food to ferment it (Smith and Reichman 1984) or selectively forage fermenting fruits (Sanchez 2004). Indeed, some anthropologists suggest that the consumption of fermented foods may have helped fuel the evolutionary development of humans' exceptionally large brains (Amato et al. 2021b)! Fermentation not only extends the shelf life of food (Levey 2004), a critical benefit before refrigeration was invented, but also increases the digestibility and nutritional density of the food, as well as providing a boost of probiotics (i.e. beneficial live microbial taxa) (Amato et al. 2021b).

Similarly, we humans can leverage what is known about gut microbial ecology to maximize health benefits via diet and other lifestyle changes. Increased fiber consumption favors the growth of fermentative microbes that produce anti-inflammatory compounds such as butyrate (David et al. 2014). Adopting a "modern paleolithic" diet can "rescue" gut microbial diversity in Westernized humans by reintroducing specialist fermentative taxa (Barone et al. 2019). Consumption of fermented foods provides an influx of specific live, probiotic microbial taxa; promotes acidic conditions; and provides nutrients that influence the characteristics of the gut microbiome and the overall health of both the microbiome and the host. A bonus is that fermented foods are often cheaper and more accessible than proprietary probiotics, helping to address social inequities (Amato et al. 2021a; Ishaq et al. 2019). For example, would you rather pay $75 for proprietary probiotic strains sold at the pharmacy, or $1 per pound for cabbage to make your own sauerkraut or kimchi?

3.6 CONCLUSION

Gut microbial ecology explores the dynamic relationships between the foods we eat, the microbes and host organs that digest those foods, and the consequences of digestion for the host's well-being. Comparative gut microbial studies across different host species and living conditions offer critical perspectives that can inform diverse practices from wildlife management and conservation to daily lifestyle choices that complement medical intervention or prevent it altogether. The direct linkages between the gut microbiome and host health and behaviors also make diet and fermentation accessible tools for public empowerment toward social equity.

REFERENCES

Amato, K.R., Yeoman, C.J., Kent, A. et al. (2013). Habitat degradation impacts black howler monkey (*Alouatta pigra*) gastrointestinal microbiomes. *The ISME Journal* 7 (7): 1344–1353. https://doi.org/10.1038/ismej.2013.16.

Amato, K.R., Leigh, S.R., Kent, A. et al. (2015). The gut microbiota appears to compensate for seasonal diet variation in the wild black howler monkey (*Alouatta pigra*). *Microbial Ecology* 69 (2): 434–443. https://doi.org/10.1007/s00248-014-0554-7.

Amato, K.R., Martinez-Mota, R., Righini, N. et al. (2016). Phylogenetic and ecological factors impact the gut microbiota of two Neotropical primate species. *Oecologia* 180 (3): 717–733. https://doi.org/10.1007/s00442-015-3507-z.

Amato, K.R., Arrieta, M.-C., Azad, M.B. et al. (2021a). The human gut microbiome and health inequities. *Proceedings of the National Academy of Sciences* 118 (25): e2017947118. https://doi.org/10.1073/pnas.2017947118.

Amato, K.R., Mallott, E.K., D'Almeida Maia, P. et al. (2021b). Predigestion as an evolutionary impetus for human use of fermented food. *Current Anthropology* 62 (S24): S207–S219. https://doi.org/10.1086/715238.

Baniel, A., Amato, K.R., Beehner, J.C. et al. (2021). Seasonal shifts in the gut microbiome indicate plastic responses to diet in wild geladas. *Microbiome* 9 (1): 26. https://doi.org/10.1186/s40168-020-00977-9.

Barone, M., Turroni, S., Rampelli, S. et al. (2019). Gut microbiome response to a modern Paleolithic diet in a Western lifestyle context. *PLoS One* 14 (8): e0220619. https://doi.org/10.1371/journal.pone.0220619.

Beasley, D.E., Koltz, A.M., Lambert, J.E. et al. (2015). The evolution of stomach acidity and its relevance to the human microbiome. *PLoS One* 10 (7). https://doi.org/10.1371/journal.pone.0134116.

Campbell, R., Hauptmann, A., Campbell, K. et al. (2022). Better understanding of food and human microbiomes through collaborative research on Inuit fermented foods. *Microbiome Research Reports* 1 (1): 5. https://doi.org/10.20517/mrr.2021.06.

David, L.A., Maurice, C.F., Carmody, R.N. et al. (2014). Diet rapidly and reproducibly alters the human gut microbiome. *Nature* 505 (7484): 559–563. https://doi.org/10.1038/nature12820.

Gillman, S.J., McKenney, E.A., and Lafferty, D.J.R. (2020). Wild black bears harbor simple gut microbial communities with little difference between the jejunum and colon. *Scientific Reports* 10 (1): 20779. https://doi.org/10.1038/s41598-020-77282-w.

Gillman, S.J., McKenney, E.A., and Lafferty, D.J.R. (2022). Human-provisioned foods reduce gut microbiome diversity in American black bears (*Ursus americanus*). *Journal of Mammalogy* 103 (2): 339–346. https://doi.org/10.1093/jmammal/gyab154.

Greene, L.K. and McKenney, E.A. (2018). The inside tract: the appendicular, cecal, and colonic microbiome of captive aye-ayes. *American Journal of Physical Anthropology* 166 (4): 960–967. https://doi.org/10.1002/ajpa.23481.

Greene, L.K., Clayton, J.B., Rothman, R.S. et al. (2019). Local habitat, not phylogenetic relatedness, predicts gut microbiota better within folivorous than frugivorous lemur lineages. *Biology Letters* 15 (6): 20190028. https://doi.org/10.1098/rsbl.2019.0028.

Greene, L.K., Williams, C.V., Junge, R.E. et al. (2020). A role for gut microbiota in host niche differentiation. *The ISME Journal* 14 (7): Article 7. https://doi.org/10.1038/s41396-020-0640-4.

Greene, L.K., McKenney, E.A., Gasper, W. et al. (2022). Gut site and gut morphology predict microbiome structure and function in ecologically diverse lemurs. *Microbial Ecology* https://doi.org/10.1007/s00248-022-02034-4.

Grime, J.P. (1977). Evidence for the existence of three primary strategies in plants and its relevance to ecological and evolutionary theory. *The American Naturalist* 111 (982): 1169–1194.

Ishaq, S.L., Rapp, M., Byerly, R. et al. (2019). Framing the discussion of microorganisms as a facet of social equity in human health. *PLoS Biology* 17 (11): e3000536. https://doi.org/10.1371/journal.pbio.3000536.

Kastl, A.J., Terry, N.A., Wu, G.D. et al. (2020). The structure and function of the human small intestinal microbiota: current understanding and future directions. *Cellular and Molecular Gastroenterology and Hepatology* 9 (1): 33–45. https://doi.org/10.1016/j.jcmgh.2019.07.006.

Lafferty, D.J.R., Gillman, S.J., Jeakle, L.K. et al. (2022). Mink (*Neovison vison*) fecal microbiomes are influenced by sex, temperature, and time postdefecation. *Journal of Mammalogy* 103 (2): 316–327. https://doi.org/10.1093/jmammal/gyab140.

Levey, D.J. (2004). The evolutionary ecology of ethanol production and Alcoholism. *Integrative and Comparative Biology* 44 (4): 284–289. https://doi.org/10.1093/icb/44.4.284.

Mallott, E.K., Skovmand, L.H., Garber, P.A., and Amato, K.R. (2022). The faecal metabolome of black howler monkeys (*Alouatta pigra*) varies in response to seasonal dietary changes. *Molecular Ecology* 31 (15): 4146–4161. https://doi.org/10.1111/mec.16559.

Martinez-Guryn, K., Leone, V., and Chang, E.B. (2019). Regional diversity of the gastrointestinal microbiome. *Cell Host & Microbe* 26 (3): 314–324. https://doi.org/10.1016/j.chom.2019.08.011.

McFall-Ngai, M., Hadfield, M.G., Bosch, T.C.G. et al. (2013). Animals in a bacterial world, a new imperative for the life sciences. *PNAS* 110 (9): 3229–3236. https://doi.org/10.1073/pnas.1218525110.

McKenney, E.A., Rodrigo, A., and Yoder, A.D. (2015). Patterns of gut bacterial colonization in three primate species. *PLoS One* 10 (5): e0124618. https://doi.org/10.1371/journal.pone.0124618.

McKenney, E.A., Greene, L.K., Drea, C.M. et al. (2017). Down for the count: *cryptosporidium* infection depletes the gut microbiome in Coquerel's sifakas. *Microbial Ecology in Health and Disease* 28 (1): 1335165. https://doi.org/10.1080/16512235.2017.1335165.

McKenney, E.A., Koelle, K., Dunn, R.R. et al. (2018a). The ecosystem services of animal microbiomes. *Molecular Ecology* 27 (8): 2164–2172. https://doi.org/10.1111/mec.14532.

McKenney, E.A., O'Connell, T.M., Rodrigo, A. et al. (2018b). Feeding strategy shapes gut metagenomic enrichment and functional specialization in captive lemurs. *Gut Microbes* 9 (3): 202–217. https://doi.org/10.1080/19490976.2017.1408762.

Randal Bollinger, R., Barbas, A.S., Bush, E.L. et al. (2007). Biofilms in the large bowel suggest an apparent function of the human vermiform appendix. *Journal of Theoretical Biology* 249 (4): 826–831. https://doi.org/10.1016/j.jtbi.2007.08.032.

Reese, A.T. and Dunn, R.R. (2018). Drivers of microbiome biodiversity: a review of general rules, feces, and ignorance. *MBio* 9 (4): eo1294-18. https://doi.org/10.1128/mBio.01294-18.

Ripple, W.J., Newsome, T.M., Wolf, C. et al. (2015). Collapse of the world's largest herbivores. *Science Advances* 1 (4): e1400103. https://doi.org/10.1126/sciadv.1400103.

Sanchez, F. (2004). The possible roles of ethanol in the relationship between plants and frugivores: first experiments with Egyptian fruit bats. *Integrative and Comparative Biology* 44 (4): 290–294. https://doi.org/10.1093/icb/44.4.290.

Schnorr, S.L., Candela, M., Rampelli, S. et al. (2014). Gut microbiome of the Hadza hunter-gatherers. *Nature Communications* 5 (1): Article 1. https://doi.org/10.1038/ncomms4654.

Smith, C.C. and Reichman, O.J. (1984). The evolution of food caching by birds and mammals. *Annual Review of Ecology and Systematics* 15: 329–351.

Smith, H.F., Fisher, R.E., Everett, M.L. et al. (2009). Comparative anatomy and phylogenetic distribution of the mammalian cecal appendix. *Journal of Evolutionary Biology* 22 (10): 1984–1999. https://doi.org/10.1111/j.1420-9101.2009.01809.x.

Smits, S.A., Leach, J., Sonnenburg, E.D. et al. (2017). Seasonal cycling in the gut microbiome of the Hadza hunter-gatherers of Tanzania. *Science* 357 (6353): 802–806. https://doi.org/10.1126/science.aan4834.

Speth, J.D. (2017). Putrid meat and fish in the eurasian middle and upper paleolithic: are we missing a key part of neanderthal and modern human diet? *PaleoAnthropology* 2017: 44–72.

Stevens, C.E. and Hume, I.D. (1998). Contributions of microbes in vertebrate gastrointestinal tract to production and conservation of nutrients. *Physiological Reviews* 78 (2): 393–427. https://doi.org/10.1152/physrev.1998.78.2.393.

Stevens, C.E. and Hume, I.D. (2004). *Comparative Physiology of the Vertebrate Digestive System*. Cambridge University Press.

Trujillo, S.M., McKenney, E.A., Hildergrand, G.V. et al. (2022). Intrinsic and extrinsic factors influence on an omnivore's gut microbiome. *PLoS One* 17 (4): e0266698.

CHAPTER 4

FERMENTATION IN THE RUMEN

EMILIO M. UNGERFELD, NATHALY CANCINO-PADILLA, AND NELSON VERA-AGUILERA
Centro Regional de Investigación Carillanca, Instituto de Investigaciones Agropecuarias INIA, Camino Cajón a Vilcún km 10, Vilcún, La Araucanía, Chile

CONTENTS

Microbial Fermentations in Nature and as Designed Processes, First Edition. Edited by Christon J. Hurst.

4.1 INTRODUCTION

Humans began a transition from a nomadic to a sedentary lifestyle about 12 000–14 000 years ago during the Late Glacial age (Vigne 2015). As they initiated plant and animal domestication, human groups gradually started to change from the hunter-gatherer lifestyle to farming (Wang et al. 2014). Most livestock species that were domesticated had been hunted for their meat and hides before (Zeder 2012). Livestock domestication commenced in the Middle East and Asia 8000–10 000 years ago, with sheep and goats being the first ruminants domesticated. Archeological evidence suggests that the first domestic ruminants were of small size (McTavish et al. 2013). Cattle domestication probably occurred first 8800–8300 years ago in the Fertile Crescent, and nearly 1500 years later in the Indus Valley. *Bos taurus* and *Bos indicus* derived from the same wild ancestor, *Bos primigenius*, or aurochs. This wild species inhabited Europe, Asia, and North Africa, and became extinct in 1627 (Ajmone-Marsan et al. 2010).

Ruminants are members of the order Artiodactyla, suborder Ruminantia, and belong to six families: Tragulidae, Moschidae, Bovidae, Giraffidae, Antilocapridae, and Cervidae, with most domestic ruminants except for reindeer belonging to the Bovidae family (Hackmann and Spain 2010). All ruminants are characterized by a forestomach divided into four chambers: the reticulum, rumen, omasum, and abomasum. The rumen is the largest compartment, and it harbors a diverse and complex microbial community that has a symbiotic relationship with the host animal (McDonald et al. 2010).

Ruminants can convert indigestible plant material into useful products. Domestic ruminants have provided humans with several products important to human societies, including milk, meat, hides, wool, and traction for soil plowing. Despite their role of providing food and other useful resources, ruminant production has implications for deforestation, erosion, water consumption, and climate change, as domestic ruminants consume 30% of the crops grown on earth and inhabit about one-third of the earth's surface (Thornton 2010), and emit the greenhouse gas methane formed in their rumens (Hristov et al. 2013; Saunois et al. 2016). Continuous human population growth and increasing demands for animal products put pressure to increase ruminant production. In turn, the expansion of ruminant production has environmental consequences on the use of natural resources and the release of greenhouse gases into the atmosphere. It is therefore equally urgent to ameliorate the negative environmental impacts of ruminant production and, in this regard, strategies to improve the efficiency and sustainability of ruminant production rest to a large extent on understanding rumen function and fermentation.

This chapter presents readers with an overview of the biochemistry of rumen fermentation and its different pathways and its physicochemical control. The main section on rumen fermentation biochemistry is preceded by shorter introductory sections explaining the main concepts of ruminants' unique digestive system and nutritional physiology, and the main types of microorganisms populating the rumen. A final section preceding the Conclusions presents concepts on the manipulation of rumen fermentation and future research directions.

4.2 NUTRITIONAL PHYSIOLOGY OF RUMINANTS

Ruminants are unique because of their digestive and metabolic adaptations, which allow them to convert fibrous feedstuffs, byproducts, and non-conventional feed resources, rich in lignocellulose and other complex polysaccharides indigestible to humans, into meat, milk, and fiber for human consumption (Robbins et al. 1995; Forbes 2007; Wanapat et al. 2013). Utilization of structural polysaccharides is due to the digestive action of the rumen microbiota, which includes fibrolytic bacteria and fungi possessing the enzymes necessary to degrade recalcitrant fibrous feedstuffs. Simple sugars released from the digestion of complex polysaccharides are then fermented to volatile fatty acids, which are absorbed through the rumen and omasum walls, and constitute the main source of energy and glucose for ruminants (McDonald et al. 2010; Naas and Pope 2020).

Ruminants also can utilize non-protein forms of nitrogen, such as urea. Again, this nutritional capacity is allowed by rumen microorganisms, which can incorporate ammonium resulting from the degradation of nitrogenous compounds in the rumen into carbon chains to synthesize their own amino acids and protein. Microbial protein so formed constitutes the main source of amino acids for ruminants when microbial cells outflow the rumen and are digested in the abomasum, the true stomach, and the duodenum and the amino acids released are absorbed in the small intestine (Wallace et al. 1997; McDonald et al. 2010; Schwab and Broderick 2017).

Ruminants absorb excess ammonium not incorporated into microbial amino acids through the rumen wall. Absorbed ammonium is converted to urea in the liver and kidneys, and urea is eliminated in urine. However, part of urea is recycled to the rumen through the rumen wall and in saliva, where it is hydrolyzed and releases ammonium, which can be again used in the synthesis of microbial amino acids. The proportion of urea recycled to the rumen increases as the dietary content of nitrogen decreases, constituting an important mechanism of nitrogen salvage in nitrogen-poor diets (McDonald et al. 2010, Schwab and Broderick 2017).

The rumen microbiota also can synthesize water-soluble vitamins, provided that their constituent macro and microminerals are in sufficient concentration in the ruminant diet. Thus, under most circumstances, the diets of ruminants do not need to be supplemented with water-soluble vitamins, although highly producing ruminants may respond to their addition to the diet (McDonald et al. 2010; Weiss 2017).

The uniqueness of ruminants´ digestive physiology has in turn resulted in the evolution of metabolic adaptations. For example, ruminants absorb little glucose from their gastrointestinal tract, and have evolved to obtain their glucose largely from propionate formed in the rumen, and to a lesser extent from glucogenic amino acids, lactate, and glycerol released from triacylglycerol hydrolysis (McDonald et al. 2010).

4.3 THE RUMEN MICROBIAL COMMUNITY

Rumen digestion and fermentation are vital not only for ruminant production but for the sheer life of the ruminant animal. More than 5000 anaerobic or facultative anaerobic microbial species inhabit and interact in the rumen, including bacteria, protozoa, fungi, methanogenic archaea, and phages (Huws et al. 2018; Cholewińska et al. 2020). The rumen microbiota composition and function are affected by ruminant species and breed, age, and, especially, diet (Henderson et al. 2015; Malmuthuge and Guan 2017).

The rumen can be conceptually divided for the study of its microbial ecology into different compartments inhabited by different, but partially overlapping, microbial communities (Cheng and McAllister 1997). Up to 25% of rumen microbial biomass is associated with the rumen liquid or fluid phase, between 70% and 80% is attached to rumen solids or particles (digesta), and up to 5% of the microbial biomass is associated with the rumen epithelium (epimural microbiota), with most of the digestion and fermentation taking place associated with digesta particles (Bickhart and Weimer 2018; Matthews et al. 2019).

The relationship between the ruminant and its rumen microbiota is a symbiotic one, in which the ruminant host animal provides an optimal environment for the growth of rumen microorganisms, specifically, temperature (maintained at 39 °C), moisture (delivered through saliva, and the consumption of water and feed), substrates (through feed consumption), removal of digestion and fermentation products through absorption and passage, and a relatively constant pH (6–7, but may vary according to the type of feed consumed). In return, rumen microorganisms provide the nutrients required for ruminant health, such as volatile fatty acid products of microbial fermentation, amino acids forming microbial proteins, and water-soluble vitamins, all resulting from microbial digestion and fermentation of feed consumed by the ruminant (Cammack et al. 2018; Matthews et al. 2019; Xu et al. 2021).

Classical rumen microbiology studies relied on the isolation and cultivation of rumen microorganisms to characterize their physiology. Those studies provided highly valuable information about the physiology and niches of culturable microbial species in the rumen ecosystem. The advent of molecular techniques revealed that most of the rumen microbial community has not been cultivated. The rumen microbial composition can now be routinely characterized through the analysis of partially conserved sections of the 16S or 18S rRNA genes (Matthews et al. 2019). More recently, the Hungate 1000 project (https://genome.jgi.doe.gov/portal/TheHunmicrobiome/TheHunmicrobiome.info.html) (Kelly 2012) has led to the isolation and sequencing of numerous rumen bacteria and archaea (Nordberg et al. 2014; Seshadri et al. 2018; Gruninger et al. 2019; Newbold and Ramos-Morales 2020). Metagenomics has allowed a thorough understanding of the digestive and metabolic potential capacity of the rumen microbiome. Further advances include binning individual rumen microbial genomes obtained from metagenomes, as well as metatranscriptomics, metaproteomics, and metabolomics. These techniques have opened a new understanding of rumen microorganisms and their ecological niches (Snelling and Wallace 2017; Huws et al. 2018; Gruninger et al. 2019; Newbold and Ramos-Morales 2020).

4.3.1 Rumen Bacteria

Bacteria are present in all ruminants and represent the largest and most diverse group of rumen microorganisms, with an abundance of 10^{10}–10^{11} cells/ml of rumen content (Gruninger et al. 2019; Cholewińska et al. 2020). There are thousands of different species of rumen bacteria, which conduct the digestion and fermentation of carbohydrates (i.e. starch, cellulose, hemicelluloses, and sugars), proteins, lipids, and other nutrients. Microbial digestion occurs through the concerted action of multiple enzymes, including amylases, cellulases, proteases, and lipases, among others (Bickhart and Weimer 2018; Huws et al. 2018). The most dominant ruminal bacterial phyla are Firmicutes, Bacteroidetes, and Proteobacteria. Bacteria can be functionally classified for their study according to the substrate they utilize, taking into account that most of them can use multiple substrates (Cammack et al. 2018; Cholewińska et al. 2020).

Fiber-degrading Bacteria Forages are sources of complex structural polysaccharides such as cellulose, hemicellulose, and pectin, with cellulose and hemicellulose often bound to lignin. Cellulolytic, hemicellulolytic, and pectinolytic bacteria secrete enzymes that degrade the cellulose, hemicellulose, and pectin, respectively, into oligo- and disaccharides (Matthews et al. 2019; Xu et al. 2021). The rate and extent of fiber digestion are affected by factors such as type, maturity, and preprocessing of forage, bacterial species and numbers, and rumen pH. Classically, *Fibrobacter succinogenes*, *Ruminococcus albus*, and *Ruminococcus flavefaciens*, which are present in relatively low numbers, are described as the predominant, highly specialized rumen bacteria responsible for cellulose digestion (Weimer 1996; Matthews et al. 2019). Recent metagenomics work has revealed the abundant *Clostridium* genus can also possess many isoforms of cellulolytic glycosyl hydrolases (Rubino et al. 2017). Hemicellulolytic bacteria include species such as *Prevotella* spp., *Butyrivibrio fibrisolvens*, *R. albus*, *R. flavefaciens*, [*Clostridium*] *polysaccharoliticum*,

Eubacterium uniforme, and *Eubacterium xylanophilum*, while *Lachnospira multipara, Succinivibrio dextrinosolvens*, [*C.*] *polysaccharoliticum*, *Clostridium longisporum*, *B. fibrisolvens, Treponema saccharophilum,* and *Treponema bryantii* are pectinolytic bacteria (Hespell et al. 1987; Hespell 1988; Stewart et al. 1997; Matthews et al. 2019; Xu et al. 2021). Importantly, there is cross-feeding among rumen microorganisms of simple sugars released from the digestion of complex carbohydrates. Cross-feeding results in multiple microbial species that are unable to hydrolyze complex structural carbohydrates being benefited by the hydrolytic action of a few highly specialized fibrolytic species (Wolin et al. 1997). The same as with other rumen processes, fiber digestion is conducted by the concerted action of a consortium of mixed microbiota rather than by a single, or a few, microbial species (Moraïs and Mizrahi 2019).

Starch- and Sugar-degrading Bacteria Adding cereal grains to ruminant diets increases the amount of starch reaching the rumen. Both starch and sugars are rapidly fermented in the rumen, although the rate and extent of starch digestion in the rumen are affected by cereal species and composition, and preprocessing and particle size (Nocek and Tamminga 1991; Kotarski et al. 1992). Some main rumen amylolytic bacteria are the *Streptococcus bovis* Group (Dekker and Lau 2016) (from here on, *S. bovis*), *Ruminobacter amylophilus*, *Selenomonas ruminantium*, and *Succinimonas amylolytica* (Stewart et al. 1997, Matthews et al. 2019, Xu et al. 2021). Compared to fiber-degrading bacteria, amylolytic, and saccharolytic bacteria are fast-growing species, and resistant to pH between 5.5 and 6 (Russell 2002).

Protein- and Amino Acid-degrading Bacteria Bacteria with proteolytic and peptidolytic activity produce enzymes that hydrolyze proteins and degrade released peptides. Proteolytic activity is widespread in rumen bacteria, except for the main specialized cellulolytic bacteria *F. succinogenes*, *R. albus*, and *R. flavefaciens*. The most abundant proteolytic species is probably *Prevotella ruminicola*. Depending on the diet, other abundant proteolytic species include *Ruminobacter amylophillus* and *B. fibrisolvens*, as well as bacteria from the genera *Eubacterium*, *Lachnospira*, *Selenomonas*, and *Succinivibrio* (Wallace et al. 1997; Cammack et al. 2018; Cholewińska et al. 2020).

Amino acids and oligopeptides released from protein breakdown can be incorporated by rumen microorganisms for the synthesis of their own proteins, or amino acids can be deaminated, and the resulting carbon skeleton further catabolized to volatile fatty acids. Deamination of amino acids is conducted both by an abundant bacterial community of proteolytic organisms having low deaminating specific activity, and by a few specialized bacteria with high deaminating activity: *Peptostreptococcus anaerobius*, *Acetoanaerobium* (*Clostridium*) *sticklandii*, and [*C.*] *aminophilum* (Wallace 1996; Wallace et al. 1997).

Bacteria-metabolizing Acylglycerols Hydrolysis of acylglycerols occurs rapidly in the rumen, mostly conducted by *Anaerovibrio lipolytica* (Lourenco et al. 2010). Lipolysis releases glycerol and free fatty acids. Glycerol is catabolized through glycolysis and unsaturated fatty acids are subjected to a biohydrogenation process consisting of successive isomerizations and reductions. Most steps of biohydrogenation of 18-carbon fatty acids are conducted by *B. fibrisolvens*, with the final step of reduction to stearic acid being conducted by *Butyrivibrio proteoclasticus* (Jenkins et al. 2008) (see Section 5.4).

Bacteria Fermenting Metabolic Intermediates In the rumen, some microbial species excrete fermentation products that are taken up and metabolized by other microorganisms (for more details, see Section 4.3). Feeding high-concentrate diets induces a fermentation shift in

S. bovis from acetate, formate, and ethanol, toward the production of lactate. Increased lactate concentration in the rumen environment in turn increases populations of lactate utilizers, mainly *Megasphaera elsdenii* and *S. ruminantium* ssp. *lactilytica* (Nagaraja and Titgemeyer 2007). Lactate metabolism to volatile fatty acids is important to control lactate accumulation, which can result in rumen acidosis (see Section 5.3).

Bacterial species such as *Fibrobacter succinogenesis*, *R. flavefaciens*, *P. ruminicola*, *S. dextrinosolvens*, *S. amylolytica*, *Wolinella succinogenes*, and *R. amylophilus* produce succinate as a major fermentation product. Succinate does not accumulate and is found at very low concentrations in the rumen, as it is taken up and metabolized to propionate by bacterial species such as *Veillonella parvula* and *Succiniclasticum ruminis* (Stewart et al. 1997; Wolin et al. 1997).

4.3.2 Rumen Methanogens

Methanogens are prokaryotes belonging to the Domain Archaea which generate adenosine triphosphate (ATP) exclusively through methane production from dihydrogen, formate, methanol, methylamines, and acetate (Thauer et al. 2008). Most rumen methanogens use dihydrogen and formate as reductants to convert carbon dioxide to methane, in a process known as hydrogenotrophic methanogenesis. In methylotrophic methanogenesis, methanol, and methylamines are reduced to methane; some methylotrophic methanogens can carry out fermentative methanogenesis from methanol in the absence of dihydrogen. Methanogens inhabit different rumen microenvironments, including fluid- and plant particle-associated methanogens, methanogens associated with protozoal surfaces or as endosymbionts, and methanogens attached to the rumen epithelium (Janssen and Kirs 2008).

Methanogens have been classified into seven orders: Methanobacteriales, Methanococcales, Methanomicrobiales, Methanopyrales, Methanosarcinales, Methanocellales, and Methanomassiliicoccales. Rumen methanogens belong to the orders Methanobacteriales, Methanomicrobiales, Methanosarcinales, and Methanomassillicoccales. Some genera commonly found are *Methanobrevibacter, Methanosphaera, Methanobacterium, Methanomicrobium*, and *Methanosarcina*. Hydrogenotrophic *Methanobrevibacter* is generally dominant, with the most abundant species being *Mbr. ruminantium* and *Mbr. gottschalkii*. Uncultured methanogens classified as "rumen cluster C" of the order Methanomassiliicoccales have been reported to be highly abundant in some animals (St-Pierre and Wright 2012; Huws et al. 2018; Gruninger et al. 2019).

4.3.3 Rumen Protozoa

Rumen protozoa are eukaryotic microorganisms belonging to the phylum Ciliophora, i.e. ciliated protozoa. They are classified into two main groups depending on the location of cilia: holotrichids (order Vestibuliferida), with ciliate covering the entire cell surface, and entodiniomorphids (order Entodiniomorphida), with cilia circumscribed to the anterior or posterior end of the cell. The family Isotrichidae is the most abundant within the order Vestibuliferida, with the three most abundant species being *Isotricha intestinalis*, *Isotricha prostoma*, and *Dasytricha ruminantium*. Most rumen entodiniomorphids belong to the family Ophryscolecidae. Abundant entodiniomorphid genera are *Entodinium, Epidinium, Enoploplastron, Ophryoscolex, Anoplodinium, Diplodinium, Eremoplastron, Diploplastron, Eudiplodinium, Metadinium, Ostracodinium*, and *Polyplastron* (Huws et al. 2020).

Rumen protozoa vary vastly in size and have been proposed to account for up to 50% of microbial biomass, although recent measurements suggest a lower contribution (Gruninger et al. 2019). They are not essential for animal life or rumen function, as they can be eliminated through chemicals or by freezing rumen contents ("defaunation"). Rumen protozoa affect ruminant production both positively and negatively. Protozoa actively phagocytize bacteria, and digest bacterial and dietary proteins, being responsible for most "intra-rumen nitrogen recycling," an inefficient process for nitrogen and energy utilization (Hartinger et al. 2018). Protozoa harbor ecto- and endosymbiotic methanogens which remove dihydrogen generated in fermentation, thus contributing to methane emissions. On the other hand, some rumen protozoa have fibrolytic activity, consume traces of oxygen, and slow down starch digestion by engulfing starch grains (Newbold et al. 2015; Gruninger et al. 2019).

4.3.4 Rumen Fungi

Rumen anaerobic fungi account for a relatively small amount of microbial biomass (7–9%) but play an important role in fiber digestion through initial colonization of plant particles, breaking down recalcitrant plant tissues both through enzymatic activity and physical disruption through hyphae growth. Eleven rumen fungi have so far been isolated and more clades have been identified through non-cultivation techniques. Rumen fungi are classified in the phylum Chytridiomycota, class Neocallimastigomycetes, and are divided into 18 different genera, including *Neocallimastix*, *Piromyces*, *Oontomyces*, *Buwchfawromyces*, *Pecoramyces*, *Liebetanzomyces*, *Feramyces*, *Agriosomyces*, *Aklioshbomyces*, *Capellomyces*, *Ghazallomyces*, *Joblinomyces*, *Khoyollomyces*, *Tahromyces*, *Anaeromyces*, *Orpinomyces*, *Caecomyces,* and *Cyllamyces* (Gruninger et al. 2019; Hanafy et al. 2020; Hess et al. 2020a; Hess et al. 2020b).

4.3.5 Rumen Viruses and Plasmids

The rumen mobilome is composed of viruses and plasmids (Gilbert and Ouwerkerk 2020). They are the least studied ruminal microorganisms (Lobo and Faciola 2021). Viruses are mobile genetic elements composed of DNA or RNA in a double or a single strand and an outer protein shell. Viruses of the rumen infect and replicate in bacteria (bacteriophages), archaea (archaeaphages), fungi (mycophages), and protozoa (protozoan viruses). Bacteriophages are the most well-characterized rumen viruses and are found in numbers of 10^7–10^9 particles/g of rumen digesta. They are, for the most part, associated with the dominant bacterial phyla Firmicutes, Bacteroidetes, and Proteobacteria (Matthews et al. 2019). Viral genomes can be integrated into chromosomes or remain extrachromosomically as plasmids. Viruses known as transducing phages may integrate new DNA in their genome. Plasmids are extrachromosomal elements that may be able to self-replicate and integrate into chromosomes. DNA sequencing has allowed identifying plasmids in bacteria from the genera *Ruminococcus*, *Selenomonas*, *Butyrivibrio*, and *Prevotella* (Gilbert and Ouwerkerk 2020).

There is little speculative understanding of the ecological role of the rumen mobilome. It is thought that it can affect the rates of microbial cell lysis, horizontal gene transfer, control of microbial populations, nutrient cycling, and fiber breakdown, and interact with the host animal in relation to immunomodulation (Gilbert and Ouwerkerk 2020; Gilbert et al. 2020).

4.3.6 Rumen Mycoplasmas

Mycoplasmas are small bacterial cells with very low guanine and cytosine content and without a cell wall. In the rumen, they are present at densities between 10^7 and 10^8 cells/g. In other environments, mycoplasmas may infect bacteria, fungi, and protozoa, but in the rumen have been found to be free-living. Their roles in rumen microbial ecology are largely unknown (Joblin and Naylor 2002).

4.4 BIOCHEMISTRY OF RUMEN FERMENTATION

4.4.1 The Rumen Environment

The rumen environment is homeostatic with regard to pH, osmolarity, reducing potential, and temperature. With most diets, the rumen pH is maintained between 5.5 and 6.5 by phosphate and bicarbonate buffers secreted in saliva during ingestive mastication and regurgitation and re-chewing of digesta, or "rumination" (McDonald et al. 2010). Osmolarity is maintained close to plasma osmolarity through ion exchange. The rumen-reducing potential is markedly negative, varying between −142 and −273 mV (Huang et al. 2018). Rumen temperature is maintained close to the host animal's temperature, i.e. 39 °C (McDonald et al. 2010), with drops for short periods after drinking cold water (Cunningham et al. 1964). As a curiosity, the Artic muskoxen consuming water as snow in the Alaskan winter maintained an average rumen temperature of 39 °C in a −40 °C environment, although there were monthly rumen temperature minimums of up to 26 °C (Crater et al. 2007).

The rumen is a strictly anaerobic environment with a highly negative reducing potential, although traces of oxygen enter the rumen in ingested feed and water, and oxygen diffuses into the rumen from the capillaries lining the rumen epithelium. Oxygen concentration rises and quickly drops after a meal (Czerkawski 1969). As the rumen microbiota includes fastidious anaerobes, oxygen-tolerant microorganisms, and facultative anaerobes (Loesche 1969; Czerkawski 1972), survival of oxygen-sensitive organisms such as methanogens is facilitated by rapid consumption of oxygen by facultative anaerobes, including bacteria and protozoa (Scott et al. 1983; Ellis et al. 1989).

Oxygen entering the rumen is rapidly consumed by acting as an electron acceptor of metabolic hydrogen generated in microbial catabolism of organic compounds (Czerkawski 1972). It has even been speculated that oxygen entering the rumen might even oxidize small amounts of the methane produced (Czerkawski 1969), although methane oxidation in the rumen seems to be minimal (Kajikawa et al. 2003). Rumen metabolism is still largely anaerobic. Carbon in organic compounds is partially oxidized to carbon dioxide and reduced to methane, short-chain fatty acids, and microbial biomass. Fermentation can be defined as an incomplete oxidation in which the ultimate electron acceptors are carbon compounds formed in the process itself as oxygen flowing into the rumen is insufficient to allow for complete oxidation of organic compounds to carbon dioxide and water. The same as respiration, fermentation is fundamentally a series of redox reactions defining a flow of electrons (Ungerfeld 2020).

Fermentation, and anaerobic respiration from nitrate and sulfate, which to some extent also occur in the rumen (Ungerfeld 2020), are exergonic processes that allow rumen microbial cells to generate ATP for cell maintenance and growth from part of the Gibbs energy released (Russell and Wallace 1997; Russell 2002; Ungerfeld and Hackmann 2020). Microorganisms use ATP so generated to reproduce at a rate matching their rumen outflow rates, resulting in their populations not washing out of the rumen while still producing microbial biomass, which provides the ruminant host with amino acids, fatty acids, vitamins, and other compounds digested and absorbed in more distal parts of the digestive system. Although the generation of ATP in fermentation and anaerobic respiration dwarfs in comparison with aerobic respiration (Voet and Voet 1995), the symbiosis between the ruminant host and its microbiota is allowed by the production of incompletely oxidized fermentation products and microbial biomass nutritionally useful for the ruminant host animal (Weimer 2015). Once absorbed, fermentation products can be completely oxidized to carbon dioxide and water in oxygen-rich animal tissues (Czerkawski 1969).

Ruminants ingest feed composed of different chemical fractions. Different microbial species differ in their capacities to utilize different substrates and produce different fermentation products

(Weimer 2015). In this section, we will examine the main pathways of metabolism of carbohydrates, amino acids, and acylglycerides, and their implications for microbial and host energy metabolism and nutrition.

4.4.2 Catabolism of Carbohydrates

Digestion of cellulose and starch yields cellobiose and maltose, respectively, with D-glucose released as the final product (McDonald et al. 2010). The foundations of the metabolism of hexoses and pentoses in the rumen were established in early *in vitro* culture experiments with labeled pure chemicals as substrates. Those experiments indicated that pure glucose as well as glucose hydrolyzed from pure cellulose were largely metabolized through glycolysis to pyruvate (Figure 4.1) (Baldwin et al. 1963; Wallnöfer et al. 1966; Hamar and Borchers 1967). A recent study, however, showed that 27% of the rumen bacterial genomes examined did not encode all the glycolytic pathway enzymes, while 11% of bacteria examined also encoded for enzymes of other pathways of glucose catabolism (Hackmann et al. 2017). Furthermore, some rumen bacteria were found to lack some glycolytic enzymes (Scardovi 1965; Kelly et al. 2010). On the other hand, a metaproteome analysis of rumen samples obtained with different sampling methods from cows, deer, and lambs eating different diets, found a prevalence of glycolytic enzymes within prokaryotic proteins (Snelling and Wallace 2017). A quantitative revisit of the fate and pathways of hexoses and pentoses metabolism in the rumen under different *in vivo* conditions and diets appears thus as necessary.

Glycolysis results in the net generation of 2 mol of ATP per mole of glucose metabolized (Voet and Voet 1995). Many microorganisms can transport disaccharides into the cells, which are then cleaved to monosaccharides using hydrolyses or phosphorylases; cleavage with phosphorylases

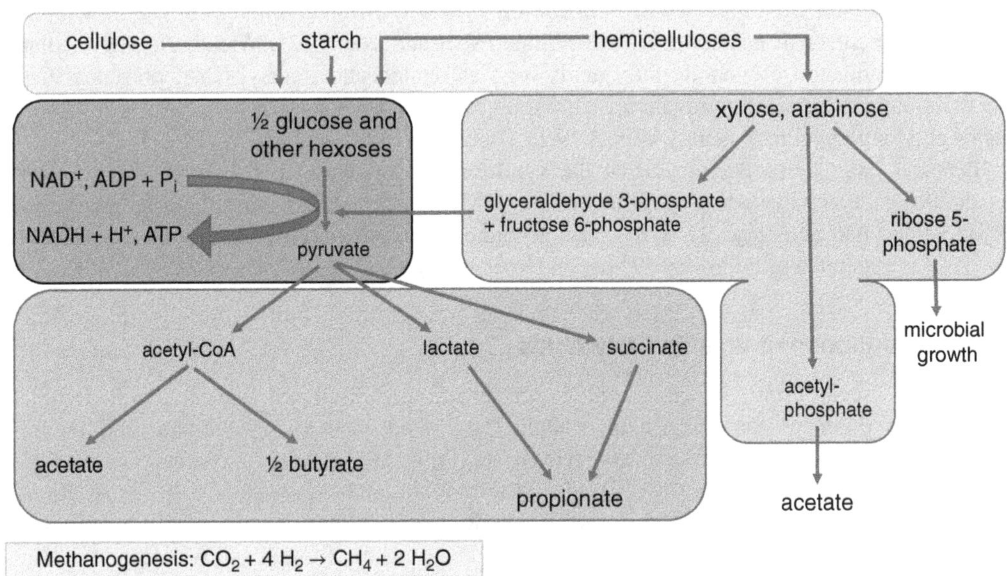

FIGURE 4.1 Overall scheme of rumen digestion of polysaccharides (inside of sky blue squircle), fermentation of hexoses and pentoses to pyruvate via glycolysis (inside of red squircle), pyruvate metabolism to volatile fatty acids (inside of purple squircle), pentose cycle and phosphoketolase cleavage (inside of green squircle), and hydrogenotrophic methanogenesis (inside of orange squircle). (*See colour plate section for colour representation of this figure.*)

yields one molecule of glucose 6-phosphate; obtaining a phosphorylated product then saves 0.5 mol of ATP per mole of glucose released. Glucose can also be phosphorylated when entering the cell if transported by the glucose phosphotransferase system, which indirectly transfers a phosphate group from phosphoenolpyruvate through a proteins cascade. This mode of transport thus saves 1 mol of ATP per mole of glucose transported, as glucose enters glycolysis as glucose 6-phosphate (Russell and Wallace 1997). An exception would occur when phosphoenolpyruvate carboxylation to oxaloacetate in the propionate randomizing pathway is catalyzed by phosphoenolpyruvate carboxykinase, as that reaction generates nucleotides triphosphate (NTP) by substrate-level phosphorylation (Schöcke and Weimer 1997); phosphoenolpyruvate carboxykinase is expressed by rumen bacteria (Asanuma and Hino 2001; Asanuma et al. 2010) and fungi (Reymond et al. 1992).

Glycolysis generates 2 mol of reduced nicotinamide adenine dinucleotide (NADH) in the oxidation of 3-phosphoglyceraldehyde to 1,3-diphosphoglycerate per mole of glucose metabolized (Voet and Voet 1995). In anaerobic environments, metabolic hydrogen in NADH cannot be disposed of with oxygen as the ultimate electron acceptor, and it is transferred in intracellular redox reactions, and between cells in electron carriers, for disposal in the formation of methane, reduced volatile fatty acids, or microbial biomass (Ungerfeld 2020) (see Sections 4.3 and 4.5).

Hemicelluloses are also abundant plant polymers containing hexoses and pentoses (Scheller and Ulvskov 2010). Pentoses are predominantly metabolized through the pentose cycle to yield fructose 6-phosphate and glyceraldehyde 3-phosphate, which enter glycolysis (Figure 4.1). They are also metabolized via phosphoketolase cleavage to yield glyceraldehyde 3-phosphate, which can also enter glycolysis, and acetylphosphate, which can yield acetate and generate 1 mol of ATP (Pazur et al. 1958; Wallnöfer et al. 1966; Russell and Wallace 1997). Pentose metabolism also provides anabolic precursors. Rumen bacteria *P. ruminicola* strains B14 and S23 and *S. ruminantium* strain D were shown to have xylose isomerase, xylulokinase, xylulose 5-phosphate epimerase, and ribose 5-phosphate isomerase activities to convert xylose to D-ribose (Matte et al. 1992), which is, in turn, a precursor of nucleotides and histidine (Voet and Voet 1995). Metabolic intermediates from the production of volatile fatty acids such as oxaloacetate, acetyl-CoA, propionyl-CoA, methylmalonyl-CoA, and pyruvate are also sources of carbon for microbial synthesis of amino acids and long-chain fatty acids (Allison 1969; Umbarger 1978; Voet and Voet 1995).

Pentoses enter the pentose cycle at the xylulose 5-phosphate step (Matte et al. 1992; Park et al. 2005), thereby bypassing the production of NADPH. If fructose 6-phosphate and glyceraldehyde 3-phosphate enter glycolysis, they will generate 1 and 2 mol of NADH per mole, respectively. No ATP is generated or hydrolyzed (Voet and Voet 1995).

4.4.3 Production of Volatile Fatty Acids

Pyruvate is a central intermediate in rumen fermentation (Russell and Wallace 1997). Pyruvate metabolism produces the three main volatile fatty acids acetate, propionate, and butyrate (Figure 4.1). The formation of acetate from pyruvate (Figure 4.2) entails an oxidative decarboxylation of pyruvate, yielding acetyl-CoA, reduced ferredoxin, and carbon dioxide if the reaction is catalyzed by pyruvate ferredoxin oxidoreductases (Ragsdale 2003). Alternatively, pyruvate can be oxidatively decarboxylated to acetyl-CoA, formate, and a proton (Russell and Wallace 1997; Himo and Eriksson 1998). Acetyl-CoA then exchanges HS-CoA with a phosphate group in a reaction catalyzed by phosphate acetyl-transferase, resulting in acetyl-phosphate. In the last step of the pathway, the phosphate group is transferred to adenosine diphosphate (ADP) and 1 mol of ATP is generated (Russell 2002). Some *Selenomonas* spp. and *Mitsuokella jalaludinii* M 9 transfer HS-CoA from acetyl-CoA to succinate formed in the propionate randomizing pathway, to form

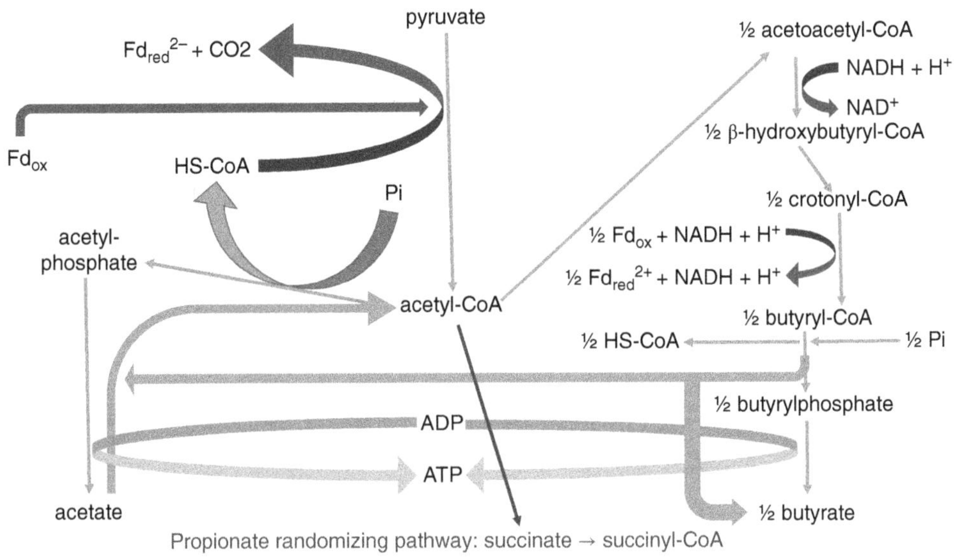

FIGURE 4.2 Metabolism of pyruvate to acetate and butyrate. Main routes of carbon compounds are marked with sky blue arrows. Alternative pathway involving butyryl-CoA: acetate CoA transferase is shown in gray arrows. Link with propionate randomizing pathway involving succinyl-CoA: acetate CoA transferase is shown in purple. Generation of ATP by substrate-level phosphorylation is shown with light green arrows. Reactions of cofactor reduction are shown in red arrows. Abbreviations: ADP, adenosine diphosphate; ATP, adenosine triphosphate; CO_2, carbon dioxide; Fe_{ox}, oxidized ferredoxin; Fe_{red}^{2-}, reduced ferredoxin; H^+, proton; HS-CoA, coenzyme A; NAD^+, oxidized nicotinamide adenine dinucleotide; NADH, reduced nicotinamide adenine dinucleotide; P_i, phosphate group. (*See colour plate section for colour representation of this figure.*)

succinyl-CoA plus acetate, in a reaction catalyzed by succinyl-CoA: acetate CoA transferase (Hackmann et al. 2017).

In contrast to acetate production, which results in the further release of reducing equivalents (whether in reduced ferredoxin or in formate), the other two major pathways of pyruvate catabolism involve the incorporation of metabolic hydrogen. Butyrate shares acetyl-CoA with acetate as its last metabolic precursor (Figure 4.2). Acetoacetyl-CoA is formed from the condensation of two acetyl-CoA units. Acetoacetyl-CoA is subsequently reduced to β-hydroxybutyryl-CoA, dehydrated to crotonyl-CoA, and reduced by NADH to butyryl-CoA in bifurcation with oxidized ferredoxin (Hackmann et al. 2017). Butyryl-CoA can exchange HS-CoA for a phosphate group generating butyryl-phosphate, which is then hydrolyzed to butyrate generating ATP, like the last steps of the acetate formation pathway. Alternatively, butyryl-CoA can transfer HS-CoA to acetate, forming butyrate and acetyl-CoA (Miller and Jenesel 1979; Russell 2002; Hackmann et al. 2017).

Propionate is produced by two different pathways that differ in the position in propionate of labeled carbon 2 in pyruvate. The so-called propionate randomizing (or succinate) pathway involves the formation of symmetrical intermediates fumarate and succinate, and a subsequent decarboxylation in the final steps of the pathway (Figure 4.3). This results in labeled carbon 2 in pyruvate being randomized into carbons 2 and 3 in propionate. The propionate randomizing pathway begins with a carboxylation of pyruvate (or phosphoenolpyruvate) to oxaloacetate, which is subsequently reduced to malate. Malate is then dehydrated to fumarate, which is reduced to succinate with NADH or reduced ferredoxin acting as an electron donor (Hackmann et al. 2017); externally taken dihydrogen (H_2) and formate can be taken up to reduce intracellular cofactors which donate electrons in fumarate reduction (Asanuma et al. 1999; Asanuma and Hino 2000),

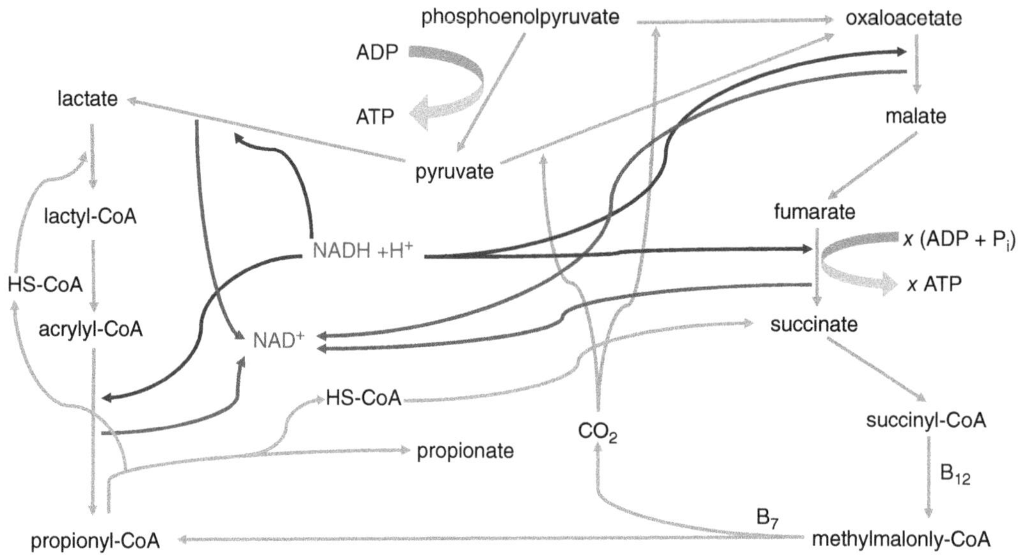

FIGURE 4.3 Propionate formation via randomizing (succinate) and non-randomizing (acrylate) pathways. Main routes of carbon compounds are marked with sky blue arrows. Reactions involving transfer of coenzyme A are shown with gray arrows. Generation of ATP by substrate level and electron transport-linked phosphorylation is shown with light green arrows. Reactions of cofactor re-oxidation are shown in red arrows. Abbreviations: ADP, adenosine diphosphate; ATP, adenosine triphosphate; B_7, biotin; B_{12}, cobalamin; CO_2, carbon dioxide; H^+, proton; HS-CoA, coenzyme A; NAD^+, oxidized nicotinamide adenine dinucleotide; NADH, reduced nicotinamide adenine dinucleotide; P_i, phosphate group. (*See colour plate section for colour representation of this figure.*)

making fumarate reduction an important reaction in intercellular metabolic hydrogen transactions. Fumarate reductase is associated with the bacterial cell membrane (Asanuma and Hino 2000) with the reduction of fumarate to succinate being associated with ATP generation through the generation of an electrochemical gradient (Ungerfeld and Hackmann 2020).

In propionibacteria, succinate formed from fumarate reduction is activated to succinyl-CoA, which undergoes a chain rearrangement to methylmalonyl-CoA co-catalyzed by vitamin B_{12} as cofactor (Figure 4.3). Methylmalonyl-CoA is decarboxylated to propionyl-CoA, and finally, propionate is formed (Russell and Wallace 1997). Importantly, succinyl-CoA decarboxylation can be catalyzed by a transcaboxylase with biotin as a cofactor to make pyruvate carboxylation to oxaloacetate energetically neutral. Similarly, propionyl-CoA hydrolysis transfers HS-CoA to succinate allowing the production of succinyl-CoA without external Gibbs energy input (Russell 2002).

Alternatively, succinate can be released to the extracellular medium, as in *F. succinogenes*, *W. succinogenes*, *S. dextrinosolvens,* or *S. amylolytica*. Bacteria such as *V. parvula* or *S. ruminis* take up succinate and convert it to propionate via the above-described pathway (van Gylswyk 1995; Stewart et al. 1997). The conversion of succinate to propionate is weakly exergonic, and the non-rumen bacterium *Propionigenium modestum* was shown to generate ATP coupled to sodium translocation by a membrane-bound methylmalonyl-CoA decarboxylase (Hilpert et al. 1984).

In the pyruvate non-randomizing (or acrylate) pathway, the labeling in carbon 2 in pyruvate appears in propionate's carbon 2, as no symmetrical intermediates are formed. In the first step of the propionate non-randomizing pathway, NADH is the electron donor in the reduction of pyruvate to lactate catalyzed by lactate dehydrogenases (Figure 4.3). Lactate is then activated to lactyl-CoA and then dehydrated to acrylyl-CoA. In the final steps, acrylyl-CoA is reduced to propionyl-CoA and, finally, propionate is formed (Russell and Wallace 1997; Russell 2002). The reduction of acrylyl-CoA occurs in the cytoplasm and is not associated with chemiosmotic ATP generation (Thauer et al. 1977; Seeliger et al. 2002). Lactate can also be released from cells and taken up by other cells and metabolized to acetate or butyrate (by lactate oxidation to pyruvate as a first step), or to succinate or propionate. Disposal of metabolic hydrogen into lactate formation is rapid, and lactate can accumulate with rapidly fermentable diets if the rates of lactate metabolism to volatile fatty acids do not match the rate of lactate production. Importantly, lactate pKa is about one unit lower than volatile fatty acids pKa, thus lactate accumulation results in a drop in rumen pH, a condition called lactic acidosis. Lactic acidosis can occur in ruminants abruptly offered high amounts of rapidly fermentable concentrates without an adaptation period to allow growth of the lactate-utilizing microbiota (see Section 5.3).

Longer chain volatile fatty acids valerate and caproate are formed through condensation of acetyl-CoA and propionyl-CoA. The formation of these more reduced fermentation products may allow for greater disposal of metabolic hydrogen, although Gibbs energy released in the CoA ester hydrolysis is utilized in the formation of carbon-to-carbon bonds rather than in ATP generation (Russell 2002).

4.4.4 Catabolism of Amino Acids and Acylglycerides

Two communities of rumen bacteria ferment amino acids in the rumen. One community of abundant bacteria, in which the genus *Prevotella* predominates, ferments amino acids at a relatively slow rate per cell. There is a second community of less abundant bacteria composed of specialized amino acids fermenters such as *P. anaerobius*, *A.* (*Clostridium*) *sticklandii*, and [*C.*] *aminophilum*. These bacteria ferment amino acids at a much faster rate, but their much lower abundance results in each population being responsible for about half of the total fermentation of amino acids (Russell and Wallace 1997, Russell 2002). Fermentation of amino acids yields little energy and begins with a deamination-releasing reducing equivalent. Branched-chain amino acids valine, leucine, and isoleucine are decarboxylated to isobutyrate, isovalerate, and 2-methylbutyrate, respectively (Allison 1978; Voet and Voet 1995; Russell 2002), which are taken up and utilized by fibrolytic bacteria for branched-chain amino acids biosynthesis (Allison et al. 1962; Allison et al. 1966; Allison 1969; Allison and Peel 1971).

In the rumen, acyl-glycerides are first hydrolyzed to glycerol and long-chain fatty acids. Glycerol constitutes a relatively small proportion of the acyl-glyceride and enters glycolysis and is mainly converted to propionate (Krueger et al. 2010), generating ATP while being redox neutral (Avila et al. 2011). Unsaturated long-chain fatty acids undergo a complex series of isomerizations and subsequent reductions until their total or partial saturation when outflowing the rumen (Jenkins et al. 2008; Or-Rashid et al. 2009). Quantitatively, unsaturated fatty acids are minor acceptors of metabolic hydrogen (Czerkawski 1972).

4.4.5 Interspecies Hydrogen Transfer and Methanogenesis

For definitions of general concepts of electron flow used in this and other sections, readers unfamiliar with electron transfer in living systems are advised to consult Ungerfeld (2020), as well as classical biochemistry textbooks. Metabolic hydrogen released in glycolysis and in pyruvate oxidative decarboxylation, and to a lesser extent from processes such as fermentation of amino acids, reduces oxidized nicotinamide adenine dinucleotide (NAD^+) to NADH, as well as oxidized to reduced ferredoxins. Much of the reoxidation of NADH and ferredoxins takes place through the transfer of electrons to protons to form dihydrogen, and as formate to a lesser extent. Most metabolic hydrogen is then transferred from the fermentative microbiota to methanogens as dihydrogen and formate (Figure 4.4) (Ungerfeld 2020). Reoxidation of NADH also takes place intracellularly in both propionate pathways, in pyruvate reduction to lactate, and in the reductions of oxaloacetate, fumarate, and acrylyl-CoA (Figure 4.3). In the butyrate formation pathway, metabolic hydrogen is incorporated in the reductions of acetoacetyl-CoA and crotonyl-CoA (Figure 4.2).

Most rumen bacteria possess hydrogenases. Under low concentrations of dihydrogen, NADH reoxidation to dihydrogen is made thermodynamically feasible through confurcation with reduced ferredoxin catalyzed by bifurcating [FeFe]-A3 hydrogenases (Greening et al. 2019). If dihydrogen concentration is elevated, this is detected by sensory hydrogenases, and ferredoxin is instead reoxidized through dihydrogen-evolving, [FeFe]-A1 hydrogenases with reduced ferredoxin as sole electron donor, while NADH is reoxidized preferentially intracellularly serving as an electron donor in the reduction of propionate and butyrate intermediates, and in ethanol formation (Figure 4.4). Changes in metabolic hydrogen disposal pathways are regulated by changes in gene expression of bifurcating-([FeFe]-A3) and fermentative, ferredoxin only-dependent ([FeFe]-A1) hydrogenases, as shown in defined cultures (Zheng et al. 2014; Greening et al. 2019) and *in vivo* (Pitta et al. 2022).

Large amounts of dihydrogen, and to a lesser extent of formate, are generated in rumen fermentation (Figure 4.4), but typically, the concentration of dihydrogen is rather low (Janssen 2010). Most dihydrogen and formate are taken up by hydrogenotrophic methanogens and incorporated into methane

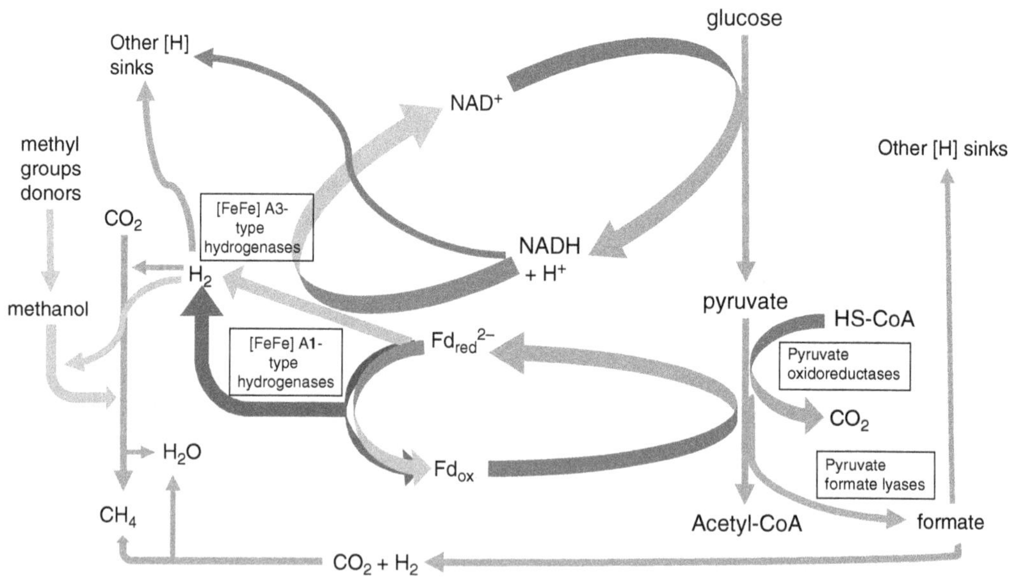

FIGURE 4.4 Main pathways of metabolic hydrogen transactions in rumen fermentation. Reduction of NAD^+ to NADH in glycolysis and of ferredoxin in pyruvate oxidative decarboxylation is depicted in sky blue arrows. Formate formation in pyruvate oxidative decarboxylation and dissociation into carbon dioxide (CO_2) and dihydrogen (H_2) previously to its incorporation into methane (CH_4) and other electron sinks is shown in gray arrows. Under low dihydrogen concentration NADH is re-oxidized by [FeFe] A3 type hydrogenases in confurcation with reduced ferredoxin (Fd_{red}^{2-}; yellow arrows). Under relatively high dihydrogen concentration NADH donates electrons in intracellular reactions and reduced ferredoxin is re-oxidized to its oxidized species (Fd_{ox}) by fermentative, dihydrogen-evolving [FeFe] A1 type hydrogenases (purple arrows). Hydrogenotrophic methanogenesis incorporates dihydrogen and carbon dioxide (sky blue arrows). Incorporation of metabolic hydrogen into methyl-reducing and fermentative methylotrophic methanogenesis is depicted in light green arrows. (*See colour plate section for colour representation of this figure.*)

formation, which is the main electron sink in rumen fermentation. Most rumen methane is formed from dihydrogen (Hungate 1967), and to a lesser extent from formate (Hungate et al. 1970), as electron donors, although recent results suggest that methylotrophic methanogenesis may be more important than previously thought (Söllinger et al. 2018; Pitta et al. 2021). The importance of interspecies dihydrogen transfer from the fermentative microbiota of bacteria, protozoa, and fungi to methanogens was revealed some decades ago by experiments in which an H_2-producing organism shifted its fermentation profile from dihydrogen and other reduced fermentation intermediates such as formate and ethanol, toward acetate, when co-cultivated with a methanogen. The production of a more oxidized product such as acetate allowed for greater ATP generation and faster growth and digestion (Latham and Wolin 1977; Bauchop and Mountfort 1981; Mountfort et al. 1982; Pavlostathis et al. 1990). Importantly, the co-cultivation of H_2-producing microorganisms with hydrogenotrophic succinate-producing bacteria could also stimulate acetate and succinate production (Iannotti et al. 1973).

In hydrogenotrophic methanogenesis, carbon dioxide is progressively reduced to methane in a series of four consecutive reductions with the one-carbon groups being transferred between methanofuran, tetrahydromethanopterin, and coenzyme M. The same process takes place when formate serves as the electron donor, as formate is oxidized by methanogens or by bacteria to CO_2 before being reduced by dihydrogen (Thauer et al. 2008). Methylotrophic methanogenesis uses methanol or methylamines as electron donors, which are released from pectins, hemicelluloses, choline, and betaine (Neill et al. 1978; Mitchell et al. 1979; Pol and Demeyer 1988). True methyl-fermenting methanogenesis conducted by methanogens of the order Methanosarcinales yields methane plus CO_2 and water from methanol, while methyl-reducing methanogenesis conducted by the more abundant in the rumen *Methanosphaera stadtmanae* utilizes dihydrogen to further reduce already reduced (compared to CO_2) methyl groups; this results in the dihydrogen threshold of *M. stadtmanae* being lower than that of the predominantly hydrogenotrophic rumen methanogen *M. ruminantium* (Feldewert et al. 2020). Rumen-predominant *Methanobrevibacter* spp. do not possess cytochromes and are thought to generate about 0.5 mol of ATP per mole of methane formed, while *Methanosarcina* spp., which possess cytochromes, are thought to generate about 1.5 mol of ATP per mole of methane. Consequently, *Methanosarcina* have higher growth yields but longer doubling times (Thauer et al. 2008).

Carbon dioxide is plentiful in the rumen, being released both in fermentation and from bicarbonate buffer in saliva (Russell and Chow 1993; Kohn and Dunlap 1998), and therefore it is not considered a limiting reactant for hydrogenotrophic methanogenesis.

4.4.6 The Role of Dihydrogen in the Control of Rumen Fermentation

The pathways of formation of acetate, and to a lesser extent of butyrate, from glucose released from digestion of cellulose and starch, determine the release of reducing equivalents, while methane and propionate incorporate them:

Acetate: $C_6H_{12}O_6 + 2\ H_2O \rightarrow 2\ CH_3COO^- + 2\ H^+ + 2\ CO_2 + 4\ [2H]$

Propionate: $C_6H_{12}O_6 + 2\ [2H] \rightarrow 2\ CH_3CH_2COO^- + 2\ H^+ + 2\ H_2O$

Butyrate: $C_6H_{12}O_6 \rightarrow CH_3CH_2CH_2COO^- + H^+ + 2\ CO_2 + 2\ [2H]$

Methane: $CO_2 + 4\ H_2 \rightarrow CH_4 + 2\ H_2O$ (only hydrogenotrophic methane is depicted)

The stoichiometry of the formation of volatile fatty acids and methane could in theory allow for very broad — virtually infinite — ranges of volatile fatty acids ratios balanced in reducing equivalents release and incorporation e.g. acetate to propionate in a 1 : 2 M ratio without any butyrate, or propionate to butyrate in a 1 : 1 M ratio with no acetate. Yet, the observed volatile fatty acids profile in the rumen lies within a much narrower range, with acetate to propionate molar ratios situating, on average, around 4:1 with high forage diets and 1.5:1 with high concentrate diets (Ungerfeld 2013). Applying classical thermodynamics, Ungerfeld and Kohn (2006) proposed that this relatively tight control of the volatile fatty acids profile could be explained through equilibrium between interconverting pools of volatile fatty acids.

Even if the observed range in volatile fatty acids profile is considerably narrower than what could be possible stoichiometrically, because changes in the volatile fatty acids profile in rumen fermentation are of much importance to the nutrition of the host animal, it is important to understand the mechanisms through which shifts in volatile fatty acids occur. A theory explaining the shift from acetate to propionate occurring when feeding concentrates was proposed by Janssen (2010). Because of having greater rates of rumen outflow, concentrates increase growth rates of methanogens, resulting in a higher concentration of dihydrogen according to the Monod law of microbial growth. An increase in dihydrogen concentration in turn thermodynamically inhibits processes releasing dihydrogen, such as acetate production, and stimulates dihydrogen-incorporating processes such as propionate production (Ungerfeld and Kohn 2006; Janssen 2010). Similarly, feeding concentrates also results in a drop in rumen pH because of rapid fermentation and the production of volatile fatty acids. Low rumen pH in turn inhibits methanogens, their maximal growth rate decreases, and dihydrogen accumulates, again inhibiting acetate production and stimulating propionate production. Importantly, the accumulation of dihydrogen signifies an increase in dihydrogen concentration but decreases dihydrogen production. Through decreasing dihydrogen production, total methane production is decreased as the supply of the main reductant for methanogenesis is decreased (Janssen 2010).

It is difficult to verify these mechanisms *in vivo*, for it would be virtually impossible to adjust rumen outflow rates or rumen pH without altering other important variables such as digestion and fermentation rates. Continuous culture experiments appropriate to test this model have been scarce, with some being confirmatory (Isaacson et al. 1975; Stanier and Davies 1981), and one obtaining opposite results (Wenner et al. 2017).

4.4.7 Microbial Growth and Protein Synthesis

Although microbial growth is an anabolic process and is thus strictly speaking not part of fermentation, this treatise would be incomplete without linking fermentation to microbial anabolism and growth, and discussing its significance in ruminant nutrition. Part of the negative Gibbs energy change associated with fermentation is used by rumen microbes to generate ATP, whose hydrolysis in turn fuels otherwise thermodynamically unfavorable anabolic reactions (Russell and Wallace 1997; Ungerfeld and Hackmann 2020). Part of Gibbs energy change resulting from fermentation pathways is used for cell maintenance, which includes maintaining ion gradients across membranes, turnover of macromolecules, and for some microbial species, motility. Energy used in cell maintenance is eventually dissipated as heat instead of used for the growth and net production of macromolecules used by the ruminant animal (Ungerfeld and Hackmann 2020).

Microbial biomass produced in the rumen is rich in highly digestible protein with a valuable profile of amino acids for the ruminant host animal. Most of the requirements of ruminants for amino acids are covered by a microbial protein produced in the rumen (Wallace et al. 1997; Broderick et al. 2010). Furthermore, ruminants can utilize non-protein nitrogen because ammonium released from the fermentation of those compounds can be incorporated into carbon chains by rumen microbes to synthesize amino acids which will be available for the ruminant host animal once microbial protein is digested in the lower gastrointestinal tract (Figure 4.5). Incorporation of ammonium into microbial amino acids not only allows replacing some true protein with urea in

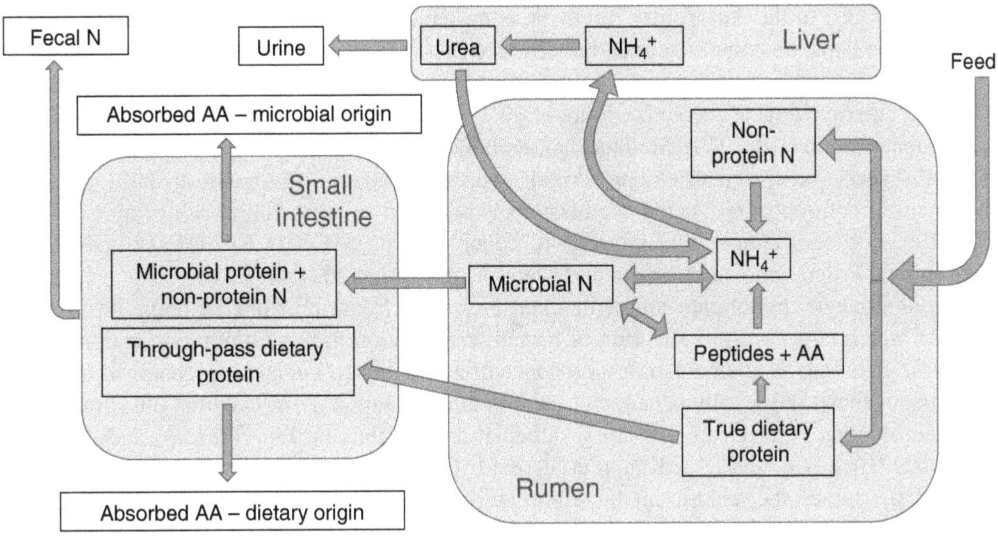

FIGURE 4.5 Simplified scheme of rumen, small intestine, and liver nitrogen metabolism. Endogenous protein is not depicted for simplicity. Liver synthesized urea is recycled to the rumen through the rumen wall and in saliva (not depicted for simplicity). Abbreviations: AA, amino acids; N, nitrogen; NH_4^+, ammonium. (*See colour plate section for colour representation of this figure.*)

ruminants diets but also salvaging nitrogen through urea recycling to the rumen through the rumen wall and saliva, which is particularly important in nitrogen-poor diets (McDonald et al. 2010). This makes microbial protein produced in the rumen not only the quantitatively most important source of amino acids for ruminants with most diets but also the most economical one. Because with most diets most amino acids absorbed in the small intestine have been de novo synthesized in the rumen, the profile of absorbed amino acids in ruminants differs from the dietary profile of amino acids.

4.5 MANIPULATION OF RUMEN FERMENTATION

The purpose of this section is not to provide a comprehensive presentation of all investigative and commercial interventions and products intended for manipulating rumen fermentation, but to delineate the main concepts, principles, and objectives being sought, when manipulating rumen fermentation. For further information, readers are directed to excellent reviews of rumen microbial ecology and manipulation of rumen microbiota (Nagaraja et al. 1997; Weimer 2015; Huws et al. 2018; Matthews et al. 2018; Newbold and Ramos-Morales 2020).

Microbial digestion and fermentation in the rumen provide ruminants with the nutritional flexibility to utilize fibrous feedstuffs and non-protein nitrogen and synthesize water-soluble vitamins. On the other hand, the efficiency of utilization of energy and nitrogen is low in ruminants compared to non-ruminants. Also, the rumen's microbial processes influence the meat and milk composition. Manipulation of rumen fermentation seeks objectives related to improving ruminant productivity, sustainability, and health, as well as the composition and quality of ruminant products.

4.5.1 Methane

Methane is the main sink of metabolic hydrogen in rumen fermentation, but also a loss of energy accounting for between 2% and 12% of gross energy ingested by ruminants (Johnson and Johnson 1995). Methane formed in the rumen is released to the atmosphere through eructation and exhalation. Once in the atmosphere, methane is a potent greenhouse gas with a 28-fold greater heat-trapping capacity compared to carbon dioxide in a 100-years horizon, with enteric methane, i.e. methane emitted by ruminants, accounting for about 30% of total anthropogenic methane. It is estimated enteric methane accounts for about 6% of total anthropogenic emissions of greenhouse gases (Beauchemin et al. 2020). Methane has a relatively short perturbation time in the atmosphere (12.4 ± 1.4 years) compared to carbon dioxide (various centuries) (Szopa et al. 2021), thus the abatement of anthropogenic methane emissions is regarded as important to achieving short-term mitigation in global temperature increase to contain it to 1.5 °C (IPCC 2021; United Nations Environment Programme and Climate and Clean Air Coalition 2021).

Various strategies to mitigate enteric methane emissions from ruminants are being investigated, many of which target the modification of rumen fermentation through diet modification or feed additives, including supplementation with concentrates or lipids, ionophores, chemical inhibitors of methanogenesis, algae, phytochemicals such as tannins, saponins, or essential oils, organic and inorganic alternative electron acceptors, defaunation (elimination of protozoa), and direct-fed microbials (Hristov et al. 2013; Knapp et al. 2014; Goopy 2019; Beauchemin et al. 2020; Terry et al. 2020). Some other enteric methane mitigation strategies not involving diet formulation or feed additives indirectly affect the rumen microbiota and fermentation. Selection for low methane-producing animals, for example, results in animals with smaller rumens, with some concerns about decreased digestibility (Løvendahl et al. 2018). Immunization against methanogens delivers antibodies to the rumen (Subharat et al. 2015), which can then bind rumen methanogens (Wedlock et al. 2013).

Importantly, the inhibition of rumen methanogenesis cannot be thought of as an isolated intervention, and even if chemical inhibitors of methanogens without direct effects on other microorganisms are used, there are also profound effects on fermentation pathways other than methanogenesis. Inhibition of rumen methanogenesis typically results in the accumulation of dihydrogen (Czerkawski 1986; Ungerfeld 2015), which has been explained based on the Monod function of microbial growth because of decreased maximal growth rate of methanogens. Accumulation of dihydrogen would in turn inhibit dihydrogen-releasing processes such as acetate production and stimulate hydrogen-incorporating processes such as propionate production, resulting in the acetate to propionate shift typically observed when inhibiting methane formation (Janssen 2010).

4.5.2 Efficiency of Nitrogen Utilization

Ruminants have the unique ability to use sources of non-protein nitrogen not available to humans or other animals. However, ruminants' efficiency in utilizing ingested nitrogen is generally lower than that of non-ruminants (Wallace et al. 1997). Excess nitrogen voided to the environment in manure, and especially in urine as urea, is converted in soils to ammonia and nitrate. Nitrate contaminates groundwater courses and can also be converted to nitrous oxide and emitted into the atmosphere (Dijkstra et al. 2013; López-Aizpún et al. 2020). Nitrous oxide is another greenhouse gas 273 times more potent than carbon dioxide at trapping heat over a 100-years horizon (Smith et al. 2021). Less frequently, nitrous oxide can also be emitted from the rumen itself by ruminants consuming nitrate (Petersen et al. 2015; de Raphélis-Soissan et al. 2018). Nitrogen voided to the environment by ruminants depends on the dietary content of nitrogen and its rumen degradability, and the animal's nitrogen utilization efficiency. In turn, the efficiency of nitrogen utilization by ruminants depends both on ruminal events and the post-absorptive metabolism of amino acids and proteins.

The most straightforward way to decrease nitrogen losses to the environment is to avoid overfeeding nitrogen (Dijkstra et al. 2013; Zhao 2019), as the single most important variable influencing nitrogen utilization efficiency is the sheer dietary content of nitrogen (Huhtanen and

Hristov 2009). In the case of confined animals fed total mixed rations, the diet should be formulated to match the animal's nutritional requirement for nitrogen and avoid exceeding it; precision feeding individual animals to cover their individual protein requirements decreases nitrogen excretion (Reynolds and Kristensen 2008). A maximum percentage of total nitrogen as non-protein nitrogen is also recommended (Kertz 2010). Manure management is another important tool to decrease nitrogen pollution from animal production (Rotz 2004). As for grazing ruminants, excess dietary nitrogen may be unusual in the tropics and subtropics but is a common problem in temperate regions during certain seasons, i.e. spring and autumn. In temperate grazing systems, nitrogen pollution can be controlled by combining agronomic, animal management, and nutritional measures. Use of excess nitrogenous fertilizers should be avoided and nitrogen-fixing legumes can be included in the sward. When grazed forages contain excess nitrogen, overstocking should be avoided (Rotz 2004), and animals can be supplemented with cereal grains, which are low in nitrogen and high in starch rapidly degradable and fermentable in the rumen, to improve microbial protein synthesis in the rumen (Dijkstra et al. 2013; Wilkinson et al. 2020).

The efficiency of utilization of dietary nitrogen can be improved by maximizing microbial protein synthesis in the rumen. The incorporation of ammonium into the synthesis of microbial amino acids is stimulated by the availability of rapidly degradable sources of energy and carbon, such as cereal grains. Increasing dietary concentrate also increases rumen outflow rates and decreases microbial cell lysis, increasing the efficiency of microbial protein synthesis (Firkins 1996).

Some processes of nitrogen metabolism are regarded as inefficient. Excess degradation of true protein and fermentation of amino acids in the rumen can overwhelm the capacity of rumen microorganisms to utilize ammonium for amino acid synthesis. This can result in high rates of ammonium absorption and synthesis of urea in the liver, and elimination of urea to the environment in urine (Wallace et al. 1997; Firkins and Mackie 2020). Feed processing, and in particular heating, decreases protein degradation in the rumen, although care should be exercised to avoid overheating and not decrease the digestion of proteins in the lower gastrointestinal tract. Phytocompounds such as tannins bind proteases and decrease rumen proteolysis, although they can also decrease intake and fiber digestibility (Firkins and Mackie 2020). Amino acid deamination can be slowed down through the use of essential oils or ionophores such as monensin and lasalocid, which inhibit a population of highly active deaminating bacteria (Wallace 1996; Hartinger et al. 2018; Firkins and Mackie 2020).

Intra-rumen recycling of microbial protein takes place as microbial cells lyse, their proteins are degraded, and part of the constituent amino acids are fermented rather than incorporated into newly synthesized protein. Protozoal predation of bacteria is quantitatively the most important cause of microbial protein turnover in the rumen (Hartinger et al. 2018), although it has been argued that its importance may be lesser in high-producing animals with relatively fast rumen outflow rates than in animals on roughage diets and low dry matter intake (Firkins et al. 2007). Elimination of protozoa generally improves the efficiency of nitrogen utilization, but there are no practical alternatives at present to keep protozoa-free animals from re-acquiring protozoa. Saponins decrease protozoal activity, but the results of saponin supplementation have been variable (Newbold et al. 2015).

Importantly, although preformed amino acids are stimulatory for microbial growth, the mixed rumen microbiota still relies on ammonium incorporation to synthesize most of its amino acids and proteins (Hackmann and Firkins 2015; Firkins and Mackie 2020). Thus, while excessive protein breakdown is inefficient, too little can also be problematic. Fibrolytic bacteria rely on ammonium as their most important source of nitrogen, and need a minimum concentration of rumen ammonium and hence dietary rumen-degradable protein, to sustain fiber breakdown (Schwab and Broderick 2017).

4.5.3 Rumen pH and Acidosis

Intensification of ruminant production has taken place through changes in genetics, animal management, health, and nutrition. In modern intensive operations, ruminants often are fed substantial amounts of concentrates based on cereal grains rich in starch. Concentrates are highly digestible and supply the rumen with rapidly fermentable carbohydrates, resulting in greater energy intake by the animal, and often help balance the energy-to-protein dietary ratio. The practice of feeding human-edible feeds to ruminants has been questioned on the basis that human-edible feed could be directly fed to human populations (i.e. feed vs. food competition), by-passing the loss of nutrients involved in the consumption and metabolism by ruminants (McDonald et al. 2010). It has been proposed that human-edible, nutrient-dense feeds, should be fed only strategically to ruminants for rectifying dietary imbalances and matching nutrient requirements that are not met by basal diets composed only of forage and byproducts (Wilkinson and Lee 2018).

Consumption of rapidly fermentable diets results in high rates of production of volatile fatty acids and lactate in the rumen (McDonald et al. 2010; Elmhadi et al. 2022). The resulting acidic load is under normal conditions neutralized by salivary buffers, ammonia produced by deamination, and volatile fatty acids absorption, passage to the omasum, and microbial utilization. Daily variation in rumen pH, e.g. after feed ingestion, is a normal condition. Acidosis is a continuum of low rumen pH that negatively affects fiber digestion, fermentation, microbial protein synthesis, as well as animal production and health (Hernandez et al. 2014).

Two types of rumen acidosis exist differing in the severity of rumen pH drops and resulting animal symptoms, and in the main metabolites causing the decrease in pH. In lactic, or acute acidosis, rumen pH can drop below 5.0 because of the accumulation of lactic acid (Nagaraja and Titgemeyer 2007). Lactate is a metabolic intermediate resulting from the disposal of metabolic hydrogen into pyruvate reduction. The predominant lactate isomer in the rumen is L(+)lactate, although the proportion of D(−)lactate increases at lower pH, with interconversions between both forms catalyzed by racemases. Some microorganisms release lactate to the extracellular medium, where it is taken up by other microbial cells, or metabolized intracellularly, and converted to propionate through propionate's non-randomizing pathway, or to butyrate or acetate. In the typical rumen fermentation, lactate does not accumulate because lactate utilizers such as *M. elsdenii* and *S. ruminantium* to a major extent, as well as others such as *Anaerovibrio l* (*lipolytica*), *Fusobacterium necrophorum*, *Peptoniphilus* (*Peptostreptococcus*) *asaccharolyticus*, *Cutibacterium* (*Propionibacterium*) *acnes*, and *V. parvula*, metabolize lactate to volatile fatty acids (Nagaraja and Titgemeyer 2007; Fan et al. 2022). When ruminants adapted to eating forage diets, which harbor relatively small populations of lactate utilizers, are abruptly changed to diets rich in rapidly fermentable carbohydrates, Gram-positive organisms such as *S. bovis* rapidly ferment carbohydrates to lactate, overwhelming the capacity of lactate utilizers to metabolize lactate to volatile fatty acids. Lactic acid has a lower pK_a ($pK_a = 3.86$) than volatile fatty acids ($pK_a \sim 4.8$), thus the decrease in rumen pH caused by lactate accumulation is more severe compared to the accumulation of volatile fatty acids. The severe drop in pH further spirals the acidotic condition by inhibiting the lactate utilizers present (Dawson et al. 1997). Most of the fermentative microbiota, including *S. bovis*, is finally inhibited, and lactobacilli establish (Russell and Wallace 1997; Nagaraja and Titgemeyer 2007).

Subacute rumen acidosis (SARA), or chronic acidosis, occurs because of a high concentration of volatile fatty acids, with no or little lactate accumulation. Subacute acidosis results from the relatively rapid production of volatile fatty acids and possibly decreased absorption, with rumen pH oscillating between 5.0 and 5.6. Lactate does not accumulate because the severity of the pH drop and the time spent at low pH still allow for the growth of lactate utilizers; depending on the time pH remains around 5.0, lactate utilizers may be eventually inhibited, and SARA transition to lactic acidosis. Subacute acidosis occurs when animals are well-managed and gradually adapted to highly fermentable diets. In this situation, the numbers of *S. bovis* remain similar to what is observed in forage-fed animals. Microorganisms such as *P. ruminicola*, *Succinivibrio* spp., and *Selenomonas* spp., among others, predominate with high-concentrate diets. *M. elsdenii* is typically the most important lactate utilizer, with *S. ruminantium* subsp. *lactilytica* having a dual role as a lactate producer and utilizer (Nagaraja and Titgemeyer 2007).

Acidosis causes important changes in the composition of the rumen microbial community. The abundance of fibrolytic bacteria, protozoa, and fungi decreases, and Gram-positive bacteria increase (Nagaraja and Titgemeyer 2007). In turn, the lysis of Gram-negative microorganisms releases lipopolysaccharides present in their cell wall, which when absorbed elicit an inflammatory response (Monteiro and Faciola 2020). Histamine produced from histidine decarboxylation is suspected to play a role in laminitis frequently observed in acidotic animals (Nagaraja and Titgemeyer 2007).

The rapid growth of *S. bovis* with resulting high rates of lactate production has been identified as a main factor triggering lactate acidosis. Therefore, *S. bovis* has been targeted with ionophores and vaccines to prevent lactic acidosis. Management practices such as step-feeding to induce a gradual adaptation of ruminants to a high-concentrate diet, and providing enough feeder space to avoid grain overeating by dominant animals, are generally effective (Dawson et al. 1997; Nagaraja and Titgemeyer 2007). There has also been much interest in using *M. elsdenii* as a direct-fed microbial to enhance lactate utilization to avoid lactate accumulation. Unlike *S. lactylitica*, *M. elsdenii* continues to metabolize lactate even in the presence of glucose or maltose, and it is also mildly acid-tolerant. *M. elsdenii* metabolizes lactate to propionate through the non-randomizing pathway to dispose of metabolic hydrogen but generates its ATP by metabolizing lactate to acetate and butyrate (Nagaraja and Titgemeyer 2007).

Acidosis causes lesions in the rumen epithelium which may serve as entry points to pathogens such as *F. necrophorum*, which can cause liver abscess and expand to cause rupture of pulmonary arteries in the bronchi. Absorption of lactic acid, and particularly the D(−) isomer, causes metabolic acidosis. Effects on the immune system have also been described (Enemark 2008).

Subacute acidosis has a high incidence in intensive operations where animals are fed high-concentrate, or even mixed, diets. Surveys indicated that 19% and 26% of early and mid-lactation cows, respectively, in the United States, and 11% and 18% in Germany and The Netherlands, in the same order, had SARA. This condition causes economic losses because of reduced feed intake and milk production, premature culling, and mortality (Enemark 2008). The incidence of SARA can be diminished by relenting starch digestion in the rumen by shifting some of the starch digestion from the rumen to the duodenum by, for example, using corn hybrids with vitreous starch and avoiding excessively fine grain grinding; care should be taken, however, to not decrease the overall tract digestibility of starch (Owens et al. 1986; Allen et al. 2021).

4.5.4 Biohydrogenation of Unsaturated Fatty Acids

The fatty acids profile of fat in ruminants' meat and milk differs from the diet's fatty acids profile, due to transformations that unsaturated fatty acids undergo in the rumen, which ultimately has implications for human health. Acylglycerides entering the rumen are hydrolyzed to glycerol and free fatty acids in a process called lipolysis. Glycerol enters glycolysis, and unsaturated, long-chain fatty acids undergo a complex series of sequential isomerizations and reductions called biohydrogenation. The consequence of biohydrogenation is that fatty acids outflowing the rumen are more saturated than fatty acids present in the diet (Harfoot and Hazlewood 1997; Jenkins et al. 2008; Lourenco et al. 2010).

The final product of biohydrogenation of 18-carbon fatty acids ($C_{18:n}$) is stearic acid ($C_{18:0}$). However, not all fatty acids complete the process, and some leave the rumen at different stages of biohydrogenation. For example, biohydrogenation of linolenic acid ($C_{18:3}$) in linseed oil partly yields linoleic ($C_{18:2}$) and oleic ($C_{18:1}$), as well as stearic acid. Biohydrogenation of 18-carbon fatty acids also results in rumen outflow of biohydrogenation intermediates unique to ruminant's fat, such as different position isomers of conjugated linoleic acid and trans fatty acids (*t*-$C_{18:1}$) (Harfoot and Hazlewood 1997; Jenkins et al. 2008).

The extent of rumen biohydrogenation of unsaturated fatty acids, i.e. how saturated is the profile of fatty acids outflowing the rumen in relation to the profile of fatty acids in the ingesta, depends on the competition between the rate of biohydrogenation and the rate of rumen outflow. Rumen biohydrogenation in this way affects the profile of fatty acids deposited in meat and milk, which has implications for consumers´ health. The consumption of saturated fat has been associated with coronary disease, and hence one of the objectives of manipulation of rumen fermentation has been to inhibit or slow down different steps of biohydrogenation to decrease the amount of saturated fatty acids leaving the rumen (Jenkins et al. 2008; Or-Rashid et al. 2009; Lourenco et al. 2010; Fievez et al. 2020).

Conversely, other fatty acids are regarded as beneficial to human health, and manipulating biohydrogenation seeks to enhance their content in ruminant fat. For example, inhibiting early steps of biohydrogenation of feed rich in linolenic acid, such as fresh grasses or linseed, would enrich ruminant products in polyunsaturated n-3 (omega-3) linolenic acid, while decreasing the percentage of saturated fatty acids. Consumption of omega-3 fatty acids is beneficial for cardiovascular disease, diabetes, cancer, rheumatoid arthritis, and mental illnesses (Shahidi and Ambigaipalan 2018). Decreasing the rates of the last step of biohydrogenation can in turn enrich ruminant fat in rumenic acid (*c*-9, *t*-11-$C_{18:2}$), a rumen intermediate of biohydrogenation which contributes to cancer prevention, decreasing atherosclerosis, stimulating the immune system, and deposition of lean tissue (Lourenco et al. 2010). Another intermediate of biohydrogenation, vaccenic acid (*t*-11-$C_{18:1}$), when absorbed in the small intestinal can be desaturated to rumenic acid in the animal's tissues. On the other hand, the formation and rumen outflow of a different biohydrogenation intermediate, conjugated linoleic acid *t*-10, *c*-12-$C_{18:2}$, is regarded as undesirable as it is associated with milk fat depression in dairy cows (Jenkins et al. 2008, Or-Rashid et al. 2009, Lourenco et al. 2010, Fievez et al. 2020).

Gram-negative *A. lipolyticus* (*lipolytica*) is generally regarded as the main rumen microorganism conducting lipolysis. Studies with pure cultures revealed that the conversion of linoleic to vaccenic acid was conducted predominantly by *B. fibrisolvens* and *Butyrivibrio* (*Clostridium*) *proteoclasticus*, the latter of which was the only bacterial species to produce stearic acid. Within 400 rumen isolates, all of those metabolizing linoleic acid to rumenic and vaccenic acid were shown to be butyrate producers belonging to the *B. fibrisolvens* group (Jenkins et al. 2008). There is evidence of the involvement of uncultivated species, in particular in the conversion of vaccenic to stearic acid (Lourenco et al. 2010). Rumen protozoa contain unsaturated fatty acids derived from the plant chloroplasts they ingest. They are also rich in rumenic and vaccenic acid from phagocytized bacteria. The biohydrogenation activity of rumen fungi is very small compared to bacteria (Jenkins et al. 2008).

Depending on the objective being sought, manipulation of the biohydrogenation process could in theory be designed to target the predominant bacteria conducting each biohydrogenation step. For example, it has been proposed that targeting *B. proteoclasticus* could enhance the content of vaccenic and rumenic acids in ruminant products (Jenkins et al. 2008). *Butyrivibrio* (*Clostridium*) *proteoclasticus* has been shown to be highly sensitive to unsaturated fatty acids (Lourenco et al. 2010).

Fatty acids composition of ruminant products can be manipulated by increasing the amount and proportion of polyunsaturated fatty acids in the diet and by altering fatty acids retention time in the rumen. Fresh forages, for example, are rich in linolenic acid. Rumen-protected oils can also be included in ruminants´ diets to ensure their unaltered absorption and improve their transfer efficiency (Or-Rashid et al. 2009; Fievez et al. 2020). The effects of dietary unsaturated fatty acids on the amount and profile of fatty acids outflowing the rumen are dual, acting not only through an increase in the amount of substrate for biohydrogenation but also through their toxicity to bacteria involved in biohydrogenation, thereby inhibiting some steps of the biohydrogenation process. In particular, highly unsaturated fatty acids with three or more double bonds present in linseed or fish oil inhibit the last step of biohydrogenation, i.e. the conversion of vaccenic to stearic acid (Lourenco et al. 2010). Care should be taken to avoid excess addition of lipids to the diet, as they can depress intake, fiber digestibility, and milk fat content (Beauchemin et al. 2020).

Phytochemicals such as essential oils and saponins can inhibit the last step of biohydrogenation and induce the accumulation of rumenic acid, but their effects have been variable, as is their composition (Lourenco et al. 2010). Plant polyphenol and in particular tannins are also being investigated to modulate different steps of rumen biohydrogenation (Toral et al. 2018; Vasta et al. 2019). Other beneficial, although variable, effects of tannins can be decreased methane production, decreased elimination of nitrogen in urine, prevention of bloat, and control of parasites. However, and depending on the dietary content and type of tannin, tannins can negatively affect palatability and feed intake, and digestion (Waghorn 2008; Patra and Saxena 2010; Patra and Saxena 2011).

4.6 CONCLUSIONS

Ruminants have been important to human societies through the provision of meat, milk, wool and hides, and draft power for tillage in agriculture. Humans adopted ruminants various millennia ago because these unique animals can transform lignocellulosic biomass and non-protein nitrogen into valuable products. These distinctive features of ruminants are the result of a symbiotic relationship that ruminants have with an anaerobic, complex microbial community that they harbor in their rumens, a large, pre-gastric fermentation vat. Ruminants provide rumen microbes with a stable environment in terms of temperature, pH, reducing potential, and osmolarity, and through feed ingestion, they ensure a more or less consistent provision of substrates for microbial digestion and fermentation. In turn, rumen microbes utilize for maintenance and growth only a small portion of the energy in feed, with most energy being absorbed through the rumen wall as incompletely oxidized volatile fatty acids that cover most of the ruminant host animal's requirements for energy and glucose. Also, microbial cells themselves constitute the main source of amino acids and other nutrients, such as water-soluble vitamins, when they pass out of the rumen to the lower gut.

Despite this extremely useful nutritional flexibility, utilization of energy and nitrogen by ruminants is poorer compared to non-ruminants, and this inefficiency has economic and environmental consequences in terms of methane and excess nitrogen released to the environment. Research in rumen microbiology and biochemistry has allowed an understanding of rumen function and designing strategies that seek to optimize ruminant production and sustainability. Many advances in our understanding of rumen fermentation and the rumen microbial community, and the interactions between microbial groups and between microbes and the host ruminant animal have taken place since the study of ruminant nutrition and rumen microbiology began, and many more exciting advances await provide solutions to pressing needs of increasing ruminant productivity and making it more sustainable. The advent of molecular microbial ecology techniques is unraveling new knowledge about the composition of the rumen microbial community and the ecological niches occupied by its different microorganisms. The integration of the new –omics techniques with classical cultivation and biochemistry techniques, and mathematical modeling based on kinetics and thermodynamic principles, will prompt our understanding of this important microbial ecosystem and its relationship with the ruminant host animal.

REFERENCES

Ajmone-Marsan, P., Garcia, J.F., and Lenstra, J.A. (2010). On the origin of cattle: how aurochs became cattle and colonized the world. *Evolutionary Anthropology: Issues, News, and Reviews* 19: 148–157. https://doi.org/10.1002/evan.20267.

Allen, M.S., Longuski, R.A., and Ying, Y. (2021). Effects of corn grain endosperm type and fineness of grind on site of digestion, ruminal digestion kinetics, and flow of nitrogen fractions to the duodenum in lactating dairy cows. *Journal of Dairy Science* 104: 7641–7652. https://doi.org/10.3168/jds.2020-18992.

Allison, M.J. (1969). Biosynthesis of amino acids by ruminal microorganisms. *Journal of Animal Science* 29: 797–807. https://doi.org/10.2527/jas1969.295797x.

Allison, M.J. (1978). Production of branched-chain volatile fatty acids by certain anaerobic bacteria. *Applied and Environmental Microbiology* 35: 872–877. https://doi.org/10.1128%2Faem.35.5.872-877.1978.

Allison, M.J. and Peel, J.L. (1971). The biosynthesis of valine from isobutyrate by *Peptostreptococcus elsdenii* and *Bacteroides ruminicola*. *Biochemical Journal* 121: 431–437. https://doi.org/10.1042/bj1210431.

Allison, M.J., Bryant, M.P., and Doetsch, R.N. (1962). Studies on the metabolic function of branched-chain volatile fatty acids, growth factors for ruminococci I. Incorporation of isovalerate into leucine. *Journal of Bacteriology* 83: 523–532. https://doi.org/10.1128/jb.83.3.523-532.1962.

Allison, M.J., Bucklin, J.A., and Robinson, I.M. (1966). Importance of the isovalerate carboxylation pathway of leucine biosynthesis in the rumen. *Applied Microbiology* 14: 807–814. https://doi.org/10.1128/am.14.5.807-814.1966.

Asanuma, N. and Hino, T. (2000). Activity and properties of fumarate reductase in ruminal bacteria. *Journal of General and Applied Microbiology* 46: 119–125. https://doi.org/10.2323/jgam.46.119.

Asanuma, N. and Hino, T. (2001). Molecular characterization, enzyme properties and transcriptional regulation of phosphoenolpyruvate carboxykinase and pyruvate kinase in a ruminal bacterium, *Selenomonas ruminantium*. *Microbiology* 147: 681–690. https://doi.org/10.1099/00221287-147-3-681.

Asanuma, N., Iwamoto, M., and Hino, T. (1999). Effect of the addition of fumarate on methane production by ruminal microorganisms in vitro. *Journal of Dairy Science* 82: 780–787. https://doi.org/10.3168/jds.S0022-0302(99)75296-3.

Asanuma, N., Kanada, K., Arai, Y. et al. (2010). Molecular characterization and significance of phosphoenolpyruvate carboxykinase in a ruminal bacterium, *Streptococcus bovis*. *The Journal of General and Applied Microbiology* 56: 121–127. https://doi.org/10.2323/jgam.56.121.

Avila, J.S., Chaves, A.V., Hernandez-Calva, M. et al. (2011). Effects of replacing barley grain in feedlot diets with increasing levels of glycerol on in vitro fermentation and methane production. *Animal Feed Science and Technology* 166–167: 265–268. https://doi.org/10.1016/j.anifeedsci.2011.04.016.

Baldwin, R.L., Wood, W.A., and Emery, R.S. (1963). Conversion of glucose-C^{14} to propionate by the rumen microbiota. *Journal of Bacteriology* 85: 1346–1349. https://doi.org/10.1128/jb.85.6.1346-1349.1963.

Bauchop, T. and Mountfort, D.O. (1981). Cellulose fermentation by a rumen anaerobic fungus in both the absence and the presence of rumen methanogens. *Applied and Environmental Microbiology* 42: 1103–1110. https://doi.org/10.1128%2Faem.42.6.1103-1110.1981.

Beauchemin, K.A., Ungerfeld, E.M., Eckard, R.J. et al. (2020). Review: fifty years of research on rumen methanogenesis: lessons learned and future challenges for mitigation. *Animal* 14: s2–s16. https://doi.org/10.1017/S1751731119003100.

Bickhart, D.M. and Weimer, P.J. (2018). Symposium review: host–rumen microbe interactions may be leveraged to improve the productivity of dairy cows. *Journal of Dairy Science* 101: 7680–7689. https://doi.org/10.3168/jds.2017-13328.

Broderick, G.A., Huhtanen, P., Ahvenjarvi, S. et al. (2010). Quantifying ruminal nitrogen metabolism using the omasal sampling technique in cattle – a meta-analysis. *Journal of Dairy Science* 93: 3216–3230. https://doi.org/10.3168/jds.2009-2989.

Cammack, K.M., Austin, K.J., Lamberson, W.R. et al. (2018). Ruminant nutrition symposium: tiny but mighty: the role of the rumen microbes in livestock production. *Journal of Animal Science* 96: 752–770. https://doi.org/10.1093/jas/skx053.

Cheng, K.-J. and McAllister, T.A. (1997). Compartmentation in the rumen. In: *The Rumen Microbial Ecosystem* (ed. P.N. Hobson and C.S. Stewart), 492–522. London: Blackie Academic & Professional. https://doi.org/10.1007/978-94-009-1453-7_12.

Cholewińska, P., Czyż, K., Nowakowski, P. et al. (2020). The microbiome of the digestive system of ruminants – a review. *Animal Health Research Reviews* 21: 3–14. https://doi.org/10.1017/S1466252319000069.

Crater, A.R., Barboza, P.S., and Forster, R.J. (2007). Regulation of rumen fermentation during seasonal fluctuations in food intake of muskoxen. *Comparative Biochemistry and Physiology Part A: Molecular & Integrative Physiology* 146: 233–241. https://doi.org/10.1016/j.cbpa.2006.10.019.

Cunningham, M.D., Martz, F.A., and Merilan, C.P. (1964). Effect of drinking-water temperature upon ruminant digestion, intraruminal temperature, and water consumption of nonlactating dairy cows. *Journal of Dairy Science* 47: 382–385. https://doi.org/10.3168/jds.S0022-0302(64)88671-9.

Czerkawski, J.W. (1969). Methane production in ruminants and its significance. *World Review of Nutrition and Dietetics* 11: 240–282. https://doi.org/10.1159/000387580.

Czerkawski, J.W. (1972). Fate of metabolic hydrogen in the rumen. *The Proceedings of the Nutrition Society* 31: 141–146. https://doi.org/10.1079/pns19720028.

Czerkawski, J.W. (1986). *An Introduction to Rumen Studies*. Exeter, UK: Pergamon Press. https://doi.org/10.1016/C2009-0-00551-0.

Dawson, K.A., Rasmussen, M.A., and Allison, M.J. (1997). Digestive disorders and nutritional toxicity. In: *The Rumen Microbial Ecosystem* (ed. P.N. Hobson and C.S. Stewart), 633–660. London: Blackie Academic & Professional. https://doi.org/10.1007/978-94-009-1453-7_14.

Dekker, J.P. and Lau, A.F. (2016). An update on the *Streptococcus bovis* group: classification, identification, and disease associations. *Journal of Clinical Microbiology* 54: 1694–1699. https://doi.org/10.1128/JCM.02977-15.

Dijkstra, J., Oenema, O., van Groenigen, J.W. et al. (2013). Diet effects on urine composition of cattle and N_2O emissions. *Animal* 7: 292–302. https://doi.org/10.1017/S1751731113000578.

Ellis, J.E., Williams, A.G., and Lloyd, D. (1989). Oxygen consumption by ruminal microorganisms: protozoal and bacterial contributions. *Applied and Environmental Microbiology* 55: 2583–2587. https://doi.org/10.1128/aem.55.10.2583-2587.1989.

Elmhadi, M.E., Ali, D.K., Khogali, M.K. et al. (2022). Subacute ruminal acidosis in dairy herds: microbiological and nutritional causes, consequences, and prevention strategies. *Animal Nutrition* 10: 148–155. https://doi.org/10.1016/j.aninu.2021.12.008.

Enemark, J.M. (2008). The monitoring, prevention and treatment of sub-acute ruminal acidosis (SARA): a review. *Veterinary Journal* 176: 32–43. https://doi.org/10.1016/j.tvjl.2007.12.021.

Fan, Y., Xia, G., Jin, Y. et al. (2022). Ambient pH regulates lactate catabolism pathway of the ruminal *Megasphaera elsdenii* BE2-2083 and *Selenomonas ruminantium* HD4. *Journal of Applied Microbiology* 132: 2661–2672. https://doi.org/10.1111/jam.15464.

Feldewert, C., Lang, K., and Brune, A. (2020). The hydrogen threshold of obligately methyl-reducing methanogens. *FEMS Microbiology Letters* 367: https://doi.org/10.1093/femsle/fnaa137.

Fievez, V., De Neve, N., and Dewanckele, L. (2020). Understanding rumen lipid metabolism to optimize dairy products for enhanced human health and to monitor animal health. In: *Improving Rumen Function* (ed. C.S. Mcsweeney and R.I. Mackie), 467–504. Cambridge, UK: Burleigh Dodds Science Publishing Limited. https://doi.org/10.19103/AS.2020.0067.15.

Firkins, J.L. (1996). Maximizing microbial protein synthesis in the rumen. *Journal of Nutrition* 126: 1347S–1354S. https://doi.org/10.1093/jn/126.suppl_4.1347S.

Firkins, J. and Mackie, R.I. (2020). Ruminal protein breakdown and ammonia assimilation. In: *Improving Rumen Function* (ed. C.S. Mcsweeney and R.I. Mackie), 383–412. Cambridge, UK: Burleigh Dodds Science Publishing Limited. https://doi.org/10.19103/AS.2020.0067.13.

Firkins, J.L., Yu, Z., and Morrison, M. (2007). Ruminal nitrogen metabolism: perspectives for integration of microbiology and nutrition for dairy. *Journal of Dairy Science* 90: E1–E16. https://doi.org/10.3168/jds.2006-518.

Forbes, J.M. (2007). A personal view of how ruminant animals control their intake and choice of food: minimal total discomfort. *Nutrition Research Reviews* 20: 132–146. https://doi.org/10.1017/s0954422407797834.

Gilbert, R. and Ouwerkerk, D. (2020). Ruminal viruses and extrachromosomal genetic elements. In: *Improving Rumen Function* (ed. C.S. Mcsweeney and R.I. Mackie), 281–320. Cambridge, UK: Burleigh Dodds Science Publishing Limited. https://doi.org/10.19103/AS.2020.0067.10.

Gilbert, R.A., Townsend, E.M., Crew, K.S. et al. (2020). Rumen virus populations: technological advances enhancing current understanding. *Frontiers in Microbiology* 11: https://doi.org/10.3389/fmicb.2020.00450.

Goopy, J. (2019). Creating a low enteric methane emission ruminant: what is the evidence of success to the present and prospects for developing economies? *Animal Production Science* 59: 1759–1776. https://doi.org/10.1071/AN18457.

Greening, C., Geier, R., Wang, C. et al. (2019). Diverse hydrogen production and consumption pathways influence methane production in ruminants. *The ISME Journal*. https://doi.org/10.1038/s41396-019-0464-2.

Gruninger, R.J., Ribeiro, G.O., Cameron, A. et al. (2019). Invited review: application of meta-omics to understand the dynamic nature of the rumen microbiome and how it responds to diet in ruminants. *Animal: An International Journal of Animal Bioscience* 13: 1843–1854. https://doi.org/10.1017/s1751731119000752.

van Gylswyk, N.O. (1995). *Succiniclasticum ruminis* gen. Nov., sp. Nov., a ruminal bacterium converting succinate to propionate as the sole energy-yielding mechanism. *International Journal of Systematic and Evolutionary Microbiology* 45: 297–300. https://doi.org/10.1099/00207713-45-2-297.

Hackmann, T.J. and Firkins, J.L. (2015). Maximizing efficiency of rumen microbial protein production. *Frontiers in Microbiology* 6: 465. https://doi.org/10.3389/fmicb.2015.00465.

Hackmann, T.J. and Spain, J.N. (2010). Invited review: ruminant ecology and evolution: perspectives useful to ruminant livestock research and production. *Journal of Dairy Science* 93: 1320–1334. https://doi.org/10.3168/jds.2009-2071.

Hackmann, T.J., Ngugi, D.K., Firkins, J.L. et al. (2017). Genomes of rumen bacteria encode atypical pathways for fermenting hexoses to short-chain fatty acids. *Environmental Microbiology* 19: 4670–4683. https://doi.org/10.1111/1462-2920.13929.

Hamar, D. and Borchers, R. (1967). Glycolytic pathway in rumen microorganisms. *Journal of Animal Science* 26: 654–657. https://doi.org/10.2527/jas1967.263654x.

Hanafy, R.A., Lanjekar, V.B., Dhakephalkar, P.K. et al. (2020). Seven new Neocallimastigomycota genera from wild, zoo-housed, and domesticated herbivores greatly expand the taxonomic diversity of the phylum. *Mycologia* 112: 1212–1239. https://doi.org/10.1080/00275514.2019.1696619.

Harfoot, C.G. and Hazlewood, G.P. (1997). Lipid metabolism in the rumen. In: *The Rumen Microbial Ecosystem* (ed. P.N. Hobson and C.S. Stewart), 382–426. London: Blackie Academic & Professional. https://doi.org/10.1007/978-94-009-1453-7_9.

Hartinger, T., Gresner, N., and Sudekum, K.H. (2018). Does intra-ruminal nitrogen recycling waste valuable resources? A review of major players and their manipulation. *Journal of Animal Science and Biotechnology* 9: 33. https://doi.org/10.1186/s40104-018-0249-x.

Henderson, G., Cox, F., Ganesh, S. et al. (2015). Rumen microbial community composition varies with diet and host, but a core microbiome is found across a wide geographical range. *Scientific Reports* 5: 14567. https://doi.org/10.1038/srep14567.

Hernandez, J., Benedito, J.L., Abuelo, A. et al. (2014). Ruminal acidosis in feedlot: from aetiology to prevention. *The Scientific World Journal* 2014: 702572. https://doi.org/10.1155/2014/702572.

Hespell, R.B. (1988). Microbial digestion of hemicelluloses in the rumen. *Microbiological Sciences* 5: 362–365.

Hespell, R.B., Wolf, R., and Bothast, R.J. (1987). Fermentation of xylans by *Butyrivibrio fibrisolvens* and other ruminal bacteria. *Applied and Environmental Microbiology* 53: 2849–2853. https://doi.org/10.1128/aem.53.12.2849-2853.1987.

Hess, M., Fliegerová, K., Paul, S. et al. (2020a). The anaerobic rumen fungi. In: *Improving Rumen Function* (ed. C.S. Mcsweeney and R.I. Mackie), 221–280. Cambridge, UK: Burleigh Dodds Science Publishing Limited. https://doi.org/10.19103/AS.2020.0067.09.

Hess, M., Paul, S.S., Puniya, A.K. et al. (2020b). Anaerobic fungi: past, present, and future. *Frontiers in Microbiology* 11: 584893. https://doi.org/10.3389/fmicb.2020.584893.

Hilpert, W., Schink, B., and Dimroth, P. (1984). Life by a new decarboxylation-dependent energy conservation mechanism with Na as coupling ion. *The EMBO Journal* 3: 1665–1670. https://doi.org/10.1002/j.1460-2075.1984.tb02030.x.

Himo, F. and Eriksson, L.A. (1998). Catalytic mechanism of pyruvate formate-lyase (pfl). A theoretical study. *Journal of the American Chemical Society* 120: 11449–11455. https://doi.org/10.1021/ja9820947.

Hristov, A.N., Oh, J., Firkins, J.L. et al. (2013). Mitigation of methane and nitrous oxide emissions from animal operations: I. A review of enteric methane mitigation options. *Journal of Animal Science* 91: 5045–5069. https://doi.org/10.2527/jas.2013-6583.

Huang, Y., Marden, J.P., Julien, C. et al. (2018). Redox potential: an intrinsic parameter of the rumen environment. *Journal of Animal Physiology and Animal Nutrition* 102: 393–402. https://doi.org/10.1111/jpn.12855.

Huhtanen, P. and Hristov, A.N. (2009). A meta-analysis of the effects of dietary protein concentration and degradability on milk protein yield and milk N efficiency in dairy cows. *Journal of Dairy Science* 92: 3222–3232. https://doi.org/10.3168/jds.2008-1352.

Hungate, R.E. (1967). Hydrogen as an intermediate in the rumen fermentation. *Archiv für Mikrobiologie* 59: 158–164. https://doi.org/10.1007/BF00406327.

Hungate, R.E., Smith, W., Bauchop, T. et al. (1970). Formate as an intermediate in the bovine rumen fermentation. *Journal of Bacteriology* 102: 389–397. https://doi.org/10.1128/jb.102.2.389-397.1970.

Huws, S.A., Creevey, C.J., Oyama, L.B. et al. (2018). Addressing global ruminant agricultural challenges through understanding the rumen microbiome: past, present, and future. *Frontiers in Microbiology* 9: 2161. https://doi.org/10.3389/fmicb.2018.02161.

Huws, S., Williams, C., and McEwan, N. (2020). Ruminal-ciliated protozoa. In: *Improving Rumen Function* (ed. C.S. Mcsweeney and R.I. Mackie), 191–220. Cambridge, UK: Burleigh Dodds Science Publishing Limited. https://doi.org/10.19103/AS.2020.0067.08.

Iannotti, E.L., Kafkewitz, D., Wolin, M.J. et al. (1973). Glucose fermentation products of *Ruminococcus albus* grown in continuous culture with *vibrio succinogenes*: changes caused by interspecies transfer of H_2. *Journal of Bacteriology* 114: 1231–1240. https://doi.org/10.1128/jb.114.3.1231-1240.1973.

IPCC (2021). Summary for policymakers. In: *Climate Change 2021: The Physical Science Basis. Contribution of Working Group I to the Sixth Assessment Report of the Intergovernmental Panel on Climate Change* (ed. V. Masson-Delmotte, P. Zhai, A. Pirani, et al.), 3–32. Cambridge, United Kingdom and New York, NY, USA: Cambridge University Press. 10.1017/9781009157896.001.

Isaacson, H.R., Hinds, F.C., Bryant, M.P. et al. (1975). Efficiency of energy utilization by mixed rumen bacteria in continuous culture. *Journal of Dairy Science* 58: 1645–1659. https://doi.org/10.3168/jds.S0022-0302(75)84763-1.

Janssen, P.H. (2010). Influence of hydrogen on rumen methane formation and fermentation balances through microbial growth kinetics and fermentation thermodynamics. *Animal Feed Science and Technology* 160: 1–22. https://doi.org/10.1016/j.anifeedsci.2010.07.002.

Janssen, P.H. and Kirs, M. (2008). Structure of the archaeal community of the rumen. *Applied and Environmental Microbiology* 74: 3619–3625. https://doi.org/10.1128/AEM.02812-07.

Jenkins, T.C., Wallace, R.J., Moate, P.J. et al. (2008). Board-invited review: recent advances in biohydrogenation of unsaturated fatty acids within the rumen microbial ecosystem. *Journal of Animal Science* 86: 397–412. https://doi.org/10.2527/jas.2007-0588.

Joblin, K.N. and Naylor, G.E. (2002). The ruminal mycoplasmas: a review. *Journal of Applied Animal Research* 21: 161–179. https://doi.org/10.1080/09712119.2002.9706368.

Johnson, K.A. and Johnson, D.E. (1995). Methane emissions from cattle. *Journal of Animal Science* 73: 2483–2492. https://doi.org/10.2527/1995.7382483x.

Kajikawa, H., Valdes, C., Hillman, K. et al. (2003). Methane oxidation and its coupled electron-sink reactions in ruminal fluid. *Letters in Applied Microbiology* 36: 354–357. https://doi.org/10.1046/j.1472-765X.2003.01317.x.

Kelly, W. (2012). The Hungate 1000. A catalogue of reference genomes from the rumen microbiome. https://genome.jgi.doe.gov/portal/TheHunmicrobiome/TheHunmicrobiome.info.html (accessed August 5 2022).

Kelly, W.J., Leahy, S.C., Altermann, E. et al. (2010). The glycobiome of the rumen bacterium *Butyrivibrio proteoclasticus* B316(T) highlights adaptation to a polysaccharide-rich environment. *PLoS One* 5: e11942. https://doi.org/10.1371/journal.pone.0011942.

Kertz, A. (2010). Review: urea feeding to dairy cattle: A historical perspective and review. *The Professional Animal Scientist* 26: 257–272. https://doi.org/10.15232/S1080-7446(15)30593-3.

Knapp, J.R., Laur, G.L., Vadas, P.A. et al. (2014). Invited review: enteric methane in dairy cattle production: quantifying the opportunities and impact of reducing emissions. *Journal of Dairy Science* 97: 3231–3261. https://doi.org/10.3168/jds.2013-7234.

Kohn, R.A. and Dunlap, T.F. (1998). Calculation of the buffering capacity of bicarbonate in the rumen and in vitro. *Journal of Animal Science* 76: 1702–1709. https://doi.org/10.2527/1998.7661702x.

Kotarski, S.F., Waniska, R.D., and Thurn, K.K. (1992). Starch hydrolysis by the ruminal microflora. *The Journal of Nutrition* 122: 178–190. https://doi.org/10.1093/jn/122.1.178.

Krueger, N.A., Anderson, R.C., Tedeschi, L.O. et al. (2010). Evaluation of feeding glycerol on free-fatty acid production and fermentation kinetics of mixed ruminal microbes in vitro. *Bioresource Technology* 101: 8469–8472. https://doi.org/10.1016/j.biortech.2010.06.010.

Latham, M.J. and Wolin, M.J. (1977). Fermentation of cellulose by *Ruminococcus flavefaciens* in the presence and absence of *Methanobacterium ruminantium*. *Applied and Environmental Microbiology* 34: 297–301. https://doi.org/10.1128/aem.34.3.297-301.1977.

Lobo, R.R. and Faciola, A.P. (2021). Ruminal phages – a review. *Frontiers in Microbiology* 12: 763416. https://www.frontiersin.org/articles/10.3389/fmicb.2021.763416/full.

Loesche, W.J. (1969). Oxygen sensitivity of various anaerobic bacteria. *Applied Microbiology* 18: 723–727. https://doi.org/10.1128/am.18.5.723-727.1969.

López-Aizpún, M., Horrocks, C.A., Charteris, A.F. et al. (2020). Meta-analysis of global livestock urine-derived nitrous oxide emissions from agricultural soils. *Global Change Biology* 26: 2002–2013. https://doi.org/10.1111/gcb.15012.

Lourenco, M., Ramos-Morales, E., and Wallace, R.J. (2010). The role of microbes in rumen lipolysis and biohydrogenation and their manipulation. *Animal* 4: 1008–1023. https://doi.org/10.1017/s175173111000042x.

Løvendahl, P., Difford, G.F., Li, B. et al. (2018). Review: selecting for improved feed efficiency and reduced methane emissions in dairy cattle. *Animal* 12: s336–s349. https://doi.org/10.1017/S1751731118002276.

Malmuthuge, N. and Guan, L.L. (2017). Understanding host–microbial interactions in rumen: searching the best opportunity for microbiota manipulation. *Journal of Animal Science and Biotechnology* 8: 8. https://doi.org/10.1186/s40104-016-0135-3.

Matte, A., Forsberg, C.W., and Gibbins, A.M.V. (1992). Enzymes associated with metabolism of xylose and other pentoses by *Prevotella* (*Bacteroides*) *ruminicola* strains, *Selenomonas ruminantium* D, and *Fibrobacter succinogenes* S85. *Canadian Journal of Microbiology* 38: 370–376. https://doi.org/10.1139/m92-063.

Matthews, C., Crispie, F., Lewis, E. et al. (2018). The rumen microbiome: a crucial consideration when optimising milk and meat production and nitrogen utilisation efficiency. *Gut Microbes* 1–18. https://doi.org/10.1080/19490976.2018.1505176.

Matthews, C., Crispie, F., Lewis, E. et al. (2019). The rumen microbiome: a crucial consideration when optimising milk and meat production and nitrogen utilisation efficiency. *Gut Microbes* 10: 115–132. https://doi.org/10.1080/19490976.2018.1505176.

McDonald, P., Edwards, R.A., Greenhalgh, J.F.D. et al. (2010). *Animal Nutrition*. Harlow, England: Prentice Hall.

McTavish, E.J., Decker, J.E., Schnabel, R.D. et al. (2013). New World cattle show ancestry from multiple independent domestication events. *Proceedings of the National Academy of Sciences* 110: E1398–E1406. https://doi.org/10.1073/pnas.1303367110.

Miller, T.L. and Jenesel, S.E. (1979). Enzymology of butyrate formation by *Butyrivibrio fibrisolvens*. *Journal of Bacteriology* 138: 99–104. https://doi.org/10.1128%2Fjb.138.1.99-104.1979.

Mitchell, A.D., Chappell, A., and Knox, K.L. (1979). Metabolism of betaine in the ruminant. *Journal of Animal Science* 49: 764–774. https://doi.org/10.2527/jas1979.493764x.

Monteiro, H.F. and Faciola, A.P. (2020). Ruminal acidosis, bacterial changes, and lipopolysaccharides. *Journal of Animal Science* 98: 1–9. https://doi.org/10.1093/jas/skaa248.

Moraïs, S. and Mizrahi, I. (2019). Islands in the stream: from individual to communal fiber degradation in the rumen ecosystem. *FEMS Microbiology Reviews* 43: 362–379. https://doi.org/10.1093/femsre/fuz007.

Mountfort, D.O., Asher, R.A., and Bauchop, T. (1982). Fermentation of cellulose to methane and carbon dioxide by a rumen anaerobic fungus in a triculture with *Methanobrevibacter* sp. strain RA1 and *Methanosarcina barkeri*. *Applied and Environmental Microbiology* 44: 128–134. https://doi.org/10.1128/aem.44.1.128-134.1982.

Naas, A.E. and Pope, P.B. (2020). Ruminal fibre digestion. In: *Improving Rumen Function* (ed. C.S. Mcsweeney and R.I. Mackie), 349–381. Cambridge, UK: Burleigh Dodds Science Publishing Limited. https://doi.org/10.19103/AS.2020.0067.12.

Nagaraja, T.G. and Titgemeyer, E.C. (2007). Ruminal acidosis in beef cattle: the current microbiological and nutritional outlook. *Journal of Dairy Science* 90 (Suppl 1): E17–E38. https://doi.org/10.3168/jds.2006-478.

Nagaraja, T.G., Newbold, C.J., van Nevel, C.J. et al. (1997). Manipulation of ruminal fermentation. In: *The Rumen Microbial Ecosystem* (ed. P.N. Hobson and C.S. Stewart), 524–632. London: Blackie Academic & Professional. https://doi.org/10.1007/978-94-009-1453-7_13.

Neill, A.R., Grime, D.W., and Dawson, R.M. (1978). Conversion of choline methyl groups through trimethylamine into methane in the rumen. *The Biochemical Journal* 170: 529–535. https://doi.org/10.1042/bj1700529.

Newbold, C.J. and Ramos-Morales, E. (2020). Review: ruminal microbiome and microbial metabolome: effects of diet and ruminant host. *Animal: An International Journal of Animal Bioscience* 14: s78–s86. https://doi.org/10.1017/s1751731119003252.

Newbold, C.J., de la Fuente, G., Belanche, A. et al. (2015). The role of ciliate protozoa in the rumen. *Frontiers in Microbiology* 6: 1313. https://doi.org/10.3389/fmicb.2015.01313.

Nocek, J.E. and Tamminga, S. (1991). Site of digestion of starch in the gastrointestinal tract of dairy cows and its effect on milk yield and composition. *Journal of Dairy Science* 74: 3598–3629. https://doi.org/10.3168/jds.S0022-0302(91)78552-4.

Nordberg, H., Cantor, M., Dusheyko, S. et al. (2014). The genome portal of the Department of Energy Joint Genome Institute: 2014 updates. *Nucleic Acids Research* 42: D26–D31. https://doi.org/10.1093/nar/gkt1069.

Or-Rashid, M.M., Wright, T.C., and McBride, B.W. (2009). Microbial fatty acid conversion within the rumen and the subsequent utilization of these fatty acids to improve the healthfulness of ruminant food products. *Applied Microbiology and Biotechnology* 84: 1033–1043. https://doi.org/10.1007/s00253-009-2169-3.

Owens, F.N., Zinn, R.A., and Kim, Y.K. (1986). Limits to starch digestion in the ruminant small intestine. *Journal of Animal Science* 63: 1634–1648. https://doi.org/10.2527/jas1986.6351634x.

Park, Y., Kim, S.-G., Park, K. et al. (2005). Fed-batch production of D-ribose from sugar mixtures by transketolase-deficient *Bacillus subtilis* SPK1. *Applied Microbiology and Biotechnology* 66: 297–302. https://doi.org/10.1007/s00253-004-1678-3.

Patra, A.K. and Saxena, J. (2010). A new perspective on the use of plant secondary metabolites to inhibit methanogenesis in the rumen. *Phytochemistry* 71: 1198–1222. https://doi.org/10.1016/j.phytochem.2010.05.010.

Patra, A. and Saxena, A. (2011). Exploitation of dietary tannins to improve rumenmetabolism and ruminant nutrition. *Journal of the Science of Food and Agriculture* 91: 24–37. https://doi.org/10.1002/jsfa.4152.

Pavlostathis, S.G., Miller, T.L., and Wolin, M.J. (1990). Cellulose fermentation by continuous cultures of *Ruminococcus albus* and *Methanobrevibacter smithii*. *Applied Microbiology and Biotechnology* 33: 109–116. https://doi.org/10.1007/bf00170581.

Pazur, J.H., Shuey, E.W., and Georgi, C.E. (1958). The conversion of d-xylose into volatile organic acids by rumen bacteria. *Archives of Biochemistry and Biophysics* 77: 387–394. https://doi.org/10.1016/0003-9861(58)90086-9.

Petersen, S.O., Hellwing, A.L., Brask, M. et al. (2015). Dietary nitrate for methane mitigation leads to nitrous oxide emissions from dairy cows. *Journal of Environmental Quality* 44: 1063–1070. https://doi.org/10.2134/jeq2015.02.0107.

Pitta, D.W., Melgar, A., Hristov, A.N. et al. (2021). Temporal changes in total and metabolically active ruminal methanogens in dairy cows supplemented with 3-nitrooxypropanol. *Journal of Dairy Science* 104: 8721–8735. https://doi.org/10.3168/jds.2020-19862.

Pitta, D., Indugu, N., Melgar, A. et al. (2022). The effect of 3-nitrooxypropanol, a potent methane inhibitor, on ruminal microbial gene expression profiles in dairy cows. Preprint: https://www.researchsquare.com/article/rs-1552496/v1. https://doi.org/10.21203/rs.3.rs-1552496/v1 (accessed July 2 2022).

Pol, A. and Demeyer, D.I. (1988). Fermentation of methanol in the sheep rumen. *Applied and Environmental Microbiology* 54: 832–834. https://doi.org/10.1128%2Faem.54.3.832-834.1988.

Ragsdale, S.W. (2003). Pyruvate ferredoxin oxidoreductase and its radical intermediate. *Chemical Reviews* 103: 2333–2346. https://doi.org/10.1021/cr020423e.

de Raphélis-Soissan, V., Nolan, J.V., Godwin, I.R. et al. (2018). Production of N_2 and N_2O from nitrate ingested by sheep. *Journal of Animal Physiology and Animal Nutrition* 102: e176–e182. 10.1111/jpn.12725.

Reymond, P., Geourjon, C., Roux, B. et al. (1992). Sequence of the phosphoenolpyruvate carboxykinase-encoding cDNA from the rumen anaerobic fungus *Neocallimastix frontalis*: comparison of the amino acid sequence with animals and yeast. *Gene* 110: 57–63. https://doi.org/10.1016/0378-1119(92)90444-t.

Reynolds, C.K. and Kristensen, N.B. (2008). Nitrogen recycling through the gut and the nitrogen economy of ruminants: an asynchronous symbiosis. *Journal of Animal Science* 86: E293–E305. https://doi.org/10.2527/jas.2007-0475.

Robbins, C.T., Spalinger, D.E., and van Hoven, W. (1995). Adaptation of ruminants to browse and grass diets: are anatomical-based browser-grazer interpretations valid? *Oecologia* 103: 208–213. https://doi.org/10.1007/BF00329082.

Rotz, C.A. (2004). Management to reduce nitrogen losses in animal production. *Journal of Animal Science* 82 (E-Suppl): E119–E137. https://doi.org/10.2527/2004.8213_supplE119x.

Rubino, F., Carberry, C., Waters, M. et al. (2017). Divergent functional isoforms drive niche specialisation for nutrient acquisition and use in rumen microbiome. *The ISME Journal* 11: 932–944. https://doi.org/10.1038/ismej.2016.172.

Russell, J.B. (2002). *Rumen Microbiology and its Role in Ruminant Nutrition*. Ithaca, NY: Cornell University.

Russell, J.B. and Chow, J.M. (1993). Another theory for the action of ruminal buffer salts: decreased starch fermentation and propionate production. *Journal of Dairy Science* 76: 826–830. https://doi.org/10.3168/jds.S0022-0302(93)77407-X.

Russell, J.B. and Wallace, R.J. (1997). Energy-yielding and energy-consuming reactions. In: *The Rumen Microbial Ecosystem* (ed. P.N. Hobson and C.S. Stewart), 246–282. London: Blackie Academic & Professional. https://doi.org/10.1007/978-94-009-1453-7_6.

Saunois, M., Jackson, R.B., Bousquet, P. et al. (2016). The growing role of methane in anthropogenic climate change. *Environmental Research Letters* 11: 120207. https://doi.org/10.1088/1748-9326/11/12/120207.

Scardovi, V. (1965). The fructose-6-phosphate shunt as a peculiar pattern of hexose degradation in the genus *Bifidobacterium*. *Annals of Microbiology and Enzymology* 15: 19–29.

Scheller, H.V. and Ulvskov, P. (2010). Hemicelluloses. *Annual Review of Plant Biology* 61: 263–289. https://doi.org/10.1146/annurev-arplant-042809-112315.

Schöcke, L. and Weimer, P.J. (1997). Purification and characterization of phosphoenolpyruvate carboxykinase from the anaerobic ruminal bacterium *Ruminococcus flavefaciens*. *Archives of Microbiology* 167: 289–294. https://doi.org/10.1007/s002030050446.

Schwab, C.G. and Broderick, G.A. (2017). A 100-year review: protein and amino acid nutrition in dairy cows. *Journal of Dairy Science* 100: 10094–10112. https://doi.org/10.3168/jds.2017-13320.

Scott, R.I., Williams, T.N., and Lloyd, D. (1983). Oxygen sensitivity of methanogenesis in rumen and anaerobic digester populations using mass spectrometry. *Biotechnology Letters* 5: 375–380. https://doi.org/10.1007/BF00131275.

Seeliger, S., Janssen, P.H., and Schink, B. (2002). Energetics and kinetics of lactate fermentation to acetate and propionate via methylmalonyl-CoA or acrylyl-CoA. *FEMS Microbiology Letters* 211: 65–70. https://doi.org/10.1111/j.1574-6968.2002.tb11204.x.

Seshadri, R., Leahy, S.C., Attwood, G.T. et al. (2018). Cultivation and sequencing of rumen microbiome members from the Hungate1000 collection. *Nature Biotechnology* 36: 359. https://doi.org/10.1038/nbt.4110.

Shahidi, F. and Ambigaipalan, P. (2018). Omega-3 polyunsaturated fatty acids and their health benefits. *Annual Review of Food Science and Technology* 9: 345–381. https://doi.org/10.1146/annurev-food-111317-095850.

Smith, C., Nicholls, Z.R.J., Armour, K. et al. (2021). The Earth's energy budget, climate feedbacks, and climate sensitivity supplementary material. In: *Climate Change 2021: The Physical Science Basis. Contribution of Working Group I to the Sixth Assessment Report of the Intergovernmental Panel on Climate Change* (ed. V. Masson-Delmotte, P. Zhai, A. Pirani, et al.). The Intergovernmental Panel on Climate Change.

Snelling, T.J. and Wallace, R.J. (2017). The rumen microbial metaproteome as revealed by SDS-PAGE. *BMC Microbiology* 17: 9. https://doi.org/10.1186/s12866-016-0917-y.

Söllinger, A., Tveit, A.T., Poulsen, M. et al. (2018). Holistic assessment of rumen microbiome dynamics through quantitative metatranscriptomics reveals multifunctional redundancy during key steps of anaerobic feed degradation. *mSystems* 3: e00038–e00018. https://doi.org/10.1128/mSystems.00038-18.

Stanier, G. and Davies, A. (1981). Effects of the antibiotic monensin and an inhibitor of methanogenesis on in vitro continuous rumen fermentations. *British Journal of Nutrition* 45: 567–578. https://doi.org/10.1079/bjn19810135.

Stewart, C.S., Flint, H.J., and Bryant, M.P. (1997). The rumen bacteria. In: *The Rumen Microbial Ecosystem* (ed. P.N. Hobson and C.S. Stewart), 10–72. London: Blackie Academic & Professional. https://doi.org/10.1007/978-94-009-1453-7_2.

St-Pierre, B. and Wright, A.-D.G. (2012). Molecular analysis of methanogenic archaea in the forestomach of the alpaca (*Vicugna pacos*). *BMC Microbiology* 12: 1. https://doi.org/10.1186/1471-2180-12-1.

Subharat, S., Shu, D., Zheng, T. et al. (2015). Vaccination of cattle with a methanogen protein produces specific antibodies in the saliva which are stable in the rumen. *Veterinary Immunology and Immunopathology* 164: 201–207. https://doi.org/10.1016/j.vetimm.2015.02.008.

Szopa, S., Naik, V., Adhikary, B. et al. (2021). Short-lived climate forcers. In: *Climate Change 2021: The Physical Science Basis. Contribution of Working Group I to the Sixth Assessment Report of the Intergovernmental Panel on Climate Change* (ed. V. Masson-Delmotte, P. Zhai, A. Pirani, et al.), 817–922. Cambridge, UK, and New York, US: Cambridge University Press. https://www.ipcc.ch/report/ar6/wg1/chapter/chapter-6.

Terry, S., Romero, C., Chaves, A. et al. (2020). Nutritional factors affecting greenhouse gas production from ruminants: implications for enteric and manure emissions. In: *Improving Rumen Function* (ed. C.S. Mcsweeney and R.I. Mackie), 505–546. Cambridge, UK: Burleigh Dodds Science Publishing Limited. https://doi.org/10.19103/AS.2020.0067.16.

Thauer, R.K., Jungermann, K., and Decker, K. (1977). Energy conservation in chemotrophic anaerobic bacteria. *Bacteriological Reviews* 41: 100–180. https://doi.org/10.1128/br.41.1.100-180.1977.

Thauer, R.K., Kaster, A.K., Seedorf, H. et al. (2008). Methanogenic archaea: ecologically relevant differences in energy conservation. *Nature Reviews. Microbiology* 6: 579–591. https://doi.org/10.1038/nrmicro1931.

Thornton, P.K. (2010). Livestock production: recent trends, future prospects. *Philosophical Transactions of the Royal Society B: Biological Sciences* 365: 2853–2867. https://doi.org/10.1098/rstb.2010.0134.

Toral, P.G., Monahan, F.J., Hervás, G. et al. (2018). Review: modulating ruminal lipid metabolism to improve the fatty acid composition of meat and milk. Challenges and opportunities. *Animal* 12: s272–s281. https://doi.org/10.1017/S1751731118001994.

Umbarger, H.E. (1978). Amino acid biosynthesis and its regulation. *Annual Review of Biochemistry* 47: 532–606. https://doi.org/10.1146/annurev.bi.47.070178.002533.

Ungerfeld, E.M. (2013). A theoretical comparison between two ruminal electron sinks. *Frontiers in Microbiology* 4: 319. https://doi.org/10.3389/fmicb.2013.00319.

Ungerfeld, E.M. (2015). Shifts in metabolic hydrogen sinks in the methanogenesis-inhibited ruminal fermentation: a meta-analysis. *Frontiers in Microbiology* 6: 1–17. https://doi.org/10.3389/fmicb.2015.00037.

Ungerfeld, E.M. (2020). Metabolic hydrogen flows in rumen fermentation: principles and possibilities of interventions. *Frontiers in Microbiology* 11: https://doi.org/10.3389/fmicb.2020.00589.

Ungerfeld, E. and Hackmann, T. (2020). Factors influencing the efficiency of rumen energy metabolism. In: *Improving Rumen Function* (ed. C.S. Mcsweeney and R.I. Mackie), 421–466. Cambridge, UK: Burleigh Dodds Science Publishing. https://doi.org/10.19103/AS.2020.0067.14.

Ungerfeld, E.M. and Kohn, R.A. (2006). The role of thermodynamics in the control of ruminal fermentation. In: *Ruminant Physiology* (ed. K. Sejrsen, T. Hvelplund, and M.O. Nielsen), 55–85. Wageningen, The Netherlands: Wageningen Academic Publishers.

United Nations Environment Programme and Climate and Clean Air Coalition (2021). *Global Methane Assessment: Benefits and Costs of Mitigating Methane Emissions*. Nairobi, Kenia: United Nations.

Vasta, V., Daghio, M., Cappucci, A. et al. (2019). Invited review: plant polyphenols and rumen microbiota responsible for fatty acid biohydrogenation, fiber digestion, and methane emission: experimental evidence and methodological approaches. *Journal of Dairy Science* 102: 3781–3804. https://doi.org/10.3168/jds.2018-14985.

Vigne, J.-D. (2015). Early domestication and farming: what should we know or do for a better understanding? *Anthropozoologica* 50: 123–150. https://doi.org/10.5252/az2015n2a5.

Voet, D. and Voet, J.G. (1995). *Biochemistry*. New York, USA: Wiley.

Waghorn, G. (2008). Beneficial and detrimental effects of dietary condensed tannins for sustainable sheep and goat production—progress and challenges. *Animal Feed Science and Technology* 147: 116–139. https://doi.org/10.1016/j.anifeedsci.2007.09.013.

Wallace, R.J. (1996). Ruminal microbial metabolism of peptides and amino acids. *Journal of Nutrition* 126: 1326S–1334S. https://doi.org/10.1093/jn/126.suppl_4.1326S.

Wallace, R.J., Onodera, R., and Cotta, M.A. (1997). Metabolism of nitrogen-containing compounds. In: *The Rumen Microbial Ecosystem* (ed. P.N. Hobson and C.S. Stewart), 283–328. London, UK: Blackie Academic & Professional. https://doi.org/10.1007/978-94-009-1453-7_7.

Wallnöfer, P., Baldwin, R.L., and Stagno, E. (1966). Conversion of c-labeled substrates to volatile fatty acids by the rumen microbiota. *Applied Microbiology* 14: 1004–1010. https://doi.org/10.1128/am.14.6.1004-1010.1966.

Wanapat, M., Kang, S., and Polyorach, S. (2013). Development of feeding systems and strategies of supplementation to enhance rumen fermentation and ruminant production in the tropics. *Journal of Animal Science and Biotechnology* 4: 32. https://doi.org/10.1186/2049-1891-4-32.

Wang, G.D., Xie, H.B., Peng, M.S. et al. (2014). Domestication genomics: evidence from animals. *Annual Review of Animal Biosciences* 2: 65–84. https://doi.org/10.1146/annurev-animal-022513-114129.

Wedlock, D.N., Janssen, P.H., Leahy, S.C. et al. (2013). Progress in the development of vaccines against rumen methanogens. *Animal: An International Journal of Animal Bioscience* 7 (Suppl 2): 244–252. https://doi.org/10.1017/s1751731113000682.

Weimer, P.J. (1996). Why don't ruminal bacteria digest cellulose faster? *Journal of Dairy Science* 79: 1496–1502. https://doi.org/10.3168/jds.S0022-0302(96)76509-8.

Weimer, P.J. (2015). Redundancy, resilience, and host specificity of the ruminal microbiota: implications for engineering improved ruminal fermentations. *Frontiers in Microbiology* 6: 1–16. https://doi.org/10.3389/fmicb.2015.00296.

Weiss, W.P. (2017). A 100-year review: from ascorbic acid to zinc—mineral and vitamin nutrition of dairy cows. *Journal of Dairy Science* 100: 10045–10060. https://doi.org/10.3168/jds.2017-12935.

Wenner, B.A., de Souza, J., Batistel, F. et al. (2017). Association of aqueous hydrogen concentration with methane production in continuous cultures modulated to vary pH and solids passage rate. *Journal of Dairy Science* 100: 5378–5389. https://doi.org/10.3168/jds.2016-12332.

Wilkinson, J.M. and Lee, M.R.F. (2018). Review: use of human-edible animal feeds by ruminant livestock. *Animal: An International Journal of Animal Bioscience* 12: 1735–1743. https://doi.org/10.1017/s175173111700218x.

Wilkinson, J.M., Lee, M.R.F., Rivero, M.J. et al. (2020). Some challenges and opportunities for grazing dairy cows on temperate pastures. *Grass and Forage Science* 75: 1–17. https://doi.org/10.1111/gfs.12458.

Wolin, M.J., Miller, T.L., and Stewart, C.S. (1997). Microbe-microbe interactions. In: *The Rumen Microbial Ecosystem* (ed. P.N. Hobson and C.S. Stewart), 467–491. London: Blackie Academic & Professional. https://doi.org/10.1007/978-94-009-1453-7_11.

Xu, Q., Qiao, Q., Gao, Y. et al. (2021). Gut microbiota and their role in health and metabolic disease of dairy cow. *Frontiers in Nutrition* 8: 701511. https://doi.org/10.3389/fnut.2021.701511.

Zeder, M.A. (2012). Pathways to animal domestication. In: *Biodiversity in Agriculture: Domestication, Evolution, and Sustainability* (ed. A.B. Damania, C.O. Qualset, P.E. Mcguire, et al.), 227–259. Cambridge, UK: Cambridge University Press. https://doi.org/10.1017/CBO9781139019514.013.

Zhao, G. (2019). Improving feed protein utilization rate in cattle through nutritional approaches. *Current Protein & Peptide Science* 20: 164–171. https://doi.org/10.2174/1389203719666180514153236.

Zheng, Y., Kahnt, J., Kwon, I.H. et al. (2014). Hydrogen formation and its regulation in *Ruminococcus albus*: involvement of an electron-bifurcating [FeFe]-hydrogenase, of a non-electron-bifurcating [FeFe]-hydrogenase, and of a putative hydrogen-sensing [FeFe]-hydrogenase. *Journal of Bacteriology* 196: 3840–3852. https://doi.org/10.1128/jb.02070-14.

CHAPTER 5

MICROBIAL DEGRADATION OF LIGNOCELLULOSE IN NATURAL AND ENGINEERED SYSTEMS – FROM THE SMALLEST TO THE BIGGEST BIOREACTOR

XAVIER GOUX[1], TONG LIU[2], MARIA WESTERHOLM[3], AND MAGDALENA CALUSINSKA[1]

[1]Environmental Research and Innovation Department, Luxembourg Institute of Science and Technology, rue du Brill 41, L-4422, Belvaux, Luxembourg

[2]Department of Forest Ecology and Management, Swedish University of Agricultural Sciences, Umeå, Sweden

[3]Department of Molecular Sciences, Swedish University of Agricultural Sciences, Uppsala, Sweden

CONTENTS

Microbial Fermentations in Nature and as Designed Processes, First Edition. Edited by Christon J. Hurst.

5.1 INTRODUCTION TO LIGNOCELLULOSE AND ITS DEGRADATION IN VARIOUS SYSTEMS

Several biological systems have evolved specialized enzymes and digestive strategies for plant biomass processing, including lignocellulose. Lignocellulose is the most abundant renewable complex of biopolymers on Earth, and its deconstruction is observed under a wide range of environmental conditions, sometimes extreme in terms of pH, temperature, redox potential, and pressure (Cragg et al. 2015). Diverse organisms, from mammals to bacteria, contribute to its efficient degradation under anaerobic or anoxic conditions, through the process commonly known as anaerobic digestion (AD). In nature, AD takes place in plenty of ecosystems such as marine and lake sediments, riverbeds, and paddy fields (Schnurer and Jarvis 2010). However, the most efficient natural lignocellulose-degrading systems are animal gut-associated habitats, including the digestive tracts of ruminants and lignocellulose-consuming insects (Godon et al. 2013). The animal gut is a complex multi-domain system where the symbiosis between prokaryotes, protists, fungi, and the host has been established over the evolutionary timescale. Being inspired by nature, man has tried to industrialize the AD process and thereby benefit from the hydrolytic potential of animal systems that digestively perform plant biomass degradation. The technique for industrializing these processes uses microorganisms to convert organic feedstock into methane-containing biogas within a closed anaerobic bioreactor ecosystem. Although significant developments have been made over the last decades to improve the industrial anaerobic digestion processes as summarized by De Vrieze (2020), currently they still are less efficient at lignocellulose degradation when compared to natural systems, especially when compared to the highly efficient herbivorous animal gut (Godon et al. 2013).

5.1.1 Past and Current Biomass Utilization: Perspectives and Challenges

In a world of finite biological resources and ecosystems, the agenda of the United Nations Sustainable Development Goals challenges the scientific community to develop disruptive and transformative technologies that would enable the replacement of petroleum-based raw materials and energy with renewable biobased feedstocks, e.g. lignocellulose (Figure 5.1). Biorefinery is

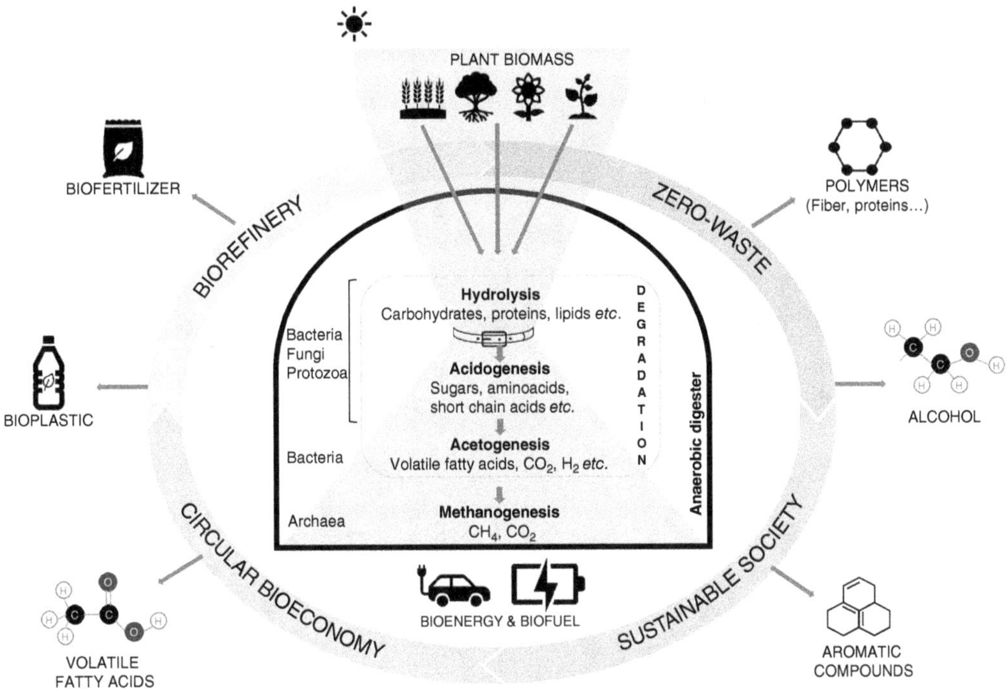

FIGURE 5.1 Schematic overview of the anaerobic digestion of plant biomass and its position within the lignocellulose biorefinery concept. (*See colour plate section for colour representation of this figure.*)

a newly defined concept that describes a multidisciplinary and complex approach of (plant) biomass processing into several marketable products (Aresta et al. 2012). In the second-generation biorefinery concept, agricultural crops and residues (e.g. grass, sugar beet leaves), and other lignocellulosic waste are processed for the production of bio-based products such as proteins, fibers, organic acids, or alcohols (Kamusoko et al. 2021). The process residues can thereafter be used as feedstock for biofuel (e.g. biogas through AD) and biofertilizer production (Martinez et al. 2018). The inclusion of biorefineries into the biomass utilization loop has the potential to further enhance carbon reutilization and contribute to the development of a circular and sustainable society.

Well before the biomass biorefinery concept emerged in the twentieth century, for millennia the microbial processing (i.e. fermentation) of (plant) biomass was used in food and feed production (Table 5.1). Milk fermentation leading to cheese production seems to have been established concomitantly with the domestication of plants and animals some 8000 years ago in the "Fertile Crescent" (Fox and McSweeney 2017). Egyptians and Sumerians profited from alcoholic and dough fermentations of plant biomass to produce wine, beer, and bread already between 4000 and 2000 BC (Ross et al. 2002). However, it was only in the nineteenth century that Louis Pasteur evidenced the essential role of microorganisms in the fermentation process. Indeed, food fermentation entirely depends on the biological activity of microorganisms, which results in oxygen consumption suppressing the growth and survival of undesirable microflora in foodstuffs, and the production of different metabolites, such as volatile fatty acids (VFAs), that lower the product pH (Ross et al. 2002). Building on this past knowledge and relying on the domestication of specific lactic acid bacteria and yeasts, microbial biomass fermentation has been industrialized over time to develop large-scale food production chains. Industrial lignocellulose fermentation is currently mainly employed in waste stabilization for compost and energy production, as well as to produce

TABLE 5.1 Main industrial processes relying on the microbial degradation of (plant) biomass: past, present, and future concepts.

Process	Products and Substrates	Main Players	Purpose	References/Remarks/Challenges
Food fermentation	Yogurt, cheese, kefir, sauerkraut, kimchi, Miso, Bahtura, Ambali, Chilra, wine, beer, sake, salami, surströmming, etc. obtained from milk and milk casein, vegetables, cereals, meat, fish, etc.	Lactic acid bacteria like *Enterococcus*, *Streptococcus*, *Leuconostoc*, *Lactobacillus*; *Pediococcus*; yeasts; molds as *Debaryomyces*, *Kluyveromyces*, *Saccharomyces*, *Geotrichium*, *Mucor*, *Penicillium*; *Rhizopus* species.	To preserve, enhance the nutritional quality (vitamins, antioxidant, etc. content), improve appearance, flavors and aroma of food. To improve the digestion, reduce the energy needed for cooking and improve the safety of food (inhibit/reduce undesirable flora).	Ross et al. (2002), Sharma et al. (2020) Remarks: Microorganisms involved in food fermentation are source of enzymes (such as amylase, proteinase, mannanase, cellulase, etc.) for industrial applications. Challenges: (i) improve microbial culture performance and activities to enhance the safety, quality, and composition of food supply, (ii) using knowledge generated by the recent molecular technics against food pathogenic organisms to design novel antimicrobial compounds.
Biowaste stabilization – composting	Organic and inorganic by-products that can be used safely and beneficially as biofertilizers and soil amendments obtained from every type of degradable organic waste.	Bacteria, molds, and yeasts.	A safe method of waste management: helps to protect underground water from becoming polluted compared to the landfilling method of waste disposal, allows a reduction of pathogenic microbes and chemical pollutants, to produce compost which helps to improve soil fertility and plant yield.	Ayilara et al. (2020) Remarks: the role of insects (and thus their gut microbiome) cannot be underestimated in composting. Challenges: (i) identification and control of the involved microorganisms to improve the composting process efficiency and speed; (ii) investigate the possibility to extract mono-nutrients from compost to avoid over-application of nutrients not needed; (iii) to develop compost process management tools (indicators and sensors); (iv) develop odor and CO_2 trapping to prevent their release.

Industrial production of volatile fatty acids (VFAs)	Acetate, propionate, butyrate, etc. obtained i.e. from wood, grass, energy crops, agricultural, and forest residues.	*Bacteroidetes*, *Proteobacteria (Enterobacter)*, *Firmicutes (Clostridium)*, etc.	VFAs can be further converted into chemicals, materials (e.g. bioplastics) and biofuels.	Sun et al. (2021) Challenges: (i) keeping low operating cost when enhancing VFA accumulation and simultaneously prevent its consumption; (ii) VFAs recovery to avoid inhibition of the process and ensure continuous system; (iii) improve pre-treatment required for the separation of cellulose and hemicellulose from lignin.
Industrial production of liquid biofuel	Bioethanol obtained from sugarcane bagasse, corn stover, wheat bran, coffee mucilage, wheat straw, oat spelt, etc.	*Zymomonas mobilis*; *Saccharomyces cerevisiae*; *Scheffersomyces stipites*.	Second generation biofuel production to replace fossil energy.	Robak and Balcerek (2020) Challenges: (i) improve strategies to remove the toxic substances (for the microbes) produced during the process; (ii) improve cell immobilization to enhance process yield.
	Biobutanol obtained from glucose, xylose, galactose, arabinose, etc. (belonging to starchy feedstocks such as corn or molasses).	Solventogenic *Clostridia* such as *Clostridium acetobutylicum*, *Clostridium beijerinckii*, *Clostridium saccharobutylicum*, *Clostridium saccharoperbutylacetonicum*, etc.	Potential alternative to fossil fuel (gasoline).	Jiang et al. (2019) Challenges: (i) optimize the biological pre-treatment and saccharification process; (ii) develop and optimize the concept of microbial co-cultivation to improve the yields and the resilience; (iii) develop continuous fermentation and recovery process to decrease the cost of production.
Industrial production of gaseous biofuel	Biohydrogen through the dark and photo fermentations obtained from agro-industrial residues (such as sugarcane bagasse, corn stalk, stover, cobs, and bran, wheat and oat straw, sorghum, rice straw, bran, and husk, etc.).	*Enterobacter*, *Bacillus*, *Clostridium*, mixed culture (anaerobic sludge from wastewater treatment plant for example), different *Rhodobacter* species.	To replace fossil fuels by producing versatile carbon rich renewable and low-cost resources.	Singh et al. (2021) Challenges: to improve the process yield, future research should be done on (i) substrate selection (to reduce pre-treatment cost); (ii) developing the use of nanomaterials as catalyst; (iii) the metabolic manipulation of the involved microbes (genetic engineering).

(*Continued*)

TABLE 5.1 (Continued)

Process	Products and Substrates	Main Players	Purpose	References/Remarks/Challenges
	Biogas obtained from numerous organic materials: sewage sludge, food waste, organic fraction of municipal solid waste, unsorted municipal waste, industrial waste (e.g. from food industry), animal and agricultural by-products, sequential and energy crops, etc.	Archaea: e.g. from genera *Methanosaeta*, *Methanosarcina*, *Methanothrix*, *Methanobacterium*, *Methanococcus*, etc.; Bacteria: e.g. from genera, *Syntrophomonas*, *Clostridium*, *Bacteroides*, *Pseudomonas*, *Bacillus*, *Streptococcus, Eubacterium,* etc.; Eukaryotes: e.g. from phylotypes Fungi, Animalia, Protista, and Plantae.	To convert organic matter into biogas including biomethane (then into energy [electricity, heat], liquid biofuel).	Brémond et al. (2021), Castellano-Hinojosa et al. (2018)) Challenges: (i) reduce biogas production cost; (ii) increase sector sustainability; (iii) set viable and sustainable support schemes; (iv) increase acceptance by local population; (v) increase the side recovery of biohydrogen and biochemicals such as acids or alcohols having a higher added value than biogas by using control strategies to redirect the microbiome toward the production of these elements.
Industrial production of plant natural products (PNPs) particularly when that type of production is accomplished by using microbial processes	Terpenoids, flavonoids, alkaloids, etc. obtained from tyrosin/ phenylalanine, glucose, glycerol.	*S. cerevisiae, Escherichia coli*	To produce chemical compounds or substances coming from plants, having a pharmacological or biological activity.	Zhou et al. (2014) Challenges: (i) identification and characterization of the enzymes involved in the PNP biosynthesis; (ii) modification of these enzymes in plants to improve their compatibility in host microorganisms; (iii) design more stable integrative and non-inducible expression systems; (iv) development of more reliable in silico model-based pathway design programs; (v) enhance the tolerance of the host microorganisms to the heterologous PNPs.

liquid biofuels (Table 5.1). In the future, different lignocellulosic (waste)streams, could also be channeled toward the production of food, feed, materials, and other (high) value-added chemicals, thus orienting microbial lignocellulosic processing toward the generation of both, nutritional (Javourez et al. 2021) and non-nutritional services within the biorefinery sector (Usmani et al. 2021). However, several challenges must be addressed before an efficient upcycling of lignocellulosic carbon can be achieved at the industrial scale. These include economic and technological aspects such as the viability of the supply chains, scale-up efforts, and the technical maturity of technologies used in the sector. In addition, a better understanding and unlocking of the microbial lignocellulolytic potential hidden in natural systems is needed to break down the lignocellulose recalcitrance (i.e. resistance to enzymatic degradation).

5.1.2 Current End-use of Lignocellulose-rich Material

The first-generation energy crops are easily degradable and have been widely used for biofuel production (Table 5.2). Typically, they are edible crops, which questions their sustainability and raises a debate on the relative merits of their usage for food or feed versus fuel production. Moreover, the environmental impact of first-generation energy crops is very high, considering the

TABLE 5.2 Composition of the most common lignocellulosic biomasses of industrial relevance.

Biomass	Cellulose (%)	Hemicellulose (%)	Lignin (%)	References
1st-generation energy crops → biofuels production (i.e. bioethanol, biogas)				
Amur silver-grass	42.0	30.1	7.0	Kikas et al. (2016)
Maize silage	28.8 ± 6.0	31.6 ± 4.7	5.8 ± 1.3	Amon et al. (2007)
Hemp	53.8	10.6	8.7	Kikas et al. (2016)
Rye	42.8	27.8	6.5	Raud et al. (2015)
Sugarcane	4.8 (wt%)	4.5 (wt%)	2.6 (wt%)	Furlan et al. (2012)
Sugarbeet	22–24	26–32	1–2	Michel et al. (1988)
Sunflower	34.0	5.2	7.7	Kikas et al. (2016)
Sweet sorghum	45.0	27.0	21.0	Kim and Day (2011)
2nd-generation energy crops → biorefinery				
Barley/wheat straw	31–38	26–38	14–19	Rocha-Meneses et al. (2017)
Corn cobs/stalk	35.0–45.70	35.0	15.0	Prasad et al. (2007)
Hardwood/barks	22–40	20–38	30–55	Rocha-Meneses et al. (2017)
Miscanthus	40	18	25	Cheng and Timilsina (2011)
Softwood/barks	18–38	15–33	30–60	Rocha-Meneses et al. (2017)
Sorghum stalk/straw	27–32	24–25	11–13	Rocha-Meneses et al. (2017)
Sugarcane bagasse	35.5–43.2	15.5–32.9	15.0–32.0	Alokika et al. (2021), Dionísio et al. (2021)
Switchgrass	28.5–45.0	29.7–32.0	6.28–7.32	Chaganti et al. (2021), Larnaudie et al. (2021), Ogden and Ileleji (2021)
Willow	57.6	14.0	19.0	Karp and Shield (2008)
Yellow poplar	41.86–43.0	15.72–17.0	27.09–28.0	Jeong et al. (2021)

soil erosion, nitrate leaching, and run-off often associated with their cultivation (Hassan et al. 2019). To overcome this, agricultural residues and short-rotation non-food grasses able to grow on marginal lands were proposed as the second-generation energy feedstock (i.e. lignocellulosic feedstock; Table 5.2). However, while the industrialization of first-generation biofuels was achieved relatively easily, the considerably higher recalcitrance of the second-generation energy crops impedes their wider utilization.

Lignocellulose is the major structural component of plant biomass and annually around 200 billion tons of lignocellulosic biomass are produced (da Silva et al. 2012). The World Bioenergy Association reported in 2016 that lignocellulosic waste accounted for 97% of the biomass supply, the remaining 3% being made up of food waste and wastewater sludge. While lignocellulose could have applications in different sectors, e.g. pharmaceutical products, food ingredients, fuels, chemicals, and bioplastics in addition to bioenergy production, currently a considerable part of lignocellulosic biomass is wasted (Usmani et al. 2021). Especially, the valorization of lignocellulosic biomass unfit for human and animal consumption including agricultural waste, wastes from forest, pulp, and paper industries, etc., would enable bio-based value chains (and biorefineries integrating such value chains) to become more sustainable and eventually to become zero-waste (Figure 5.1). To this end, there is a strong need to integrate better biomass-producing and processing sectors, with a general objective to faster, master, and reduce the cost of lignocellulose utilization, thereby supporting the establishment of sustainable bio-based value chains.

Lignocellulose Structure Lignocellulose constitutes varying proportions of lignin, hemicellulose, and cellulose fractions that are intricately linked through covalent and non-covalent bonds. Cellulose is a polysaccharide chain of glucose monomers interconnected through β-1,4 glycosidic linkages, subsequently linked together via hydrogen bonds to form microfibrils (Figure 5.2). Hemicellulose constitutes linear and branched chains of different sugar monomers linked through β-1,3 and/or β-1,4 glycosidic bonds, and typically it represents 4.5–35% of plant biomass (Table 5.2). Hemicellulosic polysaccharides differ depending on the lignocellulose source, however, they commonly include xylans, mannans, xyloglucans, and β-glucans (Figure 5.2). Xylans are often substituted with side chains of glucuronic acid or arabinose moieties, forming, respectively, glucurono- and arabinoxylans. Similarly, galactomannan comprises galactose residues branched to the mannan main chain while the glucomannan main chain is composed of mannose and glucose monomers. Lignin is composed of three phenolic building blocks (G-guaiacyl, S-syringyl, and H-p-hydroxyphenyl), and it is often ester-linked to hemicellulose, rendering the biomass recalcitrant to enzymatic hydrolysis. Lignin is often considered a by-product of many industrial sectors and more than 50 million tons of lignin are produced per year as a waste of the pulp and paper industry (Gosslink et al. 2004; Luo and Abu-Omar 2017). Nowadays, despite their multiple potential biorefining applications, hemicellulose, and especially lignin are largely underutilized and mainly considered as physical barriers limiting the enzymatic hydrolysis of cellulose.

Toward Biological Pre-treatment of Lignocellulose Due to the recalcitrance, lignocellulose hydrolysis is currently considered the rate-limiting step of many industrial processes designed to date (Himmel et al. 2007), including anaerobic digestion (Figure 5.1). Pre-treatment of lignocellulose is thus often required before industrial processing. Steam explosion with the addition of acids has been reported as one of the most efficient lignocellulose pre-treatments and is commonly applied at industrial scale. However, it is energy-consuming and its main drawback is the formation of inhibitory compounds (e.g. furfurals derived from the acid treatment of hemicellulose), that restrict further sugar utilization, e.g. via microbial processes (Heer and

(a)

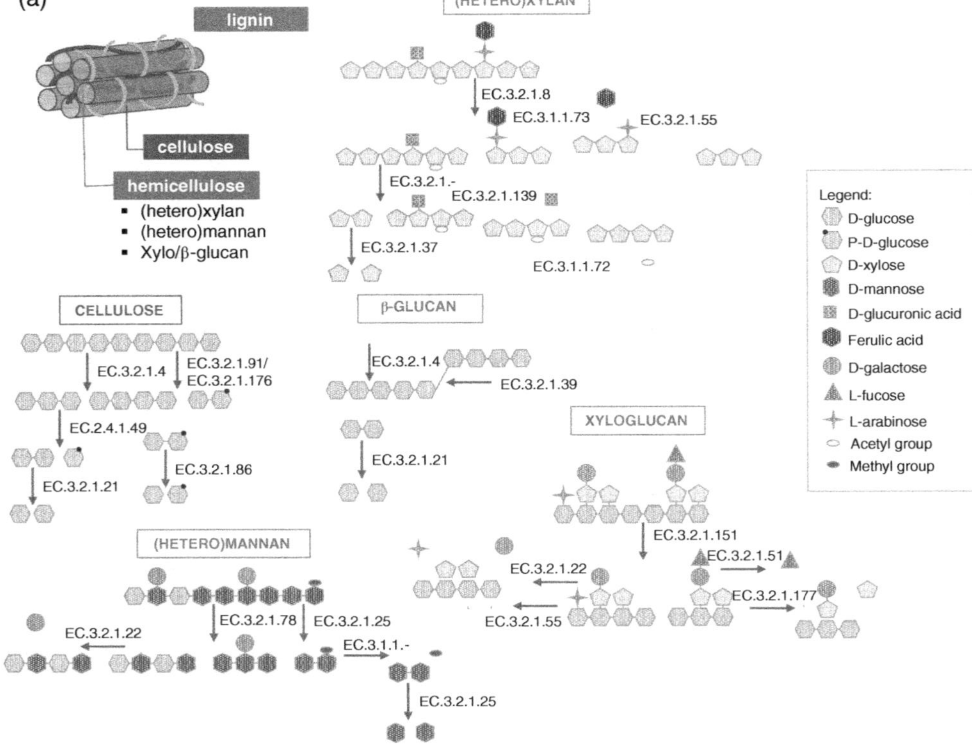

(b)

EC number	Function	Main CAZyme families
3.2.1.4	Endo-β-1,4-glucanase	GH5, 6, 7, 8, 9, 10, 12, 44, 45, 48, 51, 74, 124
3.2.1.91 3.2.1.176	Exo-β-1,4-glucanase	GH5, 6, 7, 9, 48
3.2.1.21 3.2.1.86	(phosphor)β-glucosidase	GH1, 3, 4, 5, 9, 30, 116
2.4.1.49	Cellodextrin phosphorylase	GH94
3.2.1.8	Endo-β-1,4-xylanase	GH5, 8, 10, 11, 30, 43, 51, 98, 141
3.2.1.37	β-xylosidase	GH1, 2, 3, 30, 39, 43, 51, 52, 54, 116, 120
3.1.1.72	Acetylxylan esterase	CE1, 2, 3, 4, 5, 6, 7, 12
3.1.1.73	Feruloyl esterase	CE 1
3.2.1.139	α-glucuronidase	GH 4, 67
3.2.1.55	α-L-arabinofuranosidase	GH 2, 3, 43, 51, 54, 62
3.2.1.78	endo-β-1,4-mannanase	GH 5, 26, 45, 113, 134
3.2.1.25	β-mannosidase	GH 1, 2, 5, 164
3.2.1.22	α-galactosidase	GH 4, 27, 31, 36, 57, 97, 110

FIGURE 5.2 Simplified structure of the main lignocellulosic polysaccharides and their enzymatic degradation mechanisms (a). Lignin degradation pathway is omitted here. Classification of main enzymes involved in lignocellulosic polysaccharides degradation (b). EC, enzyme commission number; CAZyme, carbohydrate-active enzyme; GH, glycoside hydrolase; CE, carbohydrate esterase. (*See colour plate section for colour representation of this figure.*)

Sauer 2008). Enzymatic (microbial) hydrolysis of lignocellulose could offer several advantages over chemical and thermal treatments in terms of sustainability issues and production cost, and especially substrate selectivity. To boost microbial hydrolysis of lignocellulose, hydrolytic enzymes including carbohydrate-active enzymes (CAZymes) and lignin-degrading oxidases have often been employed as pre-treatment (Jaramillo et al. 2015). Current strategies aim at maximizing sugar release from lignocellulose and are commonly based on e.g. iterative improvements involving corrections of individual enzyme components and the addition of new components, aiming at improving synergy and higher performance of the final enzymatic mix (Teter et al. 2014). However, the cost efficiency and implementation are still challenging and should be further reduced to promote enzymatic degradation of lignocellulosic biomasses at industrial scale.

Molecular Aspects of Lignocellulose Recalcitrance The recalcitrance of lignocellulose derives from its structural (chemical) complexity. Beyond the relatively high lignin content, the recalcitrance is also enhanced by other features, in particular acetylation and esterification with ferulic and p-coumaric acids, and cellulose tethering (non-covalent linkage) with xyloglucan (Hatfield et al. 2017). Except for cellulose, β-glucan, and cell wall glycoproteins, *O*-acetyl-substituents are present on all cell wall polymers, including on lignin (Gille and Pauly 2012). Cell wall polymer acetylation may be involved in cellulose-xylan interactions, and it was shown that acetyl-mediated steric hindrance to the binding of hydrolytic enzymes negatively influenced saccharification and further fermentation (Pauly et al. 2013). The cell wall of commelinoid monocots, including grasses such as *Miscanthus* spp., also contains large amounts of ferulic acid that are ester-linked to the cell wall polysaccharide galacto(arabino)xylan. Ferulate limits the accessibility and enzymatic digestibility of polysaccharides in grass biomass as it forms covalent linkages between polysaccharide chains (cross-linking) and polysaccharide and lignin components. In the rumen, it has been shown that reduced ferulate-mediated lignin-arabinoxylan cross-linking improved the digestibility of corn silage (Jung et al. 2011). The exact role that p-coumaroylation plays in plant cell walls remains unclear, however, it was shown that the replacement of ferulic acid with p-coumaric acids resulted in improved saccharification (Hatfield et al. 2017). The p-coumaric acid is only attached through an ester linkage to monolignols (especially S residues) and does not become cross-linked as ferulic acid. Therefore, it seems of less importance to the lignocellulose recalcitrance. Xyloglucan, binding tightly to cellulose tethers microfibrils together, leading to what is called a "tethered network." Therefore, it has been proposed that detaching xyloglucan could accelerate cellulose saccharification. Concordantly, *in vivo* expression of xyloglucan-specific endoglucanase resulted in decreased cellulose intercalation and increased enzymatic degradation of the transgenic poplar biomass (Kaida et al. 2009).

5.2 ENZYMATIC MECHANISMS OF LIGNOCELLULOSE DEGRADATION AND CARBOHYDRATE-ACTIVE ENZYMES

Over the evolutionary timescale, living organisms have been diversifying and adapting their digestive systems to become more efficient in complex biomass degradation and ingestion. They typically drive the enzymatic hydrolysis of lignocellulose by the action of carbohydrate-active enzymes (i.e. CAZymes). Fungi and bacteria excrete a large part of their hydrolytic CAZymes into the surrounding environment, where larger carbohydrate polymers are fragmented, and shorter products are further imported and metabolized inside the cell. To coordinate this cascade of reactions, many bacteria cluster their carbohydrate-active enzymes with complementary activities together in one location in the genome. A prominent example is the existence of a large diversity of so-called polysaccharide utilization loci (PULs) commonly found in genomes of members of the phylum Bacteroidetes (Lapébie et al. 2019). The PULs are defined as gene clusters encoding intracellular and cell-envelope-associated CAZymes that allow Bacteroidetes to bind onto and degrade a specific polysaccharide, and to import the released oligosaccharides inside the cell. In addition, PULs contain at least one pair of susC and susD genes encoding for a TonB-dependent transporter and a cell-surface glycan-binding protein, respectively. Similar complexes of CAZymes, often devoid of the susC and susD genes have also been detected in other bacterial phyla, including members of the phylum Spirochaetes found in the termite gut (Calusinska et al. 2020). Cellulosomes represent another example of a multi-CAZyme enzymatic complex that attaches to the plant cell wall, making the degradation more efficient as the enzymes work directly "on-site" (Bayer et al. 2007). Cellulosomes are commonly employed by bacteria of the genera *Ruminococcus*, *Clostridium*, *Acetivibrio*, and *Pseudobacteroides* (all from the phylum Firmicutes). They have been identified in different ecosystems including environments such as compost, rumen, anaerobic digesters, termite gut, human gut, soil, and sewage sludge (Artzi et al. 2017).

5.2.1 Classification of Carbohydrate-active Enzymes (CAZymes)

CAZymes are divided into four main classes, including glycoside hydrolases (GHs), carbohydrate esterases (CEs), polysaccharide lyases (PLs), and other auxiliary enzymes (AAs; (Drula et al. 2022)). GHs are the primary enzymes that cleave glycosidic bonds, and they are assisted by CEs, PLs, and AAs. Catalytic domain-containing enzymes are often associated with non-catalytic carbohydrate-binding modules (CBMs), which enhance the enzymatic association with the substrate. Typically, cellulose deconstruction is achieved through the action of endo-β-1,4-glucanases (with the enzymatic activity assigned to EC 3.2.4.1) which hydrolyze β-1,4 bonds inside cellulose and release cellobiose (Figure 5.2a). They are classified into different GH families, including GH5, GH9, GH12, GH45, and others (Figure 5.2b). Exo-β-1,4-glucanases include cellobiohydrolase I (EC 3.2.1.176) and cellobiohydrolase II (EC 3.2.1.91), respectively acting on reducing ends and non-reducing ends and releasing cellobiose and cello-oligosaccharides. Finally, β-glucosidases (EC 3.2.1.21) degrade cellobiose and cello-oligosaccharides to glucose monomers. Enzymatic hydrolysis of hemicellulose requires different sets of enzymes depending on the biomass composition. The xylan backbone in heteroxylans is depolymerized by endo-β-1,4-xylanases (EC 3.2.1.8) releasing xylo-oligosaccharides and xylobiose. Subsequently, β-xylosidases (EC 3.2.1.37) attack the extremity of the xylan chain and also target xylobiose to liberate xylose. Side-chain residues are removed by the combined action of, e.g. acetylxylan esterases (EC 3.1.1.72), feruloyl esterases (EC 3.1.1.73), α-glucuronidases (EC 3.2.1.139), and α-L-arabinofuranosidases (EC 3.2.1.55). Heteromannan degradation typically requires three types of enzymes including, endo-β-1,4-mannanases (EC 3.2.1.78), β-mannosidases (EC 3.2.1.25), and α-galactosidases (EC 3.2.1.22) in the case of galactomannans.

The process of lignin degradation is far less well understood than the degradation of lignocellulosic polysaccharides. Lignin degradation mechanisms have mainly been studied in fungi, and only a few bacteria are currently known to contribute to this process, including representatives of Actinomycetes, α- and γ-Proteobacteria (Janusz et al. 2017). Lignin degradation can also be (partially) achieved via the fungus-growing higher termites (i.e. termites that cultivate fungi in their nests, see Section 4.1.3), which rely on a symbiosis with a fungal symbiont to accomplish lignin degradation (see Section 4.1.3). The known lignin-degrading enzymes are usually divided into two categories. The first one includes lignin-modifying enzymes such as ligninolytic peroxidases and oxidoreductases (e.g. laccases). The second contains lignin-degrading auxiliary enzymes, which cannot themselves digest lignin, but support its depolymerization. In general, to degrade lignin the enzymatic machinery needs oxygen or reactive oxygen species to cleave the aromatic ring (Janusz et al. 2017), hence it does not take place in an anaerobic environment, such as an AD reactor operated under standard conditions.

5.2.2 Lytic Polysaccharide Monooxygenases (LPMOs)

As opposed to the majority of hydrolytic CAZymes, lytic polysaccharide monooxygenases (LPMOs) catalyze cellulose, hemicellulose, starch, and chitin degradation through an oxidative reaction. They have been detected in the three domains of life and even in viral genomes (Drula et al. 2022). In the CAZY database LPMOs are classified into eight AA families, including AA9, AA11, AA13, AA14, AA16, and AA17 being mainly found in fungi, AA10 in bacteria, and AA15 in viruses and metazoa. In combination with cellulose dehydrogenase (e.g. EC1.1.99.18; electron donor), they contribute to the high lignocellulose-degrading efficiency of plant cell-wall degrading fungi. The family AA10 is only present in bacteria and archaea, however, these enzymes seem to be less commonly employed by hydrolytic anaerobic bacteria, and they are mainly limited to Bacillus and some Actinobacterial species. AA15 is widespread in Eukaryotes including algae, oomycetes, and higher animals, e.g. termites and firebrats, and is active on both cellulose and chitin (Sabbadin et al. 2018). Compared to other gut-associated lignocellulolytic systems, the firebrat LPMO was shown to boost asymbiotic (i.e. without microbe assistance) cellulose digestion, rendering this animal one of the most efficient cellulose degraders in the animal world.

5.3 ANAEROBIC DIGESTION OF PLANT BIOMASS

Initially, the process of AD was designed to stabilize activated sludge from wastewater treatment (Abbasi et al. 2012). Currently, full-scale AD reactors are mainly alimented with a mixture of lignocellulosic and non-lignocellulosic polysaccharide-rich streams, including agricultural and food waste, and often energy crops. However, while part of the world's population suffers from hunger, priority should be given to food production rather than energy production from arable lands. Therefore, the production of biofuels and other added-value chemicals from lignocellulosic residues, which is a resource that does not compete with feed and food production, is gaining attention (Usmani et al. 2021).

5.3.1 The Mains Steps and Organisms Involved in the Anaerobic Degradation of Biomass

Anaerobic digestion of organic material includes four metabolic key steps, i.e. hydrolysis, acidogenesis, acetogenesis, and methanogenesis (Figure 5.1). Most of these steps are performed by different groups of organisms that coexist and cooperate, i.e. take on complementary roles. During hydrolysis, complex polymers (polysaccharides, lipids, proteins, etc.) are converted to soluble monomers (long-chain fatty acids, glycerol, amino acids, sugars, etc.) by hydrolytic bacteria (Dollhofer et al. 2017; Menzel et al. 2020). In the acidogenic step, these products are converted to VFAs, alcohols, H_2, and CO_2. During subsequent steps, these compounds are converted by the acetogens, acetate-utilizing (aceticlastic), and hydrogen or formate-utilizing (hydrogenotrophic) methanogens to methane and CO_2 (Westerholm and Schnürer 2019). Under certain operating conditions, such as in the case of the high ammonia level reactors, the aceticlastic methanogens reduce in activity due to stress and acetate is instead converted in a two-step reaction conducted by syntrophic bacteria and hydrogenotrophic methanogens (Westerholm et al. 2016).

In the first hydrolytic step, various hydrolytic microorganisms excrete extracellular hydrolytic enzymes (i.e. CAZymes) to the bulk solution or have enzymes attached to their cell wall to break down polymers into simpler substrates which are further consumed by other microbes as a source of energy and nutrients (e.g. during the fermentative growth). Some microorganisms secrete a wide diversity of enzymes, allowing them to break down different types of organic materials. Others are more specialized in the digestion of specific biomasses or lignocellulose fractions (Blair et al. 2021).

Although engineered AD is widely used today, the microbial players as well as their associations and functioning are far from being understood (Carballa et al. 2015). This is mainly due to the high diversity of involved microorganisms, which often are uncharacterized (i.e. without cultivated representatives). However, studies have shown that cellulose and hemicellulose degraders belong to a wide span of taxonomic groups including representatives within the genera *Brachybacterium, Cellulomonas* (phylum Actinobacteriota), *Acetivibrio*, *Caldicoprobacter*, *Ruminiclostridium*, the order Clostridia MBA03, the family Ruminococcaceae (phylum Bacillota) and the genera *Ruminofilibacter* (phylum Bacteroidetes), *Fibrobacter* (phylum Fibrobacteres), and *Treponema* (phylum Spirochaetes) (Izquierdo et al. 2014; Jensen et al. 2021; Jia et al. 2018; Neumann and Suen 2018; Struckmann Poulsen et al. 2022). Furthermore, genomic analyses have led to the suggestion that genes encoding enzymes needed for the degradation of cellulose existing in yet uncharacterized members of the phyla Bacteroidetes, Pseudomonadetes and the *Ca.* Bipolaricaulota, Cloacimonetes and Actinomycetes could also contribute to biomass hydrolysis (Bertucci et al. 2019; Struckmann Poulsen et al. 2022; Westerholm and Schnürer 2019).

5.3.2 Techniques Aimed to Improve the Anaerobic Digestion of Lignocellulosic Feedstock

Cellulose and hemicellulose can be relatively easily degraded under anaerobic conditions. However, as previously mentioned in this chapter complex three-dimensional structure of lignin makes it considerably more difficult to degrade by microorganisms (Khan and Ahring 2019), in particular in the absence of oxygen (Janusz et al. 2017). Due to these constraints, anaerobic microorganisms need a long time to degrade lignocellulose, which often exceeds the average time that the biomass is maintained in the reactor (i.e. the hydraulic retention time is generally in a range of 20–40 days in AD processes). Consequently, lignocellulosic material is often washed-out from AD reactors after being only poorly or even undegraded, leading to losses in methane yield and enhanced risk for methane emissions during subsequent digestate storage (Ahlberg-Eliasson et al. 2021). To improve the degree of lignocellulose degradation in AD reactors, a wide range of techniques have been developed to increase the availability of organic material to microbial degradation. Such techniques include mechanical processes (milling, ultrasound), as well as chemical (e.g. acid or alkaline hydrolysis) and thermal (e.g. steam explosion) pre-treatments, which aim to disrupt the matrix of lignin and hemicellulose before AD, thereby making those components and the cellulose more readily available for enzymatic degradation.

Certain reactor designs, such as the continuous stirred-tank reactor (CSTR) have been suggested to offer advantages for the hydrolysis of lignocellulose. However, contradictory results in this matter highlight the need to establish the correlation between feedstock characteristics and the impact of co-digestion with other organic wastes (Brown et al. 2012; Ge et al. 2016). Less investigated but still with promising potential is the serial digester set-up, e.g. of several CSTRs (Li et al. 2017b). Alternatively, combinations of different reactor technologies can be used such as a two-stage thermophilic followed by a mesophilic CSTR, a leach-bed reactor combined with a CSTR, or a plug-flow reactor (PFR) combined with CSTR (Iweh et al. 2020). Moreover, several studies have tested introducing trace oxygen amounts, so-called microaeration, to improve lignocellulose hydrolysis, specifically lignin depolymerization (e.g. reviewed in Chen et al. (2020)). According to Nguyen and Khanal, an oxygen range of 0.2–0.51 per day has a positive effect on the hydrolysis stage, by enhancing the activity and diversity of microbes able to better solubilize highly recalcitrant lignocellulose (Nguyen and Khanal 2018).

Biological pre-treatment applied as a pre-digestion step or integrated (e.g. by addition of enzymes, microorganisms, or consortia) in the AD process is an alternative method that has the advantage of low energy consumption and low chemical use compared to the physical and chemical pre-treatments (Cai et al. 2021). Several microorganisms and effects of enzyme additions have been investigated to effectively pre-treat lignocellulosic materials (Herrero Garcia et al. 2019). For instance, pre-treatment with white-rot fungi (aerobic) has shown promising results regarding lignin degradation and biogas yield from the treated biomass (Kainthola et al. 2021). Besides fungi, bacterial treatments, such as the inclusion of a pre-digestion step with hyperthermophilic bacteria *Caldicellulosiruptor bescii,* have been investigated as suitable pre-treatment approaches. This treatment brings the advantage that the pre-digestion can take place at 70–80 °C, which at the same time would function as a hygienization process of the influent feedstock (Hansen et al. 2021). Various studies have demonstrated promising potential for improved degradation of corn stover, switchgrass, and energy crops by enzymatic pre-treatment using for example laccases, peroxidases, and cellulases (Dąbkowska et al. 2019). Nevertheless, the current commercial enzymes are expensive and often quickly degraded by microbes, limiting their application in large-scale biogas plants.

5.4 OVERVIEW OF THE MOST EFFICIENT LIGNOCELLULOSE-DEGRADING SYSTEMS DEVELOPED BY NATURE

Over time, animals have been improving their biomass degrading capacities by relying on feed cutting and grinding (i.e. chewing), chemical treatment (i.e. changing pH and oxygen content), changing the size and shape of their digestive organs and adapting water content and retention time. Animal digestive tracts are usually compartmented, allowing for a physical separation of the digestion processes. Ruminants such as cows, evolved stomachs divided into four chambers (i.e. reticulum, rumen, omasum, and abomasum), to prolongate the degradation process and allow the creation of slightly different niche environments which are believed beneficial for lignocellulose degradation (Bayané and Guiot 2011). Differently, from other mammalian herbivores such as rabbits (Hofmann 1988) and horses (Frape 2008), fiber digestion in cows mainly takes place in a forestomach (reticulum, rumen, and omasum) instead of the posterior part of the digestive tract (colon and caecum). This characteristic of the ruminant digestion tract enables the indigestible plant fiber repeatedly to mix in the forestomach and be mechanically ground in the mouth, facilitating the removal of plant protective structures, thus increasing the availability of microbial digestible substrates from plant tissues. Several groups of insects that thrive on a fiber-rich diet possess gut microbiota that participates in the digestion of ingested food (Table 5.3). Some of those insects can process organic matter at loads 40–100 times higher than industrial AD systems (Godon et al. 2013). Next to being efficient in lignocellulose processing, some of these insects can emit methane, and these include termites, scarab beetle larvae, and cockroaches (Brune 2018). Similar to anaerobic digestion, methane produced in the insect gut results from a symbiotic transformation of ingested biomass, making the insects valuable models for improved industrial applications. Termites represent the best-studied group of lignocellulose-degrading insects, and within hours they can completely digest 74–99% of cellulose and 65–87% of hemicellulose, making them far more efficient than other herbivores (Brune 2014). Therefore, termites will be given the major focus in this book chapter.

TABLE 5.3 Characteristics of the most efficient insect gut lignocellulose degrading systems.

Insect Gut System	Main Syntrophic Hydrolysers	Host Main Enzymatic Contribution[a]	Main Substrate	Main Hydrolysis Product	Pre-treatment[b]	Example of Potential Application	References
Lower termite	Ciliates, and to a lesser extent hydrolytic bacteria (mainly Spirochaetes, Bacteroidetes, Elusimicrobia)	Laccase and other lignolytic enzymes, Cellulase, β-glucosidase, LPMOs, other carbohydrates targeting enzymes: chitin, alpha-glucans, peptidoglycan	Sound wood	Acetate, CO_2, H_2, CH_4	Mechanical (chewing and grinding), chemical (pH and O_2 gradient)	i) Saccharification of lignocellulosic feedstock (e.g. sugarcane, bagasse, rice husk, etc.) with the gut protein extract of *Reticulitermes flavipes* and *Heterotermes indicola* ii) Boosting the saccharification of polysaccharise with the termite superoxide dismutase	Afzal et al. (2022), Franco Cairo et al. (2022)
Higher termite	Hydrolytic bacteria (mainly Firmicutes, Fibrobacteres, Spirochaetes, and Bacteroidetes), Termitomycetes (limited to fungi-growing termites), Neocallimastigomycota[c]	Laccase and other lignolytic enzymes, Cellulase, β-glucosidase, LPMO, other carbohydrates targeting enzymes: chitin, alpha-glucans, peptidoglycan	From sound wood to soil organic matter	Acetate, CO_2, H_2, CH_4 (highest amounts are produced by soil-feeders)	Mechanical (chewing and grinding), chemical (highly alkaline hindgut), enzymatic (fungal comb)	i) Degradation of insoluble xylan polysaccharides with a multimodal GH10 hemicellulase ii) Application of multifunctional GH5_4 endoglucanases for a simultaneous degradation of cellulose and hemicellulose	Calusinska et al. (2020), Wu et al. (2021)
Scarab larvae	Firmicutes, Bacteroidetes, Proteobacteria	Cellulase, β-glucosidase	Plant root, decaying wood, soil organic matter	Volatile fatty acids, CO_2, CH_4,	Mechanical (chewing and crushing), chemical (alkaline midgut)	i) Small automated production line for lignocellulose degradation technologies ii) Gut enrichment cultures producing methane from wheat straw	Huang et al. (2010); Schroeder et al. (2022), Wang et al. (2022)

(Continued)

TABLE 5.3 (Continued)

Insect Gut System	Main Syntrophic Hydrolysers	Host Main Enzymatic Contribution[a]	Main Substrate	Main Hydrolysis Product	Pre-treatment[b]	Example of Potential Application	References
Firebrat (*Thermobia domestica*)	Microbe-free (asymbiotic) and symbiotic cellulose digestion	Oxidative degradation with LPMOs, cellulase, β-glucosidase	Crystalline cellulose (paper, cardboard, flour, insulation), chitin, hemicelluloses	Volatile fatty acids, CO_2	ND	i) Degradation of crystalline cellulose at rate comparable to rumen and termite gut, with the use of endogenous LPMOs; ii) Highest relative cellulolytic activity compared to other model insect gut systems	Sabbadin et al. (2018)

LPMO – lytic polysaccharide monooxygenase; ND – no data found.

[a] Verified experimentally and based on predicted hydrolytic potential.

[b] Pre-treatment refers here to other than enzymatic mechanisms which aid to disintegrate and digest the lignocellulose (i.e. mechanical and chemical).

[c] Based on preliminary studies (Calusinska et al. 2020; Moreira et al. 2021).

5.4.1 The Termite Gut as the Most Efficient Lignocellulose-degrading Model

The termite gut system is considered the world's smallest bioreactor (Brune 1998), at the same time being the most efficient lignocellulose-degrading system on Earth. Currently, there are around 3000 recognized termite species, classified into nine families (Krishna et al. 2013). Eight families, including Archotermopsidae, Hodotermitidae, Kalotermitidae, Mastotermitidae, Rhinotermitidae, Serritermitidae, Stolotermitidae, and Stylotermitidae constitute evolutionary basal lower termites, which mainly feed on sound wood. Higher termites are evolutionary younger and are all classified into one family Termitidae. They account for the majority (over 70%) of known species and are assigned to different feeding guilds (Donovan et al. 2001). They can consume a broader range of plant material at the different stages of humification, along the wood-soil decomposition gradient (Brune 2014). The higher taxonomic diversity of termites lies toward the humified end (i.e. soil), and in tropical ecosystems, soil-feeding termites account for 60% of the known termite species (Eggleton 2000).

Compared to the mammalian herbivores which rely exclusively on their gut microbiota for lignocellulose degradation, the insect gut system, and more specifically the termite gut is far more complex (Brune and Kühl 1996). In general, the termite digestive system is composed of three main components, including (i) the host providing mechanical (i.e. the mandibles grind up or pound the substrate into micrometer size particles) and hydrolytic pre-treatment (i.e. host excretes its own cellulases of the family GH9, mainly in the midgut), (ii) the hindgut microbiome and its lignocellulose-degrading ciliates in the case of lower termites, or bacteria in higher termites, and (iii) the fungal gardens termed combs. The fungal combs are constructed by the higher termites belonging to the subfamily Macrotermitinae and are specific structures inside the termite nest, which are composed of hydrolytic fungi Termitomyces (Aanen et al. 2002).

The termite digestive tract is compartmented and consists of the foregut, midgut, and microbe-hosting hindgut (Figure 5.3; Brune and Kühl 1996). The main innovation developed by higher termites compared to lower termites include the development of a mixed segment (at the intersection of the midgut and the hindgut) and pronounced compartmentalization of the hindgut

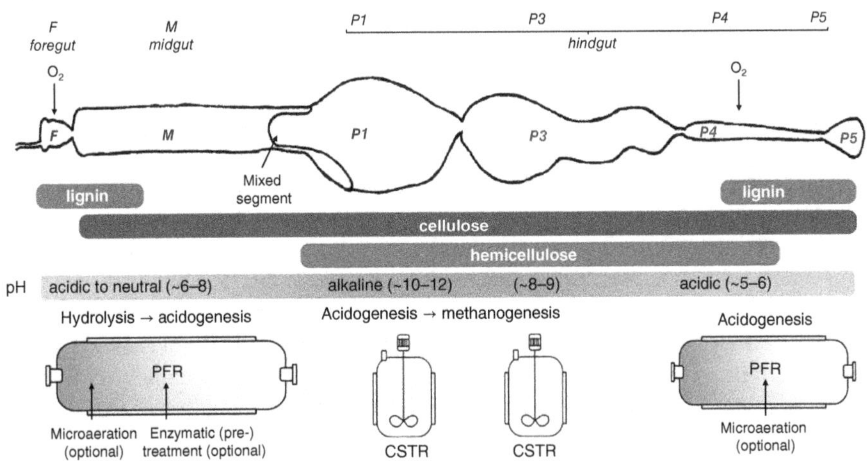

FIGURE 5.3 Characterization of the termite gut system and proposed exploitation of its potential toward improved lignocellulose utilization in engineered systems. Figure inspired by the work from Godon et al. (2013) and further modified. The highly compartmented gut model is characteristic of soil feeding termites, e.g. *Labiotermes* sp. The gut compartments involved in the degradation of the main lignocellulose fractions were identified after Marynowska et al. (2023). PFR, plug flow reactor; CSTR, continuously stirred tank reactor. (*See colour plate section for colour representation of this figure.*)

into P1 (ileum), which is often elongated, P2 (enteric valve), P3 (main paunch), P4 (colon), and P5 (rectum). The functional evolution of the termite gut microbiota has been studied for over 100 years (Scharf and Peterson 2021). A recent comparison of nearly 150 representative termite gut metagenomes revealed that the gut prokaryotic community has been remarkably conserved since termites first appeared on Earth about 150 million years ago (Arora et al. 2022).

Lower Termite Hindgut Symbionts and their Fermentative Strategies Lower termites are characterized by the presence of numerous cellulolytic flagellates in the hindgut, which can occupy more than 90% of the paunch (i.e. hindgut) volume. Flagellates belong to the phylum Parabasalia and Oxymonadida, and most of the known species do not occur elsewhere in nature, except in wood-eating *Cryptocercus* cockroaches. The expression patterns of carbohydrate targeting enzymes are different among different ciliate species (Nishimura et al. 2020), reinforcing the division of roles in wood digestion and explaining the existence of multiple species in the gut community. In higher termites, a similar labor division is assigned to the different gut prokaryotes (Arora et al. 2022; Marynowska et al. 2020). Prokaryotes in the lower termite gut reside in the lumen, remain attached to the gut epithelium or food particles, and live in endo- or exo-symbiosis with gut flagellates. Main bacterial phyla include Spirochaetes, Bacteroidetes, Firmicutes, Proteobacteria, and Elusimicrobia (Arora et al. 2022). Compared to bacteria in the higher termite gut system, lower termite gut prokaryotes are often depleted of cellulose targeting CAZymes, reflecting task partitioning between prokaryotes and eukaryotic symbionts. Typically, symbiotic prokaryotes fix nitrogen, to complement the nitrogen-limited diet of wood-feeding lower termites, and recycle hydrogen produced during the cellulolytic metabolism of a protist into acetate or methane (Ohkuma et al. 2015).

Shift into a Prokaryotic Hindgut Allowed for Diet Diversification of Higher Termites A recent analysis of the termite gut microbial communities identified 19 bacterial phyla and 16 archaeal families (Arora et al. 2022). On average archaea represent less than 1% of gut prokaryotes in wood-feeding termite species, while they reach 4.6% in fungus-growing Macrotermitinae and over 10% in some soil-feeding termites. Their abundance well correlates with methane emissions, which are considerably higher in the case of soil feeders (Brune 2018). The diversity of microbes in the higher termite gut system is much higher (in a range of hundreds to several thousand per termite species [Marynowska et al. 2020]) than the diversity of protists in lower termites (usually below 10; [Jasso-Selles et al. 2020]). The higher diversity of lignocellulolytic functions that accompany the gut microbiome diversification in the higher termite further allowed the termite host to feed on a larger range of lignocellulosic substrates and non-cellulolytic polysaccharides. It also pushed some higher termites in a direction of symbiosis with fungi in the case of Macrotermitinae (Aanen et al. 2002). The gut community of a higher termite is mainly dominated, in decreasing order, by Firmicutes, Spirochaetes, Fibrobacteres, and Bacteroidetes in the case of soil-feeding species. In turn, wood feeders mainly rely on Spirochaetes and Fibrobacteres, and to a lower extent Firmicutes (Marynowska et al. 2020). According to the profiles of sugar isomerases of the higher termite gut microbiomes, bacteria of wood feeders preferentially ferment glucose, xylose, and arabinose, the two latter found in different hemicelluloses, e.g. arabinoxylans (Marynowska et al. 2020). In turn, the soil-feeding gut prokaryotes rely on glucose, ribose, and galactosamine. In comparison to wood-feeding termites, the intestine of soil-feeding higher termites is longer, and the distinct compartments are characterized by axial and radial gradients of various physical parameters, including pH, O_2, and H_2 partial pressure (Brune and Kühl 1996; Schmitt-Wagner and Brune 1999). These innovations have transformed the soil-feeding termite gut, and especially its hindgut, into a series of complex microbial habitats densely colonized by distinct populations (Marynowska et al. 2023; Mikaelyan et al. 2017).

Externalization of Lignocellulose Digestion in Higher Termites The fascinating behavior of Macrotermitinae to cultivate hydrolytic fungi, allowed them to externalize the digestion of lignocellulose (summarized in Chouvenc et al. 2021). A (meta)genomic analysis of the termite host, gut microbiome, and the fungal comb of a fungus-growing *Macrotermes natalensis* indicated the complementarity of the three-party metabolisms to accomplish lignocellulose degradation (Poulsen et al. 2014). During the first passage through the gut of younger termite workers, crude plant material is inoculated with *Termitomyces* spores and primary feces then is deposited as a new layer of a fungal comb. The fungi symbiont can target complex carbohydrates and presumably decompose lignin (Schalk et al. 2021). Nevertheless, other studies indicate that fast and efficient lignin depolymerization already will have taken place during the passage of food through the pH-neutral gut of a young worker, thus facilitating subsequent lignin degradation via the fungus combs (Li et al. 2017a). In continuation, the older termite workers consume the comb, and the pre-treated lignocellulose is ultimately degraded by the gut microbes. The termite host also encodes its own laccases, cellulases, and other carbohydrate-active enzymes showing its contribution to lignocellulose degradation as well.

In addition to fungus gardens, some other termite lineages rely on bacterial combs, sometimes also composed of mixed bacteria-fungi communities, for external lignocellulose pre-treatment. However, this form of external symbiosis is less well studied and was only documented for a few termite species (Soukup et al. 2021). Initially, a xylophagous termite *Sphaerotermes sphaerothorax* was suggested to build two kinds of bacterial combs inside its nest and to use a similar mode of biomass digestion with the primary and secondary workers to the other Macrotermitinae species (Garnier-Sillam et al. 1989). More recently, *Cornitermes cumulans* was shown to evolve a special association with its nest microorganisms that contribute to the decomposition of plant biomass stored in the nest (Moreira et al. 2021). The profiles of lignocellulolytic enzymes were different between the nest nodules and gut microbiomes, suggesting the complementary contribution to the lignocellulose degradation of the two communities. Pre-processing of the stored plant material is mediated by bacteria and fungi residing in nest nodules, with the diversity of fungi exceeding three times those fungi presumably active in the termite gut (Moreira et al. 2021).

5.5 FROM COMPARISON TO BIOPROSPECTING: WHAT CAN WE LEARN FROM THE TERMITE GUT SYSTEM TO IMPROVE INDUSTRIAL LIGNOCELLULOSE BIOPROCESSING

Biomimicry, also called biomimetics, is an emerging trend aiming at mimicking natural systems to develop novel bio-based technologies. For maximum process efficiency, it is best to select natural systems where the process of interest takes place. Mimicking nature is nothing new and in the past rumen-inspired technologies were developed to improve lignocellulose biomass conversion in anaerobic digestion. Rumen-derived anaerobic digestion (RUDAD; [Gijzen et al. 1988]) and rumen simulating technique (RUSITEC; [Czerkawski and Breckenridge 1977]) both showed that the fermentative processes taking place in the rumen go far beyond the conditions typically applied in anaerobic reactors. Importantly, the rumen is not the most efficient lignocellulose degrading system in nature. In general, improved lignocellulose utilization by natural systems goes along with increased system complexity, which makes it even harder to mimic in engineered systems. This complexity includes not only the development of strategies aiming at the physical and chemical disintegration of lignocellulose (e.g. grinding, physicochemical gradients of O_2 and pH across the gut system, gut elongation with different feed mixing mechanisms employed in the different gut sections, etc.) but also the evolution of specific host- or organ-associated microbial communities. Termites and most of their nutritional symbionts are mutually obligate, and many microbes present in the termite gut are found nowhere else in nature. Microbes have been co-evolving together with their termite host to feed on a specific diet (Arora et al. 2022; Brune 2010). Termite gut microbial communities are vertically transmitted to the offspring, which guarantees the new generation of workers is equipped with the microflora best optimized for efficient utilization of a specific lignocellulose diet.

From an evolutionary standpoint, compared to any natural system, engineered AD is a relatively new technology of roughly 140 years old (McCarty 1982). The main development over that time was focused on the reactor engineering and process design (including the identification of key parameters such as pH, temperature, retention time, and organic loading rate), alongside the more recent development of new monitoring tools for the gas phase and the liquid digestate (i.e. for the measurement of VFAs, ammonia, etc.). At the same time, specific microbiome optimization has not been commonly taken into consideration until quite recently, when high-throughput sequencing tools started developing (De Vrieze 2020). The currently existing molecular biology tools enable scientists to gain unprecedented insights into the most efficient natural lignocellulose degrading systems. Complex phenomena used by living organisms to maximize the efficiency of lignocellulose utilization will always be out of the engineer's reach. However, modeling the AD reactor using those biologically sourced innovations, e.g. as those developed by the termite gut system, can help us advance the fundamental understanding of efficient lignocellulose processing. The inspiration from nature can be applied at different technological levels, including mimicking the whole system, or more specifically optimizing microbial communities or exploiting hydrolytic enzymes (Blair et al. 2021).

5.5.1 System-level Comparison of the Termite Gut and Anaerobic Digestion Reactors

Several comprehensive reviews comparing animal digestive strategies, including the termite gut, versus AD have been published over the last decades (Bayané and Guiot 2011; Godon et al. 2013; Rajeswari et al. 2021). The main analogy to the animal gut model for lignocellulose degradation is represented by a high concentration of VFAs, the presence of fermenting bacteria, and typical anaerobic processes including acetogenesis and methanogenesis. The main difference is the compartmentation, especially in the case of some soil-feeding higher termites, and the continuous oxygen influx, which requires efficient removal of oxygen through the activity of respiring microbes (Brune 1998). Altogether, these factors contribute to microbial community shaping, which has specifically evolved to perform different tasks in the distinct regions of the termite gut (Marynowska et al. 2023). A comparison of 190 animal digestive tracts considered efficient in lignocellulose conversion led to a conclusion that specific compartments should be considered separate "gut reactors," and the feed degradability was shown to correlate with the size of a microbial hindgut compartment (Godon et al. 2013).

According to the gut morphology and specific environmental conditions in the different compartments, industrial implementation of the termite digestive tract process was proposed to consist of a series of anaerobic reactors, including the enzymatic PFR (mimicking the host-enzyme hydrolysis) at the beginning of the process, and two CSTR reactors operating under alkaline and neutral pH conditions, respectively, with the latter hosting the specific microbiota (Godon et al. 2013). Recent employment of the omics techniques to characterize the microbiome and its lignocellulolytic strategies (e.g. [Arora et al. 2022; Marynowska et al. 2020; Marynowska et al. 2023]), could further serve to improve the proposed model. For example, the differential expression of prokaryotic CAZymes carrying different CBM domains and hydrolytic enzyme coding genes is compartment-specific in the higher (Marynowska et al. 2023) and the lower (Geng et al. 2018) termite gut systems, potentially suggesting the sequential removal of lignin, hemicellulose, and cellulose particles to be key toward efficient lignocellulose degradation. In this sense, lignin disintegration starts in the foregut, where the host enzymes contribute to its depolymerization (i.e. mimicked by the enzymatic PFR in an industrial implementation; Figure 5.3) allowing host cellulases, microbial cellulases, and hemicellulases to access the other lignocellulose fractions more easily. Lignin degradation seems to be continuing in the posterior hindgut (i.e. could be mimicked by adding another PFR at the end of the process), which is also elongated as a foregut, and in both cases, the presence of trace O_2 amounts seems essential. The effect of O_2 could be mimicked through the microaeration of the reactor, which has been shown to improve biomass hydrolysis in previous studies (Nguyen and Khanal 2018). In our model, both alkaline and neutral CSTR reactors, mimicking the P1 and P3 termite hindgut compartments, would be hosting specifically adapted microbiota. After Marynowska et al. (2023), the number and expression level of specific prokaryotic CAZymes in these two compartments indicates significant hydrolytic microbial activity taking place. Interestingly, the anterior termite hindgut is characterized by an extremely high pH (Schmitt-Wagner et al. 2003), equal to the highest values ever reported for biological systems, which brings attention to the bacterial communities living in such systems and their enzymatic machinery. Indeed, further improvements to the industrial lignocellulose bioprocessing could be made by employing specific microbes or exploiting their enzymatic potential under specifically designed conditions, as discussed below.

5.5.2 Comparison of Microbial Populations and Their Hydrolytic Potential

The structure of microbial communities and their capacity to degrade lignocellulose have been assessed in multiple studies for the termite gut (Arora et al. 2022; Marynowska et al. 2020) and using AD reactors (Calusinska et al. 2018b; Campanaro et al. 2020). However, until now a wide-ranging comparison between the two systems has not been made. Therefore, for the purpose of this chapter, we have filled this research gap by assembling information from published studies within these areas. To compare the lignocellulolytic potential of microbial communities, we relied on our analysis of the high-quality metagenome-assembled genomes (i.e. MAGs with completeness >70% and contamination <10%) from two previously published collections of MAGs, relative to the anaerobic digestion (Campanaro et al. 2020) and the termite gut systems (Hervé et al. 2020). Comparative functional ecology of microbial communities serves as a proxy to main microbes and their functional traits with regards to lignocellulose deconstruction.

Community Structure Comparison Comparison of the termite gut microbiomes has shown that the overall richness and diversity of bacteria in the higher termite gut system is three to five times lower for the plant-fiber (i.e. low diverse diet in terms of the different polysaccharides) feeding termites compared to soil-feeding species (Marynowska et al. 2020). Approximately, 1000 to 1600 different bacterial species are encountered in the soil feeders, compared with 300 on average for the plant-fiber feeders. Interestingly, these numbers match the results obtained from full-scale AD reactors. Accordingly, the farm reactors that only or mainly receive agricultural by-products (manure and plant biomass representing agricultural residue) operate with less diverse microbial communities, on average 500 species, than units co-digesting manure with more diverse feedstock (e.g. organic fraction of the municipal solid waste – OFMSW). For the latter, the microbial richness is higher, but rarely achieves the high number of species found in the soil-feeding termite gut (Calusinska et al. 2018b). In the termite gut system, Firmicutes dominate the gut community of soil-feeding termites, while they seem to be outcompeted by Fibrobacteres in wood-feeders (Marynowska et al. 2020). Many termite species seem to largely depend on Spirochaetes for lignocellulose degradation, while this phylum is rather unabundant in AD reactors (Figure 5.4b; (Calusinska et al. 2018b)). By contrast, full-scale anaerobic digestion units are typically dominated by Firmicutes and Bacteroidetes, except for reactors treating municipal wastewater, which in addition are highly enriched in Proteobacteria. In the case of the termite gut system, the diversity of microbes is also gut compartment dependent, and can largely differ between the different gut segments (Marynowska et al. 2023; Mikaelyan et al. 2017). Of course, such compartmentalization allows efficient separation between the different steps of biomass degradation, which does not take place in the full-scale AD systems, where the completely mixed CSTR reactors are typically employed. In general, even full-scale AD units operating in a mode of digester and post-digester, or similar, usually do not achieve significant functional separation and the microbial communities of the sequential reactors are very similar in terms of their structures (Calusinska et al. 2018b).

Comparison of the Community Lignocellulolytic Potential Further comparative analysis of the lignocellulolytic potential was restricted to seven main bacterial phyla known to be abundant in at least one of the two systems (i.e. AD and the termite gut), including Bacteroidetes, Chloroflexi, Elusimicrobia (dominant mainly in lower termites), Fibrobacteres, Firmicutes, Proteobacteria, and Spirochaetes. CAZymes were annotated with the dbCAN2

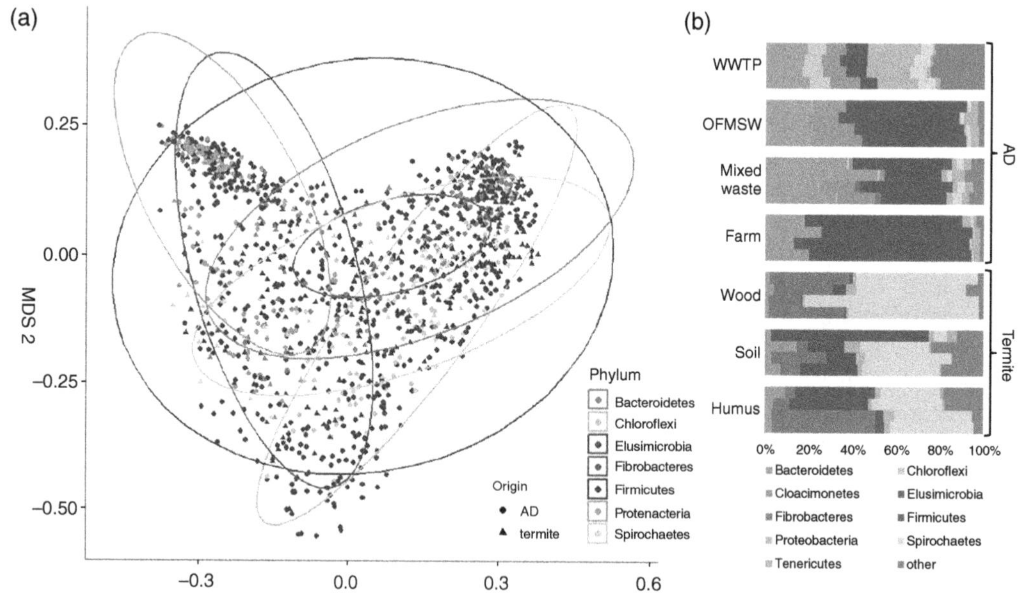

FIGURE 5.4 Comparison of microbial communities and their polysaccharide degrading potential in the anaerobic digestion (AD) reactors and the termite gut (Termite) system. Multidimensional scaling (MDS) of the calculated Jaccard similarity coefficient comparing the genomic content of the carbohydrate active enzymes (CAZymes) encoded in microbial genomes from the two systems (a). This subfigure was compiled based on microbial genomes generated in previous studies (Campanaro et al. 2020; Hervé et al. 2020). Main bacterial phyla present in the different types of AD reactors (i.e. alimented with different feedstocks), and higher termites feeding on distinct substrates (b). This subfigure was compiled based on (Calusinska et al. 2018b; Marynowska et al. 2020). WWTP, wastewater treatment plant, OFMSW, organic fraction of municipal solid waste. (*See colour plate section for colour representation of this figure.*)

(Zhang et al. 2018), using the default parameters. In the present comparison, a total of 321 CAZyme families were detected, including 140 GH families. Most of the detected GH families were shared between the microbes from the two systems, and only a minor part of the GH families was specific to any one of the studied environments (Figure 5.5a). In general, the potential to degrade polysaccharides overlaps between the different phyla, with no clear separation of MAGs between the two studied systems (Figure 5.4a). Genome comparison has revealed that members of Elusimicrobia seem to share more similarities with Proteobacteria in terms of the diversity of their CAZymes. Bacteroidetes, Spirochaetes, and Fibrobacteres, and to a lower extent also Chloroflexi (almost unrepresented in the termite gut system) are also more similar to each other in terms of their CAZyme content. Firmicutes are characterized by the largest diversity of the lignocellulolytic metabolisms in the two systems. However, a higher diversity of encoded GH genes per MAG is assigned to Bacteroidetes, Fibrobacteres, and Chloroflexi (Figure 5.5b). Collectively the largest diversity of GH coding genes is provided in the higher termite gut by Spirochaetes, Firmicutes, Fibrobacteres, and to a lower extent Bacteroidetes (Marynowska et al. 2020). In contrast, the main biomass hydrolyzers in AD reactors are associated with Bacteroidetes, Firmicutes, and Chloroflexi, the latter being mainly abundant in wastewater-treating AD reactors (Calusinska et al. 2018b). However, regarding their high CAZyme genomic content MAGs assigned to Fibrobacteres, Planctomycetes, or *Candidatus* Hydrogenedentes might also be involved in polysaccharide degradation (Bertucci et al. 2019; Campanaro et al. 2020). Nevertheless, their hydrolytic potential has hardly ever been studied in the context of an AD reactor, especially for the two latter phyla.

Specific enrichment of diverse GH families in the genomes of microbes from the two systems further highlights the presence of different lignocellulolytic strategies employed. In total, 17 GH families were more often encoded in MAGs from AD reactors, compared to 14 that were typically enriched in the termite gut microbes (Figure 5.5c). This analysis also shows that in addition to commonly employed GHs by the two microbial communities, for example, to degrade cellulose GH5, GH10, and GH9 would be more commonly employed by microbes in the termite gut, while those in the AD system would also use GH51. In general, the main endo-acting enzymes targeting the backbone of cellulose (e.g. GH5 and GH9) and xylan (e.g. GH10 and GH11) are all enriched in the termite gut microbes, compared to their relatives from the AD system.

Based on the higher diversity of encoded hydrolytic functions, many microbes in the AD system seem to be better equipped to digest a variety of plant biomass components than the termite gut prokaryotes (Figures 5.5 and 5.6). Nevertheless, comparison at the bacterial phylum level indicates that some termite gut phyla including Elusimicrobia, Spirochaetes, and Firmicutes, encode a higher diversity of GH genes per MAG, than in the case of their counterparts from the AD reactors (Figure 5.5b). Moreover, termite gut microbes seem better evolved to degrade cellulose and hemicellulose than AD microbes (Figure 5.6). At the same time, the fraction of lignocellulose in the total feedstock commonly fed to the AD reactors is (significantly) lower compared to the other plant biomass components, including α-glucans and pectin-containing bio-waste, which could shape the community potential toward the other direction. Therefore, strategies aiming at brining the termite gut isolates or adapting whole communities of the termite gut system to specific AD conditions (i.e. bio-augmentation), could further help improve the lignocellulose degradation rate in engineered systems (discussed in Section 5.3).

Collectively, this simple comparison indicates the very high biomass hydrolyzing potential of microbes from engineered lignocellulose processing environments. The challenge ahead is to explore how the biotechnological relevance of these microbes could be better tailored to specific industrial applications, including lignocellulose valorization for biogas production.

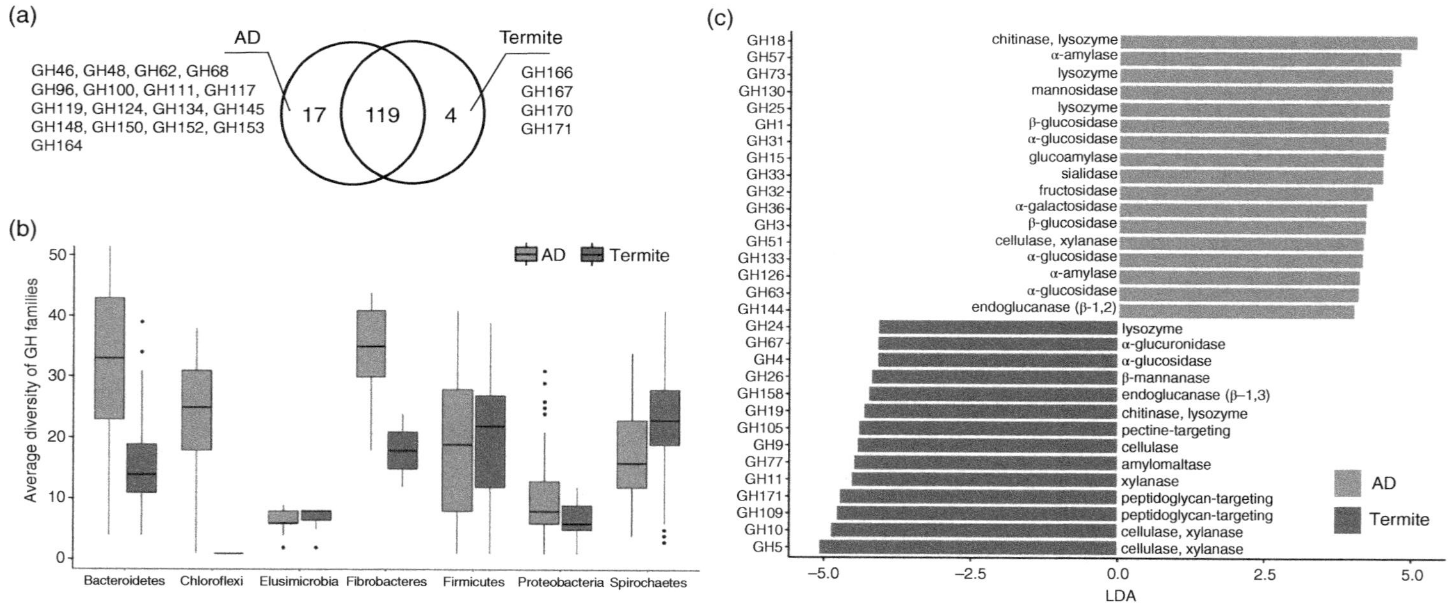

FIGURE 5.5 Glucoside hydrolases in microbial genomes from the anaerobic digestion reactors (AD) and the termite gut system (Termite). (a) Shared diversity of glucoside hydrolase (GH) families in bacterial genomes from AD and the termite gut systems. (b) Average GH family diversity per microbial genome (only genomes with >70% completeness and <10% contamination were used), displayed for the main bacterial phyla. All pair-wise comparisons are statistically significant (t.test, $p<0.05$). (c) Linear discriminant analysis (LDA) of GH gene families differentially enriched in microbial genomes (community-wise) in the two systems. This figure was compiled based on microbial genomes generated in previously published studies (Campanaro et al. 2020; Hervé et al. 2020). (*See colour plate section for colour representation of this figure.*)

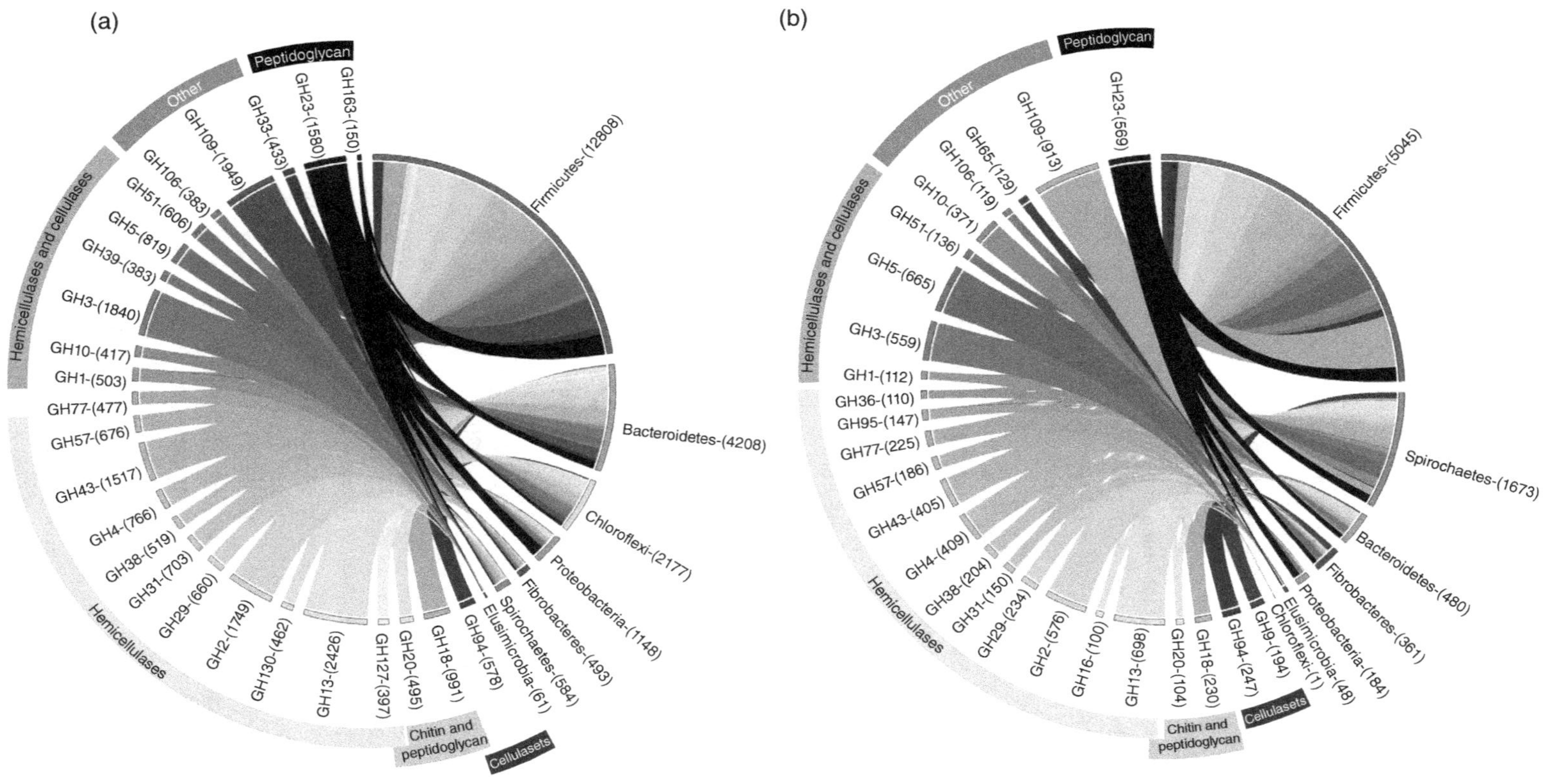

FIGURE 5.6 Comparison of the main glycoside hydrolase (GH) families targeting the different lignocellulose fractions and other non-lignocellulosic polysaccharides and their phylogenetic distribution among major bacterial phyla, for the anaerobic digestion reactors (a) and the termite gut system (b). This figure was compiled based on microbial genomes generated in previously published studies (Campanaro et al. 2020; Hervé et al. 2020). (*See colour plate section for colour representation of this figure.*)

5.5.3 Bioaugmentation and Bioprospecting with the View of Optimizing the Anaerobic Digestion of Lignocellulose

Adding beneficial microorganism(s) to improve or re-establish the performance of AD reactors is at the basis of bioaugmentation (Tale et al. 2015). Until now, most of the AD bioaugmentation strategies have proven little efficiency as they have mostly relied on reactor augmentation with synthetic mixtures composed of microbes which can be obtained and maintained in isolated cultures. Bioprospecting aims at the discovery and characterization of new bio-based products and processes with the purpose of their future commercialization. In this sense, multiple carbohydrate-active enzymes were brought to the market from microbes of the termite gut or anaerobic digestion systems.

Bioaugmentation with Mixed Communities Bioaugmentation with mixed microbial communities enriched in beneficial organisms, rather than using consortia of isolate strains, seems to improve the AD performance providing a long-lasting effect on the reactor by facilitating the establishment of the newly added organisms in the community microbiome. For example, reactor bioaugmentation with the digestate enriched in *Candidatus* Cloacimonetes bacteria, considered to be efficient propionate oxidizers and microbes involved in cellulose degradation (Limam et al. 2014), helped to improve methane production by 40% in a lab-scale anaerobic baffled reactor (Calusinska et al. 2018a). Successful trails of AD bioaugmentation with the termite gut flora were also reported. By implementing the gut content of four higher termite species in AD reactors Dumond et al. (2021) documented not only an efficient cellulose and hemicellulose degradation but also lignin removal (up to 37%) under anaerobic conditions (Dumond et al. 2021). This study confirms previous reports that the termite gut microbes are indeed active in lignin degradation, which was not always widely recognized (Li et al. 2017a). It also opens new possibilities for improved enzymatic lignin decomposition in anaerobic reactors once the model of the termite gut lignin degradation is deciphered and main enzymes and microbial strategies are identified.

A Hidden Potential of Anaerobic (Gut) Fungi The fungal kingdom is the source of the most industrially relevant enzymes used for lignocellulose bioprocessing, with fungal platforms such as Trichoderma or Aspergillus being the most prominent (Ferreira et al. 2021). However, there are also fungal superbugs suitable for bioaugmentation strategies of anaerobic lignocellulolytic processes, and these include anaerobic fungi of the phylum Neocallimastigomycota, which have been recognized as one of the key lignocellulose-degrading organisms in the cow rumen regardless their low abundance (Solomon et al. 2016). Anaerobic fungi can produce a large array of biomass-degrading enzymes (i.e. different GHs, CEs, LPMOs, pectinases, etc., Figure 5.7b,c) that synergistically degrade crude, untreated plant biomass and are competitive with optimized commercial preparations from other fungi (Solomon et al. 2016). In addition, due to mycelia growth, anaerobic fungi can also cause the mechanical disruption of lignocellulose fibers, thus reducing their durability and resistance before enzymatic lysis. Anaerobic fungi gain their energy through mixed acid fermentation, with acetate, formate, succinate, lactate, ethanol, and H_2 as the main fermentation products (Stabel et al. 2021). They are also an unprecedented reservoir of natural compounds, including novel antibiotics (Swift et al. 2021).

Neocallimastigomycota are commonly present in full-scale AD reactors, where they may account for up to 50% of the relative eukaryotic community abundance (Figure 5.7a). In general, their abundance is higher in reactors co-digesting OFMSW with manure, however, there is no consensus. Currently, we do not have a thorough understanding of anaerobic fungi and their involvement in lignocellulose degradation in full-scale AD systems. On one hand, previous reports have informed that anaerobic fungi cannot develop well in AD reactors, where they are quickly

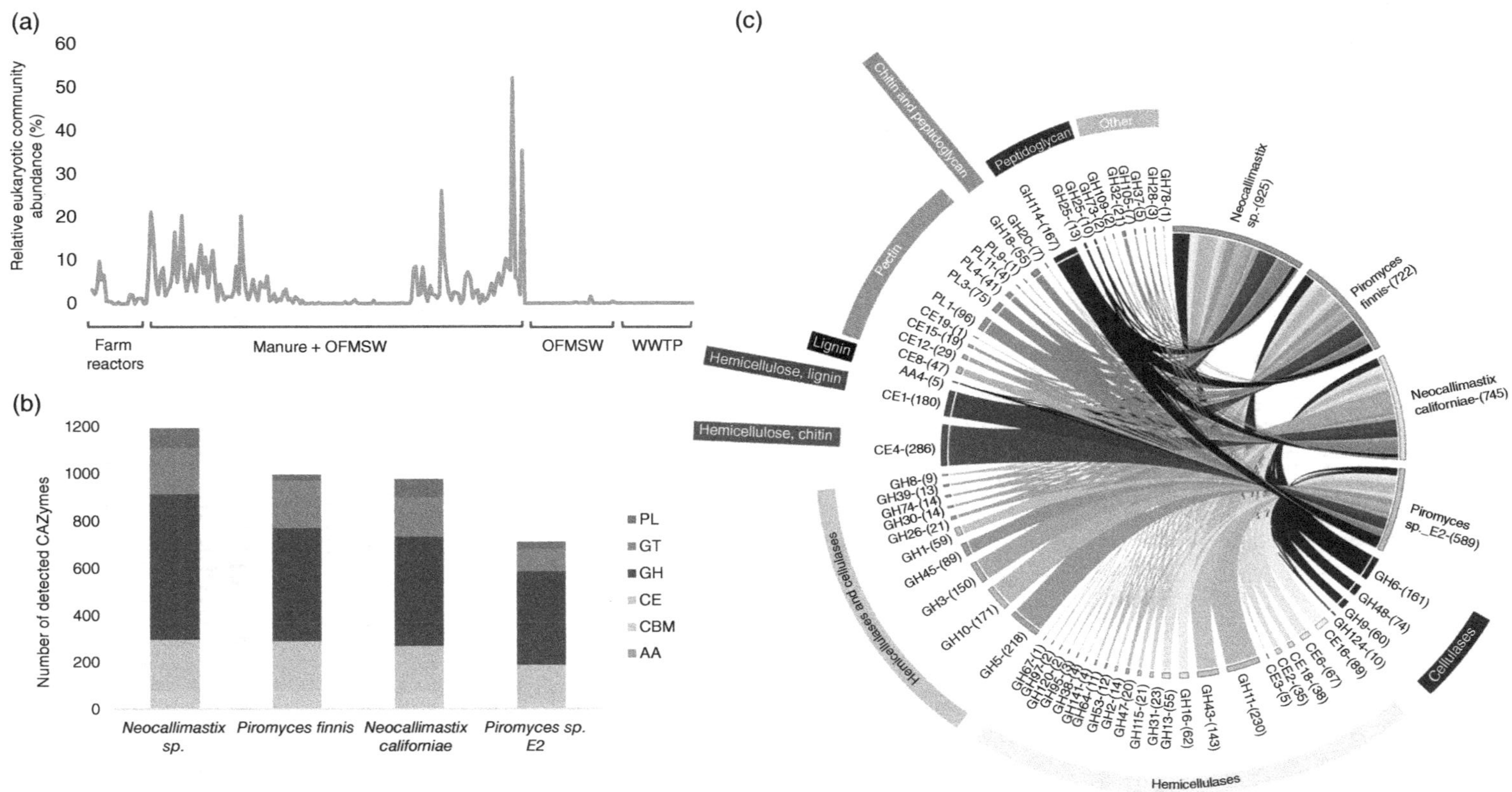

FIGURE 5.7 Anaerobic fungi in full-scale anaerobic digestion systems (a) and their lignocellulolytic potential (b and c). OFMSW, organic fraction of the municipal solid waste; WWTP, wastewater treatment plant; PL, polysaccharide lyase; GT, glycosyl transferase; GH, glycoside hydrolase; CE, carbohydrate esterase; CBM, carbohydrate-binding module; AA, auxiliary enzyme. This figure was compiled based on own, unpublished data (a), and relying on the analysis of Neocallimastigomycota genomes available in public databases (b and c). (*See colour plate section for colour representation of this figure.*)

outcompeted by cellulolytic bacteria (Dollhofer et al. 2017). On the other hand, there is strong evidence of increased methane production in reactors bioaugmented with anaerobic fungi (Peng et al. 2021), creating strong interest in their application in industrial AD systems.

Very recently, anaerobic fungi have also been suggested to be present in the higher termite gut system, where low abundant gene transcripts of putative cellulases of Neocallimastigomycota origin were detected in sequencing datasets (Calusinska et al. 2020; Moreira et al. 2021). Unfortunately, large fungal genomes, with very high AT content and the presence of manifold repeats, have hampered sequencing attempts, making genome assembly and sequence alignments difficult, and resulting in the underrepresentation of anaerobic fungi sequences in the public databases. This in consequence hinders their identification in sequencing studies of different environments.

Bioprospecting for Carbohydrate-active Enzymes The provision of lignocellulose-depolymerizing enzymes is the most prominent feature of the termite gut microbiota, and recent metatranscriptomic and metagenomic studies combined with genome re-constructions have provided more insights on the uncultured majority of the higher termite gut microbes (Calusinska et al. 2020; Marynowska et al. 2020). As shown by Marynowska et al. (2020), roughly 0.18% of the re-constructed gene transcripts from termite gut microbes, including CAZyme coding genes, were shared between the different host lineages, making each termite species gut a unique source of novel and potentially industrially relevant enzymes. When bioprospecting for enzymes from the termite gut that could be tailored to a specific waste stream, a common strategy is to investigate the adaptation of the insect gut system toward the degradation of specific biomass. Currently, it has been tackled either by exposing the living colony to an artificial diet (Calusinska et al. 2020) or by feeding a substrate of interest to the reactor inoculated with the termite hindgut fluid (Auer et al. 2017; Dumond et al. 2021). The latter example was already illustrated above (discussed in Section 5.3.1). In the former example, by imposing a *Miscanthus* fiber diet to a lab-maintained colony of a grass-feeding higher termite *Cortaritermes* sp., Calusinska et al. showed adaptation at the holobiont level (i.e. host and the gut microbiome) to the new substrate (Calusinska et al. 2020). Not only did the structure of the hindgut community change, but also the community function adapted to the new source of lignocellulose by modulating the expression of CAZyme coding genes. Specific CAZymes and even whole CAZyme clusters involved in the degradation of the two main components of *Miscanthus* biomass, which are cellulose and arabinoxylan, were more abundant in the community metatranscriptome after the new diet was introduced to the termite colony. Several microbial enzymes were identified and tested for cellulose and hemicellulose degradation, including the discovery and characterization of a multifunctional GH5_4 endoglucanase capable of simultaneously degrading cellulose, xylan, and mannan backbones. Multifunctional cellulases associated with broad substrate specificity have recently been identified as potent and versatile tools for a renewable bioeconomy (Glasgow et al. 2021). In fact, by simplifying the complexity of enzymatic cocktails while increasing substrate degradation efficiency, the use of multifunctional enzymes provides a road toward cheaper industrial enzymatic preparations.

5.6 BEYOND THE STANDARD LIGNOCELLULOSE DEGRADATION: NEW APPLICATIONS AND FUTURE CHALLENGES

Anaerobic digestion of biomass, including lignocellulose, is currently recognized as the most efficient way of biomass to bioenergy and bio-product transformation. Although the AD technology itself requires further optimization to improve the feedstock conversion rates and yields, other processes and technologies could also benefit from efficient lignocellulose degradation. A few examples are given below, and some of them could even extend or complement the range of applications of traditional AD technology.

5.6.1 Degradation of Lignocellulose-based Bioplastics

Plastic pollution is of rising concern since the high consumption of petroleum-derived plastics in today's society has led to the accumulation of plastic waste in landfills and marine environments (Sharma and Jain 2020). Currently, most of the plastics used are resistant to degradation in nature and persist for long periods if they end up in terrestrial and aquatic systems (Mohanan et al. 2020). Bioplastics have emerged as alternatives to conventional plastics, with properties that allow waste management solutions such as microbial degradation which can diminish risks of plastic accumulation in ecosystems (Bioplastics 2022). Moreover, by replacing fossil fuels with bio-based resources (e.g. starch or lactic acid) in the production of plastics, global oil use and the associated greenhouse gas emissions could be considerably lowered (Lomwongsopon and Varrone 2022). However, the actual degradability of commercially available bioplastics, which mainly originate from lignocellulose, under AD conditions varies considerably depending on the bioplastic physicochemical properties and the biogas process operating conditions (Cazaudehore et al. 2022). Consequently, for many bioplastics, the degradation time exceeds the retention time of the waste material in AD reactors. Hence, the development of eco-friendly bioplastics that are better fit for the activities of those microorganisms in anaerobic environments would enable capturing of the carbon released during the degradation of bioplastics and lower the risk for plastic residues in the digestate (Kakadellis et al. 2022).

In bioplastic degradation, the rate of the enzymatic step depends on plastic properties, such as the surface-to-volume ratio, the glass transition temperature, and the fraction of crystalline and amorphous regions (Laycock et al. 2017; Mohanan et al. 2020). The incorporation of additives into plastic compositions has been proposed as a promising strategy to enhance biodegradability. For instance, the increased degradation of bioplastic associated with the addition of fiber as a filler was suggested to be correlated with increased surface accessibility for enzymatic hydrolysis (Ryan et al. 2018). For bioplastics based on lactate, acetate, or starch, the content of starch and acetyl groups contributed to degradability (Haske-Cornelius et al. 2017). Regarding operating conditions, pre-treatments (e.g. grinding, basic treatment) that increase the surface area and allow microorganisms to more easily attach, and high operating temperature (55 °C), are parameters that have been shown to enhance biodegradability and biogas recovery from bioplastics (Kakadellis et al. 2022).

Given that bioplastics represent a minor fraction of the total feedstock in AD, there is a limited financial or energetic incentive to implement bioaugmentation of the microorganisms or enzymes that accelerate the hydrolysis of bioplastics. However, knowledge can be gained from natural systems which would advance this area and enhance its future application. Recent studies have for instance reported that certain invertebrates (e.g. mealworms, caterpillars) can degrade plastics. Microbial studies of the insect gut have identified bacteria affiliated with the genera *Bacillus*, *Citrobacter*, and *Kosakonia* to be involved in plastic degradation (Brandon et al. 2018; Yang et al. 2015). However, since the insects grind and shred the plastics before it reaches the gut microbiome, it is debated as to which extent the gut microorganisms are truly capable of degrading the plastic polymer on their own (Danso et al. 2019).

5.6.2 Improving Silage Quality and Digestibility Through Microbial Activity

Ensiling is a practice that has been applied since ancient civilizations to achieve long-term animal feed availability (Mejía-Avellaneda et al. 2022). However, even in the modern age, the herbivorous livestock digestibility of fiber-rich materials is still not optimal, due to the recalcitrant structure of lignocellulose. Lignin is one of the main compounds that hinder lignocellulose decomposition, and it has been estimated that in forage grasses 2.7 units of fiber are rendered indigestible per unit of lignin (Krämer et al. 2012). Thus, removing lignin has been considered a key step in improving silage quality and digestibility (Adesogan et al. 2019). For this purpose, a recent study introduced feruloyl esterase-producing *Lactobacillus plantarum* A1 for breaking ester bonds between lignin and hemicellulose, which resulted in increased fiber degradation during ensiling with the addition of cellulase (Xu et al. 2020). In general, bioaugmentation of the ensiling processes with lactic acid bacteria (LAB) is a common practice. Although no LABs with the capacity to degrade complex polysaccharides have been identified so far, they are believed to efficiently ferment several monosaccharides, including hexoses and pentoses, which originate from the lignocellulosic feedstock degradation (Tarraran and Mazzoli 2018). As such LABs collaborate with native lignocellulolytic bacteria and facilitate the overall lignocellulose hydrolysis process through product removal (Tarraran and Mazzoli 2018). Further genetic modifications were also applied to LABs to render them lignocellulolytic, for example by introducing genes encoding for amylases, xylanases, etc., (Artzi et al. 2017; Azman et al. 2015). Nevertheless, governmental regulations related to the use of genetically modified organisms, limit the application of genetically modified LABs in many countries.

5.6.3 Lignocellulose Bioconversion in Space

For future long space missions, lignocellulose can be the basis of fuel, energy, chemicals, and food production and recycling (Malherbe and Cloete 2002). Its generation and degradation during space missions are therefore of high relevance. Importantly, AD is considered the most promising (lignocellulosic) waste treatment technology for a Mars biomanufactory (Berliner et al. 2021). However, outer space is featured by extreme environmental conditions, including high levels of cosmic rays, vacuum, microgravity, etc. (Horneck et al. 2010). The impact of these variables on microbes and their metabolism, including lignocellulolytic activities, is difficult to predict and has not yet been well evaluated. On one hand, microgravity in space has recently been shown to promote the growth of cellulolytic bacteria belonging to the genus *Bacillus*, which significantly increased in short-term (16 hours) space conditions due to upregulated expression of an endoglucanase gene (Shakir et al. 2022). On the other hand, the question of the aging of lignocellulolytic fungi in space conditions has also been raised (Miteva-Staleva et al. 2019). Still, fungal biotechnology in space remains a promising avenue (Cortesão et al. 2020).

Lignocellulose-based materials are also important components used for the construction of different space equipment, including spacesuits (Danko et al. 2021). In this context, the safety of astronauts largely relies on the integrity of used bio-based materials, which should be resistant to any (microbial) degradation. In a closed space environment, contamination by microbes is difficult to avoid, as it can occur during the equipment manufacturing process as well as originate directly from the utilizer. Even though there is a strict material sterilization protocol used in space missions, microbes have been isolated from samples collected on the inner surface of the space station (Danko et al. 2021; Kacena et al. 1999). These included many cellulolytic bacteria and fungi, and their growth and metabolism were reported to cause damage to the structural components of the space station equipment (Novikova et al. 2006). Thus, there is a further need to evaluate, and either promote or inhibit lignocellulose degradation in the space environment.

5.6.4 Anaerobic Mixotrophy Concept for Improved Lignocellulosic Carbon Utilization

Improving fermentation technologies can lead us to more efficient carbon usage, combining the production of bio-products and renewable energy in ways that offer benefits for the environment. Multi-product system synergies can be achieved, enhancing product yields and lowering vulnerability toward price fluctuation (Galbe and Wallberg 2019). The anaerobic non-photosynthetic mixotrophic fermentation (also called acetogenic mixotrophy) refers to a simultaneous utilization of organic materials (e.g. products from the degradation of lignocellulose) and inorganic carbon sources (e.g. CO_2 and CO), and represents an intriguing concept for streamlined carbon utilization (Jones et al. 2016). Acetogens are a group of bacteria that use the Wood-Ljungdahl pathway (WLP) for energy conservation and for the reduction of CO_2 to acetyl-Coenzyme A. Some acetogenic species are also able to produce acetate from H_2 and CO_2 (Drake et al. 2008). Acetogenesis on its own is energetically challenging but the carbon yield can be optimized by using an excess of ATP generated by glycolysis in the fixation of CO_2 through the WLP (i.e. mixotrophy). Hence, in mixotrophy, two molecules of CO_2 and eight electrons produced from glycolysis are used to generate three molecules of acetyl-CoA from glucose, rather than the two molecules that are produced by typical fermentation processes (Fast et al. 2015). Several bacteria have been investigated to produce high-purity chemicals (e.g. acetate, ethanol) in mixotrophic processes, including *Blautia producta, Thermoanaerobacter kivui,* and *Clostridium ljungdahlii* (Maru et al. 2018). Furthermore, the engineering of *C. ljungdahlii* can generate acetone or mevalonate and isoprene through mixotrophic fermentation of fructose and CO_2 or syngas fixation (Diner et al. 2018; Jones et al. 2016). Future research should reveal the prospects of directly using lignocellulosic material e.g. by combining microbial pre-hydrolysis and mixotrophic fermentation for the production of bio-based compounds that can be used to replace fossil-derived compounds.

5.7 CONCLUSIONS

Lignocellulosic biomass is viable and abundant biomass worldwide, thus representing an interesting feedstock for engineered AD processes producing renewable energy, sustainable fertilizers, or green products. However, the recalcitrance of lignocellulose to the AD microbiota often hampers the degradation of the organic material and subsequently the overall yield. Currently, microbial communities are underutilized for biomass degradation applications. First, they are still poorly characterized in terms of their metabolic potential and lignocellulolytic activities, with a big part of known bacterial lineages lacking cultivable representatives. This in turn prevents their efficient exploitation in industrial applications. Second, key variables that drive the behavior and the community composition are also little understood. The termite gut system, comprising the hindgut microbiome, achieves the best lignocellulose degradation rates of any known natural and engineered systems. Thus, it could be used as a model to improve the efficiency of industrial processes relying on microbial lignocellulose processing.

Although significant developments have been made over the last century to improve the industrial transformation of biomass to bioenergy and other bio-based products, from an evolutionary standpoint, engineered anaerobic digestion is still a relatively new technology, when compared to billions of years of evolution of natural systems. However, recent development and implementation of the omic techniques aimed at the characterization of complex microbial communities could provide insights into the future optimal microbial management strategies leading to improved efficiency of microbe-dependent biomass transformative technologies. Alike the termite hindgut which contains a unique flora, specific to a termite species that specialized in a processing of a dedicated feedstock, the different full-scale AD reactors also operate with distinct microbial communities, best adapted to the applied operational conditions. However, the termite gut is more than just the microbial hindgut, and capacities such as feed cutting and grinding, chemical treatment, and adapting water content and retention time as well as the size and shape of the digestive organs largely contribute to the termite digestive strategy. Moreover, animal digestive tracts are usually compartmented, allowing for a physical separation of the different digestion processes.

Industrial implementation of the termite digestive tract (i.e. biomimicry) could consist of a series of anaerobic reactors, including a different combination of plug flow and continuously stirred tank reactors, characterized by distinct size, pH, mixing modes, oxygen levels, etc. However, up to date, bioprospecting rather than biomimicry of the termite gut system has been considered for industrial applications, and several carbohydrate-active enzymes of the host and the termite gut prokaryotic origin have been tested for their lignocellulose degrading capacities. Of special interest are the multifunctional cellulases which have been identified by other authors as potent and versatile tools for a renewable bioeconomy. Their application could simplify the complexity of enzymatic cocktails used for industrial applications, such as bioethanol production or paper and pulp processing, thus considerably lowering their cost. At the same time, increased substrate degradation efficiencies could be achieved.

ACKNOWLEDGMENTS

The authors would like to acknowledge Yves Roisin for drawing the termite gut employed in Figure 3, and Martyna Marynowska and Dominika Klimek for their help with the preparation of Figures 2 and 5. This work was supported by FNR CORE 2017 project CLOMICS (Grant number: C17/SR/11687962), FNR PoC project Microbiome (Grant number: 20/15315861), the HR2020 ERC – Grant European Research Council (grant number 948138), and the Kempe foundation (grant number JCK-2016).

REFERENCES

Aanen, D.K., Eggleton, P., Rouland-Lefevre, C. et al. (2002). The evolution of fungus-growing termites and their mutualistic fungal symbionts. *Proceedings of the National Academy of Sciences* 99 (23): 14887–14892.

Abbasi, T., Tauseef, S.M., and Abbasi, S.A. (2012). A brief history of anaerobic digestion and "biogas". In: *In: Biogas Energy* (ed. T. Abbasi, S.M. Tauseef, and S.A. Abbasi), 11–23. New York: Springer New York.

Adesogan, A.T., Arriola, K.G., Jiang, Y. et al. (2019). Symposium review: technologies for improving fiber utilization. *Journal of Dairy Science* 102 (6): 5726–5755.

Afzal, M., Shaheen, N., Shah, S.A.A. et al. (2022). Saccharification of agricultural lignocellulosic feedstocks by endogenous and symbiotic cellulases from the subterranean termites. *Biocatalysis and Agricultural Biotechnology* 39: 102265.

Ahlberg-Eliasson, K., Westerholm, M., Isaksson, S. et al. (2021). Anaerobic digestion of animal manure and influence of organic loading rate and temperature on process performance, microbiology, and methane emission from digestates. *Frontiers in Energy Research* 9: 1–16.

Alokika, A., Kumar, A., Kumar, V. et al. (2021). Cellulosic and hemicellulosic fractions of sugarcane bagasse: potential, challenges and future perspective. *International Journal of Biological Macromolecules* 169: 564–582.

Amon, T., Amon, B., Kryvoruchko, V. et al. (2007). Biogas production from maize and dairy cattle manure—influence of biomass composition on the methane yield. *Agriculture, Ecosystems & Environment* 118 (1): 173–182.

Aresta, M., Dibenedetto, A., and Dumeignil, F. (2012). *Biorefinery: From Biomass to Chemicals and Fuels*. Walter de Gruyter.

Arora, J., Kinjo, Y., Šobotník, J. et al. (2022). The functional evolution of termite gut microbiota. *Microbiome* 10 (1): 78.

Artzi, L., Bayer, E.A., and Moraïs, S. (2017). Cellulosomes: bacterial nanomachines for dismantling plant polysaccharides. *Nature Reviews Microbiology* 15 (2): 83–95.

Auer, L., Lazuka, A., Sillam-Dussès, D. et al. (2017). Uncovering the potential of termite gut microbiome for lignocellulose bioconversion in anaerobic batch bioreactors. *Frontiers in Microbiology* 8: 2623.

Ayilara, M.S., Olanrewaju, O.S., Babalola, O.O. et al. (2020). Waste management through composting: challenges and potentials. *Sustainability* 12 (11): 4456.

Azman, S., Khadem, A.F., van Lier, J.B. et al. (2015). Presence and role of anaerobic hydrolytic microbes in conversion of lignocellulosic biomass for biogas production. *Critical Reviews in Environmental Science and Technology* 45 (23): 2523–2564.

Bayané, A. and Guiot, S.R. (2011). Animal digestive strategies versus anaerobic digestion bioprocesses for biogas production from lignocellulosic biomass. *Reviews in Environmental Science and Bio/Technology* 10 (1): 43–62.

Bayer, E.A., Lamed, R., and Himmel, M.E. (2007). The potential of cellulases and cellulosomes for cellulosic waste management. *Current Opinion in Biotechnology* 18 (3): 237–245.

Berliner, A.J., Hilzinger, J.M., Abel, A.J. et al. (2021). Towards a biomanufactory on Mars. *Frontiers in Astronomy and Space Sciences* 8.

Bertucci, M., Calusinska, M., Goux, X. et al. (2019). Carbohydrate hydrolytic potential and redundancy of an anaerobic digestion microbiome exposed to acidosis, as uncovered by metagenomics. *Applied and Environmental Microbiology* 85 (15): e00895–e00819.

Blair, E.M., Dickson, K.L., and O'Malley, M.A. (2021). Microbial communities and their enzymes facilitate degradation of recalcitrant polymers in anaerobic digestion. *Current Opinion in Microbiology* 64: 100–108.

Brandon, A.M., Gao, S.-H., Tian, R. et al. (2018). Biodegradation of polyethylene and plastic mixtures in mealworms (larvae of Tenebrio molitor) and effects on the gut microbiome. *Environmental Science & Technology* 52 (11): 6526–6533.

Brémond, U., Bertrandias, A., Steyer, J.-P. et al. (2021). A vision of European biogas sector development towards 2030: trends and challenges. *Journal of Cleaner Production* 287: 125065.

Brown, D., Shi, J., and Li, Y. (2012). Comparison of solid-state to liquid anaerobic digestion of lignocellulosic feedstocks for biogas production. *Bioresource Technology* 124: 379–386.

Brune, A. (1998). Termite guts: the world's smallest bioreactors. *Trends in Biotechnology* 16 (1): 16–21.

Brune, A. (2010). Methanogenesis in the digestive tracts of insects. In: *Handbook of Hydrocarbon and Lipid Microbiology* (ed. K.N. Timmis), 707–728. Berlin, Heidelberg: Springer Berlin Heidelberg.

Brune, A. (2014). Symbiotic digestion of lignocellulose in termite guts. *Nature Reviews Microbiology* 12: 168.

Brune, A. (2018). Methanogenesis in the digestive tracts of insects and other arthropods. In: *Biogenesis of Hydrocarbons. Handbook of Hydrocarbon and Lipid Microbiology* (ed. A.J.M. Stams and D.Z. Sousa), 1–32. Cham: Springer International Publishing.

Brune, A. and Kühl, M. (1996). pH profiles of the extremely alkaline hindguts of soil-feeding termites (Isoptera: Termitidae) determined with microelectrodes. *Journal of Insect Physiology* 42 (11): 1121–1127.

Cai, Y., Zheng, Z., Schäfer, F. et al. (2021). A review about pretreatment of lignocellulosic biomass in anaerobic digestion: achievement and challenge in Germany and China. *Journal of Cleaner Production* 299: 126885.

Calusinska, M., Delfosse, P., Goux, X. et al. (2018a). Biological agent for enrichment of anaerobic digestion reactors, method of preparing such biological agent and bio-augmentation process with said agent, (Ed.) L.I.o.S.a.T. (LIST), Vol. WO/2018/122000. Luxembourg.

Calusinska, M., Goux, X., Fossépré, M. et al. (2018b). A year of monitoring 20 mesophilic full-scale bioreactors reveals the existence of stable but different core microbiomes in bio-waste and wastewater anaerobic digestion systems. *Biotechnology for Biofuels* 11 (1): 196.

Calusinska, M., Marynowska, M., Bertucci, M. et al. (2020). Integrative omics analysis of the termite gut system adaptation to Miscanthus diet identifies lignocellulose degradation enzymes. *Communications Biology* 3 (1): 275.

Campanaro, S., Treu, L., Rodriguez-R, L.M. et al. (2020). New insights from the biogas microbiome by comprehensive genome-resolved metagenomics of nearly 1600 species originating from multiple anaerobic digesters. *Biotechnology for Biofuels* 13 (1): 25.

Carballa, M., Regueiro, L., and Lema, J.M. (2015). Microbial management of anaerobic digestion: exploiting the microbiome-functionality nexus. *Current Opinion in Biotechnology* 33: 103–111.

Castellano-Hinojosa, A., Armato, C., Pozo, C. et al. (2018). New concepts in anaerobic digestion processes: recent advances and biological aspects. *Applied Microbiology and Biotechnology* 102 (12): 5065–5076.

Cazaudehore, G., Guyoneaud, R., Evon, P. et al. (2022). Can anaerobic digestion be a suitable end-of-life scenario for biodegradable plastics? A critical review of the current situation, hurdles, and challenges. *Biotechnology Advances* 56: 107916.

Chaganti, V.N., Ganjegunte, G., Meki, M.N. et al. (2021). Switchgrass biomass yield and composition and soil quality as affected by treated wastewater irrigation in an arid environment. *Biomass and Bioenergy* 151: 106160.

Chen, Q., Wu, W., Qi, D. et al. (2020). Review on microaeration-based anaerobic digestion: state of the art, challenges, and prospectives. *Science of The Total Environment* 710: 136388.

Cheng, J.J. and Timilsina, G.R. (2011). Status and barriers of advanced biofuel technologies: a review. *Renewable Energy* 36 (12): 3541–3549.

Chouvenc, T., Šobotník, J., Engel, M.S. et al. (2021). Termite evolution: mutualistic associations, key innovations, and the rise of Termitidae. *Cellular and Molecular Life Sciences* 78 (6): 2749–2769.

Cortesão, M., Schütze, T., Marx, R. et al. (2020). Fungal biotechnology in space: why and how? In: *Grand Challenges in Fungal Biotechnology* (ed. H. Nevalainen), 501–535. Cham: Springer International Publishing.

Cragg, S.M., Beckham, G.T., Bruce, N.C. et al. (2015). Lignocellulose degradation mechanisms across the Tree of Life. *Current Opinion in Chemical Biology* 29: 108–119.

Czerkawski, J.W. and Breckenridge, G. (1977). Design and development of a long-term rumen simulation technique (Rusitec). *British Journal of Nutrition* 38 (3): 371–384.

Dąbkowska, K., Alvarado-Morales, M., and Kuglarz, M. (2019). Miscanthus straw as substrate for biosuccinic acid production: focusing on pretreatment and downstream processing. *Bioresource Technology* 278: 82–91.

Danko, D., Malli Mohan, G.B., Sierra, M.A. et al. (2021). Characterization of spacesuit associated microbial communities and their implications for NASA missions. *Frontiers in Microbiology* 12: 608478.

Danso, D., Chow, J., and Streit, W.R. (2019). Plastics: environmental and biotechnological perspectives on microbial degradation. *Applied and Environmental Microbiology* 85 (19): e01095–e01019.

De Vrieze, J. (2020). The next frontier of the anaerobic digestion microbiome: from ecology to process control. *Environmental Science and Ecotechnology* 3: 100032.

Diner, B.A., Fan, J., Scotcher, M.C. et al. (2018). Synthesis of heterologous mevalonic acid pathway enzymes in *Clostridium ljungdahlii* for the conversion of fructose and of syngas to mevalonate and isoprene. *Applied and Environmental Microbiology* 84 (1): e01723–e01717.

Dionísio, S.R., Santoro, D.C.J., Bonan, C.I.D.G. et al. (2021). Second-generation ethanol process for integral use of hemicellulosic and cellulosic hydrolysates from diluted sulfuric acid pretreatment of sugarcane bagasse. *Fuel* 304: 121290.

Dollhofer, V., Callaghan, T.M., Griffith, G.W. et al. (2017). Presence and transcriptional activity of anaerobic fungi in agricultural biogas plants. *Bioresource Technology* 235: 131–139.

Donovan, S.E., Eggleton, P., and Bignell, D.E. (2001). Gut content analysis and a new feeding group classification of termites. *Ecological Entomology* 26 (4): 356–366.

Drake, H.L., Gößner, A.S., and Daniel, S.L. (2008). Old acetogens, new light. *Annals of the New York Academy of Sciences* 1125 (1): 100–128.

Drula, E., Garron, M.L., Dogan, S. et al. (2022). The carbohydrate-active enzyme database: functions and literature. *Nucleic Acids Research* 50 (D1): D571–d577.

Dumond, L., Lam, P.Y., van Erven, G. et al. (2021). Termite gut microbiota contribution to wheat straw delignification in anaerobic bioreactors. *ACS Sustainable Chemistry & Engineering* 9 (5): 2191–2202.

Eggleton, P. (2000). Global patterns of termite diversity. In: *In: Termites: Evolution, Sociality, Symbioses, Ecology* (ed. T. Abe, D.E. Bignell, and M. Higashi). Dordrecht, The Netherlands: Kluwer Academic Publishers.

Fast, A.G., Schmidt, E.D., Jones, S.W. et al. (2015). Acetogenic mixotrophy: novel options for yield improvement in biofuels and biochemicals production. *Current Opinion in Biotechnology* 33: 60–72.

Ferreira, R.G., Azzoni, A.R., and Freitas, S. (2021). On the production cost of lignocellulose-degrading enzymes. *Biofuels, Bioproducts and Biorefining* 15 (1): 85–99.

Fox, P.F. and McSweeney, P.L.H. (2017). Chapter 1 – Cheese: an overview. In: *Cheese*, 4e (ed. P.L.H. McSweeney, P.F. Fox, P.D. Cotter, et al.), 5–21. San Diego: Academic Press.

Franco Cairo, J.P.L., Mandelli, F., Tramontina, R. et al. (2022). Oxidative cleavage of polysaccharides by a termite-derived superoxide dismutase boosts the degradation of biomass by glycoside hydrolases. *Green Chemistry* 24 (12): 4845–4858.

Frape, D. (2008). *Equine Nutrition and Feeding*. Wiley.

Furlan, F.F., Costa, C.B.B., and Fonseca, G.d.C. (2012). Assessing the production of first and second generation bioethanol from sugarcane through the integration of global optimization and process detailed modeling. *Computers & Chemical Engineering* 43: 1–9.

Galbe, M. and Wallberg, O. (2019). Pretreatment for biorefineries: a review of common methods for efficient utilisation of lignocellulosic materials. *Biotechnology for Biofuels* 12 (1): 294.

Garnier-Sillam, E., Toutain, F., Villemin, G. et al. (1989). Études Préliminaires des Meules Originales du Termite XylophageSphaerotermes sphaerothorax (Sjostedt). *Insectes Sociaux* 36 (4): 293–312.

Ge, X., Xu, F., and Li, Y. (2016). Solid-state anaerobic digestion of lignocellulosic biomass: recent progress and perspectives. *Bioresource Technology* 205: 239–249.

Geng, A., Cheng, Y., Wang, Y. et al. (2018). Transcriptome analysis of the digestive system of a wood-feeding termite (*Coptotermes formosanus*) revealed a unique mechanism for effective biomass degradation. *Biotechnology for Biofuels* 11 (1): 24.

Gijzen, H.J., Zwart, K.B., Verhagen, F.J.M. et al. (1988). High-rate two-phase process for the anaerobic degradation of cellulose, employing rumen microorganisms for an efficient acidogenesis. *Biotechnology and Bioengineering* 31 (5): 418–425.

Gille, S. and Pauly, M. (2012). O-Acetylation of plant cell wall polysaccharides. *Frontiers in Plant Science* 3 (12).

Glasgow, E., Vander Meulen, K., Kuch, N. et al. (2021). Multifunctional cellulases are potent, versatile tools for a renewable bioeconomy. *Current Opinion in Biotechnology* 67: 141–148.

Godon, J.-J., Arcemisbéhère, L., Escudié, R. et al. (2013). Overview of the oldest existing set of substrate-optimized anaerobic processes: digestive tracts. *BioEnergy Research* 6 (3): 1063–1081.

Gosslink, R., De Jong, E., Guran, B. et al. (2004). Co-ordination network for lignin-standardisation, production and applications adapted to market requirements. *Industrial and Crops Product* 20: 121–129.

Hansen, J.C., Aanderud, Z.T., Reid, L.E. et al. (2021). Enhancing waste degradation and biogas production by pre-digestion with a hyperthermophilic anaerobic bacterium. *Biofuel Research Journal* 8 (3): 1433.

Haske-Cornelius, O., Pellis, A., Tegl, G. et al. (2017). Enzymatic systems for cellulose acetate degradation. *Catalysts* 7 (10): 287.

Hassan, S.S., Williams, G.A., and Jaiswal, A.K. (2019). Moving towards the second generation of lignocellulosic biorefineries in the EU: drivers, challenges, and opportunities. *Renewable and Sustainable Energy Reviews* 101: 590–599.

Hatfield, R.D., Rancour, D.M., and Marita, J.M. (2017). Grass cell walls: a story of cross-linking. *Frontiers in Plant Science* 7 (2056).

Heer, D. and Sauer, U. (2008). Identification of furfural as a key toxin in lignocellulosic hydrolysates and evolution of a tolerant yeast strain. *Microbial Biotechnology* 1 (6): 497–506.

Herrero Garcia, N., Benedetti, M., and Bolzonella, D. (2019). Effects of enzymes addition on biogas production from anaerobic digestion of agricultural biomasses. *Waste and Biomass Valorization* 10 (12): 3711–3722.

Hervé, V., Liu, P., Dietrich, C. et al. (2020). Phylogenomic analysis of 589 metagenome-assembled genomes encompassing all major prokaryotic lineages from the gut of higher termites. *PeerJ* 8: e8614.

Himmel, M.E., Ding, S.Y., Johnson, D.K. et al. (2007). Biomass recalcitrance: engineering plants and enzymes for biofuels production. *Science* 315 (5813): 804–807.

Hofmann, R.R. (1988). Anatomy of the gastro-intestinal tract. *The Ruminant Animal, Digestive Physiology and Nutrition* 14–43.

Horneck, G., Klaus, D.M., and Mancinelli, R.L. (2010). Space microbiology. *Microbiology and Molecular Biology Reviews* 74 (1): 121–156.

Huang, S.W., Zhang, H.Y., Marshall, S. et al. (2010). The scarab gut: a potential bioreactor for bio-fuel production. *Insect Science* 17 (3): 175–183.

Iweh, N.S., Koyama, M., Akizuki, S. et al. (2020). Novel wet-solid states serial anaerobic digestion process for enhancing methane recovery of aquatic plant biomass. *Science of the Total Environment* 730: 138993.

Izquierdo, J.A., Pattathil, S., Guseva, A. et al. (2014). Comparative analysis of the ability of *Clostridium clariflavum* strains and *Clostridium thermocellum* to utilize hemicellulose and unpretreated plant material. *Biotechnology Biofuels* 7 (1): 136.

Janusz, G., Pawlik, A., Sulej, J. et al. (2017). Lignin degradation: microorganisms, enzymes involved, genomes analysis and evolution. *FEMS Microbiology Reviews* 41 (6): 941–962.

Jaramillo, P.M., Rocha Gomes, H.A., Monclaro, A.V. et al. (2015). Lignocellulose degrading enzymes: an overview of the global market. In: *Fungal Bio-molecules: Sources, Applications and Recent Developments* (ed. R.L. Mach, V. Gupta, and S. . Sreenivasaprasad), 73–85. New Jersey: Wiley.

Jasso-Selles, D.E., De Martini, F., Velenovsky, J.F. IV et al. (2020). The complete protist symbiont communities of *Coptotermes formosanus* and *Coptotermes gestroi*: morphological and molecular characterization of five new species. *Journal of Eukaryotic Microbiology* 67 (6): 626–641.

Javourez, U., O'Donohue, M., and Hamelin, L. (2021). Waste-to-nutrition: a review of current and emerging conversion pathways. *Biotechnology Advances* 53: 107857.

Jensen, M.B., de Jonge, N., Dolriis, M.D. et al. (2021). Cellulolytic and xylanolytic microbial communities associated with lignocellulose-rich wheat straw degradation in anaerobic digestion. *Frontiers in Microbiology* 12: 645174.

Jeong, S.-Y., Lee, E.-J., Ban, S.-E. et al. (2021). Near infrared spectroscopy model for analyzing chemical composition of biomass subjected to Fenton oxidation and hydrothermal treatment. *Renewable Energy* 172: 1341–1350.

Jia, Y., Ng, S.-K., Lu, H. et al. (2018). Genome-centric metatranscriptomes and ecological roles of the active microbial populations during cellulosic biomass anaerobic digestion. *Biotechnology for Biofuels* 11 (1): 117.

Jiang, Y., Lv, Y., Wu, R. et al. (2019). Current status and perspectives on biobutanol production using lignocellulosic feedstocks. *Bioresource Technology Reports* 7: 100245.

Jones, S.W., Fast, A.G., Carlson, E.D. et al. (2016). CO_2 fixation by anaerobic non-photosynthetic mixotrophy for improved carbon conversion. *Nature Communications* 7 (1): 12800.

Jung, H.G., Mertens, D.R., and Phillips, R.L. (2011). Effect of reduced ferulate-mediated lignin/arabinoxylan cross-linking in corn silage on feed intake, digestibility, and milk production. *Journal of Dairy Science* 94 (10): 5124–5137.

Kacena, M., Merrell, G., Manfredi, B. et al. (1999). Bacterial growth in space flight: logistic growth curve parameters for *Escherichia coli* and *Bacillus subtilis*. *Applied Microbiology and Biotechnology* 51 (2): 229–234.

Kaida, R., Kaku, T., Baba, K. et al. (2009). Loosening xyloglucan accelerates the enzymatic degradation of cellulose in wood. *Molecular Plant* 2 (5): 904–909.

Kainthola, J., Podder, A., Fechner, M. et al. (2021). An overview of fungal pretreatment processes for anaerobic digestion: applications, bottlenecks and future needs. *Bioresource Technology* 321: 124397.

Kakadellis, S., Lee, P.-H., and Harris, Z.M. (2022). Two birds with one stone: bioplastics and food waste anaerobic co-digestion. *Environments* 9 (1): 9.

Kamusoko, R., Jingura, R.M., Parawira, W. et al. (2021). Strategies for valorization of crop residues into biofuels and other value-added products. *Biofuels, Bioproducts and Biorefining* 15 (6): 1950–1964.

Karp, A. and Shield, I. (2008). Bioenergy from plants and the sustainable yield challenge. *New Phytologist* 179 (1): 15–32.

Khan, M.U. and Ahring, B.K. (2019). Lignin degradation under anaerobic digestion: Influence of lignin modifications – a review. *Biomass and Bioenergy* 128: 105325.

Kikas, T., Tutt, M., Raud, M. et al. (2016). Basis of energy crop selection for biofuel production: cellulose vs. lignin. *International Journal of Green Energy* 13 (1): 49–54.

Kim, M. and Day, D.F. (2011). Composition of sugar cane, energy cane, and sweet sorghum suitable for ethanol production at Louisiana sugar mills. *Journal of Industrial Microbiology and Biotechnology* 38 (7): 803–807.

Krämer, M., Weisbjerg, M.R., Lund, P. et al. (2012). Estimation of indigestible NDF in forages and concentrates from cell wall composition. *Animal Feed Science and Technology* 177 (1): 40–51.

Krishna, K., Grimaldi, D.A., Krishna, V. et al. (2013). Treatise on the Isoptera of the world. *Bulletin of the American Museum of Natural History* (377): 2704.

Lapébie, P., Lombard, V., Drula, E. et al. (2019). Bacteroidetes use thousands of enzyme combinations to break down glycans. *Nature Communications* 10 (1): 2043.

Larnaudie, V., Ferrari, M.D., and Lareo, C. (2021). Life cycle assessment of ethanol produced in a biorefinery from liquid hot water pretreated switchgrass. *Renewable Energy* 176: 606–616.

Laycock, B., Nikolić, M., Colwell, J.M. et al. (2017). Lifetime prediction of biodegradable polymers. *Progress in Polymer Science* 71: 144–189.

Li, H., Yelle, D.J., Li, C. et al. (2017a). Lignocellulose pretreatment in a fungus-cultivating termite. *Proceedings of the National Academy of Sciences* 114 (18): 4709–4714.

Li, Y., Liu, C., Wachemo, A.C. et al. (2017b). Serial completely stirred tank reactors for improving biogas production and substance degradation during anaerobic digestion of corn stover. *Bioresource Technology* 235: 380–388.

Limam, R.D., Chouari, R., Mazeas, L. et al. (2014). Members of the uncultured bacterial candidate division WWE1 are implicated in anaerobic digestion of cellulose. *Microbiologyopen* 3 (2): 157–167.

Lomwongsopon, P. and Varrone, C. (2022). Contribution of fermentation technology to building blocks for renewable plastics. *Fermentation* 8 (2): 47.

Luo, H. and Abu-Omar, M.M. (2017). Chemicals from lignin. In: *Encyclopedia of Sustainable Technologies* (ed. M.A. Abraham), 573–585. Oxford: Elsevier.

Malherbe, S. and Cloete, T.E. (2002). Lignocellulose biodegradation: fundamentals and applications. *Reviews in Environmental Science and Biotechnology* 1 (2): 105–114.

Martinez, E.J., Micolucci, F., Gomez, X. et al. (2018). Anaerobic digestion of residual liquid effluent (brown juice) from a green biorefinery. *International Journal of Environmental Science and Technology* 15 (12): 2615–2624.

Maru, B.T., Munasinghe, P.C., Gilary, H. et al. (2018). Fixation of CO_2 and CO on a diverse range of carbohydrates using anaerobic, non-photosynthetic mixotrophy. *FEMS Microbiology Letters* 365 (8).

Marynowska, M., Goux, X., Sillam-Dussès, D. et al. (2020). Compositional and functional characterisation of biomass-degrading microbial communities in guts of plant fibre- and soil-feeding higher termites. *Microbiome* 8 (1): 96.

Marynowska, M., Sillam-Dussès, D., Untereiner, B. et al. (2023). A holobiont approach towards polysaccharide degradation by the highly compartmentalised gut system of the soil-feeding higher termite Labiotermes labralis. *BMC Genomics* 24: 115

McCarty, P.L. (1982). One hundred years of anaerobic treatment. One hundred years of anaerobic digestion. Anaerobic digestion. *Proceedings of the Second International Symposium on Anaerobic Digestion*, Travemünde, Federal Republic of Germany (6-11 September 1981), 3-22.

Mejía-Avellaneda, L.F., Suárez, H., Jiménez, H. et al. (2022). Challenges and opportunities for the production of lactic acid bacteria inoculants aimed for ensiling processes. *Critical Reviews in Biotechnology* 42 (7): 1028–1044.

Menzel, T., Neubauer, P., and Junne, S. (2020). Role of microbial hydrolysis in anaerobic digestion. *Energies* 13 (21): 5555.

Michel, F., Thibault, J.-F., Barry, J.-L. et al. (1988). Preparation and characterisation of dietary fibre from sugar beet pulp. *Journal of the Science of Food and Agriculture* 42 (1): 77–85.

Mikaelyan, A., Meuser, K., and Brune, A. (2017). Microenvironmental heterogeneity of gut compartments drives bacterial community structure in wood-and humus-feeding higher termites. *FEMS Microbiol ogy Ecology* 93 (1).

Miteva-Staleva, J., Kostadinova, N., and Hubenov, V. (2019). Aging in ligninolytic fungi in space conditions. *Space Research and Technology Institute-BAS, Bulgaria Konstantin Preslavsky University-Faculty of Technical Sciences Association Scientific and Applied Research* 17: 30.

Mohanan, N., Montazer, Z., Sharma, P.K. et al. (2020). Microbial and enzymatic degradation of synthetic plastics. *Frontiers in Microbiology* 11: 580709.

Moreira, E.A., Persinoti, G.F., Menezes, L.R. et al. (2021). Complementary contribution of fungi and bacteria to lignocellulose digestion in the food stored by a neotropical higher termite. *Frontiers in Ecology and Evolution* 9.

Neumann, A.P. and Suen, G. (2018). The phylogenomic diversity of herbivore-associated *Fibrobacter* spp. is correlated to lignocellulose-degrading potential. *Msphere* 3 (6): e00593–e00518.

Nguyen, D. and Khanal, S.K. (2018). A little breath of fresh air into an anaerobic system: how microaeration facilitates anaerobic digestion process. *Biotechnology Advances* 36 (7): 1971–1983.

Nishimura, Y., Otagiri, M., Yuki, M. et al. (2020). Division of functional roles for termite gut protists revealed by single-cell transcriptomes. *The ISME Journal* 14 (10): 2449–2460.

Novikova, N., De Boever, P., Poddubko, S. et al. (2006). Survey of environmental biocontamination on board the International Space Station. *Research in Microbiology* 157 (1): 5–12.

Ogden, C.A. and Ileleji, K.E. (2021). Physical characteristics of ground switchgrass related to bulk solids flow. *Powder Technology* 385: 386–395.

Ohkuma, M., Noda, S., Hattori, S. et al. (2015). Acetogenesis from H_2 plus CO_2 and nitrogen fixation by an endosymbiotic spirochete of a termite-gut cellulolytic protist. *Proceedings of the National Academy of Sciences* 112 (33): 10224–10230.

Pauly, M., Gille, S., Liu, L. et al. (2013). Hemicellulose biosynthesis. *Planta* 238 (4): 627–642.

Peng, X., Wilken, S.E., Lankiewicz, T.S. et al. (2021). Genomic and functional analyses of fungal and bacterial consortia that enable lignocellulose breakdown in goat gut microbiomes. *Nature Microbiology* 6 (4): 499–511.

Poulsen, M., Hu, H., Li, C. et al. (2014). Complementary symbiont contributions to plant decomposition in a fungus-farming termite. *Proceedings of the National Academy of Sciences of the United States of America* 111 (40): 14500–14505.

Prasad, S., Singh, A., and Joshi, H. (2007). Ethanol as an alternative fuel from agricultural, industrial and urban residues. *Resources, Conservation and Recycling* 50 (1): 1–39.

Rajeswari, G., Jacob, S., Chandel, A.K. et al. (2021). Unlocking the potential of insect and ruminant host symbionts for recycling of lignocellulosic carbon with a biorefinery approach: a review. *Microbial Cell Factories* 20 (1): 107.

Raud, M., Tutt, M., Olt, J. et al. (2015). Effect of lignin content of lignocellulosic material on hydrolysis efficiency. *Agronomy Research* 13 (2): 405–412.

Robak, K. and Balcerek, M. (2020). Current state-of-the-art in ethanol production from lignocellulosic feedstocks. *Microbiological Research* 240: 126534.

Rocha-Meneses, L., Raud, M., Orupõld, K. et al. (2017). Second-generation bioethanol production: a review of strategies for waste valorisation. *Agronomy Research* 15 (3): 830–847.

Ross, R.P., Morgan, S., and Hill, C. (2002). Preservation and fermentation: past, present and future. *International Journal of Food Microbiology* 79 (1–2): 3–16.

Ryan, C.A., Billington, S.L., and Criddle, C.S. (2018). Biocomposite fiber-matrix treatments that enhance in-service performance can also accelerate end-of-life fragmentation and anaerobic biodegradation to methane. *Journal of Polymers and the Environment* 26 (4): 1715–1726.

Sabbadin, F., Hemsworth, G.R., Ciano, L. et al. (2018). An ancient family of lytic polysaccharide monooxygenases with roles in arthropod development and biomass digestion. *Nature Communications* 9 (1): 756–756.

Schalk, F., Gostinčar, C., Kreuzenbeck, N.B. et al. (2021). The termite fungal cultivar Termitomyces combines diverse enzymes and oxidative reactions for plant biomass conversion. *Mbio* 12 (3): e03551–e03520.

Scharf, M.E. and Peterson, B.F. (2021). A century of synergy in termite symbiosis research: linking the past with new genomic insights. *Annual Review of Entomology* 66 (1): 23–43.

Schmitt-Wagner, D. and Brune, A. (1999). Hydrogen profiles and localization of methanogenic activities in the highly compartmentalized hindgut of soil-feeding higher termites (*Cubitermes* spp.). *Applied and Environmental Microbiology* 65 (10): 4490–4496.

Schmitt-Wagner, D., Friedrich, M.W., Wagner, B. et al. (2003). Phylogenetic diversity, abundance, and axial distribution of bacteria in the intestinal tract of two soil-feeding termites (*Cubitermes* spp.). *Applied and Environmental Microbiology* 69 (10): 6007–6017.

Schnurer, A. and Jarvis, A. (2010). Microbiological handbook for biogas plants. *Swedish Waste Management U* 2009: 1–74.

Schroeder, B.G., Logroño, W., Rocha, U.N.D. et al. (2022). Enrichment of anaerobic microbial communities from midgut and hindgut of sun beetle larvae (*Pachnoda marginata*) on wheat straw: effect of inoculum preparation. *Microorganisms* 10 (4): 761.

Shakir, Y., Rcheulishvili, N., Zhang, Y. et al. (2022). Effects of short-term space conditions on cellulose degradation ability and biodiversity of microorganisms. *Cellulose* 29 (9): 5061–5077.

Sharma, B. and Jain, P. (2020). Deciphering the advances in bioaugmentation of plastic wastes. *Journal of Cleaner Production* 275: 123241.

Sharma, R., Garg, P., Kumar, P. et al. (2020). Microbial fermentation and its role in quality improvement of fermented foods. *Fermentation* 6 (4): 106.

da Silva, S.S., Chandel, A.K., Wickramasinghe, S.R. et al. (2012). Fermentative production of value-added products from lignocellulosic biomass. *Journal of Biomedicine & Biotechnology* 2012: 826162–826162.

Singh, T., Alhazmi, A., Mohammad, A. et al. (2021). Integrated biohydrogen production via lignocellulosic waste: opportunity, challenges & future prospects. *Bioresource Technology* 338: 125511.

Solomon, K.V., Haitjema, C.H., Henske, J.K. et al. (2016). Early-branching gut fungi possess a large, comprehensive array of biomass-degrading enzymes. *Science* 351 (6278): 1192–1195.

Soukup, P., Větrovský, T., Stiblik, P. et al. (2021). Termites are associated with external species-specific bacterial communities. *Applied and Environmental Microbiology* 87 (2): e02042–e02020.

Stabel, M., Schweitzer, T., Haack, K. et al. (2021). Isolation and biochemical characterization of six anaerobic fungal strains from zoo animal feces. *Microorganisms* 9 (8): 1655.

Struckmann Poulsen, J., de Jonge, N., Vieira Macêdo, W. et al. (2022). Characterisation of cellulose-degrading organisms in an anaerobic digester. *Bioresource Technology* 351: 126933.

Sun, J., Zhang, L., and Loh, K.-C. (2021). Review and perspectives of enhanced volatile fatty acids production from acidogenic fermentation of lignocellulosic biomass wastes. *Bioresources and Bioprocessing* 8 (1): 68.

Swift, C.L., Louie, K.B., Bowen, B.P. et al. (2021). Anaerobic gut fungi are an untapped reservoir of natural products. *Proceedings of the National Academy of Sciences* 118 (18): e2019855118.

Tale, V.P., Maki, J.S., and Zitomer, D.H. (2015). Bioaugmentation of overloaded anaerobic digesters restores function and archaeal community. *Water Research* 70: 138–147.

Tarraran, L. and Mazzoli, R. (2018). Alternative strategies for lignocellulose fermentation through lactic acid bacteria: the state of the art and perspectives. *FEMS Microbiology Letters* 365 (15): fny126.

Teter, S.A., Sutton, K.B., and Emme, B. (2014). 7 – Enzymatic processes and enzyme development in biorefining. In: *Advances in Biorefineries* (ed. K. Waldron)), 199–233. Woodhead Publishing.

Usmani, Z., Sharma, M., Awasthi, A.K. et al. (2021). Lignocellulosic biorefineries: the current state of challenges and strategies for efficient commercialization. *Renewable and Sustainable Energy Reviews* 148: 111258.

Wang, K., Gao, P., Geng, L. et al. (2022). Lignocellulose degradation in *Protaetia brevitarsis* larvae digestive tract: refining on a tightly designed microbial fermentation production line. *Microbiome* 10 (1): 90.

Westerholm, M. and Schnürer, A. (2019). Microbial responses to different operating practices for biogas production systems. *Anaerobic Digestion* 1–36.

Westerholm, M., Moestedt, J., and Schnürer, A. (2016). Biogas production through syntrophic acetate oxidation and deliberate operating strategies for improved digester performance. *Applied Energy* 179: 124–135.

Wu, H., Ioannou, E., Henrissat, B. et al. (2021). Multimodularity of a GH10 xylanase found in the termite gut metagenome. *Applied and Environmental Microbiology* 87 (3).

Xu, D., Ding, Z., Bai, J. et al. (2020). Evaluation of the effect of feruloyl esterase-producing *Lactobacillus plantarum* and cellulase pretreatments on lignocellulosic degradation and cellulose conversion of co-ensiled corn stalk and potato pulp. *Bioresource Technology* 310: 123476.

Yang, Y., Chen, J., Wu, W.-M. et al. (2015). Complete genome sequence of *Bacillus* sp. YP1, a polyethylene-degrading bacterium from waxworm's gut. *Journal of Biotechnology* 200: 77–78.

Zhang, H., Yohe, T., Huang, L. et al. (2018). dbCAN2: a meta server for automated carbohydrate-active enzyme annotation. *Nucleic Acids Research* 46 (W1): W95–W101.

Zhou, J., Du, G., and Chen, J. (2014). Novel fermentation processes for manufacturing plant natural products. *Current Opinion in Biotechnology* 25: 17–23.

SECTION III

USING MICROBIAL FERMENTATION TO PRODUCE BEER AND WINE

CHAPTER 6

ARCHEOLOGICAL EVIDENCE FOR FERMENTED ALCOHOLIC BEVERAGES IN RITUAL FEASTS OF NEOLITHIC CHINA

LI LIU
Department of East Asian Languages and Cultures, Stanford University, Stanford, CA, USA

CONTENTS

6.1 INTRODUCTION

Alcohol fermentation techniques and drinking traditions not only form a part of the culinary practice in a society, they also play a central role in various feasting activities, which help shape sociopolitical relationships (Arnold 1999; Dietler 2006; Dietler 1990; Joffe 1998). China has a long history of cereal-based alcoholic fermentation. Alcohol types first appeared in oracle-bone inscriptions of the Late Shang dynasty (ca. 3200 to 3000 BP), recorded as *li, jiu,* and *chang* (Wen and Yuan 1983). In later texts, such as *Shangshu* (ca. fourth century BCE), the fermentation methods of these alcohol types were further described as *li* made of *nie* sprouted cereals, *jiu* made of *qu* (or Koji) moldy grain starter, and *chang* made of millet and herbs. The use of the *qu* method for alcohol production is generally thought to be unique to East Asia and was initiated in China (Huang 2000; Ling 1958). Another method was mastication, by chewing and spitting out cereals, which has been recorded in ethnographic accounts of aboriginals in Taiwan (Ling 1957).

Microbial Fermentations in Nature and as Designed Processes, First Edition. Edited by Christon J. Hurst.

Recent scientific studies of residues adhering to pottery vessels have revealed that alcoholic fermentation can be traced back to the beginning of the Neolithic (ca. 9000 cal. BP), some 6000 years before being recorded in writing. The new findings also show that the production and consumption of alcoholic beverages were closely related to the development of agriculture, ritual activities, and social complexity in China. In this chapter, I will discuss several issues relating to the origin, dispersal, and social function of alcoholic fermentation in Neolithic China. I focus on three general topics: (i) brewing ingredients and fermentation techniques; (ii) the relationship between the Neolithic revolution and the production of fermented beverages; and (iii) the development of complex society and alcohol-related ritual feasting.

6.2 METHODS: IDENTIFYING BREWING METHODS IN NEOLITHIC CHINA

The production of cereal-based fermented beverages involves two separate biochemical steps: (i) saccharification: hydrolysis of the starch in the cereal to fermentable sugars, using enzymes called amylases; and (ii) fermentation: conversion of the sugars by yeasts to ethyl alcohol and carbon dioxide. Since yeast is ubiquitous in the environment and thus readily introduced incidentally, the key step is saccharification.

In recent years, scientific analyses have revealed direct evidence of alcohol-making in prehistoric China. The methods commonly employed in these studies include chemical, microbotanical, and microbial analyses of residues extracted from pottery vessels excavated archeologically. Chemical analyses (e.g. GC-MS; HPLC; FT-IR) are effective for identifying various biomarkers in the organic remains of fermented beverages (McGovern et al. 2004, 2005). Starch grain analysis can provide taxonomical information about the starchy ingredients, and damaged starch grains that reveal features typical of fermentation and gelatinization can be indicative of the presence of alcoholic beverages (Wang et al. 2017). The brewing technique using malts can be identified by cereal husk phytoliths found on fermentation vessels, whereas the technique of using *qu* starters (often adding herbs, called *caoqu*) can be inferred when particular filamentous fungi (e.g. *Monascus, Aspergillus, Rhizopus,* and *Mucor*) and phytoliths from either leaves or stems are observed. The presence of yeast cells is also a strong indicator of alcoholic fermentation (Liu et al. 2019b, 2020a, 2020d; Wang et al. 2016) (Figure 6.1). Alcohol made with mastication may be identified if the residues contain only starches with fermentation characteristics and yeasts (Liu et al. 2020b).

The *qu* brewing method can be further divided into two techniques. One is referred to as liquid-state fermentation (LSF), using a high water content in the substrate. Fermentation is normally carried out with a vessel that has a small orifice for easy sealing and a large body for containing liquid. Such a vessel shape would help achieve an anaerobic condition to facilitate the conversion from sugars to ethanol during fermentation and facilitate storage. The second technique is so-called semi-solid-state fermentation (SSSF), using a low water content in the substrate. It requires good aeration to achieve a high oxygen level and controlled temperatures, allowing molds and yeasts to grow rapidly. This condition would be made possible by the large opening and interior space of the large vats. When yeasts grow rapidly, they actively produce heat and carbon dioxide, which in turn creates convection currents. The currents push air bubbles (carbon dioxide) and grains upward, forming a layer of substance on the mash's surface and creating an anaerobic environment for yeasts to convert sugars to ethanol. The SSSF technique has been commonly employed for making traditional *huangjiu,* yellow rice beer, in south China since antiquity (Bao and Zhou 2007).

To date, chemical, microbotanical, and microbial analyses of pottery vessels from at least two dozen Neolithic sites (ca. 9000 to 3800 cal. BP) in China have revealed scientific evidence of alcohol fermentation (Table 6.1).

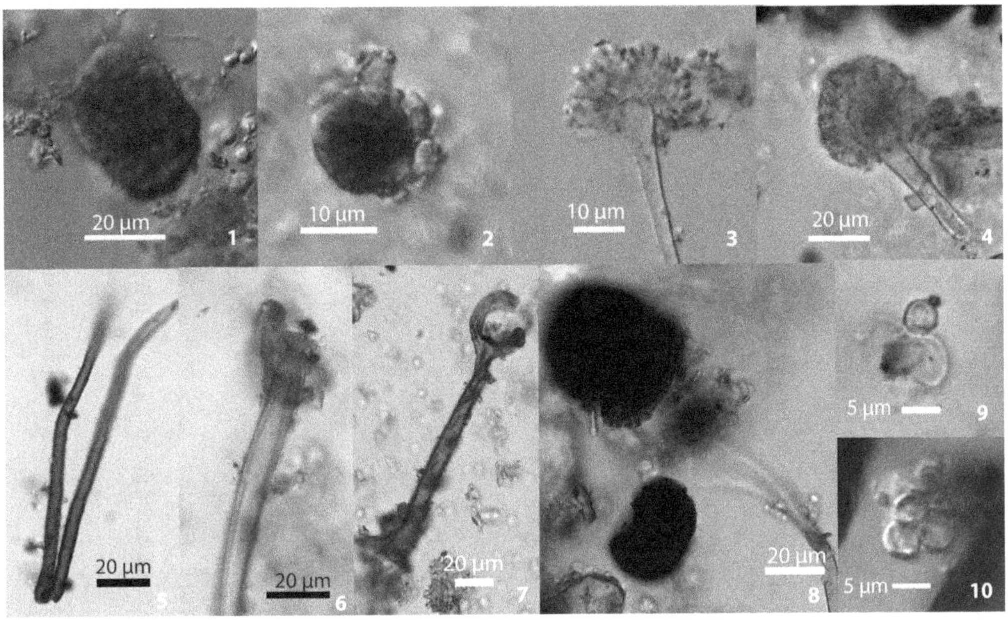

FIGURE 6.1 Examples of ancient fungal elements found on Neolithic pottery vessels in China. 1, 2: *Monascus* cleistothecia; 3, 4: *Aspergillus* vesicles; 5: cf. *Rhizopus* hyphae; 6: *Rhizopus* sporangium, showing structure of collarette and columella; 7: *Mucor* sporangium connecting to hypha; 8: *Rhizopus* or *Mucor* sporangia; 9: yeast cells in budding position; 10: a cluster of yeast cells. Source: These images appear courtesy of the author. (*See colour plate section for colour representation of this figure.*)

6.3 BREWING BEER AND NEOLITHIC REVOLUTION IN THE EARLY NEOLITHIC (ca. 9000 TO 7000 cal. BP)

It has long been debated among archeologists whether humans' thirst for beer was a driving force of cereal domestication (Braidwood et al. 1953; Hayden et al. 2013). Recent research in the Near East has shown that hunter-gatherers brewed beer with wild barley or wheat at Raqefet cave in Israel around 13 700 to 11 700 cal. BP (Liu et al. 2018b), long before the domestication of these cereals. This finding seems to support the beer hypothesis. China is one of the regions in the world where agriculture was developed independently, represented by the domestication of rice in the south and millet in the north around 10 000 to 9 000 years ago (Huan et al. 2021; Zhao et al. 2020). Therefore, it is crucial to investigate if and how the cereal domestication processes were related to fermentation technologies.

Scientific studies of residues on pottery vessels have revealed that cereal-based alcoholic fermentation is likely to have developed at the onset of the Neolithic Revolution, closely related to the beginning of rice and millet domestication in China. This development may have originated in the lower Yangzi River region of south China, represented by the Shangshan culture (11 000 to 8 600 cal. BP), in which several technological and social changes appeared simultaneously. First, people began to live in villages with increased sedentary lifeways, suggested by the presence of houses, storage pits, and burials. Second, rice domestication emerged by 10 000–9 000 cal. BP, based on the evidence from phytolith and carbonized seed remains. Third, pottery vessels of various types were made to serve different functions, including cooking, serving, and liquid storage (Huan et al. 2021; Zhejiang Provincial Institute of Archaeology and Pujiang Museum 2016; Zuo et al. 2017). Coinciding with these developments, globular jars became widespread. Such jars

TABLE 6.1 Neolithic sites revealed alcohol remains in China.

Site	Location	Archaeology Culture	Pottery Type	Alcohol Ingredients	Brewing Method	References
Early Neolithic (ca. 9000 to 7000 cal. BP), n = 6						
Qiaotou	Yiwu, Zhejiang	Middle Shangshan	Globular jars	Rice, Job's tears, tubers	*Qu* with *Aspergillus* and *Rhizopus*	Wang et al. (2021)
Xiaohuangshan	Shengzhou, Zhejiang	Middle Shangshan	Globular jars, plate	Rice, Job's tears, acorn, lily bulb	*Qu* with *Monascus*	Liu et al. (2023)
Jiahu	Wuyang, Henen	Peiligang	Globular jars	Rice, honey, fruits	Honey, fruits	McGovern et al. (2004)
Peiligang	Xinzheng, Henan	Peiligang	Globular jars, bowl, cauldron	Rice, Triticeae, foxnut, tubers	*Qu* with *Monascus*	Liu et al. (Under revision)
Guantaoyuan	Baoji, Shaanxi	Baijia-Laoguantai	Globular jars, funnel	Millet, Job's tears, Triticeae, snake gourd root, ginger, bean, tubers	*Qu* with *Aspergillus* and *Rhizopus*	Liu et al. (2019b)
Lingkou	Lintong, Shaanxi	Baijia-Laoguantai	Globular jars	Millet, rice, Job's tears, Triticeae, snake gourd root, ginger, bean, tubers	Malted broomcorn millet and rice	Liu et al. (2019b)
Middle Neolithic (ca. 7000 to 4600 cal. BP), n = 16						
Dongjiabai	Wenshang, Shandong	Middle Beixin	Globular jars	Rice, millet, Triticeae, snake gourd roots, acorn	*Qu* with *Rhizopus*; mastication	Liu et al. (2020b)
Wangyin	Yanzhou, Shandong	Early Dawenkou	*Bei* goblets	Kudzu root, snake gourd root, millet, Triticeae, tubers	*Qu* with *Aspergillus*	Liu et al. (2021b)
Xixiahou	Qufu, Shandong	Late Dawenkou	*Gaobingbei* goblets and *ping* vase	Kudzu root, snake gourd root, lily, millet, Triticeae, tubers	*Qu* with *Rhizopus* and/or *Mucor*	Liu et al. (2021b)
Yuchisi	Mengcheng, Anhui	Late Dawenkou	*Dakougang* vats, *hu* jars, *bei* cups	Rice, millet, Job's tears, Triticeae, snake gourd root	*Qu* with *Monascus*,	Liu et al. (2022b)

Lingkou	Lintong, Shaanxi	Transition to middle Neolithic (Lingkou phase)	Flat-base amphorae	Millet, rice, Job's tears, Triticeae, snake gourd root, foxnut, lily, ginger, bean, tubers	Malted broomcorn millet and rice	Liu et al. (2020c)
Lingkou	Lintong, Shaanxi	Early Yangshao	Conical-base amphorae	Millet, rice, Job's tears, Triticeae, snake gourd root, foxnut, lily, bean, tubers	Malted broomcorn millet and rice	Liu et al. (2020c)
Banpo	Xi'an, Shaanxi	Early Yangshao	Conical-base amphorae	Millet, Job's tears, Triticeae, snake gourd root, foxnut	*Qu* with *Aspergillus, Rhizopus/Mucor*	Liu et al. (2020d)
Jiangzhai	Lintong, Shaanxi	Early Yangshao	Conical-base amphorae	Millet, Job's tears, Triticeae, rice, snake gourd root, bean	*Qu* with *Aspergillus, Rhizopus/Mucor;* malted broomcorn millet and	Liu et al. (2020d)
Yangguanzhai	Yangling, Shaanxi	Middle Yangshao	Conical-base amphorae, funnels, *ping* jars	Millet, Job's tears, Triticeae, snake gourd root, yam, lily, tubers	Malted broomcorn millet	Liu et al. (2017)
Mijiaya	Xi'an, Shaanxi	Late Yangshao	Conical-base amphorae, funnels, urns	Millet, Job's tears, Triticeae, snake gourd root, lily, yam, ginger, tubers	Malted broomcorn millet	Wang et al. (2016)
Xinjie	Lantian, Shaanxi	Late Yangshao	Conical-base amphorae, funnels	Millet, Job's tears, Triticeae, rice, snake gourd root, foxnut, lily, tubers	Malted broomcorn millet and rice	Liu et al. (2018c)
Xipo	Lingbao, Henan	Middle Yangshao	Large vats, jars, bowls	Millet, Job's tears, Triticeae, rice, yam, lily, snake gourd root	*Qu* with *Monascus, Rhizopus/ Mucor, Aspergillus*	Feng et al. (2021)

(*Continued*)

TABLE 6.1 (Continued)

Site	Location	Archaeology Culture	Pottery Type	Alcohol Ingredients	Brewing Method	References
Dingcun	Mianchi, Henan	Middle Yangshao	Conical-base amphorae	Millet, Job's tears, Triticeae, rice, snake gourd root,	Malted broomcorn millet and rice; *qu* with *Monascus, Aspergillus* and *Mucor*	Liu et al. (2020a, 2021a)
Huizui	Yanshi, Henan	Late Yangshao	Pottery strainer	Millet, Job's tears, rice	Malted broomcorn millet and rice	Liu et al. (2019a)
Huizui	Yanshi, Henan	Late Yangshao	Pottery vessels, building floors	Millet, rice, Triticeae, yam, tubers	Malted cereals	Liu et al. (2018a)
Qingtai	Xingyang, Henan	Late Yangshao	Funnel, lid, painted bottle	Millet, rice, Job's tears, Triticeae, snake gourd roots, lotus roots, beans, nuts, fruits, honey	Fruits, honey	Liao et al. (2022)
Miaozigou	Inner Mongolia	Middle Yangshao	*Guan* jars and funnel	Millet, Triticeae, rice, snake gourd root, lily, ginger	Malted broomcorn millet, *qu* with *Rhizopus* or *Mucor*	He et al. (2021b)
Late Neolithic (ca. 4600 to 3900 cal. BP), n = 2						
Shimao	Yulin, Shaanxi	Late Longshan	*Gui* and *he* pitcher, *bei* cup, *weng* jar	Millet, rice, Triticeae, lily, snake gourd root, Zingiberaceae root, beans	*Qu* with *Monascus*	He et al. (2021a, 2022)
Liangchengzhen	Rizhao, Shandong	Shandong Longshan	*Guan* jars, *gui* tripos, bei cups, hu jars, swteamer rack, *pen* basin, *lei* jar	Rice, honey, possibly barley and fruit	Honey and fruits	McGovern et al. (2005)

from two sites (Qiaotou and Xiaohuangshan) of the Shangshan culture have revealed evidence of fermented beverages, dating to 9000 cal. BP (Figure 6.2). Globular jars are liquid containers, characterized by a small orifice with a restricted neck, designed for easy sealing, which would help both to prevent contamination from acid-producing bacteria in the environment, as well as to create an anaerobic condition for yeasts to convert sugars to alcohol and carbon dioxide in the LSF (Bao 2005). It is noted that globular jars were often associated with human burials, suggesting that one of the functions of the earliest alcoholic beverages may have been related to making funerary offerings to the deceased ancestors (Liu et al. 2023; Wang et al. 2021).

Based on the evidence from carbonized seeds, phytoliths, and starch grains, rice domestication technology appears to have expanded northward to the Huai River region by 9000 cal. BP (Qiu and Rao 2022; Zhang et al. 2018), and then to the Yellow River region by 8000 cal. BP (Bestel et al. 2017; Crawford et al. 2006; Jin et al. 2014; Li et al. 2020; Zhang et al. 2012, 2018). The dispersal of rice domestication was accompanied by the presence of globular jars associated with alcoholic fermentation. Similar to the Shangshan culture, such jars have often been found in burials of the Peiligang culture sites in the middle Yellow River region (Zheng 1984).

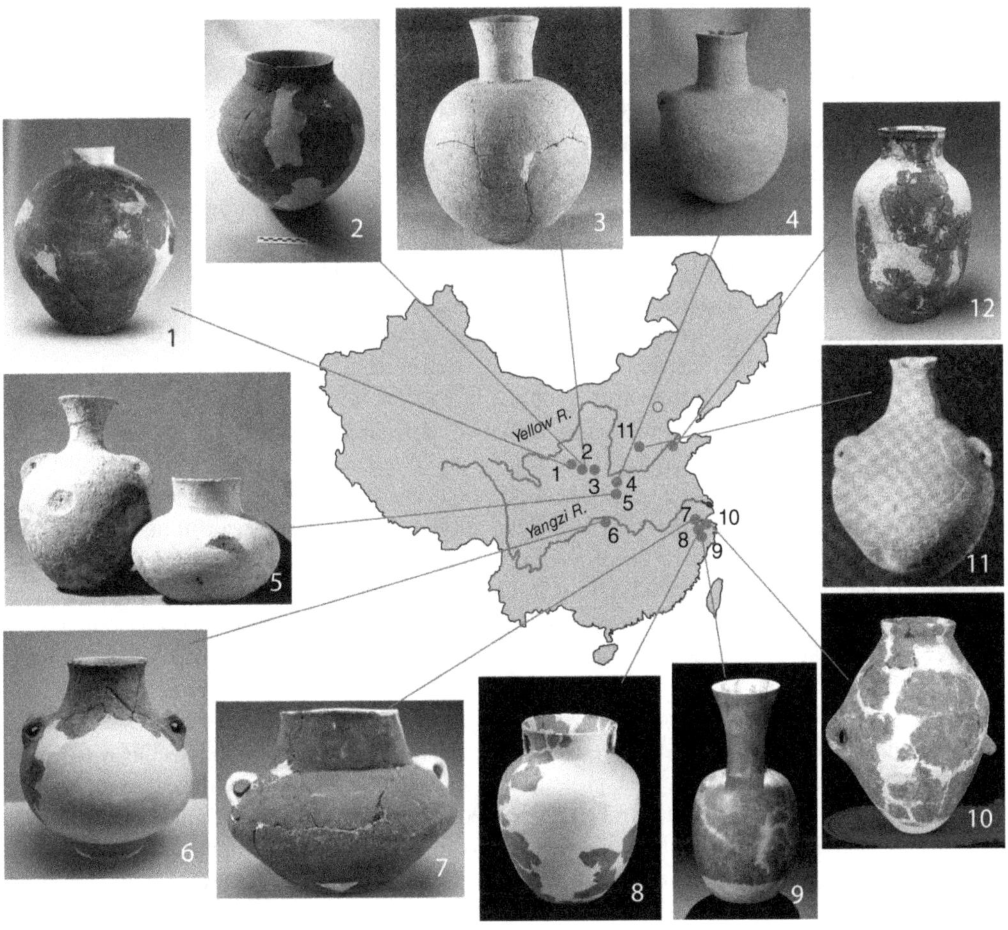

FIGURE 6.2 Globular jars unearthed from major sites in early Neolithic China, ca. 9000 to 7000 cal. BP. 1: Dadiwan; 2: Guantaoyuan; 3: Baijia-Lingkou; 4: Peiligang; 5: Jiahu; 6: Pengtoushan; 7: Kuahuqiao; 8: Shangshan; 9: Qiaotou; 10: Xiaohuangshan; 11: Cishan; 12: Houli. Source: These images appear courtesy of the author.

To date, globular jars associated with evidence of alcoholic fermentation have been identified in six early Neolithic sites, dating to ca. 9000 to 7000 cal. BP (Figure 6.2). Based on starch and phytolith analyses, the fermentation ingredients include rice (*Oryza sativa*), broomcorn millet (*Panicum miliaceum*), Job's tears (*Coix lacryma-jobi*), wild Triticeae, beans (Fabaceae), together with diverse tubers and nuts, such as foxnut (*Euryale ferox*), snake gourd root (*Trichosanthes kirilowii*), ginger (*Zingiber officinale*), lily bulb (*Lilium* sp.), and acorn (Fagaceae) (Table 6.1). Among these plants, rice and broomcorn millet are likely to have been cultivated and possibly domesticated, but others were uncertain in their domesticity. In general, the earliest alcoholic beverages were made of a mixture of domesticated cereals with many other types of wild or cultivated plants. Fermentation methods include at least three types: (i) use of malted broomcorn millet (Lingkou); (ii) preparation of *qu* starter made of moldy grains, often adding herbal plants, and the functional fungi include *Monascus, Aspergillus,* and *Rhizopus* (Qiaotou, Xiaohuangshan, Peiligang, Guantaoyuan); and (iii) use of honey and fruits as fermentation agents (Jiahu) (Table 6.1; Figure 6.2 for site location).

6.4 DEVELOPMENT OF DIVERSE BREWING METHODS IN THE MIDDLE NEOLITHIC (ca. 7000 TO 4600 cal. BP)

The middle Neolithic period witnessed the intensification of millet and rice agriculture in the Yellow and Yangzi River valleys, respectively. In the meantime, alcoholic fermentation methods became diversified, as manifested in various forms of brewing pottery vessels. We focus on the Yellow River region, where most studies on this subject have been conducted.

Archeologically, two regional systems emerged during this time. The Beixin-Dawenkou cultures (ca. 7300 to 4600 cal. BP) developed in the lower Yellow River valley of the east coast region, while the Yangshao culture (ca. 7000 to 4700 cal. BP) flourished in the middle Yellow River valley of the inland region. The Beixin-Dawenkou cultural system appears to have been more developed than the Yangshao culture in terms of social complexity expressed in mortuary practice and ceramic production (Keightley 1987; Liu and Chen 2012; Underhill et al. 2002).

In the Beixin-Dawenkou culture area, globular jars for brewing were continuously in use, but a new type of fermenter, *dakougang* wide-orifice vat, appeared around 5500 to 5000 BP, probably originated in the Lower Yangzi River region and dispersed northward (Liu et al. 2022b). Also, many types of finely made vessels related to serving and drinking emerged, including *gui* pitchers for heating and pouring, and *gaobingbei* goblets and *bei* cups for drinking (Figure 6.3). Two fermentation methods have been detected based on microfossil analysis of pottery from four sites (Dongjiabai, Wangyin, Xixiahou, Yuchisi), including (i) preparation of *qu* starter; and (ii) possibly use of human saliva (mastication) as saccharification agents. The *qu* preparation method also consists of two variants: one is the traditional LSF using globular jars, and the other is the SSSF facilitated by *dakougang* vats.

Molds for making *qu* starter included *Monascus, Aspergillus, Rhizopus,* and *Mucor*. Yeast cells, morphologically resembling *Saccharomyces cerevisiae*, are also present in the residues. The main ingredients for fermentation were rice, millet, Job's tears, Triticeae, acorn, and various tubers (e.g. snake gourd root, kudzu root, and lily), based on starch and phytolith analyses (Liu et al. 2020b, 2021b, 2022b) (Table 6.1). Interestingly, the use of broomcorn millet malts as a saccharification agent appears to have been absent in the Beixin-Dawenkou cultural system, according to current data.

The large vats and elaborate serving-drinking vessels were particularly associated with elite burials. The increased elaboration in pottery vessel forms may be related to growing competitive feasting activities in residential areas as well as in gravesites during the Dawenkou culture. Such competitive feasts would have created greater demand for prestigious foods, beverages, and vessels, a process that played a role in the development of social stratification and stimulated diversification

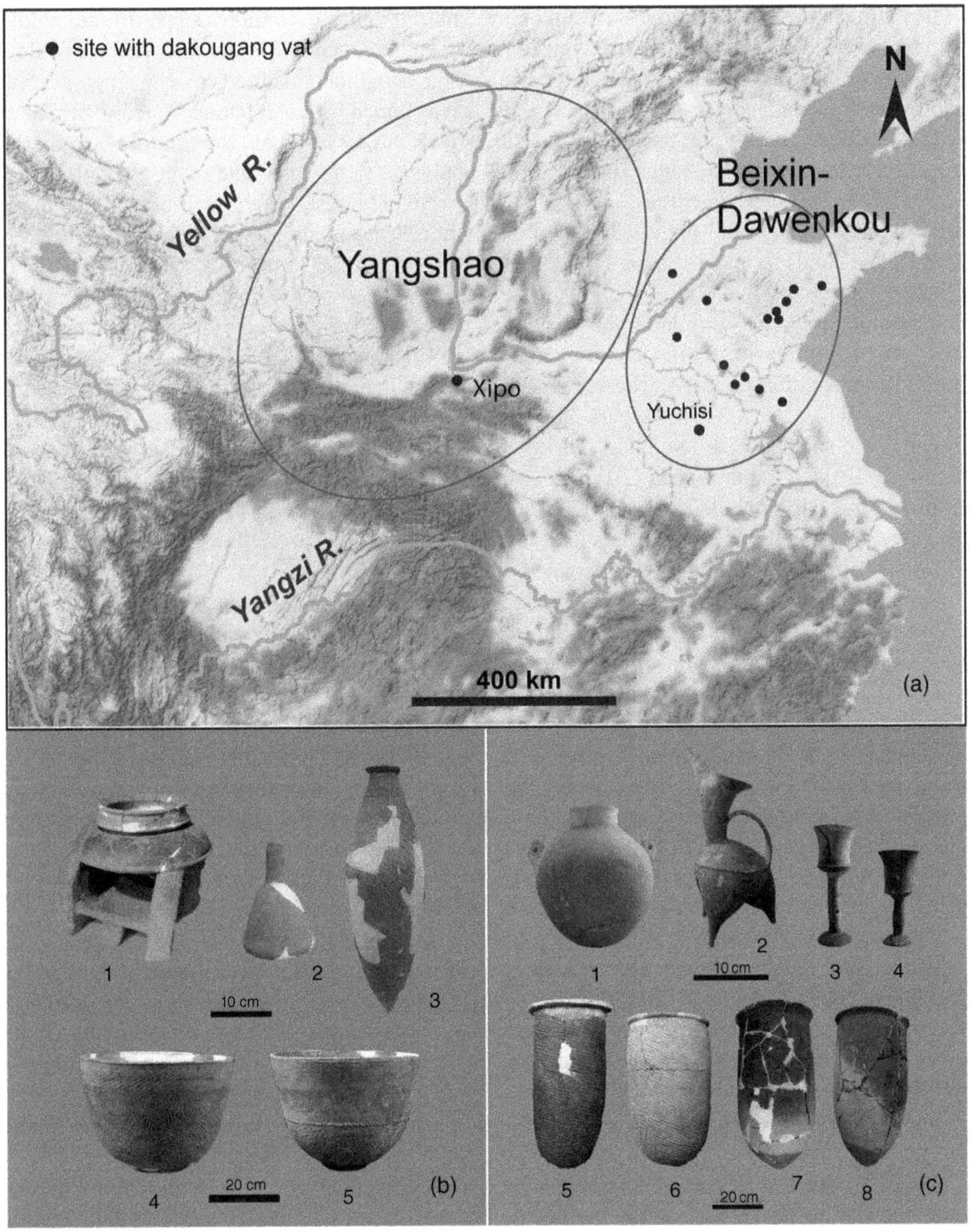

FIGURE 6.3 Regional archeological cultural development and alcohol-related pottery vessels in the Yellow River valley, the middle Neolithic period. (a) Distribution of the Beixin-Dawenkou and Yangshao cultures. (b) Alcohol-related pottery vessels in the Yangshao culture; 1: stove-cauldron; 2: funnel; 3: amphora; 4, 5: *dakougang* vats (from Xipo). (c). Alcohol-related pottery vessels in the Beixin-Dawenkou cultures; 1: globular jar; 2: *gui* pitcher; 3, 4: goblets; 5–8: *dakougang* vats (from Yuchisi). Source: These images appear courtesy of the author.

in ceramic production (Underhill 2000, 2002). An ideology of competitive funerary ritual, including libation, may have emerged in Dawenkou society, as a contributing factor in the development of social complexity in the region (Fung 2000). The alcohol-related ritual feasts appear to have emphasized individual status and social hierarchy in this region.

In sharp contrast to the Beixin-Dawenkou system, the Yangshao culture people in the middle Yellow River region produced a different ceramic assemblage. Globular jars disappeared, while a new type of brewing vessel, *jiandiping* (conical-based jar) amphora, emerged. The amphorae are formed with a small mouth, an egg-shaped body, and a conical base. Amphorae were mostly small with a few large ones during the early Yangshao period, but they became consistently large in the middle and late Yangshao (Figure 6.3). Residue analysis of amphorae and other alcohol-related vessels from 11 Yangshao culture sites shows that three previously established brewing methods were employed: (i) use of malts, (ii) use of honey and /fruits, and (iii) preparation of *qu* starter. Identifiable molds for making *qu* are predominantly *Monascus, Aspergillus, Rhizopus,* and *Mucor* (Table 6.1). Yeast cells, morphologically resembling *S. cerevisiae,* are also present. Such microbial communities are largely similar to those found in the early Neolithic in the same region, suggesting a practice of repeatedly inoculating certain microorganisms for the continuous generation of desired fermentation products. The brewing ingredients included broomcorn millet, Job's tears, Triticeae, rice, beans, various tubers and roots (e.g. snake gourd root, foxnut, lily, yam, and ginger), and, in one case (Mijiaya), barley (Table 6.1). The brewing method associated with amphorae is likely to have been LSF (Liu 2021). However, four *dakougang* vats were unearthed from two elite burials at the Xipo site of the late Yangshao culture, and the residue analysis revealed *Monascus* mold and rice starch grains in the vessels, suggesting that SSSF was also introduced to the Yangshao culture, although as an isolated case (Feng et al. 2021).

Some recurrent assemblages of certain vessel forms may have constituted a "toolkit" for the production, storage, and consumption of fermented beverages, including amphorae (for fermentation, storage, and possibly for serving), funnels (for transfer and filtration), large urns (for mashing), and portable stoves with caldrons (for heating water or fermented drinks) (Figure 6.2). The Yangshao pottery assemblage consists of very few drinking cups, in sharp contrast to the pervasive distribution of amphorae. It is interesting to note that many amphorae show vertical striations on the interior wall of their rims. Based on my experimental study, similar damages can be created by repeatedly rubbing a reed stem against a sherd in a unidirectional motion. In addition, reed stem phytoliths (Phragmite bulliform) have been uncovered from residues in some amphorae. These lines of information suggest that reed straws may have been used for communal drinking of alcoholic beverages from amphorae, referred to as siphoning or sucking (Liu 2021).

Drinking alcohol through straws has been found in the archeological record of the ancient Near East and Egypt, and it is still practiced in Africa and Vietnam today (Hiên 2007; Katz and Voigt 1986; McGovern 2009). Communal drinking through straws, known as *zajiu*, has also been recorded as a widespread type of drinking ritual among more than a dozen ethnic groups in southwestern China today, particularly well-known in the Qiang and Tibetan in northwestern Sichuan. The populations in the Tibetan Plateau of southwestern China are linguistically and genetically related to the Neolithic Proto-Sino-Tibetan populations in the Yellow River, including people of the Yangshao culture. Thus, this drinking tradition is very likely a cultural heritage of the Yangshao culture (Liu et al. 2022c).

The Yangshao culture represents a relatively egalitarian society with emerging social differentiation toward its later phase, and fermented beverages may have been served in communal feasting rituals with the function primarily of creating and reinforcing group solidarity, networking, and social identity (Liu 2021; Liu et al. 2022c).

Alcohol-related ritual feasts demonstrate very different cultural traditions in the two regions along the Yellow River valley. The Beixin-Dawenkou societies in the east coast area were characterized by an individualized sociopolitical system, while the Yangshao population in the western inlands emphasized a group-oriented social practice. These two social systems appear to have developed side by side and coexisted for thousands of years, but the ceramic assemblages together with the individualized drinking tradition from the east coast region gradually became more widespread.

6.5 INDIVIDUALIZED DRINKING TRADITION AND DEVELOPMENT OF SOCIAL COMPLEXITY IN THE LATE NEOLITHIC (4600 TO 3800 BP)

The late Neolithic societies in the Yellow River region, referred to as the Longshan culture, experienced rapid development of social complexity, characterized by increased inter-regional interactions (conflict and exchange) and social hierarchy that preceded the emergence of early state-level political organization in the Bronze Age (Liu and Chen 2012). As societies became more stratified, the alcohol-related vessels also became more elaborate in form, specialized in function, and often appeared in concentration at large sites or elite burials. Changes in the production and consumption of alcoholic beverages are manifested by the disappearance of *jiandiping* amphorae of the preceding Yangshao culture, and the widespread appearance of several ceramic types used for serving and drinking alcoholic beverages which originated in the eastern coastal regions, such as *gaobingbei* goblets, cups, and various types of pitchers (Figure 6.4). These ceramic vessels were designed for serving individualized drinking events, consistent with the increased social stratification at the time. These parallel developments suggest that the consumption of alcoholic beverages was an important part of feasting ritual activities at political and economic centers (He et al. 2021; Underhill 2018).

These developments are evident in the Shimao site in Shaanxi province, which is considered the earliest city developed in the northern Loess Plateau region. The site (430 ha in area) comprises a central palatial terrace, an inner enclosure, and an outer enclosure, all built mainly of stone. Pottery assemblages at Shimao include various forms, some (such as pitchers and cups) share features with those from the Central Plains and the lower Yellow River valley farther east (Figure 6.4), implying interactions with these regions (Sun et al. 2018, 2020).

Analysis of residue remains on pottery vessels and stone tools from Shimao has revealed their alcohol-related function. At the palatial terrace, the core locality of this urban center, people brewed beer with multiple grains and tubers, consisting of broomcorn millet, Triticeae, rice, lily, snake gourd root, Zingiberaceae (ginger or turmeric), and beans. The brewing method may have involved the preparation of *qu* starter with rice and *Monascus* mold, as well as the use of malted broomcorn millet as the saccharification agent. The ingredients and brewing method are consistent with the technological tradition of alcohol production established in the middle Yellow River valley during the early and middle Neolithic periods. Alcoholic beverages may have played an important role in the ritual feasting activities at this urban center, mediating the sociopolitical relationships and interactions between the people of Shimao and other groups. Similar to other large regional centers of the Longshan culture in the middle and lower Yellow River valley, feasting practices for forming alliances or engaging in competitions may have contributed to developing a new social order and new kinds of social relations in the region (He et al. 2021, 2022; Liu et al. 2022a).

6.6 DISCUSSION AND CONCLUSIONS

After reviewing the currently available data, I now summarize some general patterns of Neolithic alcohol production and consumption in relation to environmental conditions and social change.

The initial development of cereal-based alcoholic fermentation may have coincided with the beginning of rice and millet domestication 9000 years ago in China. The wet and warm climate in the early Holocene provided the natural conditions for the availability of diverse food resources, cultivation of plants, increased sedentary lifeways, and growing population densities. The Holocene climatic optimum generated by the strong East Asian summer monsoon also provided

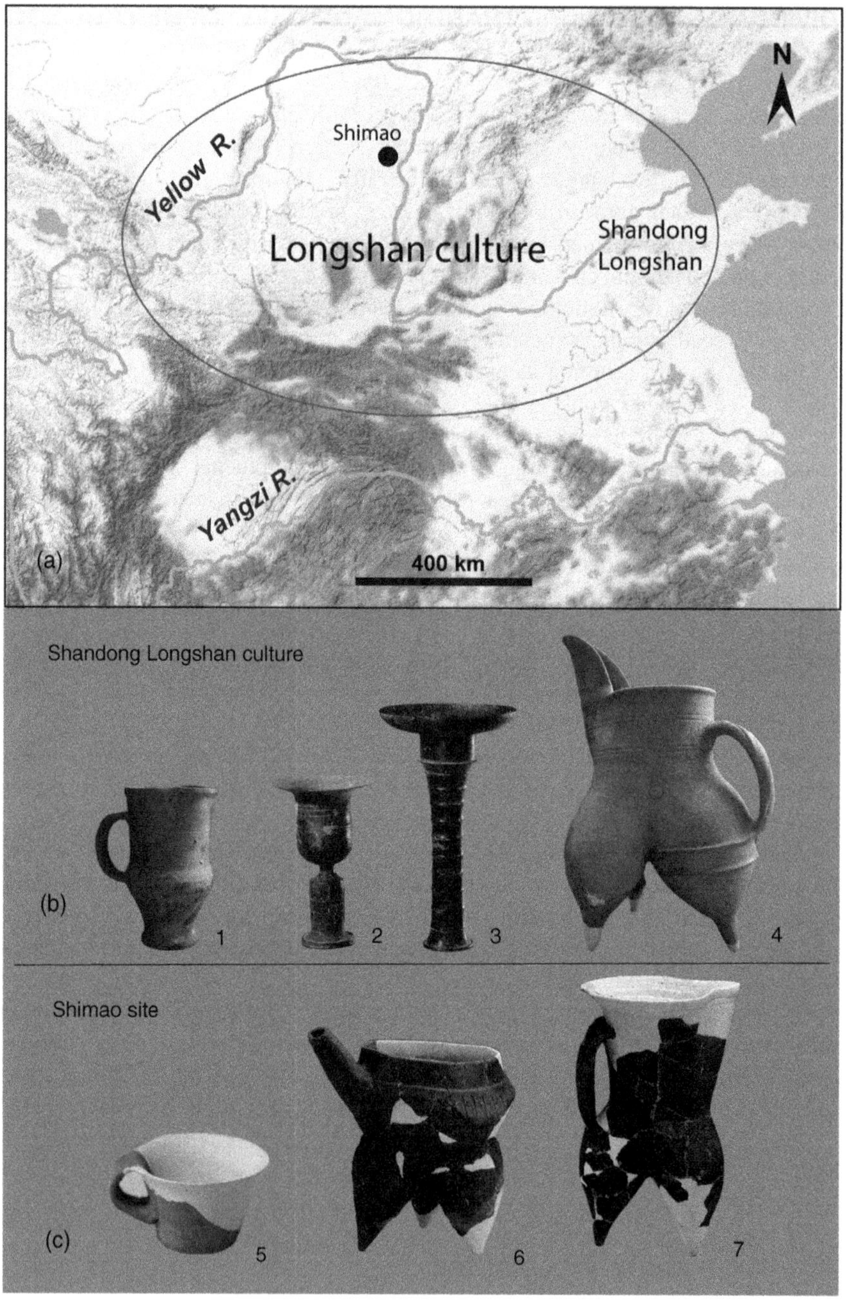

FIGURE 6.4 Examples of serving and drinking vessels during the late Neolithic Longshan culture. (a) Distribution of the Longshan culture in the Yellow River region. (b) Examples of alcohol-related vessels from the Shandong Longshan culture; 1: cup; 2, 3: goblets; 4: pitcher. (c). Examples of alcohol-related vessels from Shimao; 5: cups; 6, 7: pitchers. Source: These images appear courtesy of the author.

a natural environment suitable for the fungi to grow, facilitating the innovation of *qu* fermentation method. This is because the production of *qu* requires a humid and warm condition for a set period of incubation to stimulate the growth and metabolism of microbiota, a process often carried out in summer (Jin et al. 2017).

Rice and *Monascus* mold were essential for making *qu* starter to brew rice-based red beer, as evidenced at the Xiaohuangshan site (ca. 9000 cal. BP). Rice cultivation together with this brewing method also dispersed northward to the Yellow River region, exemplified by the findings at the Peiligang site (ca. 8000 cal. BP). In most cases, fermentation ingredients include both domesticated and wild cereals and various tubers or roots, suggesting that alcoholic beverages were made with locally available plants. However, rice or millet always were essential ingredients, indicating the close connections between cereal domestication and alcohol fermentation.

The development of Neolithic technologies (e.g. plant cultivation and ceramic production) and social systems (forming kin-based large communities) was also accompanied with increased ritual activities, such as communal feasts and provision of food offerings to the dead ancestors in mortuary practice. Pottery vessels containing fermented beverages became the key component of grave goods in many Neolithic burial sites. This phenomenon may have been related to the development of ancestral veneration, which formed the core belief system in ancient China (Keightley 1978; Liu 2000). The ritual feasting involving alcoholic beverages may have played a pivotal role in the creation and reinforcement of community solidarities, social connections, and cultural identities. These functions are particularly evident in the Yangshao culture, which demonstrated expansion to the west and southwest regions in the middle Neolithic (Liu et al. 2022c).

As the Neolithic societies became more complex, ritual feasting also focused more on social hierarchy, individual status, and competition for power. The new social values first developed in the east coast region during the middle Neolithic period, and then became widespread in the late Neolithic. Ceramic assemblages associated with alcoholic consumption consisted of diverse forms of vessels, designated for individualized serving and drinking. Such vessels were elaborately made and occurred in high concentration in political centers and elite tombs, a phenomenon which is consistent with the rapid development of ritual power controlled by political elites before the dawn of the Bronze Age dynasties in China.

Brewing methods also changed through time. It started with the LSF using small globular jars in both Yangzi and Yellow River regions. Then two regional traditions formed along the Yellow River valley, with the LSF using amphorae in the western inlands, on the one hand, and the emergence of SSSF using large vats in the east coast region. The latter method became more prevalent cross-regionally in the late Neolithic, perhaps to achieve higher levels of alcoholic contents and provide larger quantities of beverages for larger scales of feasting. This method has also become the predominant brewing technique for making traditional fermented beverages, such as *huangjiu* yellow beer, until today.

In conclusion, the emergence and development of alcoholic fermentation technologies were closely linked to the Neolithic revolution. The advance of these techniques may have been stimulated by the need for ritual activities, and in turn, it may have helped reinforce the increased social complexity through feasting events in Neolithic China.

ACKNOWLEDGEMENTS

This research was supported by the Min Kwaan Chinese Archaeology Program in Stanford Archaeology Center, Stanford University.

REFERENCES

Arnold, B. (1999). 'Drinking the Feast': alcohol and the legitimation of power in Celtic Europe. *Cambridge Archaeological Journal* 9: 71–93.

Bao, Q. (2005). Shiqian wenhua shiqi de niangjiu (er) (Wine production in the prehistoric culture period, II). *Niangjiu Keji* 7: 88–93.

Bao, Q. and Zhou, J. (2007). *Niangzao (Fermentation)*. Zhengzhou: Daxiang Press.

Bestel, S., Bao, Y., Zhong, H. et al. (2017). Wild plant use and multi-cropping at the early Neolithic Zhuzhai site in the middle Yellow River region, China. *The Holocene* 28: 195–207.

Braidwood, R.J., Sauer, J.D., Helbaek, H. et al. (1953). Symposium: did man once live by beer alone? *American Anthropologist* 55: 515–526.

Crawford, G., Chen, X., and Wang, J. (2006). Shandong Jinan Changqingqu Yuezhuang faxian Houli wenhua shiqi de tanhuadao (Discovery of the carbonized rice of the Houli culture period at Yuezhuang in Changqing district, Jinan, Shandong). In: *Archaeology of East China* (ed. D. Kaogu), 247–251. Beijing: Kexue Press.

Dietler, M. (1990). Driven by drink: the role of drinking in the political economy and the case of early Iron Age France. *Journal of Anthropological Archaeology* 9: 352–406.

Dietler, M. (2006). Alcohol: anthropological/archaeological perspectives. *Annual Review of Anthropology* 35: 229–249.

Feng, S., Liu, L., Wang, J. et al. (2021). Red beer consumption and elite utensils: the emergence of competitive feasting in the Yangshao culture, North China. *Journal of Anthropological Archaeology* 64: 101365.

Fung, C. (2000). The drinks are on us: ritual, social status, and practice in Dawenkou burials, North China. *Journal of East Asian Archaeology* 2: 67–92.

Hayden, B., Canuel, N., and Jennifer, S. (2013). What was brewing in the Natufian? An archaeological assessment of brewing tchnology in the Epipaleolithic. *Journal of Archaeological Method and Theory* 20: 102–150.

He, Y., Liu, L., Sun, Z. et al. (2021a). "Proposing a toast" from the first urban center in the north Loess Plateau, China: alcoholic beverages at Shimao. *Journal of Anthropological Archaeology* 64: 101352.

He, Y., Liu, L., and Wang, J. (2021b). Neimenggu zhongnanbu Yangshao shiqi niangjiu fenxi -- yi Miaozigou yizhi weili (Alcohol production of the Yangshao period in south-central Inner Mongolia -- example from the Miaozigou site). In: Yangshao Wenhua yu Jiu (ed. L. Liu), 214–235. Beijing: Wenwu Press.

He, Y., Liu, L., Shao, J. et al. (2022). Shaanxi Yulin Shimao yizhi Huangchengtai didian jiulei canliuwu ji xiangguan taoqi fenxi (Analysis of residues of alcoholic beverages and related pottery vessels from Huangchengtai at the Shimao site in Yulin, Shaanxi). *Kaogu yu Wenwu* 2: 146–154.

Hiên, N.X. (2007). Glutinous rice, kinship and the Têt Festival in Vietnam. In: *Kinship and Food in South East Asia* (ed. M. Janowski and F. Kerlogue), 242–265. Copenhagen: Nordic Institute of Asian Studies.

Huan, X., Lu, H., Jiang, L. et al. (2021). Spatial and temporal pattern of rice domestication during the early Holocene in the lower Yangtze region, China. *The Holocene* 31: 1366–1375.

Huang, H.T. (2000). *Science and Civilisation in China: Vol 6, Biology and Biological Technology, Part V: Fermentations and Food Science*. Cambridge: Cambridge University Press.

Jin, G., Wu, W., Zhang, K. et al. (2014). 8000-Year old rice remains from the north edge of the Shandong Highlands, East China. *Journal of Archaeological Science* 51: 34–42.

Jin, G., Zhu, Y., and Xu, Y. (2017). Mystery behind Chinese liquor fermentation. *Trends in Food Science and Technology* 63: 18–28.

Joffe, A.H. (1998). Alcohol and social complexity in ancient Western Asia. *Current Anthropology* 39: 297–322.

Katz, S.H. and Voigt, M.M. (1986). Bread and beer: the early use of cereals in the human diet. *Expedition* 28: 22–34.

Keightley, D.N. (1978). The religious commitment: Shang theology and the genesis of Chinese political culture. *History of Religion* 17: 212–214.

Keightley, D. (1987). Archaeology and mentality: the making of China. *Representations* 18: 91–128.

Li, Y., Chen, X., Gu, W. et al. (2020). Henan Xinzheng Peiligang Yizhi 2018–2019 nian fajue (The 2018–2019 season excavation at Peiligang site in Xinzheng, Henan). *Kaogu Xuebao* 4: 521–545.

Liao, J., Yang, Y., Gu, W. et al. (2022). A new filtered alcoholic beverage: residues evidence from the Qingtai site (ca. 5,500–4,750 cal. BP) in Henan Province, Central China. *Frontiers in Earth Science* 10. https://doi.org/10.3389/feart.2022.884630.

Ling, C. (1957). Zhongguo ji dongnanya de juejiu wenhua (The culture of masticating alcohols in China and Southeast Asia). *Bulletin of the Institute of History and Philology Academia Sinica* 4: 1–23.

Ling, C. (1958). Zhongguo jiu zhi qiyuan (Th origins of alcohol in China). *Bulletin of the Institute of History and Philology* 29: 883–901.

Liu, L. (2000). Ancestor worship: an archaeological investigation of ritual activities in Neolithic North China. *Journal of East Asian Archaeology* 2: 129–164.

Liu, L. (2021). Communal drinking rituals and social formations in the Yellow River valley of Neolithic China. *Journal of Anthropological Archaeology* 63: 101310.

Liu, L. and Chen, X. (2012). *The Archaeology of China: From the Late Palaeolithic to the Early Bronze Age*. Cambridge: Cambridge University Press.

Liu, L., Wang, J., Zhao, Y. et al. (2017). Yangshao wenhua de guyajiu: jiemi Yangguanzhai yizhi de taoqi gongneng (Beer in the Yangshao culture: decoding the function of pottery at the Yangguanzhai site). *Nongye Kaogu* 6: 26–32.

Liu, L., Wang, J., Chen, X. et al. (2018a). Yangshao wenhua dafangzi yu yanyin chuantong: Henan Yanshi Huizui yizhi F1 dimian he taoqi canliuwu fenxi (Large houses and feasting tradition of the Yangshao culture: starch and phytolith analyses of the residues from pottery vessels and floors of house no.1 at Huizui in Yanshi, Henan). *Zhongyuan Wenwu* 1: 32–43.

Liu, L., Wang, J., Rosenberg, D. et al. (2018b). Fermented beverage and food storage in 13,000 y-old stone mortars at Raqefet cave, Israel: investigating Natufian ritual feasting. *Journal of Archaeological Science: Reports* 21: 783–793.

Liu, L., Wang, J., Zhao, H. et al. (2018c). Shaanxi Lantian Xinjie yizhi Yangshao wenhua wanqi taoqi canliuwu fenxi: niangzao guyajiu de xinzhengju (Residue analyses on pottery from the late Yangshao culture site of Xinjie in Lantian, Shaanxi: new evidence of beer brewing). *Nongye Kaogu* 1: 7–15.

Liu, L., Wang, J., Chen, X. et al. (2019a). Huanghe zhongyou xinshiqi shidai lujiu taohu fenxi (Analysis of Neolithic pottery strainers in the middle Yellow River valley). *Zhongyuan Wenwu* 6: 55–61.

Liu, L., Wang, J., Levin, M.J. et al. (2019b). The origins of specialized pottery and diverse alcohol fermentation techniques in Early Neolithic China. *Proceedings of the National Academy of Sciences* 116: 12767–12774.

Liu, L., Li, Y., and Hou, J. (2020a). Making beer with malted cereals and qu starter in the Neolithic Yangshao culture, China. *Journal of Archaeological Science: Reports* 29: 102134.

Liu, L., Wang, J., Chen, X. et al. (2020b). Beixin wenhua xiaokou shuang'er guan de niangjiu gongneng (The fermentation function of globular jars in the Beixin culture). *Nanfang Wenwu* 5: 74–84.

Liu, L., Wang, J., and Di, N. (2020c). Cong pingdiping dao jiandiping -- Huanghe zhongyou xinshiqi shiqi niangjiuqi de yanhua he niangjiu fangfa de chuancheng (From flat-base amphorae to conical-base amphorae -- changes in fermentation vessels and brewing methods in Neolithic middle Yellow River region). *Zhongyuan Wenwu* 3: 94–106.

Liu, L., Wang, J., and Liu, H. (2020d). The brewing function of the first amphorae in the Neolithic Yangshao culture, North China. *Journal of Anthropological and Archaeological Sciences* 12: 118.

Liu, L., Li, Y., and Hou, J. (2021a). Mianchi Dingcun yizhi Yangshao wenhua de qujiu he guyajiu (Alcoholic beverages made of qu and mults discovered at the Yangshao culture site of Dingcun in Mianchi). *Zhongyuan Wenwu* 5: 75–85.

Liu, L., Wang, J., Chen, X. et al. (2021b). Shandong Dawenkou wenhua jiuqi chutan (Investigation of alcohol-related vessels of the Dawenkou culture in Shandong). *Huaxia Kaogu* 1: 49–61.

Liu, L., Levin, M., Sun, Z. et al. (2022a). Shimao yizhi chutu tao, shiqi gongneng fanying de liyi he shengji huodong (Ritual and subsistence activities reflected on pottery and stone tools from the Shimao site). *Zhongyuan Wenwu* 5: 31–51.

Liu, L., Wang, J., Chen, X. et al. (2022b). The quest for red rice beer: transregional interactions and development of competitive feasting in Neolithic China. *Archaeological and Anthropological Sciences* 14: 78.

Liu, L., Chen, J., Wang, J. et al. (2022c). Archaeological evidence for initial migration of Neolithic Proto Sino-Tibetan speakers from Yellow River valley to Tibetan Plateau. *Proceedings of the National Academy of Sciences* 19 (51): e2212006119.

Liu, L., Li, Y., Zhao, Y. et al. (Under revision). Ritual dynamics: dispersal of rice domestication and brewing of red rice beer in early Neolithic China. *Journal of Archaeological Science: Reports*.

Liu, L., Wang, H., Sun, H. et al. (2023). Serving red rice beer to the ancestors ca. 9000 years ago at Xiaohuangshan early Neolithic site in South China. *The Holocene*. https://doi.org/ 10.1177/09596836231169995.

McGovern, P.E. (2009). *Uncorking the Past: The Quest for Wine, Beer, and Other Alcoholic Beverages*. Berkeley and Los Angeles: University of California Press.

McGovern, P.E., Zhang, J., Tang, J. et al. (2004). Fermented beverages of pre- and proto-historic China. *Proceedings of the National Academy of Sciences* 101: 17593–17598.

McGovern, P.E., Underhill, A., Fang, H. et al. (2005). Chemical identification and cultural implications of a mixed fermented beverage from late prehistoric China. *Asian Perspectives* 44: 249–275.

Qiu, Z. and Rao, H. (2022). Environmental landscape and subsistence strategy of the Shunshanji culture: a review. *Frontiers in Plant Science* 13: 966635.

Sun, Z., Shao, J., Liu, L. et al. (2018). The first Neolithic urban center on China's north Loess Plateau: the rise and fall of Shimao. *Archaeological Research in Asia* 14: 33–45.

Sun, Z., Shao, J., Di, N. et al. (2020). Shaanxi Shenmu Shimao yizhi Huangchengtai fajue qude zhongyao shouhuo (Important gains from excavations in Huangchengtai at the Shimao site in Shenmu, Shaanxi). *Zhongguo Wenwubao* 249.

Underhill, A.P. (2000). An analysis of mortuary ritual at the Dawenkou site, Shandong, China. *Journal of East Asian Archaeology* 2: 93–128.

Underhill, A.P. (2002). *Craft Production and Social Change in Northern China*. New York: Kluwer Academic/ Plenum Publishers.

Underhill, A.P. (2018). Urbanization and new social contexts for consumption of food and drink in northern China. *Archaeological Research in Asia* 14: 7–19.

Underhill, A., Feinman, G., Nicholas, L. et al. (2002). Regional survey and the development of complex societies in southeastern Shandong, China. *Antiquity* 76: 745–755.

Wang, J., Liu, L., Ball, T. et al. (2016). Revealing a 5,000-y-old beer recipe in China. *Proceedings of the National Academy of Sciences* 113: 6444–6448.

Wang, J., Liu, L., Georgescu, A. et al. (2017). Identifying ancient beer brewing through starch analysis: a methodology. *Journal of Archaeological Science: Reports* 15: 150–160.

Wang, J., Jiang, L., and Sun, H. (2021). Early evidence for beer drinking in a 9000-year-old platform mound in southern China. *PLoS One* 16: e0255833.

Wen, S. and Yuan, T. (1983). *Yinxu Buci Yanjiu: Kexue Jishu Pian (A Study of Oracle Bones from Yinxu: Science and Technolgy)*. Chengdu: Sichuansheng shehui kexueyuan chubanshe.

Zhang, J., Lu, H., Gu, W. et al. (2012). Early mixed farming of millet and rice 7800 years ago in the middle Yellow River region, China. *PLoS One* 7: e52146.

Zhang, J., Cheng, Z., Lan, W. et al. (2018). Henan Wuyang Jiahu yizhi zhiwu kaogu yanjiu de xin jinzhan (New progress of archaeobotanical research at the Jiahu site in Wuyang, Henan). *Kaogu* 4: 100–110.

Zhao, Z., Zhao, C., Yu, J. et al. (2020). Beijing Donghulin yizhi zhiwu yicun fuxuan jieguo ji fenxi (Flotation results and analysis of plant remains from the Donghulin site in Beijing). *Kaogu* 7: 99–106.

Zhejiang Provincial Institute of Archaeology & Pujiang Museum (2016). *Pujiang Shangshan (Shangshan in Pujiang)*. Beijing: Wenwu Press.

Zheng, N. (1984). 1979 nian Peiligang yizhi fajue baogao (excavation of the Peiligang site in 1979). *Kaogu Xuebao* 1: 23–51.

Zuo, X., Lu, H., Jiang, L. et al. (2017). Dating rice remains through phytolith carbon-14 study reveals domestication at the beginning of the Holocene. *Proceedings of the National Academy of Sciences* 114: 6486.

CHAPTER 7

BREWS OF THE PAST: BIOARCHAEOLOGY OF MICROBIAL FERMENTATION

KEITH THOMAS
Brewlab Ltd, Sunderland, United Kingdom
Department of Health Sciences and Well Being, University of Sunderland, Sunderland, United Kingdom

CONTENTS

7.1 INTRODUCTION

Why bother with the past? In the rapidly moving and increasingly popular field of food fermentation, it is easy to focus on the future and ignore what was eaten in the past. Industrial fermentations today are increasingly technical employing advanced techniques to analyze and predict fermentations and define product character. Analytical chemistry, molecular biology, and even robotic sensory detection are all driving more informed fermentations resulting in safer and ever more novel foods. Moreover, the overlap with the heavily funded fermented plant and synthetic meat industries provides strong momentum to the industry and an increasingly long-term view (Ismail et al. 2020; Aschemann-Witzel et al. 2021).

Foods of the past are, in many cases, considered lost and forgotten or mentioned in passing. Even if considered they are often associated with poor quality production, periods of famine, and

Microbial Fermentations in Nature and as Designed Processes, First Edition. Edited by Christon J. Hurst.

potential health hazards. Beer, for example, is noted as a historic means of avoiding disease from contaminated water but is non-essential with today's provision of potable supplies.

Aside from academic interest, there is, nevertheless, an arena where past ferments are of relevance — the production of traditional foods, increasingly those with associations to health benefits (Kapp and Sumner 2019; Cuamatzin-García et al. 2022).

That said the definition of traditional is difficult. The term is ubiquitous, a marketing tag on many foods implying some degree of historical authenticity and credibility. In many cases, this is supported by reference to a recipe, ingredients, or to processing features associated with the product and, in some foods, a direct continuity of production. Occasionally, a microbiology association is claimed or physically involved, often for probiotic input but in a few cases as a recreation of past products.

However, an often-asked question is "how do we know what past fermentations actually tasted like, how they were produced, and how they compare to today's foods?" While this has a superficial interest there are supplementary questions with a more valuable focus to consider. These include "did historic beers contain beneficial microorganisms?" and specifically, "did they contain probiotic microbes and prebiotic constituents?"

Other questions relate to whether the ingredients and processing provided nutritional or health benefits and, as a health and social concern, whether fermented beverages really were protection against food pathogens. While these questions have some academic relevance, and commercial potential there is continual interest in recreating past beverages, in part for commercial potential but also as an investigative challenge (McGovern 2009; Dineley 2011; Schwartz 2013; Jenner 2016).

Contemporary fermentations are carefully managed and often conducted using purified microorganism inoculations, typically a single species for beer, wine, and spirits and a dual mix for yogurt. Natural (spontaneous) fermentation is an active area of research and application today and is likely to benefit from an understanding of the past when products evolved under natural conditions rather than marketing influences (Michel et al. 2016).

Can bioarchaeology and its numerous tools address these issues and provide valuable information toward their appraisal with relevance to health practitioners and contemporary food producers? Food residues can of course be analyzed, and interpretations of ingredients and processing made (Cheng 2018; Wang et al. 2016, 2017; Farag et al. 2019). In some instances, microorganisms have been isolated from samples (Aouizerat et al. 2019; Thomas et al. 2021) but modern, next-generation DNA sequencing provides a wider window on the range of microorganisms present — both alive and dead (Hofreiter et al. 2015; Shapiro and Hofreiter 2014; Thomas et al. 2021). Knowing these profiles allows an application to contemporary production such as using comparable microorganisms sourced from culture collections or occasionally from recovered samples.

It is important at this stage to note the difficulties in handling and interpreting data from bioarcheological analysis. Molecular deterioration is inevitable, and some degree of extrapolation may be needed to estimate the actual levels of a compound. As a simple example, ethanol is readily oxidized to acetic acid requiring levels of both to be determined and considered. Lipids also oxidize and proteins fragment while phenols polymerize with time. It is clear that changes in metabolites and flavors are complex and depend on the recipe and environmental factors making it difficult to determine how beer has deteriorated (Lehnhardt et al. 2019).

Nucleic acids are similarly affected, and the degradation of DNA is a well-recognized issue with an inverse and exponential correlation being noted between the length of DNA fragments and their abundance (Schwarz et al. 2009; Allentoft et al. 2012). Recovered DNA is likely to be of short fragment length, possibly less than 300 base pairs, and more difficult to match to barcoding databases (Rawlence et al. 2014). It is also suggested that particular nucleotide sequence patterns are more likely to deteriorate than others leading to biases in the identification of particular microbiological groups (Ziesemer et al. 2015). Damage to DNA bases such as from deamination may introduce miscoding and thus identification errors so requiring assessment and caution in interpretations (Hofreiter et al. 2001; Willerslev et al. 2004) including downstream analysis of the bioinformatic data obtained (Ginolhac et al. 2011).

The potential for external and contemporary contamination provides a further source of the difficulty. Recommended protocols for handling artifacts in dedicated facilities are a typical requirement of working while the inclusion of controls is an important quality assurance (Cooper and Poinar 2000; Willerslev et al. 2004; Lydolph et al. 2005; Korlević et al. 2015; Drieu et al. 2020). Controls may include samples from soils adjacent to an artifact source and from the outside of vessels (Hendy et al. 2018) although laboratory blanks carry risks of unintended contamination (Cooper and Poinar 2000).

To illuminate work in these areas this review will focus on beer as an example beverage with a large industrial and social impact. Nevertheless, methods are universal, and the continued resurrection of more niche fermented products continues to provide interest and challenge.

Indeed, it is relevant to note the features of beer in contrast to other historically recognized fermented foods, for example, dairy products. Evidence of dairy fermentation in China has been placed at 1980 to 1213 BCE through the identification of milk peptides (Yang et al. 2014; Greco et al. 2018) and at 5700 BCE in Croatia and 7000 BCE in Northern Europe from lipid residue analysis (Salque et al. 2013; McClure et al. 2018). Unlike beer, milk is a rapidly fermenting food that readily spoils if left at warm temperatures due to its high levels of sugars and amino acids. Controlled fermentation, however, produces yogurt, kafir lassi, and many regional products primarily through the metabolism of lactic acid bacteria (Shiby and Mishra 2013). A major difference between these foods and traditional beers is the fundamental involvement of lactic acid in milk fermentations and yeast in beer. Today though, increasing interest in naturally fermented beers is providing interest in historic brewing which inevitably would have involved a mix of microorganisms.

7.2 HISTORIC ALE AND BEER

The origins of beer fermentations may go back further than milk with various claims of earliest use including suggestions of brewing from 8000 BCE covering locations varying from Mesopotamia to Egypt (Hornsey 2003). Because of the common cereal ingredients in beer and bread, there has been continual commentary over the commonality of brewing and baking (Curtis 2001; Ishida 2002) but it is very clear from visual artifacts that brewing was extensively practiced in Mesopotamia and Egypt from early neolithic millennia (Geller 1992).

As with dairy fermentations, bioarchaeology has focused on the analysis of residues from production locations and particularly from storage and drinking vessels (Samuel 1996; McGovern 2009; Perruchini et al. 2018; Aouizerat et al. 2019). Reference texts have provided some basic information on recipes and processes, but these have not always been authenticated by bioarchaeology. For example, the potential use of dates as a beer ingredient has validity due to their extensive cultivation in the middle east (Faltings 1991; Farag et al. 2019). However, the absence of date stones in some beer residues and brewing sites suggests this may be incorrect or variable, at least in part (Samuel 1996; McGovern 2009).

While analysis of neolithic artifacts has been relatively extensive much less attention has been addressed to artifacts from more interim periods — the Middle Ages for example. Written records are, inevitably, more extensive but more detailed bioarchaeology would be of interest in this period as it covers the introduction of hops to brewing so providing a shift away from more malt-based

FIGURE 7.1 Embossed and screw top Victorian beer bottles.

beverages (Burton 2013). The industrialization of brewing which followed in subsequent centuries moved beer into a commercial beverage and initiated early industrial processes (Hornsey 2003; Anderson 2005). As technologies progressed beers developed greater shelf life along with wider distribution in bottles, particularly after the invention of the screw top design in 1879 (Lott 1901; Cornell 2006). These developments allowed for the prospect of survival of samples remaining intact to the present day and thus an ability to reach back and conduct detailed bioarchaeology on actual samples (Jeandet et al. 2015; Roullier-Gall et al. 2017; Kajala et al. 2018; Thomas et al. 2021). A tantalizing prospect for analysis and recreation (Figure 7.1).

To explore the prospects of bioarchaeology to the evolution of beers it is worth considering beers and brewing in specific periods or golden ages, representing different characteristics and to identify their features.

7.3 GOLDEN AGES OF BREWING

Assigning a quality index to brewing periods has strong subjective biases, particularly if viewed from a contemporary position. This is because it is not possible to compare the satisfaction of drinkers across centuries, or millennia since the characteristics of beers will differ extensively. Records of sales and consumption are not comparable and marketing analysis has difficulty in evaluating the success of commercial trade in different areas. Moreover, the varying stature of beer compared to other fermented drinks, particularly wine and spirits, further complicates any standardization on quality. With this proviso, any judgment is limited to identifying periods when advances produced beers of identifiable and differing character.

As mentioned, brewing archeology has an extensive chronicle ranging from early Egyptian, Mesopotamian, and Abyssinian production to recent industrial developments. This time range represents several distinct periods of development some of which can be charted by historical records and modern techniques illustrating the application of specific technologies such as the introduction of hops (Burton 2013), the appearance of lager yeast (Gallone et al. 2016) as well as genetic changes in the brewing yeast genome (Gonçalves et al. 2016).

It is inevitable that brewing microbiology has altered during this extensive period making it difficult and speculative to ascertain the microbial communities present in different eras. Nevertheless, some detailed studies have been conducted and indicate major features of beers produced in specific periods, the ingredients used, and their production processes. For example, black and spruce beers from Dantzig (Baker 2012) coming of age ales (Cornell 2013), bock bier (Flissebaalje 2013), Devon white ale (Swift 2014), Lambic beer (Mulder 2018), and audit ales (Compton-Davey 2008). Further studies have reported recreations of past beers and assessments of their character (Dineley 1996; Schwartz 2013; Wang et al. 2017; Dineley and Dineley 2022), albeit using modern ingredients.

Health fermentations are increasingly popular today as evidenced by the explosion of interest in the kombucha and low-alcohol markets (Amarasinghe et al. 2018; Villarreal-Soto et al. 2018; Bellut and Arendt 2019). Knowledge and input from historic fermentations can thus be very relevant to these developments by providing details of beverage microbiology before industrialization and brand marketing produced standardized beverages.

While brewing through the ages has followed similar pathways of production, a sequence of development may be identified based on historical commentary and technical developments.

- Raw and infused grain fermentation. Egyptian, Mesopotamian >1000 BCE
- Herbal infusions. Viking. Gruit. 500–1200 CE
- Development of high-alcohol beers. Monastical. 1000–1700 CE
- Use of hops for flavor and to increase stability and life span. 1300 CE onward
- Temperature and gravity control to optimize and standardize extract from mashing and achieve target alcohol levels. 1750 CE
- Gravity control to manage fermentation effectively and achieve consistency in alcohol targets. 1750 CE
- Bottling to allow wider distribution of sales. 1880 CE
- Hygiene control. (Pasteur) to manage quality and shelf life. 1860 CE
- Yeast purity control. (Hansen) to assist consistency in flavor profiles. 1896 CE
- Filtration and packaging management to increase stability and shelf life. 1900 CE

From this sequence certain Golden Ages of brewing may be postulated where beer character became optimized. These may, arguably, be:

- Mashed grain fermentation (Egyptian, Mesopotamia) supplemented with herbs and fruit.
- Gruit ales supplemented with herbal infusions.
- Monastic ales.
- Early hopped ales. Elizabethan.
- Industrialization of production with scientific input. Georgian, Victorian.

A comparative scaling of these developments may be constructed as in Figure 7.2 to show changes through the periods.

Each of these periods will exemplify beer characters different from each other but accepted as exemplary to drinkers of each period. Characterizing these is a task of the bioarcheologist. It is highly likely that beers from early periods would be difficult to sell today and quite likely be

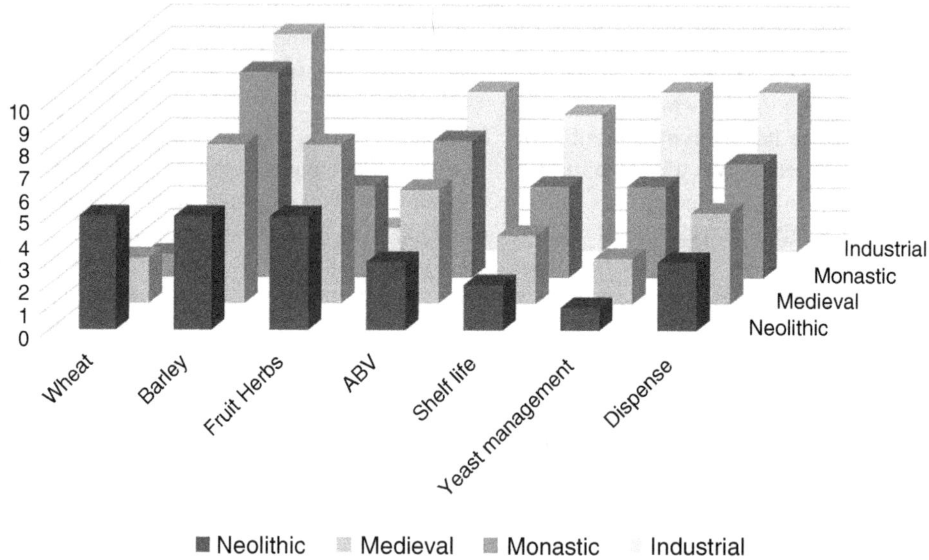

FIGURE 7.2 Profiles of major features of brewing through periods of development. Wheat – proportion used as a major ingredient: barley — proportion used as a major ingredient; fruit & herbs — prevalence in recipes, particularly as gruit; ABV — general level of alcohol achieved: shelf life — likely drinkable period (days to months); yeast management — awareness of value of yeast repatching; dispense — sophistication of draught and packaged product. (*See colour plate section for colour representation of this figure.*)

rejected by drinkers. However, the rise in popularity of sour beers in recent years (Dysvik et al. 2019, 2020) suggests that a proportion of the drinking public may be appreciative of beverages of the past.

7.4 EARLY BEERS. EGYPTIAN AND MESOPOTAMIAN

Extensive evidence of beer production and culture is available from written and pictorial texts, particularly in Egyptian records (Forman et al. 1992; Geller 1992). Two of the oldest brewery sites investigated in Egypt, Hierakonpolis and Abydos, contained cereal grains of emmer wheat and barley within structures commensurate with brewery operations. Residues on the inner surfaces of 4001 vessels were identified as containing carbonized sugar and carboxylic acids — intermediates of brewing reactions (Geller 1992). The association of oxalic acid residues has been particularly associated with beer production reflecting their common presence as "beer stone" (Michael et al. 1992).

Early analysis of artifacts suggests that these beers were produced by crushing cereal grains, infusing them in warm water, and allowing a semi-solid fermentation. Such beers would have short shelf lives and be low in alcohol unless supplemented with sugars from fruit. Evidence of pitting from analysis of cereal grains indicated that a warm temperature incubation is likely to have occurred and allowed enzymatic digestion as encountered in standard mashing processes (Samuel 1996; Wang et al. 2017). The use of barley and emmer wheat is noted by Samuel (1996) and would have an impact on beer character. Samuel also suggests that the grains had lower levels of protein compared to grains today although it is unclear whether this is a feature of the varieties or of growing conditions.

Seventeen types of beer are listed on medical papyri (Strouhal 1992) suggesting variations possibly based on different added ingredients. Identification of these in archeological samples would be a valuable collaboration of recipes. The potential of isolating indigenous yeasts from very early brewing artifacts, as demonstrated by Aouizerat et al. (2019) would be particularly rewarding.

7.5 GRUIT ALES

Although middle eastern brewing overlapped with the Greek and Roman periods the focus moved toward wine as the beverage of choice. This may have been due to improvements in irrigation allowing better cultivation of grapes but also an endorsement of wine as a status beverage compared to beer (Murray 2000).

In Europe beer became more extolled in the Viking period and developed the practice of flavorings with herbal and fruit mixtures commonly termed gruit. The provenance of gruit is unclear with several possible compositions providing different impacts on beer (Verberg 2018). A simple mixture of flavoring herbs is a basic option but the documentation of gruithuis establishments providing gruit for breweries and of recipes with complex infusions of herbs and malt suggests that gruit was a more extensive product possibly involving fermentation technology (Gray 1828; Ruis 2016). Gruit ingredients themselves such as bog myrtle, juniper, and yarrow (Figure 7.3) are variably common and available to brewers suggesting that they may well have been transformed. Moreover, the taxing of gruit and its commercial potential indicate strong benefits to breweries such that it may have been fundamental to the character of medieval beers (Verberg 2018).

Aside from the flavor contribution from herbs the production process such as described by Gray (1828)

> …infusing 6 or 8 lb. of malt in a gallon and a half of water, covering it warm by the fireside, stirring it often: when in full fermentation it is to be boiled down to a thick paste

suggest that there may be additional benefits from gruit addition. These include nutritional compounds for yeast vitality, and possible inoculations of yeast and bacteria to assist fermentation and fortification of the wort (Verberg 2018, 2022). This may be akin to the well-established use of

FIGURE 7.3 Example ingredients of gruit. (a) Bob myrtle/sweet gale; (b) Juniper berries; (c) Yarrow.

TABLE 7.1 Major components of gruit and their characteristics.

Herb	Growth Habitat	Flavor	Other Properties
Bog Myrtle (*Myrica gale*)	Acid bogs, sandy soil. Coastal zones.	Bitter, astringent taste, slightly sweet, resinous scent.	Listed abortifacient.
Marsh Rosemary (*Rhododendron tomentosum*)	Wet, mildly acid, peaty soil, shade loving.	Strong aromatic fragrance, minty, citrus.	Contains ledol, possible narcotic, poisonous in high doses.
Yarrow (*Achillea millefolium*)	Broad distribution, Grassland, open forests. Drought tolerant.	Sweet scent similar to chrysanthemum. Some astringency.	Contains salicylic acid. Possibly poisonous if >500 mg taken over prolonged period.
Juniper (*Juniperus* species)	Shrubs from dry rocky locations, woodlands, and open areas.	High terpenoid and aromatic flavors.	Some juniper pollens are allergy stimulants.
Laurel berries (bay) (*Laurus nobilis*)	Evergreen shrub in oceanic climates.	Astringent taste, terpine, and pinene fragrances.	Common culinary herb.
Laserwort/snake weed (*Laserpitium siler*)	Mountain perennial umbellifer. Light, mildly alkaline soils.	Spicy with bitter-tasting fruit. Seeds have anis aroma.	Seeds may be diuretic, emmenagogue, and stomachic.

sour mashes in German brewing practice (Peyer et al. 2017; Dysvik et al. 2020). The inclusion of plant resins, as well as some herbs, with antibacterial properties, may have given beers an extended shelf life.

The possibility of gruit providing medicinal benefits from the herbal infusion is a further possibility that could be investigated by bioarchaeology. The range of potential pharmacologically active compounds from gruit ingredients would support this (Table 7.1). To date, it appears that residues of gruit have not been assessed with advanced techniques but the opportunity should be taken if samples should become available. The wide geographical spread of gruit use is likely to show differences but the comparative chemical analysis would be of considerable interest.

7.6 MONASTIC BREWING AND EARLY-HOPPED ALES

Much medieval brewing would have been conducted in domestic settings and probably interlinked with baking (Verberg 2018). With the rise of technology in monasteries brewing became more structured with larger scales of production and a wider range of styles (van Lilsteren 1996).

The more fundamental change to brewing arrived in the fourteenth century with the use of hops. Initially recorded in the eighth century in Germany hops eventually spread throughout much of Europe and the UK (Behre 1999; Wubs-Mrozewicz 2005; Verberg 2022). Although also used as food and for fiber production hops are progressively recorded as ingredients in beer. Their use in brewing is likely to be twofold in providing flavor, particularly bitterness, and aroma, but also imparting preservation qualities and so extending shelf life.

This latter not only allowed beer to be transported further but stored for maturation. It may also have provided a further hurdle of protection against pathogenic, food spoilage microorganisms particularly gram-positive bacteria (Bocquet et al. 2018). This preservation effect results particularly from the action of the hop bitterness agent iso-humulone an isomer of humulone that is generated from humulone by heat and thus develops in boiling wort. Humulone is concentrated in lupulin glands in hop flowers, although β acids and other isomers are also present and effective (Bocquet et al. 2018; Michel et al. 2020). Levels of humulone in ancient hops are unknown and likely to be low compared to varieties today which typically range from 5–15%. Analysis of historic samples would be of considerable interest in the interpretation of recipes.

Iso-humulone's mode of action is as a proton ionophore agent of bacterial plasma membranes leading to cytoplasmic leakage (Simpson 1993). Such effect has been extensively demonstrated against contemporary bacteria but it is particularly relevant that some bacteria have developed resistance due to the production of efflux pump proteins among other mechanisms (Simpson 1993). These are typically lactic acid bacteria that possess one or more of the plasmid-encoded genes, *hit*A, *hor*A, or *hor*C (Sami et al. 1997; Hayashi et al. 2001; Suzuki et al. 2006). In many natural fermentations such as wine and Lambic brewing yeast are found to co-exist with lactic acid bacteria even in hopped products (Bartle et al. 2019; De Roos et al. 2020).

It is relevant to speculate whether the microbiology of medieval and Tudor beers were mixed fermentations of yeast and lactic acid bacteria. To date, no evidence has been presented to allow such interpretations and it would be of interest to determine when the *hit*A, *hor*A, and *hor*C genes appeared. However, molecular analysis of residues offers the possibility to determine whether these genes were present at early stages of hop usage as has been conducted for more recently recovered beers detailed below (Kajala et al. 2018).

Analysis of lactic acid bacteria genomes may also allow the identification of probiotic features and an indication of potential health benefits in such beers. Further bioarcheological analysis of medieval beers would allow the determination of other ingredients, particularly the composition of gruit. Written records indicate complex mixtures of herbs but may well be incomplete in disclosure (Verberg 2018).

Monastic brewing did not cease with the end of the Middle Ages but has continued particularly in the Trappist monasteries of Belgium and Holland producing highly distinctive ales and sour Lambic beers. It is tempting to speculate that some of these may be the oldest continuous brewing operations. Recent studies have identified biochemical and microbiological features of their production indicating the extensive impact of mixed culture fermentations (De Roos et al. 2018, 2019, 2020).

7.7 VICTORIAN AND EDWARDIAN HIGH-HOPPED BEERS

Technical developments between 1750 and 1900 allowed extensive expansion of the commercial brewing industry in many countries. In particular applications of thermometers and saccharometers allowed monitoring of production leading to greater consistency (Martin 2007). Mechanical handling via steam power allowed increases in volume, electricity allowed refrigeration to be applied to bulk storage while chemistry and microbiology established the basis for quality assurance (Anderson 2006).

As understanding of the biochemistry of mashing, boiling, and fermentation developed brewing became more predictable and controllable allowing more assured innovations in beer styles and dispense. Packaging in bottles was increasingly practiced and prompted international trade such as the extensive sales of Porter and Stouts from the UK to Scandinavia and Pale ales to India (Anderson 2005).

The survival of significant artifacts from this period allows the relatively accessible opportunity to analyze samples including intact products as well as residues. Bottles were initially stoppered with cork to prevent leakage and loss of carbonation but were later commonly sealed with lead. Survival of intact bottles is rare but has occurred and is particularly valuable when recovered from circumstances with constant and low-temperature conditions such as cellars, excavations, and shipwrecks (Jeandet et al. 2015; Londesborough et al. 2015; Kajala et al. 2018; Roullier-Gall et al. 2017; Thomas et al. 2021).

Considering the brewing industry in this period what information may be obtained to further elucidate the microbiology of these beers? A particularly pertinent issue is to obtain evidence of the range of microorganisms present in the light of the awareness of contamination as identified by Pasteur (1876) and the selection and purification of yeast by Hansen in 1884 (Hansen 1892). Brewers were increasingly aware of the benefits of limiting spoilage microorganisms and so producing longer shelf life beers without spoilage taints.

The benefits of pure yeast cultures were, however, a more debatable point. Hansen purified yeast for the Carlsberg brewery in 1883 and the use of single-strain yeasts was successfully promoted on the Continent thereafter (Claussen 1904; Anderson 2007a). In the UK, resistance from brewers was strong, partly due to a conservative mindset but also a belief that long-matured "stock" ales benefited from the combined action of a yeast consortium (Anderson 2007b). *Brettanomyces* yeast was considered an essential contributor to stock ales and despite some successful trials many UK beers retained their traditional yeasts. This could have altered with greater promotion of pure strains but an entrenchment of the industry after WW1 limited this until the 1980's in many breweries.

Microbiology and metagenomic analysis of some recovered samples from UK breweries have indicated that *Brettanomyces* was clearly present, even in more routine running beers (Claussen 1904; Thomas et al. 2021) so confirming this assertion. Metagenomics of beers from two different breweries indicated the presence of other fermenting yeast species including *Debaryomyces hansenii*, *Saccharomyces boullardi* (no longer a valid species name), and *Saccharomyces byanus* suggesting a broad mix was present (Thomas et al. 2021). The isolation of *Debaryomyces* yeast is particularly interesting as the species is not recognized as a standard brewing yeast but has been identified in other isolates (Jenner 2016) and, incidentally in Belgium Lambic brewing (De Roos et al. 2018, 2019, 2020).

The involvement of bacteria in such beers has also been confirmed. Kajala et al. (2018) isolated *Lactobacillus backii* (now renamed *Loigolactobacillus backii*) and *Pediococcus damnosus* from an 1820s shipwreck beer and identified several genomic differences with comparable species isolates from culture collections. For example, 15 genome functions were lacking in the *L. backii* genome but also that some additional features were present related to energy metabolism, potassium uptake, and response to heavy metals. Further differences were observed with the isolated *P. damnosus* strains with additional gene clusters for capsular exopolysaccharide synthesis (Kajala et al. 2018).

A wider range of lactic acid bacteria and some acetobacter were identified by DNA metagenomics in three 1895 bottles of McEwan's stout shown in Figure 7.4 and as listed in Table 7.2. Interestingly lactic acid bacteria were not identified in samples of a 1902 Bass brewery presentation beer. Instead, six species of *Staphylococcus* were found to be present, the most prevalent of which was *Staphylococcus equorum* the species most associated with horses (Thomas et al. 2021).

The presence of bacterial contaminants is quite likely in Victorian and Edwardian beers despite awareness of hygiene. In part, this may reflect the use of wood for fermentation and that casks as the porous structure would allow colonization and protection from cleaning. The use of wood as a common material in contact with beer also opens the potential of timber saprophytes contributing to the microflora. This is perhaps unlikely as these would in the main be aerobes but there may be the potential for flavor contribution from mold contamination in the cask wood.

(a) (b)

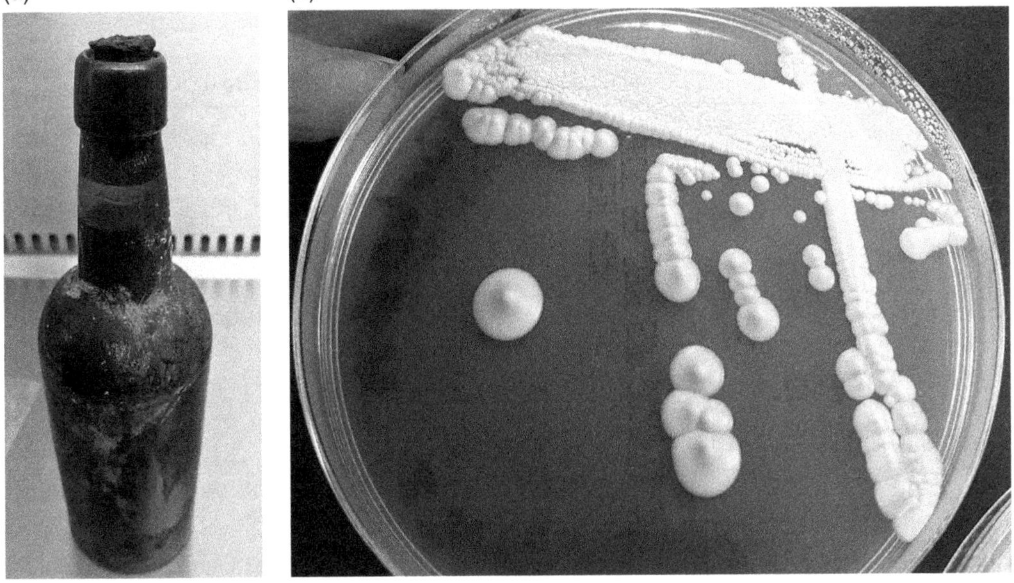

FIGURE 7.4 (a) Bottle #2 from the Wallachia shipwreck in preparation for analysis. (b) Yeast colonies isolated from Wallachia shipwreck bottle.

TABLE 7.2 Metagenomic profile of bacteria in three bottles retrieved from the Wallachia shipwreck (Thomas et al. 2021).

Bacteria Species. Percentage of Reads >1%					
Wallachia Bottle #1		Wallachia Bottle #2		Wallachia Bottle #3	
Acetobacterium tundrae	1.10	*Aminomonas* spp.	2.01	*Acetobacterium tundrae*	12.53
Acidaminococcus spp.	1.62	*Chroococcidiopsis* sp.	5.28	*Aminomonas* spp.	4.10
Actinotignum urinale	2.27	*Cyanothece* spp.	1.65	*Bacillus amyloliquefaciens*	1.00
Bacillus licheniformis	1.45	*Desulfobacter psychrotolerans*	1.55	*Bacillus subtilis*	3.32
Bacteroides sp.	3.96	*Fusobacterium* sp.	9.21	*Corynebacterium kroppenstedtii*	5.57
Finegoldia magna	7.59	*Fusobacterium* spp.	2.88	*Desulfovibrio* sp.	8.69
Fusobacterium sp.	2.47	*Lacticaseibacillus camelliae*	0.84	*Gramella* sp.	3.05
Kocuria rosea	1.26	*Paucilactobacillus vaccinostercus*	17.56	*Gluconacetobacter liquefaciens*	2.10
Lactobacillus spp.	1.39	*Parabacteroides distasonis*	1.65	*Halomonas xinjiangensis*	8.40
Paucilactobacillus vaccinostercus	32.01	*Pediococcus damnosus*	54.74	*Secundilactobacillus malefermentans*	30.99
Lactococcus lactis	1.56			*Leuconostoc citreum*	8.18

(Continued)

TABLE 7.2 (Continued)

Bacteria Species. Percentage of Reads >1%					
Wallachia Bottle #1		Wallachia Bottle #2		Wallachia Bottle #3	
Mogibacterium pumilum	2.34			*Marinilabilia* sp.	6.75
Pediococcus damnosus	10.75			*Oscillospira* spp.	1.02
Pediococcus inopinatus	11.08			*Stenotrophomonas maltophilia*	1.45
Shigella sonnei	2.46			*Xanthomonas* sp.	1.11
Staphylococcus epidermidis	3.26				
Stenotrophomonas maltophilia	1.96				
Varibaculum cambriense	1.18				

Nevertheless, evidence of DNA from *Talaromyces rotundus* and *Stereum sanguinolentum* were found in Victorian and Edwardian beer (Thomas et al. 2021) suggesting that metabolites could transfer into the beer. The presence of wood biochemical markers, 5-carboxyvanillic acid, and castalin in the 170-year-old Baltic champagne confirms the potential flavor impact of wooden casks (Jeandet et al. 2015).

Metabolomic analysis offers a further window of information with particular relevance to flavor compounds. Walther et al. (2015) analyzed archival samples of Carlsberg lager beers from 1880, 1890, and 1900 in comparison to contemporary beers. The NMR profiles of metabolites indicated distinctive metabolic signatures of the historic and contemporary samples reflecting different processing at different ages. Differences in the profiles of carbohydrates indicated that the 1900 beer had an improved stability compared to the earlier beers while higher levels of maltotriose and maltotetrose in these early beers compared to the contemporary sample suggests a less optimized raw material processing (Walther et al. 2015).

A comparable analysis was conducted by Pryde et al. (2011) using gas chromatography and sensory techniques on a sample of Mackinlay whisky from the Shackleton Antarctic expedition in 1909. Metabolomic profiles indicated a signature of a whisky matured in "first-fill" sherry casks of American white oak with a complex, the light-peated character of a malt rather than a blended spirit – clearly different from the heavy robust character expected of that age. Metagenomic analysis of other historic cask samples may provide further data in this prospect, but the chemical signature of wood has been identified in historic samples of champagne from a Baltic shipwreck (Jeandet et al. 2015). Further chemical analysis in this study also suggested that malolactic fermentation was uncontrolled and that concentrated grape sugar was used to supplement sweetness.

The value of characterizing the microbial biome of historic beers is emphasized by the current interest in brewing with mixed cultures and novel microorganisms (Piraine et al. 2021; Michel et al. 2016). Awareness of the advantages of mixed cultures is increasing in the brewing community but also other fermentations such as kombucha and kimchi (Park et al. 2019; Lee et al. 2020; Villarreal-Soto et al. 2020). This does relate to a push for novelty and diversification but also has a strong health impetus, partly focused on probiotic potential (Kapp and Sumner 2019; Vargas et al. 2021). The suggested presence of *Saccharomyces boulaardi* and various lactic acid bacteria in the 1902 Bass Ale sample has relevance to this and could be the basis of brewing recreated beers.

A detailed analysis of mixed microbes has been presented in the report by Londesborough et al. (2015), noted above. Another analysis of lactic acid bacteria isolated from an 1840 shipwreck bottle assessed the potential of the isolates to resist the action of hop-derived iso-humulone present in contemporary beers (Kajala et al. 2018). Of the bacteria isolated, the *P. damnosus* strains demonstrated the presence of all three hop resistance genes, *hit*A, *hor*A, and *hor*C, while the other, *L. backii,* contained *horA*. Nevertheless, the isolated bacteria were unable to grow in contemporary lager beer and were slower to grow in wheat beer compared to comparable strains from contemporary isolations. It is noted, however, that the challenge against today's beers brewed with hops containing high levels of α acids may not be representative of that experienced in the beer from 1840 which possessed higher levels of β acids due to different hop varieties and processing. It was further suggested that these bacteria possessed additional stress resistance systems to assist survival in the beer for an extended period as suggested by other investigations for lactic acid bacteria (Bergsveinson et al. 2016; Behr et al. 2016).

A further potential is the identification of microorganisms or genes which may relate to biotechnology applications. The extended survival of microorganisms may be associated with specific stress resistance genes and mechanisms that may be of value to identify. A specific example of this is the presence of high levels of arsenic and lead in the 1895 Wallachia beers (Thomas et al., unpublished data). In this context, it is further relevant that the 1840s lactic acid bacteria investigated had elevated levels of copper homeostasis and arsenic efflux genes associated with resistance to arsenic toxicity (Kajala et al. 2018). High levels of iron and copper were also noted in the 170-year-old Champaign bottles analyzed by Jeandet et al. (2015) possibly arising from metal nails used in assembling barrels. The survival of microorganisms in such conditions may have relevance to potential use in the remediation of contamination.

7.8 CONCLUSIONS

How relevant is historic brewing today? Novel microbial organisms and fermentations are certainly an active area of research (Yeo and Liu 2014) both for beverage production and biotechnology applications (Wang et al. 2010; Bellut and Arendt 2019; Krogerus et al. 2022). The sources of potential microorganisms are extensive ranging from environmental isolates to genetic recombinants (Wang et al. 2010; Iattici et al. 2020; Krogerus et al. 2022) and increasingly subject to extensive screening (Gamero et al. 2016).

The potential of historic isolates in such searches has additional benefits as their heritage includes information on the conditions during their use as well as indications of how they might have evolved and adapted. Bioarchaeology is increasingly active in revealing this information through biochemical and microbiological detail. In particular, advanced techniques can be used to extract a wealth of information from increasingly small sample sizes and to provide insights into the character of past foods and how these may be recreated. One increasingly prevalent application is in identifying the complex community structure of many mixed fermentations involving yeasts and bacteria as would have been typical in traditional products.

While recreations have an academic interest there is commercial potential in such products – either as novel products or, potentially, due to enhanced nutritional benefits such as improved probiotic and prebiotic features. It is also relevant to refer back to the questions posed earlier "how do we know what past fermentations actually tasted like, how they were produced, and how they compare to today's foods?"

An increasing number of brewing recreations seek to address this issue (Schwartz 2013; Dineley and Dineley 2022) but there are few cases where actual taste comparisons can be made as age differences are typically too long. The singular case of the recreation of an 1825 porter retrieved from a shipwreck being tasted by an 87-year-old drinker with a memory of the porter being a rare exception (https://brewlab.co.uk/the-original-flag-porter-story/).

The greater detail in recipes of more recent times along with some indication of microbiology management (Thomas 2001) does allow the opportunity for more exact recreation. Chemical signatures, moreover, can provide information to translate into flavor character and offer a means of making recreations more authentic (Pryde et al. 2011; Jeandet et al. 2015; Roullier-Gall et al. 2017; Thomas et al. 2021).

The identification of beneficial components, particularly pre and probiotics is a further area of investigation and is supported by the presence of specific yeast and lactic acid bacteria in some retrieved samples. Fiber, vitamins, and antioxidants are other potential components in beer linked with dietary health (Bamforth 2002).

In contrast, the presence of toxic metals and other contaminants suggests that beers of the past contained more hazardous content. Some of this may have been from accidental additions (Coping 2009) but others as means to enhance flavor or narcosis.

Fortunately, there is limited evidence of pathogenic growth in fermented products due to the presence of acid, alcohol, and inhibitors although survival is possible (Roering et al. 1999; McClay and Thomas 2021; Wong and Thomas 2021). The use of certain selective microorganisms could nevertheless be hazardous, particularly ones identified from studies but without recognition of their safety. Strains of *Bacillus cereus* for example may be used in some fermented products such as Natto but are also associated with food poisoning (Jovanovic et al. 2021). The use of relatively unknown microorganisms or combinations from wild isolations should be viewed with caution.

Similarly, reconstructions of gruit mixtures as flavor additions to historic beer recreations may incur an excess of pharmacologically active ingredients leading to health impacts. While acute hazards are unlikely continued low-level use over a period could cause an accumulated effect. For example, the presence of salicylic acid in yarrow and heather or mycotoxins in fungi.

One issue to highlight is the effect of time on the potential isolation of past microorganisms and their residues. Organic matter degrades and microorganisms die with time. Living microorganisms have been most easily isolated from intact samples of up to 200 years of age although the potential for older sources possibly through spore survival has been reported (Aouizerat et al. 2019).

The likely mechanism of survival of yeast and non-sporulating bacteria is the cannibalism of dead cells and slow metabolism due to cold and inert environmental conditions. This will, inevitably, have time limitations, and with the window of successful retrieval slowly closing it becomes increasingly imperative to identify sources. Some of these will be fortuitous such as from shipwrecks and archeological retrieval but others may be identified by auditing archives in heritage trusts and museums. The microbial past may be shelved at present but awaiting recovery.

REFERENCES

Allentoft, M.E., Collins, M., Harker, D. et al. (2012). The half-life of DNA in bone: measuring decay kinetics in 158 dated fossils. *Proceedings of the Royal Society B: Biological Sciences* 279: 4724–4733. https://doi.org/10.1098/rspb.2012.1745.

Amarasinghe, H., Weerakkody, N.S., and Waisundara, V.Y. (2018). Evaluation of physicochemical properties and antioxidant activities of kombucha "Tea Fungus" during extended periods of fermentation. *Food Science & Nutrition* 6: 659–665. https://doi.org/10.1002/fsn3.605.

Anderson, R. (2005). The transformation of brewing: an overview of three centuries of science and practice. *Brewery History* 121: 5–24.

Anderson, R. (2006). The sword and the armour. Science and practice in the brewing industry, 1837–1914. *Brewery History* 123: 55–83.

Anderson, R. (2007a). One yeast or two? Pure yeast and top fermentation. Part 1: a question of flavour and condition. *The Brewer and Distiller International* 3: 26–28.

Anderson, R. (2007b). One yeast or two? Pure yeast and top fermentation part II: *Brettanomyces* and beyond. *The Brewer and Distiller International* 3: 38–40.

Aouizerat, T., Gutman, I., Paz, Y. et al. (2019). Isolation and characterization of live yeast cells from ancient vessels as a tool in bio-archaeology. *MBio* 10. https://doi.org/10.1128/mBio.00388-19.

Aschemann-Witzel, J., Gantriis, R.F., Fraga, P. et al. (2021). Plant-based food and protein trend from a business perspective: markets, consumers, and the challenges and opportunities in the future. *Critical Reviews in Food Science and Nutrition* 61: 3119–3128. https://doi.org/10.1080/10408398.2020.1793730.

Baker, S. (2012). A note on black, spruce and Dantzig beers. *Brewery History* 151: 45–47.

Bamforth, C.W. (2002). Nutritional aspects of beer—a review. *Nutrition Research, Preventive Nutrition* 22: 227–237. https://doi.org/10.1016/S0271-5317(01)00360-8.

Bartle, L., Sumby, K., Sundstrom, J. et al. (2019). The microbial challenge of winemaking: yeast-bacteria compatibility. *FEMS Yeast Research* 19: foz040. https://doi.org/10.1093/femsyr/foz040.

Behr, J., Geissler, A.J., Schmid, J. et al. (2016). The identification of novel diagnostic marker genes for the detection of beer spoiling *Pediococcus damnosus* strains using the BlAst diagnostic gene findEr. *PLoS One* [WWW Document], n.d. https://journals.plos.org/plosone/article?id=10.1371/journal.pone.0152747 (accessed 7 December 2022).

Behre, K.-E. (1999). The history of beer additives in Europe — a review. *Vegetation History and Archaeobotany* 8: 35–48.

Bellut, K. and Arendt, E.K. (2019). Chance and challenge: non-*Saccharomyces* yeasts in nonalcoholic and low alcohol beer brewing – a review. *Journal of the American Society of Brewing Chemists* 77: 77–91. https://doi.org/10.1080/03610470.2019.1569452.

Bergsveinson, J., Friesen, V., and Ziola, B. (2016). Transcriptome analysis of beer-spoiling *Lactobacillus brevis* BSO 464 during growth in degassed and gassed beer. *International Journal of Food Microbiology* 235: 28–35. https://doi.org/10.1016/j.ijfoodmicro.2016.06.041.

Bocquet, L., Sahpaz, S., and Rivière, C. (2018). An overview of the antimicrobial properties of hop. In: *Natural Antimicrobial Agents, Sustainable Development and Biodiversity* (ed. J.-M. Mérillon and C. Riviere), 31–54. Cham: Springer International Publishing https://doi.org/10.1007/978-3-319-67045-4_2.

Burton, K.D. (2013). The citie calls for beere: the introduction of hops and the foundation of industrial brewing in early modern London. *Brewery History* 150: 6–16.

Cheng, A. (2018). Review: shaping a sustainable food future by rediscovering long-forgotten ancient grains. *Plant Science* 269: 136–142. https://doi.org/10.1016/j.plantsci.2018.01.018.

Claussen, N.H. (1904). On a method for the application of Hansen's pure yeast system in the manufacture of well-conditioned English stock beers. *Journal of the Institute of Brewing* 10: 308–311. https://doi.org/10.1002/j.2050-0416.1904.tb04656.x.

Compton-Davey, J.A.R. (2008). Audit ales: a short history. *Brewery History* 128: 2–48.

Cooper, A. and Poinar, H.N. (2000). Ancient DNA: do it right or not at all. *Science* 289: 1139. https://www.proquest.com/openview/15abea590f53f2f3c8a04ac5e770f22b/1/index.

Coping, M. (2009). Death in the beer-glass: the Manchester arsenic-in-beer epidemic of 1900-1 and the long-term poisoning of beer. *Brewery History* 132: 31–57.

Cornell, M. (2006). Half and half to mother-in-law: a history of beer, 1837–1914. *Brewery History* 123: 44–54.

Cornell, M. (2013). 'A glass of something very treble extra': coming-of-age ale, a long forgotten British beer style. *Brewery History* 146: 2–10.

Cuamatzin-García, L., Rodríguez-Rugarcía, P., El-Kassis, E.G. et al. (2022). Traditional fermented foods and beverages from around the world and their health benefits. *Microorganisms* 10: 1151. https://doi.org/10.3390/microorganisms10061151.

Curtis, R.J. (2001). *Ancient Food Technology*. Leiden: Brill.

De Roos, J., Vandamme, P., and De Vuyst, L. (2018). Wort substrate consumption and metabolite production during Lambic beer fermentation and maturation explain the successive growth of specific bacterial and yeast species. *Frontiers in Microbiology* 9: 1–20. https://doi.org/10.3389/fmicb.2018.02763.

De Roos, J., Van der Veken, D., and De Vuyst, L. (2019). The interior surfaces of wooden barrels are an additional microbial inoculation source for lambic beer production. *Applied and Environmental Microbiology* 85: e02226-18. https://doi.org/10.1128/AEM.02226-18.

De Roos, J., Verce, M., Weckx, S. et al. (2020). Temporal shotgun metagenomics revealed the potential metabolic capabilities of specific microorganisms during Lambic beer production. *Frontiers in Microbiology* 11: 1–17. https://doi.org/10.3389/fmicb.2020.01692.

Dineley, M. (1996). Finding magic in stone age real ale. *British Archaeology* 19: 6–13.

Dineley, M. (2011). Experiment or demonstration? Making fermentable malt sugars from the grain and a discussion of some of the evidence for this activity in the British Neolithic. In: *Experimentation and Interpretation. The Use of Experimental Archaeology in the Study of the Past* (ed. D.C.E. Millson), 96–108. Oxford: Oxbow Books.

Dineley, M. and Dineley, G. (2022). The oldest mash forks in Britain and other archaeological evidence for ale in the neolithic. *Brewery History* 189: 53–61.

Drieu, L., Rageot, M., Wales, N. et al. (2020). Is it possible to identify ancient wine production using biomolecular approaches? *STAR: Science & Technology of Archaeological Research* 6: 16–29. https://doi.org/10.1080/20548923.2020.1738728.

Dysvik, A., Liland, K.H., Myhrer, K.S. et al. (2019). Pre-fermentation with lactic acid bacteria in sour beer production. *Journal of the Institute of Brewing* 125: 342–356. https://doi.org/10.1002/jib.569.

Dysvik, A., La Rosa, S.L., De Rouck, G. et al. (2020). Microbial dynamics in traditional and modern sour beer production. *Applied and Environmental Microbiology* 86: e00566-20. https://doi.org/10.1128/AEM.00566-20.

Faltings, D. (1991). Die Bierbrauerei im AR1. *Zeitschrift fur Agyptische sprache und altertumskunde* 118: 104.

Farag, M.A., Elmassry, M.M., Baba, M. et al. (2019). Revealing the constituents of Egypt's oldest beer using infrared and mass spectrometry. *Scientific Reports – Nature* 9: 16199. https://doi.org/10.1038/s41598-019-52877-0.

Flissebaalje, T. (2013). Bockbeer in the Netherlands: a historical overview. *Brewery History* 152: 33–42.

Forman, W., Stroughal, E., and Zemina, M. (1992). *Life in Ancient Egypt*. London: Opus Publishing Ltd.

Gallone, B., Steensels, J., Prahl, T. et al. (2016). Domestication and divergence of *Saccharomyces cerevisiae* beer yeasts. *Cell* 166: 1397–1410.e16. https://doi.org/10.1016/j.cell.2016.08.020.

Gamero, A., Quintilla, R., Groenewald, M. et al. (2016). High-throughput screening of a large collection of non-conventional yeasts reveals their potential for aroma formation in food fermentation. *Food Microbiology* 60: 147–159. https://doi.org/10.1016/j.fm.2016.07.006.

Geller, J.R. (1992). From prehistory to history: beer in Egypt. In: *The Followers of Horus* (ed. R. Friedman and B. Adams), 19–26. Oxford: Oxbow Books.

Ginolhac, A., Rasmussen, M., Gilbert, M.T.P. et al. (2011). mapDamage: testing for damage patterns in ancient DNA sequences. *Bioinformatics* 27: 2153–2155. https://doi.org/10.1093/bioinformatics/btr347.

Gonçalves, M., Pontes, A., Almeida, P. et al. (2016). Distinct domestication trajectories in top-fermenting beer yeasts and wine yeasts. *Current Biology* 26: 2750–2761. https://doi.org/10.1016/j.cub.2016.08.040.

Gray, S.F. (1828). *The Operative Chemist*. London: Hurst, Chance & Co.

Greco, E., El-Aguizy, O., Ali, M.F. et al. (2018). Proteomic analyses on an ancient Egyptian cheese and biomolecular evidence of brucellosis. *Analytical Chemistry* 90: 9673–9676. https://doi.org/10.1021/acs.analchem.8b02535.

Hansen, E.C. (1892). Recherches faites dans la pratique de l'industrie de la fermentation Comptes Rendus des Meddelelser Fra Carlsberg Laboratoriet III., 2.

Hayashi, N., Ito, M., Horiike, S. et al. (2001). Molecular cloning of a putative divalent-cation transporter gene as a new genetic marker for the identification of *Lactobacillus brevis* strains capable of growing in beer. *Applied Microbiology and Biotechnology* 55: 596–603. https://doi.org/10.1007/s002530100600.

Hendy, J., Colonese, A.C., Franz, I. et al. (2018). Ancient proteins from ceramic vessels at Çatalhöyük West reveal the hidden cuisine of early farmers. *Nature Communications* 9: 4064. https://doi.org/10.1038/s41467-018-06335-6.

Hofreiter, M., Jaenicke, V., Serre, D. et al. (2001). DNA sequences from multiple amplifications reveal artifacts induced by cytosine deamination in ancient DNA. *Nucleic Acids Research* 29: 4793–4799. https://doi.org/10.1093/nar/29.23.4793.

Hofreiter, M., Paijmans, J.L.A., Goodchild, H. et al. (2015). The future of ancient DNA: technical advances and conceptual shifts. *BioEssays* 37: 284–293. https://doi.org/10.1002/bies.201400160.

Hornsey, I.S. (2003). *A History of Beer and Brewing*. Royal Society of Chemistry.

Iattici, F., Catallo, M., and Solieri, L. (2020). Designing new yeasts for craft brewing: when natural biodiversity meets biotechnology. *Beverages* 6: 3. https://doi.org/10.3390/beverages6010003.

Ishida, H. (2002). Insight into ancient Egyptian beer brewing using current folkloristic methods. *Technical Quarterly – Technical Quarterly – Master Brewers Association of the Americas* 39: 81–88.

Ismail, I., Hwang, Y.-H., and Joo, S.-T. (2020). Meat analog as future food: a review. *Journal of Animal Science and Technology* 62: 111–120. https://doi.org/10.5187/jast.2020.62.2.111.

Jeandet, P., Heinzmannb, S.S., Roullier-Gallb, C. et al. (2015). Chemical messages in 170-year-old champagne bottles from the Baltic Sea: revealing tastes from the past. *Proceedings of the National Association of Sciences* 112: 5893–5898. https://www.pnas.org/doi/abs/10.1073/pnas.1500783112.

Jenner, M. (2016). Recreating 'Imperial extra double stout'. *Brewery History* 166: 70–73.

Jovanovic, J., Ornelis, V.F.M., Madder, A. et al. (2021). *Bacillus cereus* food intoxication and toxicoinfection. *Comprehensive Reviews in Food Science and Food Safety* 20: 3719–3761. https://doi.org/10.1111/1541-4337.12785.

Kajala, I., Bergsveinson, J., Friesen, V. et al. (2018). *Lactobacillus backii* and *Pediococcus damnosus* isolated from 170-year-old beer recovered from a shipwreck lack the metabolic activities required to grow in modern lager beer. *FEMS Microbiology Ecology* 94: https://doi.org/10.1093/femsec/fix152.

Kapp, J.M. and Sumner, W. (2019). Kombucha: a systematic review of the empirical evidence of human health benefit. *Annals of Epidemiology* 30: 66–70. https://doi.org/10.1016/j.annepidem.2018.11.001.

Korlević, P., Gerber, T., Gansauge, M.-T. et al. (2015). Reducing microbial and human contamination in DNA extractions from ancient bones and teeth. *BioTechniques* 59: 87–93. https://doi.org/10.2144/000114320.

Krogerus, K., Eerikäinen, R., Aisala, H. et al. (2022). Repurposing brewery contaminant yeast as production strains for low-alcohol beer fermentation. *Yeast* 39: 156–169. https://doi.org/10.1002/yea.3674.

Lee, S.H., Whon, T.W., Roh, S.W. et al. (2020). Unraveling microbial fermentation features in kimchi: from classical to meta-omics approaches. *Applied Microbiology and Biotechnology* 104: 7731–7744. https://doi.org/10.1007/s00253-020-10804-8.

Lehnhardt, F., Gastl, M., and Becker, T. (2019). Forced into aging: analytical prediction of the flavor-stability of lager beer. A review. *Critical Reviews in Food Science and Nutrition* 59: 2642–2653. https://doi.org/10.1080/10408398.2018.1462761.

van Lilsteren, V.T. (1996). From herbs to hops; outlines of the brewing process in medieval Europe. *Halve Maen; Magazine of the Dutch Colonial Period in America* LXIX (3): 41–46.

Londesborough, J., Dresel, M., Gibson, B. et al. (2015). Analysis of beers from an 1840s' shipwreck. *Journal of Agricultural and Food Chemistry* 63: 2525–2536. https://doi.org/10.1021/jf5052943.

Lott, F.E. (1901). Bottled beers. *Journal of the Federated Institutes of Brewing* 7: 190–211. https://doi.org/10.1002/j.2050-0416.1901.tb00169.x.

Lydolph, M.C., Jacobsen, J., Arctander, P. et al. (2005). Beringian paleoecology inferred from permafrost-preserved fungal DNA. *Applied and Environmental Microbiology* 71: 1012–1017. https://doi.org/10.1128/AEM.71.2.1012-1017.2005.

Martin, P.A. (2007). Hydrometry and slide rules in brewing and distilling. *Brewery History* 124 (5): 14–37.

McClay, C. and Thomas, K. (2021). The persistence of pathogens in fruit juice and cider. *Aspects of Applied Biology* 123: 91–94.

McClure, S.B., Magill, C., Podrug, E. et al. (2018). Fatty acid specific δ13C values reveal earliest Mediterranean cheese production 7,200 years ago. *PLoS One* 13: e0202807. https://doi.org/10.1371/journal.pone.0202807.

McGovern, P. (2009). *Uncorking the Past: The Quest for Wine, Beer and Other Alcoholic Beverages*. Berkeley: University of California Press.

Michael, R.H., McGovern, P.E., and Badler. (1992). Chemical evidence for ancient beer. *Nature* 360: 24. https://doi.org/10.1038/360024b0.

Michel, M., Meier-Dörnberg, T., Jacob, F. et al. (2016). Review: pure non-*Saccharomyces* starter cultures for beer fermentation with a focus on secondary metabolites and practical applications. *Journal of the Institute of Brewing* 122: 569–587. https://doi.org/10.1002/jib.381.

Michel, M., Cocuzza, S., Biendl, M. et al. (2020). The impact of different hop compounds on the growth of selected beer spoilage bacteria in beer. *Journal of the Institute of Brewing* 126: 354–361. https://doi.org/10.1002/jib.624.

Mulder, R. (2018). Lambic: the need for a new historical narrative. *Brewery History* 175: 2–13.

Murray, M.A. (2000). Fruits, vegetables, pulses and condiments. In: *Ancient Egyptian Materials and Technology* (ed. P.T. Nicholson and I. Shaw), 609. Cambridge: Cambridge University Press.

Park, S.-E., Seo, S.-H., Kim, E.-J. et al. (2019). Changes of microbial community and metabolite in kimchi inoculated with different microbial community starters. *Food Chemistry* 274: 558–565. https://doi.org/10.1016/j.foodchem.2018.09.032.

Pasteur, M.L. (1876). *Études sur la bière: ses maladies, causes qui les provoquent, procédé pour la rendre inaltérable; avec une théorie nouvelle de la fermentation*. Paris: Gauthier-Villars.

Perruchini, E., Glatz, C., Hald, M.M. et al. (2018). Revealing invisible brews: a new approach to the chemical identification of ancient beer. *Journal of Archaeological Science* 100: 176–190. https://doi.org/10.1016/j.jas.2018.05.010.

Peyer, L.C., Zarnkow, M., Jacob, F. et al. (2017). Sour brewing: impact of *Lactobacillus Amylovorus* FST2.11 on technological and quality attributes of acid beers. *Journal of the American Society of Brewing Chemists* 75: 207–216. https://doi.org/10.1094/ASBCJ-2017-3861-01.

Piraine, R.E.A., Leite, F.P.L., and Bochman, M.L. (2021). Mixed-culture metagenomics of the microbes making sour beer. *Fermentation* 7: 174. https://doi.org/10.3390/fermentation7030174.

Pryde, J., Conner, J., Jack, F. et al. (2011). Sensory and chemical analysis of 'Shackleton's' Mackinlay scotch whisky. *Journal of the Institute of Brewing* 117: 156–165. https://doi.org/10.1002/j.2050-0416.2011.tb00455.x.

Rawlence, N.J., Lowe, D.J., Wood, J.R. et al. (2014). Using palaeoenvironmental DNA to reconstruct past environments: progress and prospects. *Journal of Quaternary Science* 29: 610–626. https://doi.org/10.1002/jqs.2740.

Roering, A.M., Luchansky, J.B., Ihnot, A.M. et al. (1999). Comparative survival of *Salmonella typhimurium* DT 104, *Listeria monocytogenes*, and *Escherichia coli* O157:H7 in preservative-free apple cider and simulated gastric fluid. *International Journal of Food Microbiology* 46: 263–269. https://doi.org/10.1016/S0168-1605(98)00198-6.

Roullier-Gall, C., Heinzmann, S.S., Garcia, J.-P. et al. (2017). Chemical messages from an ancient buried bottle: metabolomics for wine archeochemistry. *NPJ Science of Food* 1: 1. https://doi.org/10.1038/s41538-017-0001-5.

Ruis, F. (2016). A note on the essence of gruit. *Brewery History* 166: 50–53.

Salque, M., Bogucki, P.I., Pyzel, J. et al. (2013). Earliest evidence for cheese making in the sixth millennium BC in northern Europe. *Nature* 493: 522–525. https://doi.org/10.1038/nature11698.

Sami, M., Yamashita, H., Hirono, T. et al. (1997). Hop-resistant lactobacillus brevis contains a novel plasmid harboring a multidrug resistance-like gene. *Journal of Fermentation and Bioengineering* 84: 1–6. https://doi.org/10.1016/S0922-338X(97)82778-X.

Samuel, D. (1996). Archaeology of ancient Egyptian beer. *Journal of the American Society of Brewing Chemists* 54: 3–12.

Schwartz, G. (2013). A pint of the past: assessing authenticity in modern versions of ancient and historic ales. *Brewery History* 154: 17–30.

Schwarz, C., Debruyne, R., Kuch, M. et al. (2009). New insights from old bones: DNA preservation and degradation in permafrost preserved mammoth remains. *Nucleic Acids Research* 37: 3215–3229. https://doi.org/10.1093/nar/gkp159.

Shapiro, B. and Hofreiter, M. (2014). A Paleogenomic perspective on evolution and gene function: new insights from ancient DNA. *Science* 343: 1236573. https://doi.org/10.1126/science.1236573.

Shiby, V.K. and Mishra, H.N. (2013). Fermented milks and Milk products as functional foods—a review. *Critical Reviews in Food Science and Nutrition* 53: 482–496. https://doi.org/10.1080/10408398.2010.547398.

Simpson, W.J. (1993). Cambridge prize lecture. Studies on the sensitivity of lactic acid bacteria to hop bitter acids. *Journal of the Institute of Brewing* 99: 405–411. https://doi.org/10.1002/j.2050-0416.1993.tb01180.x.

Strouhal, E. (1992). *Life in Ancient Egypt*. Cambridge: Cambridge University Press.

Suzuki, K., Iijima, K., Sakamoto, K. et al. (2006). A review of hop resistance in beer spoilage lactic acid bacteria. *Journal of the Institute of Brewing* 112: 173–191. https://doi.org/10.1002/j.2050-0416.2006.tb00247.x.

Swift, A. (2014). Devon white ale. *Brewery History* 156: 51–60.

Thomas, K. (2001). Yeasts of yesterday. *The Brewer International* 1: 29–32.

Thomas, K., Ironside, K., Clark, L. et al. (2021). Preliminary microbiological and chemical analysis of two historical stock ales from Victorian and Edwardian brewing. *Journal of the Institute of Brewing* 127: 167–175. https://doi.org/10.1002/jib.641.

Vargas, B.K., Fabricio, M.F., and Záchia Ayub, M.A. (2021). Health effects and probiotic and prebiotic potential of Kombucha: a bibliometric and systematic review. *Food Bioscience* 44: 101332. https://doi.org/10.1016/j.fbio.2021.101332.

Verberg, S. (2018). The rise and fall of gruit. *Brewery History* 174: 46–79.

Verberg, S. (2022). Reconstructing medieval gruit beer: separating facts from fiction. In: *Beer and Brewing in Medieval Culture and Contemporary Medievalism* (ed. J.A. Geck, R. O'Neill, and N. Phillips). Cham: Palgrave Macmillan https://doi.org/10.1007/978-3-030-94620-3_3.

Villarreal-Soto, S.A., Beaufort, S., Bouajila, J. et al. (2018). Understanding kombucha tea fermentation: a review. *Journal of Food Science* 83: 580–588. https://doi.org/10.1111/1750-3841.14068.

Villarreal-Soto, S.A., Bouajila, J., Pace, M. et al. (2020). Metabolome-microbiome signatures in the fermented beverage, Kombucha. *International Journal of Food Microbiology* 333: 108778. https://doi.org/10.1016/j.ijfoodmicro.2020.108778.

Walther, A., Ravasio, D., Qin, F. et al. (2015). Development of brewing science in (and since) the late 19th century: molecular profiles of 110–130 year old beers. *Food Chemistry* 183: 227–234. https://doi.org/10.1016/j.foodchem.2015.03.051.

Wang, J., Wang, Z.-Y., He, X.-P. et al. (2010). Construction of amylolytic industrial brewing yeast strain with high glutathione content for manufacturing beer with improved anti-staling capability and flavor. *Journal of Microbiology* 20: 1539–1545. https://doi.org/10.4014/jmb.1004.04007.

Wang, J., Liu, L., Ball, T. et al. (2016). Revealing a 5,000-y-old beer recipe in China. *PNAS* 113: 6444–6448. https://doi.org/10.1073/pnas.1601465111.

Wang, J., Liu, L., Georgescu, A. et al. (2017). Identifying ancient beer brewing through starch analysis: a methodology. *Journal of Archaeological Science: Reports* 15: 150–160. https://doi.org/10.1016/j.jasrep.2017.07.016.

Willerslev, E., Hansen, A.J., and Poinar, H.N. (2004). Isolation of nucleic acids and cultures from fossil ice and permafrost. *Trends in Ecology & Evolution* 19: 141–147. https://doi.org/10.1016/j.tree.2003.11.010.

Wong, V.S.Q. and Thomas, K. (2021). Does acid inhibit pathogens in fruit juice? *Aspects of Applied Biology* 123: 101–108.

Wubs-Mrozewicz, J. (2005). Hopped beer as an innovation. The Bergen beer market around 1200–1600 in the European context. In: *Trade, Diplomacy and Cultural Exchange* (ed. H. Brand), 152–168. Hilversum: Uitgeverij Verloren.

Yang, Y., Shevchenko, A., Knaust, A. et al. (2014). Proteomics evidence for kefir dairy in early bronze age China. *Journal of Archaeological Science* 45: 178–186. https://doi.org/10.1016/j.jas.2014.02.005.

Yeo, H.Q. and Liu, S.-Q. (2014). An overview of selected specialty beers: developments, challenges and prospects. *International Journal of Food Science & Technology* 49: 1607–1618. https://doi.org/10.1111/ijfs.12488.

Ziesemer, K.A., Mann, A.E., Sankaranarayanan, K. et al. (2015). Intrinsic challenges in ancient microbiome reconstruction using 16S rRNA gene amplification. *Scientific Reports* 5: 16498. https://doi.org/10.1038/srep16498.

CHAPTER 8

AN ACCIDENTAL ART FORM: SPONTANEOUS FERMENTATION OF BEER

CASSANDRA SUTHER
Department of Food Science, University of Massachusetts Amherst, Amherst, MA USA

CONTENTS

8.1 INTRODUCTION

Musicians, painters, and brewers share a common trait. They are all artists. Reader, the claim that "art is science" holds true for me. Painters use ratios and color theory while musicians must understand arithmetic and frequencies. Alternately, brewing science requires a precise knowledge of biochemical reactions, which, in turn, creates something artful and elegant. However, restriction stifles creativity. Improvisation elevates an art form, akin to Mile Davis's jazz solos or the works of Russian painter Vasily Kandinsky. The same principle also applies to certain forms of brewing.

Beer brewing is a complicated process both home and commercial brewers alike must master to achieve a good product. The control and care taken to brew your favorite beer are demanding, with an almost comical number of changing variables. These factors consist of a type of grain, water quality, hops variety, and, importantly, microbes (Lodolo et al. 2008). But what happens when the latter is left to chance?

Microbial Fermentations in Nature and as Designed Processes, First Edition. Edited by Christon J. Hurst.

One could define spontaneous fermentation as an accidental art form. The term scientifically refers to the process of leaving the inoculate (or the introduction of the required microbes) up to the organisms that happen to be in the surrounding environment (De Roos and De Vuyst 2019). Spontaneously fermented alcohol is thought to be one of the oldest and most important beverages ever produced. Historians suggest most yeast and bacteria from original brews came off plant materials or what was carried in the air (Keersmaecker 1996). Thanks to the introduction of modern brewing and microbiology, we now have a fair understanding of the specific bacteria and mechanisms involved in the art form.

This work will review the process and biology behind spontaneous beer fermentation. In completion, I hope you reexplore brewing and what fermentation means to me as an art form.

8.2 FERMENTATION AND ITS SPONTANEITY

To understand spontaneous fermentation, you must first understand fermentation. In relation to beer, it could simply be described as the conversion of sugars from grains into other compounds, comprising alcohol, esters, and CO_2 (Briggs et al. 2004). Yeast, commonly *Saccharomyces*, is the major microbe involved in this process and is required in the production of nearly all alcoholic drinks (Lodolo et al. 2008). However, some varieties of beer, like sour beer, also incorporate bacteria into their carefully planned inoculation (Sparrow 2005). These bacteria will produce distinct byproducts yeast forgoes, including the acids found in cheese and yogurt. Pitches, or the microbes thrown into the beer, are carefully planned by brewers to deliver the best flavor profiles (Garshol 2021). Spontaneous fermentation depends on good faith (or an understanding of the beerhouse microbiome).

A microbiome is a term used to describe the community of microorganisms found within a certain habitat (Suther et al. 2020). This includes not only our human microbiome (gut, lung, and skin) but also environmental microbiomes (soil, water, and plant) (Aagaard et al. 2014; Islam et al. 2020; Sokolowska et al. 2018; Suther et al. 2020). What may be the most interesting aspect of spontaneous fermentation is how much it relied on the ecosystem of the brewer's equipment and facilities (we can call it the house microbiota) (De Roos et al. 2019). The microbiome of each brewery will have similar microbes; however, strains and amounts will vary. As such, each beer could be pictured as a representation of the microbial ecology of that environment and may even relate to seasonal changes. Thanks to modern-day DNA sequencing and organism identification, we can accurately profile the microbes found within breweries (Spitaels et al. 2014). Due to this understanding, modern companies produce their own commercial versions of spontaneous beer, the most common being the lambic or coolship style (Keersmaecker 1996).

The lambic style of beer originated in Brussels, Belgium, specifically near the Senne Valley (Spitaels et al. 2014). The valley is believed to harbor local ambient microbes that are important for the fermentation and maturation process of the lambic beer wort. In fact, the term lambic, at least in Belgium, only refers to the beer brewed next to the Senne Valley (similar to Champagne only coming from the Champagne Province in France). Outside of Belgium, no such requirement exists. However, out of respect for the Belgium brewers, commercial companies outside of Brussels have used different names to describe the beer style. Allagash Brewing Company (Portland, ME) was the first brewery to bring the beverage to the US market in 2009. The company coined the name "Coolship" to refer to the style. A coolship itself is a large, metal vessel vital for production (explained in more detail below) (Keersmaecker 1996). The Dutch and Flemish word for such equipment is called a "koelschip" and is where Allagash derived the name.

8.3 MUCH A BREW ABOUT NOTHING

The biological process behind beer brewing is akin to a theatrical production, while spontaneous fermentation adds a layer of improv. Brewers act as directors in this improvisation comedy filled with bacteria and yeast. In these next sections, I will introduce you to each step involved in the complete fermentation of the popular lambic or coolship-style beer.

8.3.1 Act 1: Preparation

The preparation phase starts with the mixture and heating (termed mashing) of the ingredients used in brewing, including water, grains, and hops (Briggs et al. 2004). Most beer contains malted barley (sugar source), which provides the necessary endogenous enzymes to break down starches into easily fermentable sugars (Briggs et al. 2004). However, spontaneous beer contains a high amount of unmalted wheat (30–40%) in addition to malted barley (Sparrow 2005). This addition allows for a wort high in a complex sugar called maltooligosaccharides. This addition is important for the later maturation phase utilized by the yeast *Brettanomyces* (De Roos and De Vuyst 2019). It should be noted, however, different microbes other than *Brettanomyces* may break down maltooligosaccharides early in fermentation, including *Saccharomyces kudriavzevii* and several different acetic acid bacteria (Bongaerts et al. 2021).

The addition of hops adds an important bittering attribute to the beer (Briggs et al. 2004). Without hops or a different type of bitter replacement (like the gruit styles which are bittered with botanicals), all conventional beer would taste like sticky sweet grain soda (an image given to me by an old biochemistry professor). As such, hops contribute the desirable sensory attributes of bitterness to the sweetness of the malt while supping a reasonable amount of aroma. Conveniently, they also work as a natural antimicrobial (Vriesekoop et al. 2012).

This particular antimicrobial activity comes from their hydrophobic bitter acids, α-acids (humulone), and β-acids (lupulone) (Vriesekoop et al. 2012). This is a desired property of hops. Brewers deal with contamination in the same way a scientist does in a lab. Growth from unintended microbes can give beer unintended sensory attributes and produce tastes that differ between batches. Although the lambic and coolship styles are known for their uncontrolled bacterial inoculation, hops still prevent a range of undesirable or harmful bacteria from contaminating the fermentation. Spontaneous beer uses older, gentler hops (De Roos and De Vuyst 2019). This mild flavor is due to the creation of humulinic acids from the bitter humulone. This gentler flavor is used to better complement the complex flavors given off by the many microbes found in these wild ales. It also helps provide higher microbial stability compared to the humulone found in typical beer (De Roos and De Vuyst 2019). A longer boil is required (three to five hours) due to off-flavors given from the older hop cones (mainly oxidized resins) (De Roos and De Vuyst 2019).

Once mashing is complete, the sugar sweet liquid is filtered off and collected. This hot mixture is called the wort (Briggs et al. 2004). The next step, and the most unique, is the transfer of the hot wort to an open, high surface-area vessel called a koelschip (or coolship). The wort cools overnight exposed to the microorganisms in the air and the residual in the coolship (Bokulich et al. 2012). Traditionally, lambic was only brewed between the colder months of October to March (Bongaerts et al. 2021). The cold temperatures allowed for a rapid cooling that encourage microbes to grow, as too high a temperature is detrimental to bacterial growth (Vriesekoop et al. 2012). Year-round brewers have overcome nature by pre-chilling the wort before transferring it to the coolship (Bongaerts et al. 2021). However, some companies like Allagash keep to tradition.

As mentioned, Allagash Brewing Company could be described as the flagship coolship brewer in the US. While most famous for their Allagash White (a Belgian-style wheat beer I frequently order), Allagash makes several different batches of their coolship in the colder months of November through January every year. Currently residing near Boston, Massachusetts, I took a day trip up to Portland, Maine, to visit Allagash, where these wild ales are made. Staff were kind enough to let me look at their coolship vessel, located near their own Senne Valley river (a small man-made pond on their property). The vessel itself was inside a small, single-room building with two circulating fans, a hanging light, and four propped open square stained-glass windows. They told me that the wooden ceiling was crucial for harboring important microbes. Nothing is cleaned, nothing is touched, and nothing disrupts the microbial community in that room.

Allagash waits for overnight lows of around 30 °F. This cold temperature allows the wort to cool within a perfect time frame, not too short but also not too long on the coolship. Through trial and error, Allagash found too long a time on their coolship can lead to undesirable flavors, however, the exact reason remains unknown. Because science (especially the undefined by nature) can be so unpredictable, they told me tradition and superstition can become a major part of coolship brewing (Figure 8.1).

The next morning (around 16 hours later), the wort is transferred to wooden (mostly oak) barrels. Alcohol has a long history of being aged in wooden casks. This is how many whiskeys or wines get their traditional oaky flavor (Mosedale 1995). The oak flavor itself comes from oak lactones (which give a coconut, wood flavor), vanillin (as the name suggests, typically associated with vanilla), and guaiacol (responsible for a smokey, burned flavor) (Jordão and Cosme 2022). This flavor is lost after several uses of the barrel. Another significant aspect of wooden barrels is the physical integrity and porosity of their surface. These are difficult to sanitize and, therefore, harbor diverse microorganisms (De Roos et al. 2019). Often, distillers or winemakers will discard old barrels, or, to the delight of our cause, sell them to brewers. In the case of spontaneous beer,

FIGURE 8.1 Allagash Brewing Company coolship room. Windows are propped open at all times and the wooden ceiling is important for harboring organisms. The fans help circulate airflow. The beer sits overnight and is then drained back into the main brewing building to be placed into barrels. Source: These images appear courtesy of the author.

the flavor profiles given from fresh wooden barrels are not necessarily desired. What is desired, however, is the left-over harboring of microorganisms in old barrels which contribute to the flavor and aroma of the finished beer.

The next act begins.

8.3.2 Act 2: Initial Growth

At the draw of the theater curtain, our cooled, inoculated wort sits, filled to the brim, in stacks of large, old, discarded oak barrels. This prevents as much oxygen as possible to minimize the growth of aerobic microorganisms (Bongaerts et al. 2021). However, these barrels are filled with pores (and exist on earth), so O_2 is still present. This begins our second act and our first fermentation (Figure 8.2).

The first phase of bacterial growth is thought to be the most unimportant but showcases the true spontaneous nature of lambic or coolship beer production. It takes place over the first few weeks and starts with the increase of resistant bacteria found within the *Enterobacteriaceae family* and oxidative yeast (Martens et al. 1991; De Roos et al. 2018). If lactic acid had been added to the wort at the beginning, like so often the case, this phase has less of a presence due to decreased pH. The genera most often encountered during this first fermentation phase are *Enterobacter* and *Klebsiella*. Specifically, previous studies have identified *Klebsiella oxytoca, Klebsiella aerogenes, Klebsiella pneumonia,* and *Pantoea agglomerans* but also *Enterobacter ludwigii, Enterobacter cloacae, Enterobacter mori, Escherichia coli,* and *Serratia ureilytica* within the first one to two months of fermentation (Bongaerts et al. 2021; Martens et al. 1991). Yeasts like *Hanseniaspora* have also been reported (Bongaerts et al. 2021). One study looked at specific areas of the brewer's house and linked many *Enterobacteria* to the cellar and attic air (Spitaels et al. 2014).

FIGURE 8.2 Allagash Brewing Company aging barrels. "CS" refers to "Coolship." Apple refers to the fresh pomades added while lavender refers to the lavender-smoked malt used. Beer was poured into barrels to begin aging on March 27th, 2019, November 20th, 2020, or January 4th, 2022. Source: These images appear courtesy of the author. (*See colour plate section for colour representation of this figure.*)

Acetic acid bacteria are found during this initial fermentation phase as well, however, the presence of acetic acid bacteria throughout the spontaneous beer production process has long been considered unpredictable. The metabolic outcomes of acetic acid bacteria, enterobacteria, and yeasts will decrease pH, monosaccharides, and oxygen while producing several organic acids (De Roos and De Vuyst 2019). Increases in ethanol and carbon dioxide concentrations make the environmental conditions harsher for the undesirable *Enterobacteria* and those microbes will fade from the scene.

8.3.3 Act 3: Alcoholic Fermentation

Our next act, the alcoholic fermentation phase, may last several weeks to months. It has a slow start and partially begins in the initial fermentation phase. Important to this act, the wooden barrels purchased from other alcohol production sites harbor an important microbe, *Saccharomyces*.

Everyone's favorite character is *Saccharomyces*. It has multiple fan clubs and wiki pages dedicated to its origins, growth curve, and appearance. This is thanks to the involvement of *Saccharomyces* in the production of sourdough, cocoa, and alcohol production (Parapouli et al. 2020). Beer can be simplistically classified based on the type of *Saccharomyce*s utilized. *Saccharomyces cerevisiae* is used for ales (top fermenters), while *Saccharomyces pastorianus* is used for lagers (bottom fermenters) (Briggs et al. 2004). Wild ales contain a variety of yeast and are not typically placed into the ale/lager hierarchy.

For spontaneous beer, the main yeast strain at the beginning of alcoholic fermentation is *S. cerevisiae*. It has the job of converting fermentable carbohydrates (maltose and maltotriose) into ethanol and carbon dioxide. They also contribute flavor (alcoholic flavor and sweet esters). Near the end of this act, the main yeast changes from *S. cerevisiae* to *S. kudriavzevii* or *S. pastorianus*. However, other, less prominent, yeast has been reported in this phase, including *Candida, Hanseniaspora, Kregervanrija, Naumovozyma, and Wickerhamomyces* (Spitaels et al. 2014).

Temperature plays an important role in any biological process, so it's thought that temperature drops favor yeast that are more resistant to the cold, such as *S. kudriavzevii* and *S. pastorianus* (Bongaerts et al. 2021). Seasonal changes can lead to different beer outcomes because of this factor. *S. kudriavzevii* may also metabolize the increased amounts of maltooligosaccharides, mentioned above, unlike *S. cerevisiae* (De Roos et al. 2020). This may also encourage its shift as the more dominant yeast species.

8.3.4 Act 4: Acidification

I see the acidification of the wort as the climax of our production. Both acetic and lactic acid bacteria present in the wort encourage the pH to typically drop below pH 3.5 over the course of several months (De Roos and De Vuyst 2019). Acetic and lactic acid bacteria produce their name's-sake organic acids, acetic and lactic acid. Lactic acid is produced from the fermentation of saccharides (sugars) while acetic acid is produced from the oxidation of ethanol into acetaldehyde, which is further metabolized into acetic acid (or vinegar acid) (Narendranath et al. 2001). However, at this point, the oxygen level in the wort is low, restricting much acetic acid production (Now, if this wasn't the case, the final product might be better used as a coffee pot cleaner than a crafted beverage). The ratio balance between these two acids will determine the outcome of the beer. Lactic acid gives a soft, fresh yogurt-like taste, while acetic acid has a sharp, strong vinegar taste (hence, the coffee pot cleaner) (Hartwig and McDaniel 1995). However, as acetic acid is rather volatile, it contributes to much of the aroma of spontaneous beer (De Roos and De Vuyst 2019).

The acetic acid bacteria species most isolated during this fermentation phase include *Acetobacter pasteurianus* and *Acetobacter lambici* (De Roos and De Vuyst 2018). These species are thought to be better adapted to the harsher conditions involved in the initial fermentation phase, including their possession of extra genes coding for acid and ethanol tolerance. They also help aid in the breakdown of maltooligosaccharides, unlike other acetic acid bacteria species (Bouchez and De Vuyst 2022).

Lactic acid bacteria contribute core flavor to the brew (being much more desired than acetic bacteria) and mainly include those in the *Lactobacillus* and *Pediococcus* genus (Thompson Witrick et al. 2017). Some of the main Lactobacillus species include those regularly found in milk, cheese, and yogurt including *Lactobacillus brevis* (Bokulich et al. 2012). Others found include *Lactobacillus curvatus*, *Lactobacillus fructivorans,* and *Lactobacillus lindneri* (De Roos and De Vuyst 2018). *Pediococcus damnosus* has a bit of a bad reputation, acting as a major soilage bacteria in non-sour beer. Its prominence in spontaneous fermented beer is due to its resistance to unfavorable growth environments, although this is thought to be a strain-specific trait rather than species-specific (Snauwaert et al. 2015). On an unpleasant note, *P. damnosus* may also produce slimy and ropy beer through the production of β-D-glucan (Gindreau et al. 2001).

8.3.5 Act 5: Maturation

This final act is the longest, taking a minimum of six months to three years (Bongaerts et al. 2021). At this point, lactic and acetic bacteria, along with the *Brettanomyces* yeast genus, are still present. The overall composition of lactic and acetic bacteria is similar to the previous phase. However, this is where the *Brettanomyces* genus shines.

Just like *P. damnosus*, *Brettanomyces* is thought of as a spoilage bacterium in traditional breweries. This is due to the production of secondary metabolites the yeast produces when performing alcoholic fermentation (Serra Colomer et al. 2019). These metabolites give the beer a special flavor termed the "Brett" flavor, which has been characterized by leathery and medicinal notes. Other (*arguably strange*) descriptions of Brett flavor have included, "vomit," "barnyard", "clove," "horsy," "mousy," "smoky," and "horse blanket" (Steensels et al. 2015). Outside of spontaneously fermented beer, *Brettanomyces* is gaining attention as a novel craft beer microbe to harness its funky flavors (Serra Colomer et al. 2019).

In spontaneous beer, the *Brettanomyces* species commonly present during the maturation phase are *B. bruxellensis*, *B. anomalus,* and *Brettanomyces custersianus* (Bongaerts et al. 2021). *Brettanomyces* are prevalent during this phase because of their ability to break down maltooligosaccharides that cannot be used by most other yeast species. As mentioned earlier, higher amounts of unmalted barley are used in the crafting of the wort to allow for this maturation to occur. Because of this, higher ethanol and lower sugar levels are found in spontaneous beer compared to traditional lagers or ales (De Roos and De Vuyst 2019).

After three long years, your beer is now flat, cloudy, and shockingly sour. Did we come all this way for disappointment? Luckily no. The last step before you're likely to pour a bottle is to fine-tune a mixture of different maturation ages. Most often, a young (one-year) ale is mixed with older brews (two- and three-year) and poured straight into a bottle (Spitaels et al. 2015). Since the younger ale is not fully fermented, it contains sugars available for the yeast to produce carbon dioxide for the delightful pop and fizz when you open any bottle of beer. The lambic-linked name of this blend is called a gueuze. Outside of Belgian, a blend may be mentioned on the label. Another popular end product is a fruit blend. Artfully, whole fruit can be added to a blend, which provides extra sugar, which aids in a second fermentation (Keersmaecker 1996). Common fruits include raspberries, peaches, and strawberries.

TABLE 8.1 Microbial cast list: in order of major appearance.[a]

Group of locationally ubiquitous pathogen bacteria	Including, *Enterobacter* spp., *Klebsiella* spp. and *Serratia* spp. (Bongaerts et al. 2021). Known to cause infections but unlikely to survive the final environment of beer.
Background yeast	Including, *Hanseniaspora, Candida, Hanseniaspora, Kregervanrija, Naumovozyma,* and *Wickerhamomyces* (Spitaels et al. 2014). Many are found within the barrel but also on the coolship and in the air of the brewer's house.
Saccharomyces	*S. cerevisiae* → *S. kudriavzevii* or *S. pastorianus* Meaning sugar fungus. Of high importance in food production (Lodolo et al. 2008).
Acetobacter	A genus of acetic acid bacteria. Acetic acid bacteria are known for their ability to convert ethanol to acetic acid (Bouchez and De Vuyst 2022).
Pediococcus	A genus of lactic acid bacteria. Known as spoilage bacteria to some, a hero to others. Some strains produce slime (Gindreau et al. 2001).
Lactobacillus	A genus of lactic acid bacteria. Commonly found in cheese, yogurt, and your cherished probiotic supplement (Hartwig and McDaniel 1995; Suther et al. 2020).
The artist previously known as *Dekkera – Brettanomyces*	Including *B. bruxellensis*, *B. anomalus,* and *Brettanomyces custersianus.* Associated with the Senne valley near Brussels, Belgium (Bongaerts et al. 2021). Known to give off strange flavors, including "goat-like" and "Band-Aid."

[a] Give or take, this is a spontaneous process.

Typically, final blends are at the discretion of the brewmaster, as they decipher which blends of aged beer produce the finest flavor profile. In preparation for this paper, I talked to Jason Perkin, a brewmaster of over 20 years at Allagash Brewing Company, about his process. He explained it's mostly sensory-driven with small batches. It's unlikely he'll want to pair a more acidic end product with tart fruit, like cherry, and instead draws from batches that come out milder. For Allagash's version of a gueuze, he said they start with a base, 50% three years, 25% two years, and 25% one year. From this, he'll mix varying amounts until a desirable blend is achieved. After, residual sugar is taken to ensure enough sugar is available for secondary fermentation and CO_2 production. Shockingly, he informed me about 50% of coolship beer is tossed before bottling due to off-flavors (that are intentionally uncontrolled).

Table 8.1 includes a cast list of tonight's entertainment.

8.4 CURTAIN CALL

Months and seasons have passed, and you now have a fully mature spontaneously fermented beer! You're welcome to applaud your local brewer and ambient microbes. The difference in flavor is rather distinct compared to your traditional beer.

Art has always been subjective based on personal preference, but it is hard to dispute the scientific methodologies used to assess and critique artistic pieces. If the distinct taste and aromas of spontaneous beers are not welcomed on your palate, the artful way in which they are made certainly deserves your respect. Nonetheless, I implore you to find a lambic or coolship style beer in your area. Pop it open and drink slowly. Art is a personal experience.

REFERENCES

Aagaard, K., Ma, J., Antony, K.M. et al. (2014). The placenta harbors a unique microbiome. *Science Translational Medicine* 6: 237ra65.

Bokulich, N.A., Bamforth, C.W., and Mills, D.A. (2012). Brewhouse-resident microbiota are responsible for multi-stage fermentation of American coolship ale. *PLoS One* 7: e35507.

Bongaerts, D., De Roos, J., and De Vuyst, L. (2021). Technological and environmental features determine the uniqueness of the Lambic beer microbiota and production process. *Applied and Environmental Microbiology* 87: e0061221.

Bouchez, A. and De Vuyst, L. (2022). Acetic acid bacteria in sour beer production: friend or foe? *Frontiers in Microbiology* 13: 957167.

Briggs, D.E., Brookes, P.A., Stevens, R. et al. (2004). *Brewing: Science and Practice*. Cambridge, United Kingdom: Woodhead Publishing.

De Keersmaecker, J. (1996). The mystery of lambic beer. *Scientific American* 275: 74–80.

De Roos, J. and De Vuyst, L. (2018). Acetic acid bacteria in fermented foods and beverages. *Current Opinion in Biotechnology* 49: 115–119.

De Roos, J. and De Vuyst, L. (2019). Microbial acidification, alcoholization, and aroma production during spontaneous lambic beer production. *Journal of the Science of Food and Agriculture* 99: 25–38.

De Roos, J., Vandamme, P., and De Vuyst, L. (2018). Wort substrate consumption and metabolite production during Lambic beer fermentation and maturation explain the successive growth of specific bacterial and yeast species. *Frontiers in Microbiology* 9: 2763.

De Roos, J., Van der Veken, D., and De Vuyst, L. (2019). The interior surfaces of wooden barrels are an additional microbial inoculation source for Lambic beer production. *Applied and Environmental Microbiology* 85 (1): e02226–e02218.

De Roos, J., Verce, M., Weckx, S. et al. (2020). Temporal shotgun metagenomics revealed the potential metabolic capabilities of specific microorganisms during Lambic beer production. *Frontiers in Microbiology* 11: 1692.

Garshol, L.M. (2021). Pitch temperatures in traditional farmhouse brewing. *Journal of the American Society of Brewing Chemists* 79: 181–186.

Gindreau, E., Walling, E., and Lonvaud-Funel, A. (2001). Direct polymerase chain reaction detection of ropy *Pediococcus damnosus* strains in wine. *Journal of Applied Microbiology* 90: 535–542.

Hartwig, P. and McDaniel, M.R. (1995). Flavor characteristics of lactic, malic, citric, and acetic acids at various pH levels. *Journal of Food Science* 60: 384–388.

Islam, W., Noman, A., Naveed, H. et al. (2020). Role of environmental factors in shaping the soil microbiome. *Environmental Science and Pollution Research International* 27: 41225–41247.

Jordão, A.M. and Cosme, F. (2022). The application of wood species in enology: chemical wood composition and effect on wine quality. *Applied Sciences* 12: 3179.

Lodolo, E.J., Kock, J.L., Axcell, B.C. et al. (2008). The yeast *Saccharomyces cerevisiae* – the main character in beer brewing. *FEMS Yeast Research* 8: 1018–1036.

Martens, H., Dawoud, E., and Verachtert, H. (1991). Wort enterobacteria and other microbial populations involved during the first month of lambic fermentation. *Journal of the Institute of Brewing* 97: 435–439.

Mosedale, J.R. (1995). Effects of oak wood on the maturation of alcoholic beverages with particular reference to whisky. *Forestry: An International Journal of Forest Research* 68: 203–230.

Narendranath, N.V., Thomas, K.C., and Ingledew, W.M. (2001). Effects of acetic acid and lactic acid on the growth of *Saccharomyces cerevisiae* in a minimal medium. *Journal of Industrial Microbiology & Biotechnology* 26: 171–177.

Parapouli, M., Vasileiadis, A., Afendra, A.S. et al. (2020). *Saccharomyces cerevisiae* and its industrial applications. *AIMS Microbiology* 6: 1–31.

Serra Colomer, M., Funch, B., and Forster, J. (2019). The raise of Brettanomyces yeast species for beer production. *Current Opinion in Biotechnology* 56: 30–35.

Snauwaert, I., Stragier, P., De Vuyst, L. et al. (2015). Comparative genome analysis of *Pediococcus damnosus* LMG 28219, a strain well-adapted to the beer environment. *BMC Genomics* 16: 267.

Sokolowska, M., Frei, R., Lunjani, N. et al. (2018). Microbiome and asthma. *Asthma Research and Practice* 4: 1.

Sparrow, J. (2005). *Wild Brews: Beer Beyond the Influence of Brewer's Yeast*. Boulder, CO: Brewers Publications.

Spitaels, F., Wieme, A.D., Janssens, M. et al. (2014). The microbial diversity of traditional spontaneously fermented Lambic beer. *PLoS One* 9: e95384.

Spitaels, F., Van Kerrebroeck, S., Wieme, A.D. et al. (2015). Microbiota and metabolites of aged bottled Gueuze beers converge to the same composition. *Food Microbiology* 47: 1–11.

Steensels, J., Daenen, L., Malcorps, P. et al. (2015). Brettanomyces yeasts--from spoilage organisms to valuable contributors to industrial fermentations. *International Journal of Food Microbiology* 206: 24–38.

Suther, C., Moore, M.D., Beigelman, A. et al. (2020). The gut microbiome and the big eight. *Nutrients* 12.

Thompson Witrick, K., Duncan, S.E., Hurley, K.E. et al. (2017). Acid and volatiles of commercially-available lambic beers. *Beverages* 3: 51.

Vriesekoop, F., Krahl, M., Hucker, B. et al. (2012). 125th anniversary review: bacteria in brewing: the good, the bad and the ugly. *Journal of the Institute of Brewing* 118: 335–345.

CHAPTER 9

JAPANESE TRADITIONAL FERMENTATION USES SOLID-STATE FUNGAL CULTIVATION TO PRODUCE SAKE

NORIO TAKESHITA[1] AND KEN ODA[2]

[1]Microbiology Research Center for Sustainability (MiCS), Faculty of Life and Environmental Sciences, University of Tsukuba, Tsukuba, Japan

[2]National Research Institute of Brewing, 3-7-1 Kagamiyama, Higashi-hiroshima, Hiroshima 739-0046, Japan

CONTENTS

9.1 INTRODUCTION TO TRADITIONAL JAPANESE FERMENTATIONS PERFORMED USING *ASPERGILLUS ORYZAE*

"Washoku" means traditional Japanese cuisine and the dietary culture (https://www.maff.go.jp/j/shokusan/gaisyoku/pamphlet/pdf/washoku_english.pdf): it was added to UNESCO Intangible Cultural Heritages in 2013 and attracts increasing attention globally (Ministry of Agriculture, Forestry and Fisheries. (n.d.).). Shoyu (soy sauce) and miso (soybean paste) are used as fundamental seasonings for Japanese dishes. In addition, the traditional Japanese alcoholic beverage sake is primarily paired with Japanese foods. The filamentous fungus *Aspergillus oryzae* has been used in the production of traditional fermented foods such as shoyu and miso and drinks such as sake for more than 1000 years in Japan, making it the most important filamentous fungus in the country (Figure 9.1) (Machida et al. 2008). Impressively, *A. oryzae* has even been heralded as the "national microorganism" of Japan (Kitamoto 2015; Ichishima 2016). The safety of *A. oryzae* is guaranteed by the long history of its use in food fermentation industries including molecular genomic and

Microbial Fermentations in Nature and as Designed Processes, First Edition. Edited by Christon J. Hurst.

FIGURE 9.1 An electron microscopy image of an *A. oryzae* conidiophore and a photo of rice *koji*.

metabolomic analyses (Barbesgaard et al. 1992; Machida et al. 2005), which is supported by the World Health Organization (Machida 2002). The fungal species *A. oryzae* has been used commercially as a host for homologous and heterologous protein production as well in modern biotechnology (Christensen et al. 1988; Gouka et al. 1997).

One of the distinctive features of the use of *A. oryzae* in Japanese traditional fermentation is the use of solid-state cultivation (rice grain, soybean, and wheat bran). Since filamentous fungi often secrete more enzymes in solid-state culture than in submerged culture (Diaz-Godinez et al. 2001; Elinbaum et al. 2002; Oda et al. 2006), several commercial enzymes are produced using *A. oryzae* in solid-state culture in addition to traditional fermentation methods. For example, glucoamylase and acid protease encoding genes, *glaB* and *pepA*, respectively, are known as the solid-state-specific genes of *A. oryzae* (Hata et al. 1998; Ishida et al. 2000; Kitano et al. 2002). These gene expressions are controlled by solid-state specific transcriptional factor flbC (Tanaka et al. 2016). Transcriptome and proteome analyses have revealed the ability of *A. oryzae* to produce heterologous proteins in solid-state culture (Hata et al. 1998; Akao et al. 2002).

"Rice *koji*" is a solid culture of *A. oryzae*, also called koji mold, on steamed rice grains (Figure 9.1). Multiple parallel fermentation, wherein saccharification of rice by *A. oryzae* and alcohol fermentation by the budding yeast (*Saccharomyces cerevisiae*) occur simultaneously, leading to the formation of a variety of ingredients in Japanese sake with its characteristic tastes. Sake contains more than 280 metabolites that affect its quality. The metabolite composition of sake depends on the combination of raw materials and sake-making parameters (e.g. rice races, rice polishing ratio, water quality, *koji* mold, yeast strains, sake mash fermentation methods) used during manufacturing (Yazawa et al. 2019).

Japanese people normally consume varieties of eating rice as food, while sake is made with a special variety called sake rice, which is especially suitable for sake brewing. Features of sake rice are large grains generally with a distinct "white core" (or *shinpaku* in Japanese) which contains most of the starch content, a low protein content, and high solubility during the brewing process (Okuda 2019). The distribution of nutrients in a grain of rice is not uniform. Proteins and lipids are most abundant near the surface layer of the rice grains and polishing removes most of that surface layer. After polishing, the grains have a higher percentage of white starch. Inadequate polishing results in miscellaneous taste and off-flavor of the produced sake (Bao 2019). The Japanese high-quality sake, *daiginjo-shu*, is made from highly polished rice with a polishing ratio of less than 50%, meaning that less than half of the grain's original mass remains after the friction has been used to remove the outermost material (Anzawa et al. 2013).

9.2 PREPARATION OF RICE KOJI BY CONTROLLED FUNGAL GROWTH

Koji mold is used for making a variety of traditional Japanese foods. Strains of koji have been commercialized since the Muromachi period (1333–1573), which is earlier than yeast strains. Let me explain how to make Rice koji. The first step in the process of making koji is to cook the rice by steaming it and then cool down the steamed rice before bringing it into a koji room. On a lengthy table, the rice is spread out thinly, adjustment made to the water content and the temperature maintained at around 30 °C. The rice then is dusted with koji mold spores. To ensure that the spores adhere uniformly to the steamed rice, the mass is thoroughly mixed. To maintain the rice's constant temperature and humidity, the rice is then piled high and covered with cloth. The dispersed spores grow their hyphae and begin to germinate. As a result of the heat produced by the koji metabolism, the temperature of the koji increases during this process. The koji is mixed and the pile is spread out once more to maintain the proper temperature, moisture, and hyphae growth.

The koji extends hyphae not only on the rice surface but also into the rice grains. While the koji mold is growing on the steamed rice, it produces enzymes. The degree of hyphal penetration into the steamed rice, called "*haze-komi,*" highly correlates with the digestibility and quality of *koji* (Ito et al. 2017), since the hyphae growing into the rice secrete amylases and digest the starch. To enhance *haze-komi*, koji rice is often dried by raising the temperature and lowering the humidity of the koji room during the fungal growth period as the hyphae penetrate the rice. The *haze-komi* is believed empirically to determine the distribution and composition of enzymes in rice *koji*, the step of fermentation, and the qualities of sake. Recently, the hyphal penetration into steamed rice was visualized by β-glucuronidase (GUS)-expressing *A. oryzae* (Wisman et al. 2020). The correlation of fungal metabolism and growth with the spread of glucose during fermentation was shown by mass spectrometry imaging (MSI) of glucose (Wisman et al. 2020). Moreover, the spread of secreted peptidase was also shown by MSI (Wisman et al. 2022).

We investigated the mycelial distribution of GFP-tagged *A. oryzae* in rice *koji* made with different types of rice, such as sake rice and eating rice, with a 50 or 90% polishing rate to remove abundant proteins and lipids near the surface (Yasui et al. 2020). Our analyses indicate that *A. oryzae* hyphae grew more deeply into 50% polished rice (half of the original grain volume had been removed) than 90% polished rice (10% of the original grain volume had been removed, Figure 9.2). Since proteins and lipids are abundant near the surface of a grain of rice (Okuda 2019), 50% of polished rice consists of mainly starch. *A. oryaze* mycelia likely grow near the surface in

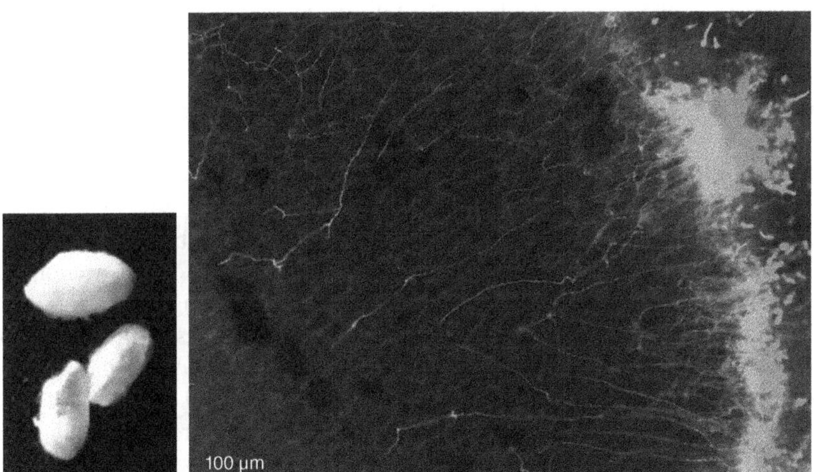

FIGURE 9.2 Haze komi and invasive growth of *A. oryzae* (green) into steamed rice (red). (*See colour plate section for colour representation of this figure.*)

90% polished rice due to a sufficient supply of nutrients near the surface, while the mycelia grow deeply into 50% polished rice to search for nutrients and water as well. That is believed by the experience of sake brewers as evaluated by several methods (Ito et al. 2017), which correlates with our results. We visualized fungal hyphae at the cellular level in rice *koji*. Our data support that hyphae penetrate between rice cells and grow inside rice cells.

We compared *A. oryaze* gene expression in 50% polished rice and 90% polished rice (Yasui et al. 2020). Gene expression of α-amylase and glucoamylase was higher in rice koji with 50% polished rice than 90% rice. Gene expression of fatty acid synthase, fatty acid unsaturation enzyme, phytase, acid/alkaline phosphatase, vitamin B (B1; thiamine, B5; pantothenic acid, B7; biotin) synthesis, and biosynthesis was higher in rice koji with 50% polished rice than 90% rice. On the other hand, gene expression of acidic proteases was decreased in rice koji with 50% polished rice than 90% rice. In total, 22 carboxypeptidase genes and 15 lipase genes did not show a significant trend. From these imaging and gene expression analyses, in rice koji with 90% polished rice, mycelium growth preferentially uses the relatively abundant proteins and lipids near the rice surface and does not break deep into the rice, whereas in rice koji with 50% polished rice, mycelium breaks more deeply into the rice. The surface of the rice is scraped to expose the rice cells, making it easier for the mycelium to penetrate between the cells, and differences in the moisture content and evaporation rate of the rice interior may also affect the penetration of the mycelium.

Water quality is important for sake brewing. The most harmful mineral in water is iron. If sake has too much iron in it, it becomes reddish brown and tastes and smells bad. Water used to make sake must have an iron concentration of less than 0.02 ppm, which is less than the regulations for tap water (0.3 ppm or less). Fortunately, Japan has several excellent water sources and places where there are lots of sake brewers.

9.3 FERMENTATION OF THE RICE AS A MASH TO PRODUCE ETHANOL

Alcohol is produced by budding yeast, *S. cerevisiae*, using sugar. Since starch has longer sugar molecules, yeast cannot directly convert it to alcohol. The enzymes in the koji transform the starch in the fermentation mash into sugar. Alcohol can be created from sugar by yeast. Nowadays, most sake breweries employ trusted sake yeast strains that have been carefully chosen from wild yeast. Dr. Kikuji Yabe first isolated sake yeast from a fermentation mash in 1895. This particular sake yeast was first made available to sake breweries across Japan by the Brewing Society of Japan. Sake is produced steadily and at a high standard thanks to the use of sake yeast. Different sake yeast produce various flavors and fragrances. The right yeast strains are used by sake breweries.

Since a large amount of yeast is required for creating a fermentation mash, a fermentation starter, called *moto,* is prepared to cultivate the yeast. Steamed rice, rice koji, and water are placed in a tank to cultivate the yeast. Lactic acid must be added to the fermentation starter to inhibit the growth of harmful bacteria. We can create a fermentation starter that is a pure culture of yeast by the judicious management of microorganisms and precise temperature control.

The steamed rice, rice koji, water, and fermentation starter are combined in a tank to create the fermentation mash (Figure 9.3). Each of these materials is divided into three sections of progressively increasing volume, and each portion is added one at a time taking three days. This approach works well to avoid lowering the mash's acidity and yeast density as well as to avoid bacterial contamination. After proceeding through the three steps, the fermentation mash is mixed every day. The temperature of the fermentation tank rises daily by about 1 °C until around 15 °C depending on fermentation progress and there is a desire to keep the temperature from exceeding about 15 °C. The fermentation mash's alcohol content will rise to 18% after three weeks of fermentation.

FIGURE 9.3 A tank to create the fermentation mash and fresh sale. (*See colour plate section for colour representation of this figure.*)

After fermentation, the fermentation mash contains a lot of solid ingredients such as undissolved rice and rice koji as well as the alcoholic liquid now called sake. The mash is pressed using a filter press or a conventional press to collect the fermented liquid sake. The solid substance, known as sake kasu, is removed. We then have fresh sake with a lovely scent after the pressing procedure. The sake then may be allowed to age before it is marketed.

9.4 SUMMARY

Humankind has been producing fermented foods long before correctly recognizing the existence of microorganisms. These processes were done by utilizing microorganisms that match the land topography and climate. Microorganisms can have an impact on local culture, religious beliefs, and more in addition to dietary habits. Valuably, *A. oryzae* has been bred in the brewing industry for more than a thousand years and is vital to Japanese society and business. Microbial biotechnology is built on methods for fermentation, which have been improved over time. There is still much to be learned or exploited about this fungus.

REFERENCES

Akao, T., Gomi, K., Goto, K. et al. (2002). Subtractive cloning of cDNA from *Aspergillus oryzae* differentially regulated between solid-state culture and liquid (submerged) culture. *Current Genetics* 41: 275–281.

Anzawa, Y., Nabekura, Y., Satoh, K. et al. (2013). Polishing properties of sake rice Koshitanrei for high-quality sake brewing. *Bioscience, Biotechnology, and Biochemistry* 77: 2160–2165.

Bao, J. (2019). *Rice Milling Quality. Rice Chemistry and Technology*, 4e, 339–369. Elsevier Inc. in cooperation with AACC International.

Barbesgaard, P., Heldt-Hansen, H.P., and Diderichsen, B. (1992). On the safety of *Aspergillus oryzae*: a review. *Applied Microbiology and Biotechnology* 36: 569–572.

Christensen, T., Woeldike, H., Boel, E. et al. (1988). High level expression of recombinant genes in *Aspergillus oryzae*. *Nature Biotechnology* 6: 1419–1422.

Diaz-Godinez, G., Soriano-Santos, J., Augur, C. et al. (2001). Exopectinases produced by *Aspergillus niger* in solid-state and submerged fermentation: a comparative study. *Journal of Industrial Microbiology & Biotechnology* 26: 271–275.

Elinbaum, S., Ferreyra, H., Ellenrieder, G. et al. (2002). Production of *Aspergillus terreus* alpha-l-rhamnosidase by solid state fermentation. *Letters in Applied Microbiology* 34: 67–71.

Gouka, R.J., Punt, P.J., and van den Hondel, C.,.A. (1997). Efficient production of secreted proteins by *Aspergillus*: progress, limitations and prospects. *Applied Microbiology and Biotechnology* 47: 1–11.

Hata, Y., Ishida, H., Ichikawa, E. et al. (1998). Nucleotide sequence of an alternative glucoamylase-encoding gene (*glaB*) expressed in solid-state culture of *Aspergillus oryzae*. *Gene* 207: 127–134.

Ichishima, E. (2016). Development of enzyme technology for *Aspergillus oryzae*, *A. sojae*, and *A. luchuensis*, the national microorganisms of Japan. *Bioscience, Biotechnology, and Biochemistry* 80: 1681–1692.

Ishida, H., Hata, Y., Kawato, A. et al. (2000). Identification of functional elements that regulate the glucoamylase-encoding gene (*glaB*) expressed in solid-state culture of *Aspergillus oryzae*. *Current Genetics* 37: 373–379.

Ito, K., Gomi, K., Kariyama, M. et al. (2017). Quantitative evaluation of haze formation of *koji* and progression of internal haze by drying of *koji* during *koji* making. *Journal of Bioscience and Bioengineering* 124: 62–70.

Kitamoto, K. (2015). Cell biology of the koji mold *Aspergillus oryzae*. *Bioscience, Biotechnology, and Biochemistry* 79: 863–869.

Kitano, H., Kataoka, K., Furukawa, K., and Hara, S. (2002). Specific expression and temperature-dependent expression of the acid protease-encoding gene (*pepA*) in *Aspergillus oryzae* in solid-state culture (rice-*koji*). *Journal of Bioscience and Bioengineering* 93: 563–567.

Machida, M. (2002). Progress of *Aspergillus oryzae* genomics. *Advances in Applied Microbiology* 51: 81–106.

Machida, M., Asai, K., Sano, M. et al. (2005). Genome sequencing and analysis of *Aspergillus oryzae*. *Nature* 438: 1157–1161.

Machida, M., Yamada, O., and Gomi, K. (2008). Genomics of *Aspergillus oryzae*: learning from the history of *Koji* mold and exploration of its future. *DNA Research* 15: 173–183.

Ministry of Agriculture, Forestry and Fisheries. (n.d.). Government of Japan. The Washoku Way: Japan's nuanced approach to food, https://www.maff.go.jp/j/shokusan/gaisyoku/pamphlet/pdf/washoku_english.pdf (accessed 19 April 2023).

Oda, K., Kakizono, D., Yamada, O. et al. (2006). Proteomic analysis of extracellular proteins from *Aspergillus oryzae* grown under submerged and solid-state culture conditions. *Applied and Environmental Microbiology* 72: 3448–3457.

Okuda, M. (2019). Rice used for Japanese sake making. *Bioscience, Biotechnology, and Biochemistry* 83: 1428–1441.

Tanaka, M., Yoshimura, M., Ogawa, M. et al. (2016). The C2H2-type transcription factor, FlbC, is involved in the transcriptional regulation of *Aspergillus oryzae* glucoamylase and protease genes specifically expressed in solid-state culture. *Applied Microbiology and Biotechnology* 100: 5859–5868.

Wisman, A.P., Tamada, Y., Hirohata, S. et al. (2020). Mapping *haze-komi* on rice *koji* grains using β-glucuronidase expressing *Aspergillus oryzae* and mass spectrometry imaging. *Journal of Bioscience and Bioengineering* 129: 296–301.

Wisman, A.P., Minami, M., Tamada, Y. et al. (2022). Visualization of dipeptidyl peptidase B enzymatic reaction in rice koji using mass spectrometry imaging. *Journal of Bioscience and Bioengineering* 134: 133–137.

Yasui, M., Oda, K., Masuo, S. et al. (2020). Invasive growth of *Aspergillus oryzae* in rice koji and increase of nuclear number. *Fungal Biology and Biotechnology* 7: 8.

Yazawa, H., Tokuoka, M., Kozato, H. et al. (2019). Investigation of relationship between sake-making parameters and sake metabolites using a newly developed sake metabolome analysis method. *Journal of Bioscience and Bioengineering* 128: 183–190.

SECTION IV

USING MICROBIAL FERMENTATION TO PRODUCE FOOD AND FODDER

CHAPTER 10

SAFETY DEMONSTRATION OF FOOD AND FEED CULTURES

FRANÇOIS BOURDICHON, ALESSANDRA FONTANA, VANIA PATRONE, AND LORENZO MORELLI
Facoltà di Scienze Agrarie, Alimentarie Ambientali, Università Cattolica Del Sacro Cuore, Piacenza, Italy

CONTENTS

10.1 INTRODUCTION

Fermented foods have been consumed for more than 10 000 years, making food fermentation probably one of the oldest food technologies implemented by man. Presently, fermented food products are estimated to represent a third of our food intake.

Microbial Fermentations in Nature and as Designed Processes, First Edition. Edited by Christon J. Hurst.

The scientific knowledge behind its role and mechanisms of action have only been studied in the past 150 years. Food microbiology has provided many answers behind the roles, modes of action, nutrition, and health effects of fermented foods. Nevertheless, one of the major topics of concern remains the safety demonstration of microbial food cultures. Despite this very long history of traditional use, what is considered fermented in a part of the world is considered spoiled, if not unsafe, in another one. As an example, in late 2017, China blocked the importation of cheeses from Europe due to the presence of microbial food cultures not present in the Chinese 2010 positive list: *Penicillium roquefortii*, *Penicillium camemberti*, and the presence in their genomes of mycotoxin genes. This is not unheard of per se, as the Biohazard Panel of the European Food Safety Authority has excluded filamentous fungi from the Qualified Presumption of Safety. The United States (GRAS – Generally Recognized as Safe) also has a respective procedure in place for the microbial risk assessment of microbial species voluntarily added to the food chain. The International Dairy Federation (IDF), in collaboration with the European Food and Feed Cultures Association, has been working for the past 20 years on an inventory of microbial species with technological properties in fermented foods.

Safety demonstration of food and feed cultures through inoculation into a food matrix for use in the food chain is aimed to avoid trade barriers between countries when a history of safe use cannot be established for an indigenous fermented food product on the international market. It is as well important to avoid the pitfalls of cross-over fermentation, when changing the traditional food matrix, and when possible, to also avoid deleterious microbial metabolic activities.

10.2 FERMENTED FOOD PRODUCTS AND INTERNATIONAL TRADE

In Early September 2017, China banned the importation of European cheeses. The ban started in July and August in some major cities and was officially extended nationally in early September. Specifically, soft cheeses such as Stilton Blue Cheese (UK), Danish Blue Cheese (DK), Camembert, Brie, and Roquefort (FR), and lastly Gorgonzola (IT) were implicated in the ban.

Food adulteration? Ongoing foodborne outbreaks? *Penicillium roqueforti* and *P. camemberti* as microbial food cultures were the common factors and reason for the ban of these famous, tasty(?) cheeses with a long history of safe use. The decision was based upon the Chinese 2010 list of food cultures allowed in foods according to the Ministry of Health. Chinese health regulations permit only a few types of food cultures in fermented food products. While there is an exemption for "cultures that are traditionally used for food production," that exemption does not apply to imported goods.

The ban was short-lived and lasted two months. Following a round of meetings between European Commission representatives and Chinese quarantine and health officials, China reversed the ban in late October, effective immediately. Since then, multiple technical meetings have occurred between Chinese and European microbiologists to update the regulatory status of microbial food cultures in China. A new GB standard (Guobiao – 国标· –, Chinese standardization scheme) is presently going (early 2023) through public consultation before final publication. In these new sets of standards, the list of 2010 and its following updates are expected to be taken out, while the technical details of safety demonstration and safety dossier will be detailed.

Recently, China has been working on updating the food cultures found in indigenous traditional fermented foods, going through the same process that Western countries (Europe, United States) had done in previous decades (Yao et al. 2021). International trade of fermented food products, research in the understanding of the mechanisms of action of food cultures to get out of the historical black box approach, consumer acceptance, and regulatory hurdles for the use of microorganisms in the food chain are all contributing factors that have put traditional food processes and cultural usage in the spotlight and raised, necessarily or not, various concerns about the conditions of use and safety of the traditional indigenous fermented foods, and their constitutive food cultures, whether inoculated or not.

10.3 FOOD FERMENTATION. FOOD AND FEED CULTURES

Fermentation is probably one of the oldest food technologies used by man, dating back to Paleolithic times. For millennials, the approach was exclusively empiric, with little if no awareness of the role played by microbial communities. The importance of food fermentation is still underestimated in Western societies, where the focus on Dairy fermented products provides a consumer bias in the understanding of the incredible variety of fermented food products, which are expected to represent a third of our food intake (Steinkraus 2002).

Despite its role in everyday life, there is not presently any consensual definition (and therefore understanding) of what food fermentation precisely means.

In his 2002 review, Steinkraus proposed five roles to be considered for fermentation in food processing:

1. Enrichment of the human diet through the development of a wide diversity of flavors, aromas, and textures in food
2. Preservation of substantial amounts of food through lactic acid, alcoholic, acetic acid, alkaline fermentations, and high salt fermentations
3. Enrichment of food substrates biologically with vitamins, protein, essential amino acids, and essential fatty acids
4. Detoxification during food fermentation processing
5. A decrease in cooking times and fuel requirements

These roles were readdressed by the IDF in the 2012 publication of the second update of the inventory of food cultures (Bourdichon et al. 2012a, b):

1. Preservation of food through the formation of inhibitory metabolites such as organic acids (lactic acid, acetic acid, formic acid, propionic acid), ethanol, bacteriocins, etc., often in combination with a decrease of water activity (by drying or use of salt)
2. Improving hygiene through inhibition and even elimination of food pathogens
3. Detoxification of food (e.g. removal of cyanogens, goitrogens, etc.)
4. Improving wholesomeness through improved digestibility of polymers after removal of flatulence-causing compounds
5. Enrichment of food substrates with essential nutrients (vitamins, proteins, essential amino acids, fatty acids, etc.) and enhanced bioavailability of food components through catabolism of the food matrix
6. Organoleptic properties through effects on flavor, texture, and color: flavor compounds are produced by microorganisms during the initial fermentation step or subsequent ripening

Most recently, the International Scientific Association of Probiotics and Prebiotics (ISAPP) published a review paper on fermented foods as well (Marco et al. 2021), where food fermentation per se was not defined, but as done by Steinkraus and the IDF, the outcome, namely fermented food products, as: "foods made through desired microbial growth and enzymatic conversions of food components."

10.4 HISTORY OF (SAFE) USE

One of the points of focus regarding the safety demonstration of microbial food cultures originates either in Europe or in North America for the evaluation of novel foods products, either through the European Novel Food Regulation (Regulation [EC] No. 258/97 of the European Parliament and of the Council of 27 January 1997 concerning novel foods and novel food ingredients) or the United States – Food and Drug Administration (US – FDA) GRAS regulation. When it comes to the novel food process evaluation, Europe considers 1997 as a milestone, whereas US-FDA considers the year 1958. And science? Considering the history of microbial use can be complicated, the wording used in microbiology that mentions the "origin" of a micro-organism instead of its "isolation" is not helpful. Recently, the phrasing has been adapted to better represent this field of science, but education is still needed.

In December 2004, during the second scientific colloquium of the European Food Safety Authority (EFSA) about the Qualified Presumption of Safety of Micro-organisms in Food and Feed (QPS), it was agreed that "a long history of apparent safe use of a given micro-organism or micro-organisms for the making of a given food or feed product suggests a very high safety level for the consumption of such products." Even so, the application in food of a microbial strain belonging to a species with an accepted history of safe use always remains subject to consideration regarding the potential presence of undesirable traits that are of safety concern (Leuschner et al. 2010).

While "history of use" is found on 21 pages out of 144 of the QPS scientific colloquium report, the concept is not defined in the global report, leaving the appreciation of what is historical and what is not to other stakeholders. This point was tentatively addressed by the IDF expert group, following a survey, and lack a clear definition of the different regulations in place. One definition of "history of safe use" proposes "significant human consumption of food over several generations and in a large, genetically diverse population for which there exist adequate toxicological and allergenicity data to provide reasonable certainty that no harm will result from consumption of the food" (Health Canada 2003). A history of the "apparent safe use" concept is not limited to use over a long period; it also includes factors such as the level of exposure to a micro-organism. A history of apparently safe use does not, in itself, constitute a risk assessment. Furthermore, no assurances can be made that something is absolutely safe. An appropriate history of safe use does, however, provide evidence in support of a "reasonable certainty of no harm." The history of use, in a nutshell, replaces the exposure assessment part of the risk assessment framework proposed in the risk analysis guidelines of the FAO; it is, therefore, necessary but far from sufficient to consider that any individual isolate of the evaluated microorganism's species is safe (FAO 2007).

The traditional use of microbial food cultures was tentatively placed in the QPS colloquium in the three following categories:

1. Spontaneous fermentation processes i.e. without any microbial food cultures added intentionally, and so-called back-slopping processes where an undefined mixture of micro-organisms, naturally present in a product, is recycled. Products of such processes include, e.g. olives, cream, and sourdough-based bread;

2. Processes based on deliberately added, but undefined microbial mixtures, which were not originally part of the natural flora in the raw material. An example of the use of an undefined microbial mixture is Kefir;
3. Processes using defined microbial food cultures, which are identifiable at the strain level.

Microbial food cultures per se do not have a clear definition either and even a clear regulatory status.

There is presently an ISO standard in place, ISO 27205:2010 [IDF 149:2010] standard for Fermented milk products – Bacterial starter cultures – Standard of identity, which "specifies characteristics of industrial bacterial starter cultures, which are principally lactic acid bacteria (LAB), but which also include *Bifidobacterium* and *Propionibacterium* used for the manufacture of fermented milk products such as yoghurt, sour cream, cultured butter and cheese," but does not apply to bacterial cultures which are added as an ingredient to foods only because of their probiotic properties. (https://www.iso.org/standard/44069.html). This standard is presently under revision (Chaired by the IDF) to extend its scope to non-dairy and non-bacterial food cultures. Part of the safety demonstration of the microbial food cultures is also to define them. This is a small yet significant gap in the QPS approach with the QPS internal assessment of EFSA that addresses the microorganisms voluntarily added into the food chain, and not all of them are defined as microbial food cultures, as they do not necessarily play a technological role in food fermentation.

Probiotic cultures are specifically addressed in the FAO/WHO 2001 & 2002, compiled in 2006 reports (FAO/WHO 2001, 2002, 2006). From the concept and terms used until now, a probiotic is the only one which is precisely defined and consensually recognized: "live microorganisms which when administered in adequate amounts confer a health benefit on the host." Yet even with 20 years of experience plus definition, and clear guidelines, the use (and abuse) of probiotics remains a concern when it comes to their safety and precaution about use, adding more substance to the global concerns on microbial food cultures. Most particularly, the abuse of probiotics for specific uses, and medical purposes, which is already mentioned in the 2001 report as biotherapeutic agents. The concept of biotherapeutic agents is clearly regulated, at least in the United States and Europe (Cordaillat-Simmons et al. 2020). Either probiotic or biotherapeutic, these microorganisms fall into the scope of the safety demonstration of both GRAS and QPS, and the lack of precaution from many (sometimes significant) operators has placed doubt on their safe use (Pararajasingam and Uwagwu 2017).

10.5 SAFETY CONSIDERATIONS

According to the FAO/WHO expert panel in the 2002 report, probiotics have been proposed to be theoretically responsible for four types of side effects (Marteau 2001):

1. Systemic infections
2. Deleterious metabolic activities
3. Excessive immune stimulation in susceptible individuals
4. Gene transfer

These side effects were re-evaluated by the IDF expert group in the 2012 Publication set (Bourdichon et al. 2012a, b) with small modifications.

There has been indeed growing evidence of safety considerations in the past 30 years, as shown in the following timeline, from 1994–2021 with a selection of relevant milestones:

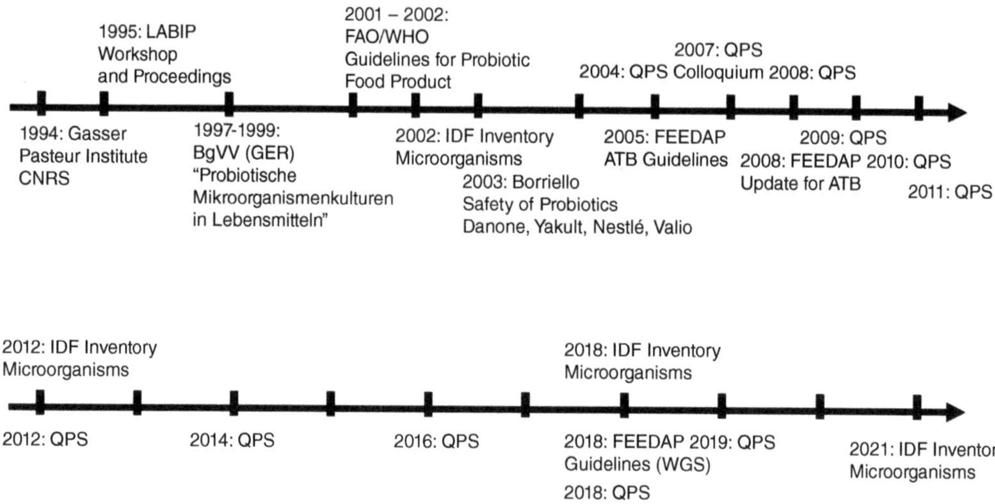

In his 1994 review in the "Bulletin de L'Institut Pasteur," Pr. Francis Gasser addressed one of the first reviews on the evidence of the safety of LAB and their clinical isolations (Gasser 1994). This initial work was followed by a specific workshop of LABIP – Lactic Acid Bacteria Industrial Platform and position from the BgVV – Bundesinstitut für Gesundheitlichen Verbraucherschutz und Veterinärmedizin – of Germany in 1999 (BgVV 1999). The FAO/WHO experts reports followed in 2001 and 2002 with a proposal for the safety assessment of probiotic candidate strains, while the IDF published its first edition of the inventory of food cultures to establish the history of safe use and answer regulatory expectations (Mogensen et al. 2002a,b). A joint expert group at the initiative of industrial key players reviewed the available evidence in 2003 (Borriello et al. 2003). The QPS initiative started with the second EFSA scientific colloquium in December 2004 and has been regularly updated and extended since then (Leuschner et al. 2010; Herman et al. 2018). The key milestones of the QPS are detailed on the EFSA website (https://www.efsa.europa.eu/en/topics/topic/qualified-presumption-safety-qps#milestones)

Among the major points of concern, the gap of understanding between clinical and food microbiologists when it comes to the assessment of LAB is particularly "interesting." None of the commonly used food cultures, even when causing relative harm, falls into the scheme of Koch's 1890 postulate, or even the recently proposed molecular ones (Falkow 1988)

As originally stated, the four criteria are:

1. The microorganism must be found in diseased but not healthy individuals
2. The microorganism must be cultured from the diseased individual
3. Inoculation of a healthy individual with the cultured microorganism must recapitulate the disease
4. The microorganism must be re-isolated from the inoculated, diseased individual and matched to the original microorganism.

As for LAB, postulates 1, 3, and 4 do not apply: through fermented food products, one can reasonably consider that the global world population is exposed, hence as previously mentioned the lack of exposure assessment relevance and the alternative consideration of the history of safe use. Nevertheless, LAB are sometimes classified as pathogens of concern and infectious substances by clinical microbiologists (e.g. Government of Canada: https://www.canada.ca/en/public-health/services/laboratory-biosafety-biosecurity/pathogen-safety-data-sheets-risk-assessment/lactobacillus.html), despite 10 000 years at least of history of safe use. *Lacticaseibacillus rhamnosus* (Former taxon *Lactobacillus rhamnosus*) is even classified as class 2 in Germany (https://www.baua.de/DE/Angebote/Rechtstexte-und-Technische-Regeln/Regelwerk/TRBA/pdf/TRBA-466.pdf?__blob=publicationFile&v=11), after reports of opportunistic infections in the 1990s by Professors Klein and Reuter (Freien Universität Berlin, Germany). The same point for *Leuconostoc mesenteroides*, used in the food culture for sauerkraut and blue cheese, is considered by clinical microbiologists as an emerging pathogen with probably 100% exposure to the population. Maybe the conclusion is not appropriate and the issue is elsewhere…

10.6 OPPORTUNISTIC INFECTIONS BY LACTIC ACID BACTERIA

Correlation is not causation and the isolation of a microorganism where the medical dogma expects microbial sterility can lead to quite "interesting" conclusions (Mofredj et al. 2007). When considering the prevalence of opportunistic infections by LAB, one can start by considering the available evidence in the scientific literature. The review of the opportunistic case report as presently done by the EFSA Biohazard Panel is based on a literature database web search using keywords the Appendix C — Search strategies for the maintenance and update of a list of QPS-recommended biological agents (EFSA 2018).

As an example, for *Bifidobacterium* spp.:

***Bifidobacterium* spp.**

String for species	
"*Bifidobacterium adolescentis*" OR "*Bifidobacterium animalis*" OR "*Bifidobacterium bifidum*" OR "*Bifidobacterium breve*" OR "*Bifidobacterium longum*" OR "*B. adolescentis*" OR "*B. animalis*" OR "*B. bifidum*" OR "*B. breve*" OR "*B. longum*"	
Outcome	**String**
1. Antimicrobial/Antibiotic/Antimycotic	"antimicrobial resistan*" OR "antibiotic resistan*" OR "antimicrobial susceptibil*"
2. Infection/Bacteremia/Fungemia/Sepsis	"infection*" OR "abscess*" OR "sepsis*" or "septic*" OR "bacteremia" OR "bacteraemia" OR "toxin*"
3. Type of disease	"endocarditis" OR "abscess" OR "meningitis"
4. Mortality/Morbidity	"clinical*" OR "death*" OR "morbidit*" OR "mortalit*" OR "disease*" OR "illness*"
5. Disease Risk	"opportunistic" OR "virulen*"

Source: EFSA et al. (2018), p. 26.

This screening approach has its limitation. Considering these research keywords for *Lactobacillus* spp. (former taxon), reported cases such as Pararajasingam and Uwagwu 2017

(Case report of Probiotic Infection) would not be identified. Opportunistic infections of *Lactobacillus* spp. (former taxon) and *Bifidobacterium* spp. are also not restricted to endocarditis and meningitis. Several cases of peritonitis, hepatitis, canaliculitis, cellulitis, cholangitis, cholecystitis, chondritis, colitis, empyema, enteritis, onset of fever, gastroenteritis, esophagitis, osteomyelitis, pyelonephritis, urosepsis, vaginitis, etc. have been reported. The picture is presently clearly underestimated in its reporting and its description as done by the EFSA QPS Panel.

10.7 ANTIBIOTIC RESISTANCE

The Antibiotic resistance of Food cultures is a point specifically addressed in the Guidelines of the EFSA's Panel on Additives and Products or Substances used in Animal Feed (FEEDAP) in the first guidelines of 2005, 2008, and most recently the Guidance on the characterization of microorganisms used as feed additives or as production organisms (FEEDAP 2018).

The proposed microbiological cut-off values (mg/l) for specific microbial species have been established based on the data generated from the European Project ACE-ART (Assessment and Critical Evaluation of Antibiotic Resistance Transferability in the food chain; https://cordis.europa.eu/project/id/506214/reporting/fr) (Domig et al. 2008) (Table 10.1).

ISO 10932|IDF 223:2010 – Milk and milk products – Determination of the minimal inhibitory concentration (MIC) of antibiotics applicable to bifidobacteria and non-enterococcal LAB is the international standard of reference used for phenotypic antibiotic susceptibility testing (https://www.iso.org/standard/46434.html).

This standard was last reviewed and confirmed in 2020. Considering the outcome and comments of the systematic review and the technical discussions within the IDF, the existing standard could be improved with respect to several areas:

- Update the standard with respect to up-date taxonomy.
- Removal of the currently included supplier of antibiotic panels that are no longer on the market.
- More sustainable by use of fewer antibiotics and plastics and supporting a longer shelf life of the medium.
- Clear guidance for antibiotic susceptibility testing of food cultures will reduce the burden of antibiotic resistance in the food chain.
- Increase the quality of MIC data by a better description of how to read MIC data (in line with EUCAST reading guidance) and how to evaluate MIC data of test strains and control strains.
- Include the options to use the equipment for standardizing the inoculum, reading devices to assist the reading of MIC values with automated solutions for data transfer, or other options for automatization.
- Extension with more "reference" strains in the interlaboratory trial (e.g. include *Lactobacillus delbrueckii* subsp. *lactis*)

The standard includes testing of 16 antibiotics some of which are not clinically relevant for LAB. The range of antibiotics included in the standard are overlapping with the EFSA FEEDAP (2005) guidance: 14 of the 16 antibiotics were requested by EFSA at the time of development of the standard. Considering the lack of clinical relevance, only nine antibiotics of the present standard are required and considered relevant in EFSA's present assessment.

From a regulatory perspective, it would be more efficient to provide the guidelines for setting the MIC values and introduce new species of interest, particularly in the context of major changes in taxonomy for *Lactobacillaceae* and *Leuconostocaceae* (Zheng et al. 2020)

TABLE 10.1 Microbiological cut-off values (mg/l).

	Ampicillin	Vancomycin	Gentamicin	Kanamycin	Streptomycin	Erythromycin	Clindamycin	Tetracycline	Chloramphenicol	Tylosin	Ciprofloxacin	Colistin	Fosfomycin
Lactobacillus obligate homofermentative[a]	2	2	16	16	16	1	4	4	4	n.r.	n.r.	n.r.	n.r.
Lactobacillus acidophilus group	1	2	16	64	16	1	4	4	4	n.r.	n.r.	n.r.	n.r.
Lactobacillus obligate heterofermentative[b]	2	n.r.	16	64	64	1	4	8[c]	4	n.r.	n.r.	n.r.	n.r.
Lactobacillus reuteri	2	n.r.	8	64	64	1	4	32	4	n.r.	n.r.	n.r.	n.r.
Lactobacillus facultative heterofermentative[d]	4	n.r.	16	64	64	1	4	8	4	n.r.	n.r.	n.r.	n.r.
Lactobacillus plantarum/pentosus	2	n.r.	16	64	n.r.	1	4	32	8	n.r.	n.r.	n.r.	n.r.
Lactobacillus rhamnosus	4	n.r.	16	64	32	1	4	8	4	n.r.	n.r.	n.r.	n.r.
Lactobacillus casei/paracasei	4	n.r.	32	64	64	1	4	4	4	n.r.	n.r.	n.r.	n.r.
Bifidobacterium	2	2	64	n.r.	128	1	1	8	4	n.r.	n.r.	n.r.	n.r.
Pediococcus	4	n.r.	16	64	64	1	1	8	4	n.r.	n.r.	n.r.	n.r.
Leuconostoc	2	n.r.	16	16	64	1	1	8	4	n.r.	n.r.	n.r.	n.r.
Lactococcus lactis	2	4	32	64	32	1	1	4	8	n.r.	n.r.	n.r.	n.r.
Streptococcus thermophilus	2	4	32	n.r.	64	2	2	4	4	n.r.	n.r.	n.r.	n.r.
Bacillus	n.r.	4	4	8	8	4	4	8	8	n.r.	n.r.	n.r.	n.r.
Propionibacterium	2	4	64	64	64	0.5	0.25	2	2	n.r.	n.r.	n.r.	n.r.
Enterococcus faecium	2	4	32	1024	128	4	4	4	16	4	n.r.	n.r.	n.r.
Corynebacterium and other Gram-positive	1	4	4	16	8	1	4	2	4	n.r.	n.r.	n.r.	n.r.
Enterobacteriaceae	8	n.r.	2	8	16	n.r.	n.r.	8	n.r.	n.r.	0.06	2	8

n.r.: not required.

[a] Including *L. delbrueckii, L. helveticus.*

[b] Including *L. fermentum.*

[c] For *L. buchneri* the cut-off for tetracycline is 128.

[d] Including the homofermentative species *L. salivarius.*

Source: EFSA et al. (2018), p. 8.

10.8 FOOD FERMENTATION AS END USE OF MICROBIAL FOOD CULTURES: HISTORY OF SAFE USE

The history of safe use, despite a lack of a definition, is the major aspect pushed forward when setting the safety of microbial food cultures. As expected by the QPS panel of EFSA, this is a key component of the body of knowledge of a microbial species of interest.

The IDF initiative on the inventory of microbial food cultures uses the demonstration of food usage of a microbial species in a food matrix as the key demonstrator of the history of safe use (Bourdichon et al. 2012b, c).

The history of the safe use of a food culture is linked to those food matrices in which it has been recognized to be used traditionally. This microbial food culture–food matrix approach is the major point of difference in the rationale of the IDF list of food cultures when compared to the EFSA QPS List (Bourdichon et al. 2019).

There is a trend recently in food fermentation research to assess the use of species historically isolated in a type of food matrices in other food types. Dank et al. (2021) introduced most recently the concept of cross-over fermentations, "processes in which a microorganism from one traditional fermentation process is introduced onto a new substrate and/or to a new partner." The traditional use, however, which grants a history of safe use, is not applicable anymore and a specific safety demonstration for this new food usage must be demonstrated (Capozzi et al. 2017). Even if a microorganism originates from an established traditional/artisanal fermented food, a thorough safety examination must be performed also at the strain level before it is used as a food culture (Capozzi et al. 2020).

10.9 BENEFIT – RISK ASSESSMENT OF FOOD AND FEED CULTURES

The explanation and building of rationale for use of microbial food cultures in the food chain highlight the importance of doing a benefit–risk assessment when considering the potential role of a microbial strain of interest, or when investigating the diversity of the food microbiota for the technological role of the present microbial species.

A specific focus is expected on the (body of) knowledge around the microbial species present. These microorganisms should not be considered as safe by design: more understanding and sound evidence are needed about their introduction into the food chain. They do not harbor necessarily per se specific unwanted properties such as those assessed when it comes to the hazard characterization of microorganisms of concerns such as food-borne pathogens, but it does not mean that only a little consideration is needed, particularly when considering the use of non-usual food matrices, or a specific population with medical conditions and underlying factors prone to adverse effects.

Intrinsic as well as extrinsic factors must be properly considered and addressed: the interaction with the ecological niche, essentially the food matrices they are usually associated with, but also the human host. The approach and understanding of microbial ecology are particularly different between food microbiologists and clinical microbiologists. Due to their isolation in human samples and correlation with various health situations, some of them severe if not fatal, these food culture species are eventually considered by clinical microbiologists as opportunistic pathogens, leading to regulatory approaches often provoking barriers to trade for fermented food products. Depending upon the competent authorities, the ministry of health or the ministry of agriculture, both covering the area of food safety (similar to WHO and FAO), the management decision can be particularly different, despite being based on the same evidence.

The usual segmentation of the risk assessment (FAO 2007) into hazard identification, hazard characterization exposure assessment, and risk characterization have their limits when it comes to microbial food cultures and probiotics. For the latter, guidelines have been in place since 2001 and 2002, reinforced by the ISAPP, and more recently by the International Probiotics Association (IPA) (Hill et al. 2014; Binda et al. 2020). Food Cultures per se do not have such strict guidelines, not to mention a regulatory hesitance between additives or ingredients (Laulund et al. 2017). As mentioned earlier, and addressed, e.g. by the IDF in the initiative of the inventory of food cultures, or the EFSA BIOHAZARD Panel for the QPS evaluation, the establishment of the history of safe use (when it will be clearly agreed upon what "history" means in this evaluation) is the approach used by food safety and regulatory authorities to "mimic" the exposure assessment. The gap between these approaches is mostly due to a lack of mutual understanding in assessing what constitutes a history of safe use.

What about the hazard characterization part? The role of the host and precisely the gut microbiota is a new area of investigation for food safety authorities such as the European Food Safety Authority (EFSA) (Merten et al. 2020) and the Food and Agriculture Organisation of the United Nations (FAO) – Sustainable Development Goals 2 & 3 (FAO 2019). We are not all equal when it comes to infectious diseases, and this might not be all about our immune systems. Whether one does consider the potential risk of opportunistic infections of microbial food cultures, or the use as biotherapeutic agents for specific medical purposes, the sensitivity of the host as a halobiont must be properly considered in the conditions of use of the microbial food cultures, and or the fermented food products. The diet has been studied most recently for its influence and capacity to modulate the gut microbiome (Wastyk et al. 2021), which in turn can impact other effectors of the human immune system. All is not bad about a slightly oriented microbial stimulation, but considerations of use need to be properly addressed. The EFSA Panel on Additives and Products or Substances used in Animal Feed (FEEDAP) Guidance on the characterization of microorganisms used as feed additives or as production organisms (FEEDAP 2018) should indeed be followed, but they are only providing a part of the work that must be done to ensure safe use of microbial food cultures, and the guidance does not apply to undefined food cultures and spontaneous fermentation processes. There is a need to compile and harmonize the global knowledge of microbial food cultures, as the science of food fermentation and "positive" microorganisms has grown very enthusiastically in recent years, while somehow still adequately emphasizing essential points of precaution.

10.10 UNDERLYING FACTORS AND SPECIFIC HEALTH CONDITIONS OF CONCERN VS. OPPORTUNISTIC INFECTIONS

Among the one hundred thousand plus case reports found in the literature for the genera *Lactobacillus*, *Leuconostoc*, *Lactococcus*, *Weissella*, and *Bifidobacteria*, circa 50–80 of them were directly correlated with the consumption of probiotic food products (D'Agostin et al. 2021; Topcuoglu et al. 2015), where in some cases, the clinical team found clear evidence through Whole Genome Sequencing (WGS) connecting the probiotic strain of the food product and the clinical isolate (Pararajasingam and Uwagwu 2017). In several reports, the clinical isolate had not been compared to the probiotic strain consumed, and a definitive conclusion could therefore not be done to consider the causative agent being the food product (Tommasi et al. 2008). With the implementation of molecular typing, the major breakthrough of Next-Generation Sequencing (NGS) since 2012, and awareness of the health care practitioners, one might expect epidemiology to become more effectively done and allow proper conclusions (Aroutcheva et al. 2016). In recent case reports, the authors usually have presented a review of the previously published reported cases, but none of the reviews have provided the same exhaustivity as the approach presently proposed to review several databases, global web search by general keywords, and investigations of the hits, searching for cross references. This would confirm the limits and pitfalls of presuming to be exhaustive when partially using only one source of information. On top of the difficulty involved when trying to gather the available yet scattered evidence, LAB infection cases are underreported by design by health care practitioners, because of inadequate screening techniques (aerobic incubation on a blood agar plate) that cannot be efficient for the isolation of the specific growth conditions of these microorganisms, most specifically *Bifidobacterium*. Nevertheless, some publications about endocarditis, the most severe condition of infection caused by LAB, see to provide an estimate of 0.02–0.4% of global endocarditis being caused by LAB. Considering the broad exposure of human populations to LAB, we could anticipate concluding there to be a significant and concerning pathogenic potential when underlying factors are often if not always identified, such as diabetes mellitus, liver cirrhosis, history of valves replacement, invasive procedures, poor dental hygiene, immune-compromised status among the most reported ones (Rossi et al. 2019). Nevertheless, the risk, despite being low, is far from being negligible, and specific considerations, education, and awareness by the clinical microbiologists as well as the producers are needed.

10.11 OPPORTUNISTIC INFECTIONS AND MICROBIAL FOOD CULTURES: GUIDELINES TO HEATH CARE PRACTITIONERS

Probiotic strains are expected to be introduced into the food chain for the generally healthy population. In the case of using a microbial strain for a specific targeted population for medical purposes, this is not the definition of a probiotic strain per se but a therapeutic agent, as mentioned in the 2002 FAO/WHO Expert report and presently in place, e.g. in the United States (Cordaillat-Simmons et al. 2020; Rouanet et al. 2020). In the late 2000s, Probiotics gained interest for their use in specific medical conditions, e.g. *Clostridioides difficile* infection following the initiative and publication from a UK health care practitioners' team (Hickson et al. 2007). Since then, the use (and marketing) of probiotics has boomed for this type of condition. Considering the specific population and its poor health conditions, the concomitant use of vancomycin as the first line treatment of *C. difficile* infection but facing natural resistance from most LAB, the benefit-risk assessment should not have been overlooked. Several inappropriate applications of probiotics were reported, mostly through case reports, which highlights the need to have a clear consideration of the use of microbial cultures before their use (Liu et al. 2020; Pace et al. 2020). Of all misuses, the PROPATRIA study in 2004–2008 was the most "advanced" in the lack of reflection upon the considerations of use (Bongaerts and Severijnen 2016), while it is recommended for such populations to consider a low microbial diet (Lund 2014).

By definition, there cannot be a clinical study of a probiotic strain in medical settings or specific at-risk populations. These are biotherapeutic agents and do not fall under the regulatory considerations of a food product but a drug. Probiotic strains, as defined by the FAO/WHO expert report, are designed for a generally healthy population and regulated as food products. On top of having to be produced in a different manufacturing setting, not under a Food Safety Management System (FSMS) but according to cGMP (Current Good Manufacturing Practices), it is the responsibility of the manufacturing producer to provide clear guidelines for use of the biotherapeutic agent, not to mention the antibiotic resistance profile of the microorganisms of interest (Castro-González et al. 2019)

It is indeed a common practice in a clinical setting to use vancomycin as a first line treatment of antibiotic prophylaxis against bacterial infection. While this does make sense against *C. difficile*, it is counterproductive for LAB which are most commonly resistant to this antibiotic, therefore leading to a gut disbalance, overgrowth of LAB in this ecological niche, translocation through the gut barrier into the vena portae and liver infection. This is why liver cirrhosis, diabetes mellitus, and organ failure are reported as associated with opportunistic infection from LAB, which is of particular risk when considering the concomitant use of antibiotics along with probiotic consumption. When considering probiotic prophylaxis, and promoting medical purposes, it is the responsibility of the marketing company to inform, based on the available evidence, about the risk of opportunistic infection, the need for correct identification, and the anticipated intravenous ampicillin treatment in case of lactobacillemia, as empirically done and reported in the case reports.

A better screening methodology would also be needed, to identify viable but noncultivable cells, and avoid the specific growth conditions on agar media and the bias of culture-based techniques. As such, screening techniques based on molecular methodologies seem more relevant and efficient to clearly identify, in the onset of fever and sepsis, the causative agent(s) (Peri et al. 2021).

10.12 FOOD FERMENTATION AS END USE OF FOOD AND FEED CULTURES

The one health approach is particularly under scrutiny during the COVID-19 pandemic. One of the major concerns linked to the use of microbial organisms is the issue of antibiotic resistance and the potential for the transfer of genes. In the mid-2000s, the ACE-ART European project proposed MIC values for the most commonly used LAB. These requirements remain still today a major topic of surveillance, either for food cultures (Zarzecka et al. 2020) or probiotics (Daniali et al. 2020).

The present ACE-ART cut-offs are still used and required by the EFSA FEEDAP Panel.

In the FAO/WHO expert report, 2001 and 2002, in the IDF initiative, or the EFSA BIOHAZARD QPS evaluation, the end use is necessarily considered when it comes to assessing the introduction of a microbial food culture in the food chain and, ultimately, the human host. Food fermentation and the mechanism of action in the fermentation process have been driven mostly by a "black box" approach, where the evidence was more than limited when available. Sir John Lister in the late nineteenth century established the role of *Lactococcus lactis* (At that time, *Bacterium lactis*) in the lactic acid fermentation of milk, although his objective was not to understand food fermentation (Santer 2010) but to prove a point vs. the infectivity of bacteria in humans.

The major hurdle remains a lack of consensus on the approach to studying food fermentation. Some 20 years ago, the Food and Agriculture Organisation initiated work on the global diversity of fermented foods (Battcock and Azam-Ali 1998; Haard et al. 1999; Deshpande et al. 2000), while the World Health Organisation focused on the potential role of microbial food cultures in biopreservation to address among other control measures safe food production (Motarjemi et al. 1996). The IDF, in collaboration with the European Food and Feed Cultures Associations took the lead on a similar initiative for dairy food products and lastly enlarged their perspective to include all types of food matrices (Mogensen et al. 2002a,b; Bourdichon et al. 2012a,b,c,d, 2018). The International Scientific Association on Probiotics and Prebiotics also investigated this topic (Marco et al. 2021).

Despite all these numerous publications, expert reports, and global symposia, there is still a lack of alignment on defining what is understood and should in fine detail be regulated and supervised as a food fermentation and its role in the food chain. This topic is far from being solely a linguistic issue, there are a lot of implications in the classification of fermented foods and microbial food cultures. Presently in Europe, there is a status quo on the classification, where it is not obvious to know if a food culture is an additive or an ingredient and whether or not a food product is a novel food product.

Precise fermentation is one of the new concepts recently proposed for the production of a food component by a microorganism. The organisms most generally are genetically engineered and the marketing is not very scientific. This is a new application of industrial-engineered fermentation, not of the type applied to medical and pharma ingredients. A question that we need to ask is how should this be assessed under the EU 178/2002 regulation, and whether or not it should be classified as a GMO component.

Cross-over fermentation has recently been conceptualized by the Dutch scientists of Wageningen University (Dank et al. 2021), this approach has also been promoted in other sectors such as the wine industry (Capozzi et al. 2021). But what about the history of use when changing the usual food matrix? Should this still be considered as a traditional use or be considered a novel food? For the time being, the consumer trend and lack of concern have provided a marketing opportunity to skip the regulatory assessment, until an issue arises. Alternative protein sources developed by new fermentation processes mean new epitopes, hence new allergies. The use of pea protein for "alternative cheeses," is only one of many examples that might look promising, but remains to be properly addressed. The lack of consideration could endanger this new area of food production and should be considered by the industrial players before regulatory authorities or consumer associations ask for better supervision.

10.13 CONCLUSION

Despite their variety and wide consumption around the world with full integration in the global heritage of the different cultures, indigenous fermented foods remain still today unelucidated in their manufacturing process and internal mode of action.

After more than 10000 years of daily consumption and despite the considerable advances in science, the approach used for food fermentation remains very empirical. With the breakthroughs in microbiology achieved during the last 150 years, one would have expected the interactions between microbes and food to be largely understood and easily reproducible. It is far from being the case today, and the mystery is about to last for a while.

Fermented food products represent nevertheless a huge opportunity to ensure global food security. Several international organizations have been participatory in this approach during the last 20 years: Food and Agriculture Organisation, "Office International de l'Elevage," World Health Organisation, IDF, and the academic world. They are also major business opportunities, with major worldwide companies have built their business on the science of "good bacteria" (but also fungi and yeasts).

The major barrier to trade on fermented foods and microbial food cultures remains today is the lack of harmonization in regulation, not to say bluntly the lack of regulation. Regulation is expected to be science-based, and the science of food microbiology as such has not reached full maturity yet. There is a need to widen the scope, get away from the Western-oriented vision, and have an exhaustive approach to fermentation. The recent trend from the Western world on plant-based fermentation should not obliterate our successful reliance upon those plants, among which are cereals, vegetables, seeds, and legumes which have been the most used matrix for food fermentation from the beginning. Dairy is "just" a part of global diversity.

While scientists explore or rediscover the use of microbial food cultures, the demonstration of safety should be fully integrated into new product development. Lack of clear regulation and therefore lack of clear supervision still does not allow food business operators to do an integrated and adaptive risk analysis and fully understand the opportunities and limits of microbial food cultures in the food chain. Standardizations of processes and methods in the coming years are expected to provide a rationalized framework for continuing to produce safe food for everyone.

REFERENCES

Aroutcheva, A., Auclair, J., Frappier, M. et al. (2016). Importance of molecular methods to determine whether a probiotic is the source of *Lactobacillus* bacteremia. *Probiotics and Antimicrobial Proteins* 1: 31–40. https://doi.org/10.1007/s12602-016-9209-y.

Battcock, M. and Azam-Ali, D. (1998). Fermented frutis and vegetables. A global perspective. FAO agricultural services bulletin No. 134. https://www.fao.org/3/x0560e/x0560e00.htm (accessed 19 April 2023).

Binda, S., Hill, C., Johansen, E. et al. (2020). Criteria to qualify microorganisms as "Probiotic" in foods and dietary supplements. *Frontiers in Microbiology* 11: 1662. https://doi.org/10.3389/fmicb.2020.01662.

Bongaerts, G.P. and Severijnen, R.S. (2016). A reassessment of the PROPATRIA study and its implications for probiotic therapy. *Nature Biotechnology* 34 (1): 55–63. https://doi.org/10.1038/nbt.3436. PMID: 26744983.

Borriello, S.P., Hammes, W.P., Holzapfel, W. et al. (2003). Safety of probiotics that contain lactobacilli or bifidobacteria. *Clinical Infectious Diseases* 36 (6): 775–780.

Bourdichon, F., Casaregola, S., Farrokh, C. et al. (2012a). Food fermentations: microorganisms with technological beneficial use. *International Journal of Food Microbiology* 154 (3): 87–97. https://www.sciencedirect.com/science/article/pii/S0168160511007586.

Bourdichon, F., Berger, B., and Casaregola, S. (2012b). A safety assessment of microbial food cultures with history of use in fermented dairy products. *Bulletin of IDF*, 455, 2–12. https://www.fil-idf.org/wp-content/uploads/2016/05/Presentation-food-fermentations-Francois-Bourdichon-05-09-2012-v2.pdf (accessed 19 April 2023).

Bourdichon, F. and Berger, B., Casaregola, S. et al. (2012c). Building an inventory of microbial food cultures with a technological role in fermented food products. Bulletin of IDF 455, 13–21. https://shop.fil-idf.org/collections/publications (accessed 19 April 2023).

Bourdichon, F. and Boyaval, P., Casaregola, S. (2012d). The 2012 Inventory of Microbial Species with Technological Beneficial Role in Fermented Food Products. Bulletin of IDF 455, 22–61. https://shop.fil-idf.org/collections/publications (accessed 19 April 2023).

Bourdichon, F., Alper, I., Bibiloni, R. et al. (2018). Inventory of microbial food cultures with safety demonstration in fermented food products. Update of the Bulletin of the IDF 455–2012. Bulletin of IDF 495 -2018. https://shop.fil-idf.org/collections/publications (accessed 19 April 2023).

Bourdichon, F., Laulund, S., and Tenning, P. (2019). Inventory of microbial species with a rationale: a comparison of the IDF/EFFCA inventory of microbial food cultures with the EFSA Biohazard Panel qualified presumption of safety. *FEMS Microbiology Letters* 366 (5): fnz048. https://doi.org/10.1093/femsle/fnz048. PMID: 30839062.

Bundesinstitut für Gesundheitlichen Verbraucherschutz und Veterinärmedizin (1999). Final report of the BgVV Working Group "Probiotic Microorganism Cultures in Food". Bundes Federal Riskbewertung Büro. http://www.bfr.bund.de/cm/349/probiotika_en.pdf (accessed 19 April 2023).

Capozzi, V., Fragasso, M., Romaniello, R. et al. (2017). Spontaneous food fermentations and potential risks for human health. *Fermentation* 3 (4): 49. https://doi.org/10.3390/fermentation3040049.

Capozzi, V., Fragasso, M., and Russo, P. (2020). Microbiological safety and the management of microbial resources in artisanal foods and beverages: The need for a transdisciplinary assessment to conciliate actual trends and risks avoidance. *Microorganisms* 8 (2): 306. https://doi.org/10.3390/microorganisms8020306. PMID: 32098373; PMCID: PMC7074853.

Capozzi, V., De Simone, N., Fragasso, M. et al. (2021). Wine microbial diversity and cross-over applications: emerging results and future perspectives. Poster presented at Macrowine virtual (June 23–30). *Infowine: Internet Journal of Viticulture and Enology*.

Castro-González, J.M., Castro, P., Sandoval, H. et al. (2019). Probiotic lactobacilli precautions. *Frontiers Microbiology* 12 (10): 375. https://doi.org/10.3389/fmicb.2019.00375. PMID: 30915041.

Cordaillat-Simmons, M., Rouanet, A., and Pot, B. (2020). Live biotherapeutic products: the importance of a defined regulatory framework. *Experimental and Molecular Medicine* 52: 1397–1406. https://doi.org/10.1038/s12276-020-0437-6.

D'Agostin, M., Squillaci, D., Lazzerini, M. et al. (2021). Invasive infections associated with the use of probiotics in children: a systematic review. *Children* 8 (10): 924. https://doi.org/10.3390/children8100924. PMID: 34682189; PMCID: PMC8534463.

Daniali, M., Nikfar, S., and Abdollahi, M. (2020). Antibiotic resistance propagation through probiotics. *Expert Opinion on Drug Metabolism & Toxicology* 12: 1207–1215. https://doi.org/10.1080/17425255.2020.1825682. Epub 2020 Sep 27. PMID: 32938241.

Dank, A., van Mastrigt, O., Yang, Z. et al. (2021). The cross-over fermentation concept and its application in a novel food product: the dairy miso case study. *LWT* 142: https://doi.org/10.1016/j.lwt.2021.111041.

Deshpande, S.S., Salunkhe, D.K., Oyewole, O.B. et al. (2000). Fermented grain legumes, seeds and nuts: a global perspective. *FAO Agricultural Services Bulletin*, No. 142. https://agris.fao.org/agris-search/search.do?recordID=XF2000397085 (accessed 19 April 2023).

Domig, K.J., Mayrhofer, S., Huys, G. et al. (2008). The ACE-ART database – antibiotic susceptibility of lactic acid bacteria and bifidobacteria in relation to their biological, geographical and chronological origin. *International Journal of Probiotics and Prebiotics* 3 (4): 281–286.

EFSA Panel on Biological Hazards (BIOHAZ), Ricci, A., Allende, A. et al. (2018). Statement on the update of the list of QPS-recommended biological agents intentionally added to food or feed as notified to EFSA 8: suitability of taxonomic units notified to EFSA until March 2018. *EFSA Journal* 16 (7): 5315. https://doi.org/10.2903/j.efsa.2018.5315.

Falkow, S. (1988). Molecular Koch's postulates applied to microbial pathogenicity. *Reviews of Infectious Diseases* (Suppl 2): S274–S276. https://doi.org/10.1093/cid/10.supplement_2.s274. PMID: 3055197.

FAO (2007). Food safety risk analysis: A guide for national food safety authorities. https://www.fao.org/3/a0822e/a0822e.pdf (accessed November 30, 2021).

FAO (2019). Microbiome: the missing link? https://www.fao.org/documents/card/fr/c/ca6767en (accessed on November 30, 2021).

FAO and WHO (2001). Report on Joint FAO/WHO expert consultation on evaluation of health and nutritional properties of probiotics in food including powder milk with live lactic acid bacteria. Cordoba, Argentina (1–4 October 2001). https://www.fao.org/3/y6398e/y6398e.pdf (accessed November 30, 2021).

FAO and WHO (2002). Guidelines for the evaluation of probiotics in Food. Joint FAO/WHO Workgroup on Drafting Guidelines for the Evaluation of Probiotics in Food. London, Ontario, Canada (April 30 and May 1, 2002). https://www.scribd.com/doc/110123033/Guidelines-for-the-Evaluation-of-Probiotics-in-Food# (accessed November 30, 2021).

FAO and WHO (2006). Health and nutritional properties and guidelines for evaluation – Report of a Joint FAO/WHO Expert Consultation on Evaluation of Health and Nutritional Properties of Probiotics in Food including Powder Milk with Live Lactic Acid Bacteria. https://www.fao.org/publications/card/fr/c/7c102d95-2fd5-5b22-8faf-f0b2e68dfbb6 (accessed November 30, 2021).

FEEDAP Panel, EFSA (2005). Opinion of the Scientific Panel on additives and products or substances used in animal feed (FEEDAP) on the updating of the criteria used in the assessment of bacteria for resistance to antibiotics of human or veterinary importance. *The EFSA Journal* 223: 1–12. https://www.efsa.europa.eu/en/efsajournal/pub/223.

FEEDAP Panel, EFSA (2018). Guidance on the characterisation of microorganisms used as feed additives or as production organisms. *EFSA Journal* 16 (3): 5206.

Gasser, F. (1994). Safety of lactic acid bacteria and their occurrence in human clinical infections [opportunistic bacteria, various infections]. *Bulletin de l'Institut Pasteur (France)* 92 (1): 45–67.

Haard, N.F., Odunfa S.A., and Lee, C.H. (1999). Fermented cereals. A global perspective. FAO Agricultural Services Bulletin No. 138. Rome. https://www.fao.org/3/x2184e/x2184e00.htm (accessed 19 April 2023).

Health Canada (2003). *Amendment (Schedule No. 948) to Division 28 of the Food and Drug Regulations, Sections B.28.001–003.* Food and Drugs Regulations.

Herman, L., Chemaly, M., Cocconcelli, P.S. et al. (2018). The qualified presumption of safety assessment and its role in EFSA risk evaluations: 15 years past. *FEMS Microbiology Letters* 366 (1): fny260. https://academic.oup.com/femsle/article/366/1/fny260/5237703.

Hickson, M., D'Souza, A.L., Muthu, N. et al. (2007). Use of probiotic *Lactobacillus* preparation to prevent diarrhoea associated with antibiotics: randomised double blind placebo controlled trial. *BMJ* 335 (7610): 80. https://doi.org/10.1136/bmj.39231.599815.55. Epub 2007 Jun 29. PMID: 17604300.

Hill, C., Guarner, F., Reid, G. et al. (2014). Expert consensus document. The International Scientific Association for Probiotics and Prebiotics consensus statement on the scope and appropriate use of the term probiotic. *Nature Reviews Gastroenterology and Hepatology* 11 (8): 506–514. https://doi.org/10.1038/nrgastro.2014.66.

Laulund, S., Wind, A., Derkx, P.M.F. et al. (2017). Regulatory and safety requirements for food cultures. *Microorganisms* 5 (2): 28. https://doi.org/10.3390/microorganisms5020028.

Leuschner, R.-G.K., Robinson, T., Hugas, M. et al. (2010). Qualified presumption of safety (QPS): a generic risk assessment approach for biological agents notified to the European Food Safety Authority (EFSA). *Trends in Food Science and Technology* 21: 425–435.

Liu, D., Zeng, L., Yan, Z. et al. (2020). The mechanisms and safety of probiotics against toxigenic. *Clostridium difficile. Expert Review of Anti-Infective Therapy* (10): 967–975. https://doi.org/10.1080/14787210.2020.1778464. Epub 2020 Jun 16. PMID: 32520637.

Lund, B.M. (2014). Microbiological food safety and a low-microbial diet to protect vulnerable people. *Foodborne Pathogen Diseases* 11 (6): 413–424. https://doi.org/10.1089/fpd.2013.1679. Epub 2014 Mar 12. PMID: 24621274.

Marco, M.L., Sanders, M.E., Gänzle, M. et al. (2021). The International Scientific Association for Probiotics and Prebiotics (ISAPP) consensus statement on fermented foods. *Nature Reviews Gastroenterology & Hepatology* 18 (3): 196–208. https://doi.org/10.1038/s41575-020-00390-5.

Marteau, P. (2001). Safety aspects of probiotic products. *Näringsforskning* 45 (1): 22–24. https://doi.org/10.3402/fnr.v45i0.1785.

Merten, C., Schoonjans, R., Di Gioia, D. et al. (2020). Editorial: Exploring the need to include microbiomes into EFSA's scientific assessments. *EFSA Journal* 18 (6): e18061. https://doi.org/10.2903/j.efsa.2020.e18061.

Mofredj, A., Bahloul, H., and Chanut, C. (2007). *Lactococcus lactis*: un pathogène opportuniste? [*Lactococcus lactis*: an opportunistic bacterium?]. *Médecine et Maladies Infectieuses* 37 (4): 200–207. French. https://doi.org/10.1016/j.medmal.2007.01.005. Epub 2007 Mar 1. PMID: 17336016.

Mogensen, G., Salminen, S., O'Brien, J. et al. (2002a). Food microorganisms – health benefits, safety evaluation and strains with documented history of use in foods. *Bulletin of International Dairy Foundation* 377: 4–9.

Mogensen, G., Salminen, S., O'Brien, J. et al. (2002b). Inventory of microorganisms with a documented history of use in food. *Bulletin of International Dairy Foundation* 377: 10–19.

Motarjemi, Y. and Nout, M.J.R. (1996). Food fermentation: a safety and nutritional assessment. Joint FAO/WHO Workshop on Assessment of Fermentation as a Household Technology for Improving Food Safety. *Bulletin of the World Health Organization* 74 (6): 553–559.

Pace, F., Macchini, F., and Massimo Castagna, V. (2020). Safety of probiotics in humans: a dark side revealed? *Digestive and Liver Disease* 52 (9): 981–985. https://doi.org/10.1016/j.dld.2020.04.029.

Pararajasingam, A. and Uwagwu, J. (2017). *Lactobacillus*: the not so friendly bacteria. *BMJ Case Reports* https://doi.org/10.1136/bcr-2016-218423.

Peri, A.M., Harris, P.N.A., and Paterson, D.L. (2021). Culture-independent detection systems for bloodstream infection. *Clinical Microbiology and Infection* https://doi.org/10.1016/j.cmi.2021.09.039.

Rossi, F., Amadoro, C., and Colavita, G. (2019). Members of the *Lactobacillus* Genus Complex (LGC) as opportunistic pathogens: a review. *Microorganisms* 7 (5): 126. https://doi.org/10.3390/microorganisms7050126.

Rouanet, A., Bolca, S., Bru, A. et al. (2020). Live biotherapeutic products, a road map for safety assessment. *Frontiers in Medicine* 7: 237. https://doi.org/10.3389/fmed.2020.00237. PMID: 32637416; PMCID: PMC7319051.

Santer, M. (2010). Joseph Lister: first use of a bacterium as a 'model organism' to illustrate the cause of infectious disease of humans. *Notes and Records of the Royal Society Journal of the History of Science* 64 (1): 59–65. https://doi.org/10.1098/rsnr.2009.0029.

Steinkraus, K.H. (2002). Fermentations in world food processing. *Comprehensive Reviews in Food Science and Food Safety* 1: 23–32.

Tommasi, C., Equitani, F., Masala, M. et al. (2008). Diagnostic difficulties of *Lactobacillus casei* bacteraemia in immunocompetent patients: a case report. *Journal of Medical Case Reports* 2: 315. https://doi.org/10.1186/1752-1947-2-315. PMID: 18826603.

Topcuoglu, S., Gursoy, T., Ovalı, F. et al. (2015). A new risk factor for neonatal vancomycin-resistant Enterococcus colonisation: bacterial probiotics. *The Journal of Maternal-Fetal & Neonatal Medicine* 12: 1491–1494. https://doi.org/10.3109/14767058.2014.958462.

Wastyk, H.C., Fragiadakis, G.K., Perelman, D. et al. (2021). Gut-microbiota-targeted diets modulate human immune status. *Cell* 184 (16): 4137–4153.e14. https://doi.org/10.1016/j.cell.2021.06.019.

Yao, S., Wang, P., Bai, F. et al. (2021). Research on the inventory of microbial food cultures in Chinese traditional fermented foods (2nd edition). *Food and Fermentation Industries* 48 (01): 1–30. https://doi.org/10.13995/j.cnki.11-1802/ts.029332.

Zarzecka, U., Zadernowska, A., and Chajęcka-Wierzchowska, W. (2020). Starter cultures as a reservoir of antibiotic resistant microorganisms. *LWT Food Science and Technology* 127: 109424.

Zheng, J., Wittouck, S., Salvetti, E. et al. (2020). A taxonomic note on the genus *Lactobacillus*: description of 23 novel genera, emended description of the genus *Lactobacillus* Beijerinck 1901, and union of *Lactobacillaceae* and *Leuconostocaceae*. *International Journal of Systematic and Evolutionary Microbiology* 70 (4): 2782–2858. https://doi.org/10.1099/ijsem.0.004107. Epub 2020 Apr 15. PMID: 32293557.

CHAPTER 11

STARTER CULTURES AND THEIR ROLE IN FERMENTED FOODS

MARTA LARANJO
MED – Mediterranean Institute for Agriculture, Environment and Development & CHANGE-Global Change and Sustainability Institute, IIFA – Institute for Advanced Studies and Research, Universidade de Évora, Évora, Portugal

CONTENTS

11.1 INTRODUCTION

Starters are pure or mixed cultures of microorganisms that are used to boost and conduct the fermentation of foods (Laranjo et al. 2017; Laranjo et al. 2019b).

Fermented foods (FF) are defined as "foods and beverages made through desired microbial growth and enzymatic conversions of food components" (Marco et al. 2021). These include sausages, yogurts and cheeses, table olives, or wine and beer, just to cite some of the most common examples.

Fermentation has long been used not only as a food preservation methodology but also for enhancing food quality attributes. Fermented foods have been part of the human diet since ancient times and account for nearly a third of the world's food consumption (Xiang et al. 2019).

Microbial Fermentations in Nature and as Designed Processes, First Edition. Edited by Christon J. Hurst.

But what exactly is fermentation? In biochemical terms, fermentation is the metabolic pathway by which energy (ATP) is extracted from organic compounds (carbohydrates) through the action of enzymes in the absence of oxygen. Nevertheless, fermentation requires metabolically active microorganisms.

According to David Rains Wallace from "The Fermentation Association," "Fermentation may have been a greater discovery than fire." (https://fermentationassociation.org/david-rains-wallace-fermentation-may-have-been-a-greater-discovery-than-fire)

11.2 FERMENTATION

Most sugars are fermentable. Different fermentation processes can be classified either by the substrate of fermentation or by its products.

Based on the end-product formed, fermentation can be classified into main four types, namely lactic acid fermentation, alcohol fermentation, acetic acid fermentation, and butyric acid fermentation.

There are two kinds of lactic acid fermentation, namely homolactic and heterolactic fermentation. In homolactic fermentation, one molecule of glucose is converted to two molecules of lactic acid in the end. Heterolactic fermentation, in contrast, yields carbon dioxide and ethanol in addition to lactic acid. Streptococci as well as some lactobacilli can perform homolactic fermentation, while leuconostocs and other lactobacilli undergo heterolactic fermentation.

Alcoholic fermentation or ethanol fermentation is a biological anaerobic process by which sugar is converted into alcohol and carbon dioxide. Byproducts of this fermentation process include heat, carbon dioxide, water, and alcohol. Alcoholic fermentation is mostly performed by yeasts and *Zymomonas*.

In acetic acid fermentation, the first step is anaerobic fermentation (alcoholic fermentation of sugars into ethanol by yeasts), while the second step is aerobic fermentation (oxidation of ethanol into acetic acid by acetic acid bacteria). *Clostridium aceticum, Moorella thermoacetica, Acetobacter* spp., and *Komagataeibacter* spp., among others, are usually responsible for this type of fermentation.

The general principle of butyric acid fermentation includes glycolysis in which bacteria use sugar and oxidize it to pyruvate, which is then oxidized to acetyl-CoA. One part of Acetyl-CoA later forms acetic acid, and the rest generates acetoacetyl-CoA, which is reduced to butyryl CoA, which is then converted to butyric acid. This type of fermentation is performed by *Clostridium butyricum*.

Other less common food fermentation processes include mixed acid fermentation and propionic fermentation.

Mixed acid fermentation is carried out by enteric bacteria, namely *Escherichia*, *Salmonella*, *Shigella*, *Klebsiella*, and *Enterobacter*. Its end-products are ethanol, butanediol, succinic acid, lactic acid, acetic acid, and formic acid, although not all of them are produced by all microorganisms.

Propionic fermentation is performed by *Propionibacterium* and *Clostridium propionicum* and yields both propionic and acetic acid.

According to the major microbial metabolites, industrial-fermented foods may be classified into five distinct classes* (Table 11.1):

Fermentation processes have several positive or desirable effects on fermented foods, which are here shortly described. Fermentation plays an important role in biopreservation because it improves the shelf-life of foods. Moreover, it develops distinctive sensory properties, namely flavor, aroma, and texture. Additionally, fermentation also eliminates naturally occurring toxins

TABLE 11.1 Classification of fermented foods based on their main fermentation products.

Main Fermentation Products	Fermented Foods
Lactic acid	Yogurt Cheese Pickles Meat
Ethanol + CO_2	Beer Wine Distilled spirits Bread
Acetic acid	Vinegar
Esters	Coffee Cocoa Tea
Ammonia	Shoyu Tempeh Miso Natto

Source: Adapted from Vinicius De Melo Pereira et al. (2020).

and antinutritional components that are present in raw materials. Furthermore, fermentation has the ability to recycle food byproducts and enrich fermented foods with desirable microbial metabolites, such as L-lactic acid or amino acids. Finally, fermentation also delivers beneficial microbes to fermented foods.

For all the above-mentioned reasons, fermentation enhances the nutritional and functional attributes of fermented foods.

11.3 FERMENTED FOODS

Thousands of different types of fermented foods are eaten around the world, with fermented foods and beverages comprising between 5% and 40% of the human diet in some populations (Borresen et al. 2012).

In our review, we will consider 12 main groups or categories of fermented foods (Table 11.2).

Numerous works have been published on fermented meat or dairy products, while limited research works have been published on other groups of fermented foods.

For example, in the fermented fish category, there are only three types of fermented fish known in Europe: surströmming (lightly salted fermented Baltic Sea herring), rakfisk (Norwegian fermented freshwater fish), and hákarl (Icelandic fermented shark). However, other fermented fish products are consumed mostly in the Orient including: fesikh (Egyptian fermented fish); kusaya (Japanese fermented then dried fish); pla ra (Fermented fish and rice flour seasoning common throughout Southeast Asia); and shiokara (Japanese seafood fermented in highly salted viscera).

Plant-based fermented foods, such as kimchi or sauerkraut, are good alternative protein sources to shift from a diet mainly based on animal proteins, as has been largely recommended to reduce environmental impact, but also reduce the emergence of antimicrobial resistance and prevent the further loss of biodiversity (Sun et al. 2022).

TABLE 11.2 Categories of fermented foods (FF): examples and recent literature.

Category of FF	Examples of FF	Recent References
Meat	Chorizo, ham, linguiça, paio, catalão, salsichón, salsichão	Dias et al. (2020), Franciosa et al. (2022), Lorenzo et al. (2017)
Fish	Surströmming, rakfisk, hákarl, fesikh, kusaya, pla ra, shiokara	Riebroy et al. (2008), Saithong et al. (2010), Skåra et al. (2015)
Dairy	Yogurt, cheese, kefir, ayran	Bintsis and Papademas (2022), Ferroukhi et al. (2023)
Plant-based	Table olives, kimchi, pickles, sauerkraut	Papadaki and Mantzouridou (2016), Peñas et al. (2010)
Legumes and pulses	Tempeh, miso, shoyu, natto, tofu	Saeed et al. (2022)
Cereal based	Bread, sake, tarhana, seitan	Blandino et al. (2003)
Vinegar	Distilled white vinegar, apple cider vinegar, balsamic vinegar, white wine vinegar, red wine vinegar, rice vinegar, malt vinegar, red rice vinegar, champagne vinegar, sherry vinegar, black vinegar, cane vinegar, beer vinegar	Budak et al. (2014), Chinnici et al. (2016)
Coffee		Vinicius De Melo Pereira et al. (2020)
Tea		Wang et al. (2022)
Cocoa		Díaz-Muñoz et al. (2023)
Non-alcoholic beverages (ethanol level below 0.5% [v/v]) (except dairy)	Kombucha, ginger beer, water kefir	Júnior et al. (2022), Lynch et al. (2021), Moretti et al. (2022), Yang et al. (2022)
Alcoholic beverages	Wine, beer, cider, drinking spirits	Aponte et al. (2022), Chaïb et al. (2022), Coton et al. (2010), Han and Du (2023)

The legumes and pulses FF group includes several soy-based food products, such as miso, shoyu, natto, or tempeh (Jang et al. 2021).

According to the European Commission, "geographical indications establish intellectual property rights for specific foods, whose qualities are specifically linked to the area of production" (European Commission n.d.). There are three types of geographical indications: *Protected Designation of Origin* (PDO), *Protected Geographical Indication* (PGI), and *Traditional Speciality Guaranteed (TSG)* (Dias and Mendes 2018). TSG describes the composition of a given food product highlighting the way a product is manufactured. A TSG registration protects foods against forgery and misuse.

eAmbrosia is the legal European database that contains all registered and protected agricultural products and foodstuffs, wine, and spirit drinks across the European Union (https://ec.europa.eu/info/food-farming-fisheries/food-safety-and-quality/certification/quality-labels/geographical-indications-register). It further provides product specifications.

TABLE 11.3 Main microbial groups of starter cultures.

Microbial Group	Species	References
Lactic acid bacteria (LAB)	*Lactobacillus* spp. *Lactococcus* spp. *Enterococcus* spp. *Pediococcus* spp. *Oenococcus* spp.	Laranjo et al. (2017), Laranjo et al. (2019b)
Gram-positive catalase-positive cocci (GCC+)	*Staphylococcus* spp. *Micrococcus* spp. *Kocuria* spp.	Laranjo et al. (2017), Laranjo et al. (2019b)
Yeasts	*Debaromyces* spp.	Laranjo et al. (2017), Laranjo et al. (2019b)
Molds	*Penicillium*	Laranjo et al. (2017), Laranjo et al. (2019b)

11.4 STARTERS

As already introduced at the beginning of this chapter, "starters are pure or mixed cultures of microorganisms that are used to boost and conduct the fermentation of foods" (Laranjo et al. 2017; Laranjo et al. 2019b).

Distinct "types" of starters can be considered, depending on both their origin and the associated processing technology: autochthonous starters, allochthonous, commercial starters, and back-slopping (Mota-Gutierrez and Cocolin 2021).

Autochthonous starters are microbial cultures isolated from and re-used on the same raw matrix, apart from the geographical origin (Di Cagno et al. 2013). However, some authors defend that true autochthonous starter cultures should be used also within the same origin (e.g. same manufacturing unit) (Chessa et al. 2021; Tangwatcharin et al. 2019). This ensures the establishment of these inoculated autochthonous starter cultures throughout the ripening process. Otherwise, starter microorganisms may not survive due to competitive exclusion from the native microbiota (Eisenbach et al. 2019; Li et al. 2023).

Commercial or allochthonous starters are used to ferment many diverse food products, independent of the raw matrices they were isolated from Di Cagno et al. (2013).

Back-slopping can be defined as the inoculation of a particular food with a small amount of a previously successfully fermented final product (Shrivastava and Ananthanarayan 2015).

The more traditional or artisanal fermentation process usually relies on spontaneous fermentation by the autochthonous microbiota. However, an also common traditional technology is the use of back-slopping.

According to their taxonomic position, starter cultures are traditionally divided into four main groups (Table 11.3): lactic acid bacteria (LAB), Gram-positive catalase-positive cocci (GCC+), yeasts, and molds.

11.5 THE ROLE OF STARTERS IN FOOD SAFETY

The use of selected protective starters, mainly LAB, may be an important practice and is used to control microbiological hazards or inhibit potential foodborne pathogens (Fraqueza et al. 2021).

Within microbiological hazards, bacteria are the most relevant ones. Foodborne pathogens include *Escherichia coli*, *Salmonella* spp., *Staphylococcus aureus*, *Listeria monocytogenes*, *Campylobacter* spp., *Yersinia enterocolitica*, *Clostridium botulinum*, or *Clostridium perfringens* (Fraqueza et al. 2021)

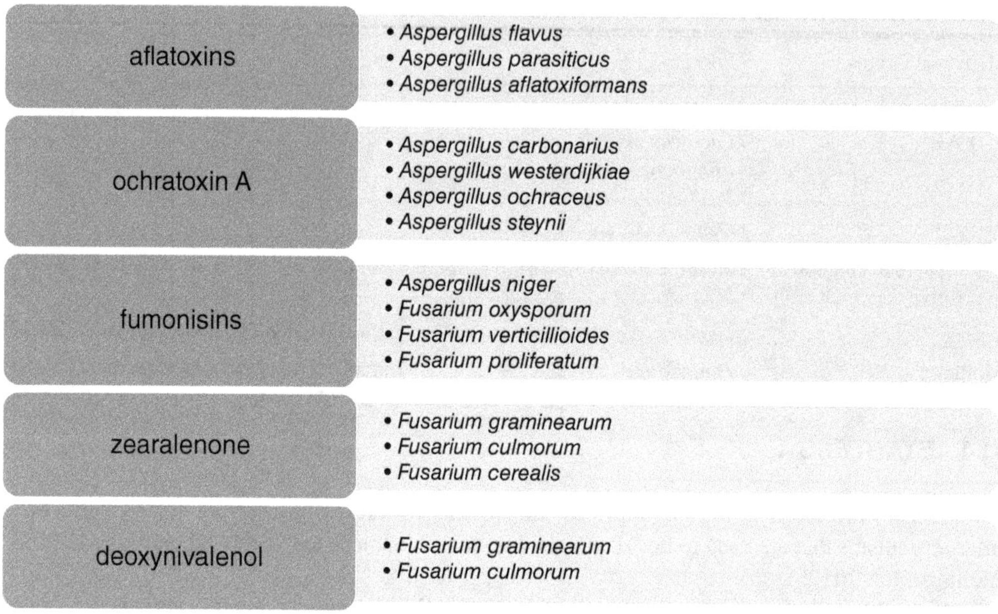

FIGURE 11.1 Major mycotoxins and fungal species that produce them.

Regarding fungi, most species are not pathogenic, but different *Aspergillus* and *Fusarium* species produce mycotoxins (Figure 11.1).

Nevertheless, some fungi are further responsible for food spoilage, such as *Botrytis cinerea*, *Aspergillus niger*, *Aspergillus fumigatus*, *Aspergillus terreus*, *Penicillium* spp., *Cladosporium* spp., or *Alternaria* spp., causing enormous losses in fresh fruits or vegetables (Cheung et al. 2020; Laranjo et al. 2019a).

Hepatitis A and E virus, norovirus, rotavirus, and astrovirus are some of the most common foodborne viruses (Bosch et al. 2016).

Other biological hazards include the nematode species *Trichinella* spp., *Taenia solium*, *Taenia saginata*, and *Toxoplasma gondii* (Todd 2014; Zolfaghari Emameh et al. 2018).

The spontaneous fermentation of raw materials requires knowledge of the phenomena that control the process to avoid contamination by spoiling and pathogenic microorganisms (Adams and Mitchell 2002).

L. monocytogenes and *E. coli* can adapt to the acid and saline environment of fermented food (Mataragas et al. 2015; McLeod et al. 2016).

Several types of chemical hazards may affect fermented foods, but biogenic amines are the product of microbial metabolism and their concentration can thus be decreased using starter cultures (Fraqueza et al. 2021).

Biogenic amines (BAs) are small nitrogen compounds formed from amino acids by decarboxylation, or by amination and transamination of aldehydes and ketones (Elias et al. 2018). The BA are responsible for nausea, headache, or anaphylactic shock in sensitive individuals (Elias et al. 2018), but constitute a potential chemical hazard, mainly because they are precursors of carcinogenic nitrosamines. Nevertheless, the presence of BA in FF is considered normal because some "gastronomic" LAB, namely some enterococci, lactobacilli, and pediococci, involved in the fermentation process, can decarboxylate amino acids and can therefore produce biogenic amines.

11.6 THE ROLE OF STARTERS IN FOOD SENSORY PROPERTIES

The role of starters in the safety of fermented foods has been largely addressed. However, starters are also used in the manufacturing of fermented foods due to their ability to improve or enhance the sensory or organoleptic characteristics of these food products.

For example, in meat products starter staphylococci enhance and fix the red cured color of sausages (Gøtterup et al. 2008; Škaljac et al. 2022), while yeasts are responsible for the release of volatile compounds that improve the aroma and flavor, for which these food products are highly appreciated (Pastink et al. 2008). Moreover, LAB are known for improving the texture, color, and flavor of sausages (Hu et al. 2022).

In cheese, LAB are responsible for flavor development (Pastink et al. 2008). Moreover, the use of starter cultures in the manufacturing of goat cheese has been shown to have a pronounced effect on texture and volatile components (Jia et al. 2021).

The use of different yeast strains as starter cultures in the fermentation of cocoa can induce the development of aroma and flavor profiles (Toker et al. 2020).

11.7 STARTERS PROMOTE INNOVATION IN FERMENTED FOODS

The food industry is always looking out for the latest developments to introduce innovation into its products. This ensures that they reach new markets and new consumers every day.

However, the first innovation about the use of introduced starters in fermented foods is the standardization of food products. The use of starters of a defined microbial composition and in known concentrations allows the fermentation to be conducted always in a similar, controlled process, that guarantees final products of known, constant quality. This is very important and mandatory to build consumer loyalty.

While some food industry owners see innovation as an enemy of tradition, others embrace it due to its power for standardizing and enhancing food quality and safety, expanding shelf-life, and improving environmental sustainability. According to West (2020), "Consumers often contrast authentic and traditional with modern and industrial …," and this does not have to necessarily be so.

The COST Action PIMENTO (Promoting Innovation of ferMENTed fOods) has recently established a network of scientists and food industry specialists that aim to place Europe at the spearhead of innovation on microbial foods, promoting health, regional diversity, and local production (https://fermentedfoods.eu).

11.8 THE HEALTH BENEFITS OF FERMENTED FOODS

Fermented foods play a central role in our modern food systems and can make significant contributions toward sustainability (Lortal et al. 2020). The place of fermented foods in the diet needs to be clarified considering their health benefits and their impact on gut microbiota. In recent years, the intake of fermented foods has been assessed in several studies in an attempt to link their consumption with human health (Li et al. 2021).

Although numerous studies have focused on and reviewed the health benefits of fermented foods (Castellone et al. 2021; Diez-Ozaeta and Astiazaran 2022; Leeuwendaal et al. 2022; Rul et al. 2022), particularly of dairy products, yogurt still currently remains the only FF for which an approved health claim exists (EFSA 2010).

Although fermented foods are recognized as the primary source of LAB, these microorganisms are present not only in foods but also in the human and animal gut and the environment, being a true example of the tripartite concept of One Health (Bell et al. 2018). Nevertheless, the link between LAB-fermented foods and the human gut from the perspective of health promotion still needs to be further studied and better elucidated (De Filippis et al. 2020).

De Angelis et al. (2020) have recently highlighted how diet can influence the functions of the human gut microbiome. Another recent study has reported that a diet rich in fermented foods steadily increased microbiota diversity and decreased inflammatory markers (Wastyk et al. 2021).

Finally, starter cultures not only directly confer health benefits via ingestion of fermented foods, but starters can also act as carriers of live microorganisms with specific functional roles, namely probiotics, prebiotics, and postbiotics.

The original definition of probiotics dates back to 1974 (Parker 1974), however, the role of these microorganisms was first described by the Nobel prize winner Metchnikoff (1907). The current definition of probiotics has been written by a joint FAO/WHO Working Group as: "Live microorganisms which when administered in adequate amounts confer a health benefit on the host" (Reid 2016).

The new concept of prebiotics defined it as "a substrate that is selectively utilized by host microorganisms conferring a health benefit". This definition expanded the well-established concept of prebiotics to non-carbohydrates, and body parts other than the gut (Gibson et al. 2017).

Postbiotics are dead microorganisms or their components or metabolites that confer health benefits to the host (Vinderola et al. 2022).

A lot of room for innovation is available here. In a recent work, Di Cagno et al. (2020) highlighted the potential of yeasts designed for the manufacturing of a functional fruit beverage with an improved volatile profile and probiotic properties, which can modulate the gut microbiota.

11.9 CONCLUSIONS

Autochthonous or allochthonous starter cultures definitely play an important role in fermented foods contributing to their safety but also to their quality, extending their shelf-life and enhancing their aromas and flavors.

Moreover, traditional spontaneous fermentation processes eventually coupled with the use of autochthonous starter cultures contribute to a more sustainable diet. Fermented foods represent a sustainable production alternative, moreover if we consider plant-based protein alternatives to meat.

ACKNOWLEDGMENTS

This work was supported by Portuguese National Funds through FCT – Fundação para a Ciência e a Tecnologia under Project UIDB/05183/2020 (MED) and written within the framework of the PIMENTO COST ACTION CA20128 – Promoting Innovation of ferMENTed fOods.

REFERENCES

Adams, M. and Mitchell, R. (2002). Fermentation and pathogen control: a risk assessment approach. *International Journal of Food Microbiology* 79 (1): 75–83.

Aponte, M., Esposito, F., Sequino, G. et al. (2022). Stuck or sluggish fermentations in home-made beers: beyond the surface. *International Journal of Food Microbiology* 383: 109956.

Bell, V., Ferrão, J., Pimentel, L. et al. (2018). One health, fermented foods, and gut microbiota. *Food* 7 (12): 195.

Bintsis, T. and Papademas, P. (2022). The evolution of fermented milks, from artisanal to industrial products: a critical review. *Fermentation* 8 (12): 679.

Blandino, A., Al-Aseeri, M.E., Pandiella, S.S. et al. (2003). Cereal-based fermented foods and beverages. *Food Research International* 36 (6): 527–543.

Borresen, E.C., Henderson, A.J., Kumar, A. et al. (2012). Fermented foods: patented approaches and formulations for nutritional supplementation and health promotion. *Recent Patents on Food, Nutrition & Agriculture* 4 (2): 134–140.

Bosch, A., Pintó, R.M., and Guix, S. (2016). Foodborne viruses. *Current Opinion in Food Science* 8: 110–119.

Budak, N.H., Aykin, E., Seydim, A.C. et al. (2014). Functional properties of vinegar. *Journal of Food Science* 79 (5): R757–R764.

Castellone, V., Bancalari, E., Rubert, J. et al. (2021). Eating fermented: health benefits of LAB-fermented foods. *Foods* 10 (11): 2639.

Chaïb, A., Philippe, C., Jaomanjaka, F. et al. (2022). Phage–host interactions as a driver of population dynamics during wine fermentation: betting on underdogs. *International Journal of Food Microbiology* 383: 109936.

Chessa, L., Paba, A., Daga, E. et al. (2021). Autochthonous natural starter cultures: a chance to preserve biodiversity and quality of pecorino Romano PDO cheese. *Sustainability* 13 (15): 8214.

Cheung, N., Tian, L., Liu, X. et al. (2020). The destructive fungal pathogen *Botrytis cinerea* – insights from genes studied with mutant analysis. *Pathogens* 9 (11): 923.

Chinnici, F., Durán-Guerrero, E., and Riponi, C. (2016). Discrimination of some European vinegars with protected denomination of origin as a function of their amino acid and biogenic amine content. *Journal of the Science of Food and Agriculture* 96 (11): 3762–3771.

Coton, M., Romano, A., Spano, G. et al. (2010). Occurrence of biogenic amine-forming lactic acid bacteria in wine and cider. *Food Microbiology* 27 (8): 1078–1085.

De Angelis, M., Ferrocino, I., Calabrese, F.M. et al. (2020). Diet influences the functions of the human intestinal microbiome. *Scientific Reports* 10 (1): 4247.

De Filippis, F., Pasolli, E., and Ercolini, D. (2020). The food-gut axis: lactic acid bacteria and their link to food, the gut microbiome and human health. *FEMS Microbiology Reviews* 44 (4): 454–489.

De Melo, V., Pereira, G., De Carvalho Neto, D.P. et al. (2020). A review of selection criteria for starter culture development in the food fermentation industry. *Food Reviews International* 36 (2): 135–167.

Di Cagno, R., Coda, R., De Angelis, M. et al. (2013). Exploitation of vegetables and fruits through lactic acid fermentation. *Food Microbiology* 33 (1): 1–10.

Di Cagno, R., Filannino, P., Cantatore, V. et al. (2020). Design of potential probiotic yeast starters tailored for making a cornelian cherry (*Cornus mas* L.) functional beverage. *International Journal of Food Microbiology* 323: 108591.

Dias, C. and Mendes, L. (2018). Protected designation of origin (PDO), protected geographical indication (PGI) and traditional speciality guaranteed (TSG): a bibiliometric analysis. *Food Research International* 103: 492–508.

Dias, I., Laranjo, M., Potes, M.E. et al. (2020). Autochthonous starter cultures are able to reduce biogenic amines in a traditional Portuguese smoked fermented sausage. *Microorganisms* 8 (5): 686.

Díaz-Muñoz, C., Van de Voorde, D., Tuenter, E. et al. (2023). An in-depth multiphasic analysis of the chocolate production chain, from bean to bar, demonstrates the superiority of *Saccharomyces cerevisiae* over Hanseniaspora opuntiae as functional starter culture during cocoa fermentation. *Food Microbiology* 109: 104115.

Diez-Ozaeta, I. and Astiazaran, O.J. (2022). Fermented foods: an update on evidence-based health benefits and future perspectives. *Food Research International* 156: 111133.

EFSA (2010). Scientific opinion on the substantiation of health claims related to live yoghurt cultures and improved lactose digestion (ID 1143, 2976) pursuant to article 13(1) of regulation (EC) no 1924/2006. *EFSA Journal* 8 (10): 1763.

Eisenbach, L., Geissler, A.J., Ehrmann, M.A. et al. (2019). Comparative genomics of Lactobacillus sakei supports the development of starter strain combinations. *Microbiological Research* 221: 1–9.

Elias, M., Fraqueza, M.J., and Laranjo, M. (2018). Biogenic amines in food: presence and control measures. In: *Biogenic Amines (BA): Origins, Biological Importance and Human Health Implications* (ed. J. Stadnik), 129–176. New York: Nova Science Publishers, Inc.

European Commission. (n.d.) Geographical indications and quality schemes explained. https://agriculture.ec.europa.eu/farming/geographical-indications-and-quality-schemes/geographical-indications-and-quality-schemes-explained_en (accessed 19 Apirl 2023).

Ferroukhi, I., Bord, C., Lavigne, R. et al. (2023). Exploring alternative salting methods to reduce sodium content in blue-veined cheeses. *International Dairy Journal* 138: 105555.

Franciosa, I., Ferrocino, I., Corvaglia, M.R. et al. (2022). Autochthonous starter culture selection for Salame Piemonte PGI production. *Food Research International* 162: 112007.

Fraqueza, M.J., Laranjo, M., Elias, M. et al. (2021). Microbiological hazards associated with salt and nitrite reduction in cured meat products: control strategies based on antimicrobial effect of natural ingredients and protective microbiota. *Current Opinion in Food Science* 38: 32–39.

Gibson, G.R., Hutkins, R., Sanders, M.E. et al. (2017). Expert consensus document: the International Scientific Association for Probiotics and Prebiotics (ISAPP) consensus statement on the definition and scope of prebiotics. *Nature Reviews. Gastroenterology & Hepatology* 14 (8): 491–502.

Gøtterup, J., Olsen, K., Knøchel, S. et al. (2008). Colour formation in fermented sausages by meat-associated staphylococci with different nitrite- and nitrate-reductase activities. *Meat Science* 78 (4): 492–501.

Han, Y. and Du, J. (2023). A comparative study of the effect of bacteria and yeasts communities on inoculated and spontaneously fermented apple cider. *Food Microbiology* 111: 104195.

Hu, Y., Li, Y., Li, X.-a. et al. (2022). Application of lactic acid bacteria for improving the quality of reduced-salt dry fermented sausage: texture, color, and flavor profiles. *LWT* 154: 112723.

Jang, C.H., Oh, J., Lim, J.S. et al. (2021). Fermented soy products: beneficial potential in neurodegenerative diseases. *Food* 10 (3): 636.

Jia, R., Zhang, F., Song, Y. et al. (2021). Physicochemical and textural characteristics and volatile compounds of semihard goat cheese as affected by starter cultures. *Journal of Dairy Science* 104 (1): 270–280.

Júnior, J.C.d.S., Meireles Mafaldo, Í., de Lima Brito, I. et al. (2022). Kombucha: formulation, chemical composition, and therapeutic potentialities. *Current Research in Food Science* 5: 360–365.

Laranjo, M., Elias, M., and Fraqueza, M.J. (2017). The use of starter cultures in traditional meat products. *Journal of Food Quality* 2017: 1–18.

Laranjo, M., Fernández-León, A.M., Agulheiro-Santos, A.C. et al. (2019a). Essential oils of aromatic and medicinal plants play a role in food safety. *Journal of Food Processing and Preservation* 46 (2).

Laranjo, M., Potes, M.E., and Elias, M. (2019b). Role of starter cultures on the safety of fermented meat products. *Frontiers in Microbiology* 10 (853).

Leeuwendaal, N.K., Stanton, C., O'Toole, P.W. et al. (2022). Fermented foods, health and the gut microbiome. *Nutrients* 14 (853): 1527.

Li, K.J., Brouwer-Brolsma, E.M., Burton-Pimentel, K.J. et al. (2021). A systematic review to identify biomarkers of intake for fermented food products. *Genes & Nutrition* 16 (1): 5.

Li, X.-m., Deng, J.-y., Nie, W. et al. (2023). Study on the safety and excellent probiotic properties of fermentation strains isolated from traditional dry-cured duck for starter development. *Food Bioscience* 51: 102244.

Lorenzo, J.M., Munekata, P.E.S., and Domínguez, R. (2017). Role of autochthonous starter cultures in the reduction of biogenic amines in traditional meat products. *Current Opinion in Food Science* 14: 61–65.

Lortal, S., El Mecherfi, K.-E., Mariotti, F. et al. (2020). Aliments fermentés & bénéfices santé : un défi pour la recherche. *Cahiers de Nutrition et de Diététique* 55 (3): 136–148.

Lynch, K.M., Wilkinson, S., Daenen, L. et al. (2021). An update on water kefir: microbiology, composition and production. *International Journal of Food Microbiology* 345: 109128.

Marco, M.L., Sanders, M.E., Gänzle, M. et al. (2021). The International Scientific Association for Probiotics and Prebiotics (ISAPP) consensus statement on fermented foods. *Nature Reviews Gastroenterology & Hepatology* 18 (3): 196–208.

Mataragas, M., Bellio, A., Rovetto, F. et al. (2015). Risk-based control of food-borne pathogens listeria monocytogenes and *Salmonella enterica* in the Italian fermented sausages cacciatore and Felino. *Meat Science* 103: 39–45.

McLeod, A., Måge, I., Heir, E. et al. (2016). Effect of relevant environmental stresses on survival of enterohemorrhagic *Escherichia coli* in dry-fermented sausage. *International Journal of Food Microbiology* 229: 15–23.

Metchnikoff, I.I. (1907). *The Prolongation of Life: Optimistic Studies*. London: G.P. Putnam & Sons.

Moretti, A.F., Moure, M.C., Quiñoy, F. et al. (2022). Water kefir, a fermented beverage containing probiotic microorganisms: from ancient and artisanal manufacture to industrialized and regulated commercialization. *Future Foods* 5: 100123.

Mota-Gutierrez, J. and Cocolin, L. (2021). Current trends and applications of plant origin lactobacilli in the promotion of sustainable food systems. *Trends in Food Science & Technology* 114: 198–211.

Papadaki, E. and Mantzouridou, F.T. (2016). Current status and future challenges of table olive processing wastewater valorization. *Biochemical Engineering Journal* 112: 103–113.

Parker, R.B. (1974). Probiotics, the other half of the antibiotics story. *Animal Nutrition & Health* 29: 4–8.

Pastink, M.I., Sieuwerts, S., de Bok, F.A.M. et al. (2008). Genomics and high-throughput screening approaches for optimal flavour production in dairy fermentation. *International Dairy Journal* 18 (8): 781–789.

Peñas, E., Frias, J., Gomez, R. et al. (2010). High hydrostatic pressure can improve the microbial quality of sauerkraut during storage. *Food Control* 21 (4): 524–528.

Reid, G. (2016). Probiotics: definition, scope and mechanisms of action. *Best Practice & Research Clinical Gastroenterology* 30 (1): 17–25.

Riebroy, S., Benjakul, S., and Visessanguan, W. (2008). Properties and acceptability of Som-fug, a Thai fermented fish mince, inoculated with lactic acid bacteria starters. *LWT – Food Science and Technology* 41 (4): 569–580.

Rul, F., Béra-Maillet, C., Champomier-Vergès, M.-C. et al. (2022). Underlying evidence for the health benefits of fermented foods in humans. *Food & Function* 13 (9): 4804–4824.

Saeed, F., Afzaal, M., Shah, Y.A. et al. (2022). Miso: a traditional nutritious & health-endorsing fermented product. *Food Science & Nutrition* 10 (12): 4103–4111.

Saithong, P., Panthavee, W., Boonyaratanakornkit, M. et al. (2010). Use of a starter culture of lactic acid bacteria in plaa-som, a Thai fermented fish. *Journal of Bioscience and Bioengineering* 110 (5): 553–557.

Shrivastava, N. and Ananthanarayan, L. (2015). Use of the backslopping method for accelerated and nutritionally enriched idli fermentation. *Journal of the Science of Food and Agriculture 95* (10): 2081–2087.

Škaljac, S., Jokanović, M., Tomović, V. et al. (2022). Color characteristics and content of polycyclic aromatic hydrocarbons of traditional dry fermented sausages throughout processing in controlled conditions. *Polycyclic Aromatic Compounds* 1–11.

Skåra, T., Axelsson, L., Stefánsson, G. et al. (2015). Fermented and ripened fish products in the northern European countries. *Journal of Ethnic Foods* 2 (1): 18–24.

Sun, Z., Scherer, L., Zhang, Q. et al. (2022). Adoption of plant-based diets across Europe can improve food resilience against the Russia–Ukraine conflict. *Nature Food* 3 (11): 905–910.

Tangwatcharin, P., Nithisantawakhup, J., and Sorapukdee, S. (2019). Selection of indigenous starter culture for safety and its effect on reduction of biogenic amine content in Moo som. *Asian-Australasian Journal of Animal Sciences* 32 (10): 1580–1590.

Todd, E.C.D. (2014). Foodborne diseases: Overview of biological hazards and foodborne diseases. *Encyclopedia of Food Safety* 221–242.

Toker, O.S., Palabiyik, I., Pirouzian, H.R. et al. (2020). Chocolate aroma: factors, importance and analysis. *Trends in Food Science & Technology* 99: 580–592.

Vinderola, G., Sanders, M.E., and Salminen, S. (2022). The concept of postbiotics. *Food* 11 (8): 1077.

Wang, R., Sun, J., Lassabliere, B. et al. (2022). UPLC-Q-TOF-MS based metabolomics and chemometric analyses for green tea fermented with *Saccharomyces boulardii* CNCM I-745 and Lactiplantibacillus plantarum 299V. *Current Research in Food Science* 5: 471–478.

Wastyk, H.C., Fragiadakis, G.K., Perelman, D. et al. (2021). Gut-microbiota-targeted diets modulate human immune status. *Cell* 184 (16): 4137–4153.e4114.

West, H.G. (2020). Crafting innovation: continuity and change in the "living traditions" of contemporary artisan cheesemakers. *Food and Foodways* 28 (2): 91–116.

Xiang, H., Sun-Waterhouse, D., Waterhouse, G.I.N. et al. (2019). Fermentation-enabled wellness foods: a fresh perspective. *Food Science and Human Wellness* 8 (3): 203–243.

Yang, J., Lagishetty, V., Kurnia, P. et al. (2022). Microbial and chemical profiles of commercial Kombucha products. *Nutrients* 14 (3): 670.

Zolfaghari Emameh, R., Purmonen, S., Sukura, A. et al. (2018). Surveillance and diagnosis of zoonotic food-borne parasites. *Food Science & Nutrition* 6 (1): 3–17.

CHAPTER 12

AFRICAN FERMENTED FOODS AND BEVERAGES. POTENTIAL IMPACT ON HEALTH

VICTORIA BELL[1], JOSÉ GUINA[2], AND TITO H. FERNANDES[3,4]

[1]Faculty of Pharmacy, University of Coimbra, Pólo das Ciências da Saúde, Azinhaga de Santa Comba, 3000-548 Coimbra, Portugal

[2]ISEUNA, Polytechnic University, N13, Nampula, Mozambique

[3]CIISA – Centre for Interdisciplinary Research in Animal Health, Faculty of Veterinary Medicine, University of Lisbon, Pólo da Ajuda, 1300-477 Lisboa, Portugal

[4]Former CEIL, Lúrio University, Marrere, Nampula, Mozambique

CONTENTS

Microbial Fermentations in Nature and as Designed Processes, First Edition. Edited by Christon J. Hurst.

12.1 INTRODUCTION

Centuries ago, believed over 9000 years ago, African civilizations embarked upon spontaneously fermenting foods and beverages with the deliberate purpose to stockpile fresh food raw materials to extend preservation and endure seasonal changes in food availability. In the era of the Neolithic Period, with the beginning of agriculture, there were already traces of many fermented products such as milk, sorghum, millet, fish, meat, bones, hides, skins, hooves, gall bladder, fat, intestines, caterpillars, locusts, frogs, and even cow urine. Beer was not invented but discovered in Egypt (circa 10 000 BCE), as fermentation was a fortuitous by-product of the collection of wild grain.

Nearly every culture in the world developed a single staple crop such as cereals and potatoes, which formed the bulk of its diet. North Africa's climate and geography are very different from those of Sub-Saharan Africa, the main agricultural challenge in the regions being access or lack of water, and arable land. Humans' emergence in Africa as foragers adopting subsistence farming as a way of living occurred quite late relative to some other geographical regions. Agriculture's progress southward from the north of the continent was quite slow, more to support the development of animal husbandry, while domesticated crops did not arrive in South Africa until the seventeenth century, and they came with European sailors, not overland.

According to the UN's World Food Programme (WFP), presently in 2022, there are 44 million people in 38 countries in the "emergency" phase of food insecurity and as many as 38 million Africans are living under the threat of famine, facing the effects of severe food insecurity, including malnutrition, starvation, and poverty, conflicts, or environmental challenges being the largest drivers of the scarcity of food. While many people may not be "hungry" in the sense that they are suffering physical discomfort, they may still be food insecure (Table 12.1).

Subsistence small-scale farming for food security has been a policy focus for many African governments but any agricultural development is highly dependent on stressors such as climate changes, serious and sustained droughts, cyclones, floods, and water availability, namely in Sub-Saharan Africa. Therefore, it is imperative to improve the interpretation of the interconnections among climate, food, sustainable livelihoods, policies, actions, and practical initiatives.

Food security is dependent not only on nutritious food production but also on consistent food access and adequate food utilization. The overall consumption of animal products (dairy, meat, eggs, fish, and seafood) in Africa is well below the global average. Most traditional African dishes, based on hearty vegetable soups and stews, contain starchy items, some meat or fish, and a variety of spices and herbs. Many African dishes consist of maize flour and center around curried dishes.

TABLE 12.1 The most at-risk of food insecurity among the 54 African countries.

Country	% Hungry Population	Total Population (million)
Somalia	68%	12.3
Burkina Faso	61%	19.8
South Sudan	60%	11.0
Mali	60%	19.1
Sierra Leone	55%	8.2
Niger	55%	22.4
Lesotho	50%	2.1
Guinea	48%	12.2
Benin	47%	11.5

Africa is also home to an extensive and diverse functional and medicinal botanical life, which is an essential element of the indigenous African healthcare systems.

Thus, food security, currently firmly aggravated by the pandemic and the war in Europe, continues to be a major concern in Africa and needs to be tackled to eradicate poverty and reduce inequalities through sustainable development. Fermentation portrays an important function in food security in that it can recycle and transform inedible by-products and food wastes into valuable nutrient-rich and healthy food sources.

Fermented foods have a historical tradition in Africa, probably introduced by Arab merchant travelers. Foods and beverages that are prepared via a fermentation process constitute an essential segment of human nutrition in most food cultures around the world, although not illustrated in food guidelines. Fermentative pickled fruits and leafy vegetables are very popular in many regions of Africa, Europe, Asia, America, and the Middle East and can address the problems related to the world's balanced diet.

Africa has a rich tradition in fermented foods and beverages of plant and animal sources, essentially homemade under unhygienic substandard environments for own consumption or offered for sale in informal small markets. Indigenous knowledge concerning these foods is found in elderly rural women.

African fermented foods can be grouped into three clusters: (i) the foodstuffs; (ii) the relishes, condiments, and sauces; and (iii) the beverages. However, the absence of writing culture in most of Africa makes their origin difficult to trace. The most documented of the fermented foods is sour milk but over 5000 fermented products have long been a part of the human diet, and their traditional and cultural value varies among countries.

Africa has a rich tradition in fermented foods and beverages of plant and animal sources, essentially homemade under unhygienic substandard environments for own consumption or offered for sale in informal small markets. Indigenous knowledge concerning these foods is found in elderly rural women.

Traditional fermentation processes in Africa provide a means of food preservation, stalling the rottenness of perishable foods, extending the shelf life, and reducing food waste. Indeed, microbiota metabolize the raw materials, improving organoleptic properties, adding nutritional value, and repairing biosafety through the synthesis of anti-microbial metabolites.

African fermented foods and beverages are so specific that they identify cultures and civilizations. There is a difference in gut microbiota from distinct origins. The enteric microbial community of infants from rural Africa is dominated by bacteria able to utilize cellulose and xylan (hemicellulose) contrary to the gut microbiota of European children. As justified by different diets, African children's microbiota is based on foods rich in starch, fiber, and plant polysaccharides whereas European children's microbiota represents their being accustomed to protein and fat rich foods.

Research has shown that the fermentation of sorghum raises the digestibility of its protein from a value as low as 45% to as high as 86%. Despite the awareness of the favorable impact of fermented foods and beverages on well-being and diseases, their many healthcare benefits or suggested systematic consumption have not been generally contemplated in most world food guidelines.

Knowledge about the specific multispecies microbial biodiversity of each food and beverage ecosystem is essential to understand the eventual impact on human welfare and health, but progress has been slow, namely with traditional African products. The understanding of food microorganisms' contribution to health is still in its infancy and much remains to be discovered.

The present global regulatory frameworks are confusing and limiting and do not accommodate the increased rate of digitalization which influences our communities and markets. Science is relentlessly developing, and many exciting new technologies will continue to transform the developing world and improve human welfare. Foods prepared by fermentation will increase in amount and its use will spread to different parts of the world, as they contribute to the requirements of the vast diversity of gut microbiota, now known to have an indirect impact on general and even mental health.

In Africa, the potential risk of microbiological contamination is vast, raising the question of balance between food security and food safety. The overall advantages of indigenous or ethnic fermented foods and beverages are accepted, the doubt is if they are safe for human consumption and approved as an everyday food.

Differences in flavor compounds and bacterial communities of traditional and modern fermentation are poorly understood. Furthermore, many challenges remain regarding the establishment of dietary guidelines integrating education, agriculture, health, environment, economy, and industry.

The purpose of this chapter is an illustration of some African fermented foods and beverages, their potential health implications, and eventual hazards. Identifying social determinants and formulating appropriate strategic responses is not an easy task and no suggestion is made on possible master plans for improving their nutritional composition and potential well-being and general health benefits.

12.2 SOME STAPLE FERMENTABLE FOODS IN AFRICA

Africa, the second-most populous, vast, and diverse continent, has numerous varieties of gastronomic traditions with very distinct regions of unique cultures and cuisines. Trying to figure out the most popular foods in the continent is an impossible task since African foods are as diverse as its people and they are typically overlooked by scientists, policymakers, and the world at large.

The vast potential of each vegetable in Africa has been mainly screened for nutraceutical and pharmaceutical potential and less to help curb malnutrition, boost food security, foster rural development, or create sustainable land care in Africa. The limited information is not perfectly aggregated and is fragmentary and widely dispersed in many different journals and books.

While the customary African diet is miscellaneous, vegetables are the largest staple. Cereals (maize, sorghum, millet), cassava, sweet potatoes, African spinach, okra, cabbages, beans, lentils, carrots, aubergine, pumpkin, onions, tomatoes, tropical fruits, spices, and herbs, are common ingredients in African dishes.

The microbiota in any fermenting food substrate is an outcome of the hygio-sanitary conditions of the raw material, the local environment settings, the tools used, and the handlers, leading to a wide range of fermented foods and beverages.

12.2.1 Maize (*Zea mays*)

In Africa, maize is mostly grown in small-holder farming systems. It is common and relevant with a double harvest happening in a year. Maize being a C4 plant, like sugar cane and sorghum, is more efficient than others (e.g. rice, wheat, potato, soybean) and presently there are many challenges when attempting to improve crops through C3-to-C4 engineering, namely in warmer climates.

Maize is produced in all African countries albeit mainly in Nigeria, South Africa, Egypt, and Ethiopia, and is common in drought or humid areas. The desired genetic enhancements for early and extra-early maturing germplasm, for striga (*Striga hermonthica*) resistance and management, higher levels of lysine and tryptophan, more resistance or tolerance to drought, and low soil nitrogen needs, have been researched and used in the continent.

The introduction of insect-resistant Bt genetically modified (GM) maize, cotton, and soybeans, which include a gene from the *Bacillus thuringiensis* (Bt) bacterium, can provide insect protection. We also can try to create plants that are more resistant to pesticides or produce pesticides themselves, changing the way that pesticides are used in agriculture and perhaps helping small-scale farmers to improve yields and control pests without chemical insecticides. However, many countries still ban this technology, with several pros and cons, and continue to adopt the indigenous white maize, which requires the use of more land, water, fertilizers, pesticides, and herbicides, which are not only energy expensive but also detrimental to the environment.

12.2.2 Cassava (*Manihot esculenta*)

Cassava (Figure 12.1) is produced not only for its root vegetable, but also for its leaves (*matapa*), and for the production of biofuels, foods (*gari*), feedstock, starch, beverages, flour, glucose syrup, and glue. Traditionally, cassava root is peeled, cleaned, and boiled to soften the tissue, then the microbial starter culture is added and incubated to ferment (*fufu*) for four to six days to effect sufficient detoxification of the roots, eliminating cyanide.

FIGURE 12.1 Cassava roots and leaves.

Every part of Africa has developed its own process to ferment and use cassava and their designations are extensive, providing over 50% of the average daily energy intake in some countries. The cassava fermented foods are quite similar except the plant varieties and ingredients are different and there is diversity in the associated microorganisms in cassava fermentation.

12.2.3 Fermented Beans

Africans consume reduced amounts of protein of animal origin and their main sources of essential amino acids are beans, legumes, nuts, and seeds, which usually means their diet is lacking one or more of the nine essential amino acids. All beans primarily lack the essential sulfur amino acid methionine.

Generally, beans may have numerous anti-nutritional factors such as condensed tannins, trypsin inhibitor activity, α-galactoside oligosaccharides, polyphenols, lectins, and pyrimidine glucosides. Processing food by fermentation may be positively associated with enhanced nutritional quality by reducing the levels of these anti-nutritional factors.

These beans and other seeds are mainly fermented to yield condiments and convert substrates into products that contribute to the chemical composition and taste of the final product. *Bacillus subtilis* seems to be the most predominant microorganism followed by *Escherichia, Proteus, Pediococcus, Micrococcus, Staphylococcus,* and *Streptococcus,* but surely the microbiota profile will be much wider.

Some of the most popular condiments based on fermented vegetables in West Africa comprise *daddawa* and *soumbala* from soybean (*Glycine max*), *iru* or *dawadawa* from locust bean (*Parkia biglobosa*), *ugba* from African oil bean seed (*Pentaclethra macrophylla*), *ogiri* from melon seeds (*Citrullus lanatus* subsp. *vulgaris*), and *owoh* from cotton seeds (*Gossypium hirsutum*).

Cowpea (*Vigna unguiculata*) is a valuable legume widely cultivated. Fermentation of this bean seems to significantly improve its nutritional value, by improving amino acid digestibility and mineral (P, Mg, and K) bioavailability.

12.2.4 Moringa (*Moringa oleifera*)

There are many valuable evergreen trees and shrubs in Africa producing the three vital "F's" that desert dwellers require, Fruit, Fodder, and Fuel. Rural populations in arid regions use these trees to meet their daily requirements of biomass or biomass-based products, such as food, fuel (firewood), fodder, fertilizer (organic manure, forest litter), building materials (poles), and medical herbs.

All parts of moringa are sources of food or medicine. It is a plant that is often called the miracle tree used for centuries due to its medicinal properties and health benefits (Figure 12.2). Rich in vitamin C, 7 times more than oranges, and in potassium 15 times more than bananas, 10 times more vitamin A than carrots, 25 times more iron than spinach, 17 times more calcium than milk, 9 times more protein than yogurt, it is known to have some 540 compounds that can treat or prevent 300 health issues.

There has been considerable research to prove it has antifungal, antiviral, antidiabetic, antidepressant, anti-inflammatory, and anticholesterol properties. However, most tests have mainly taken place on animals as opposed to humans. The concentrations of crude protein, small peptides, and amino acids present in moringa leaves may increase significantly after fermentation. Antinutritional factors in moringa, such as tannins and glucosinolates, could affect the palatability, digestion, and absorption of nutrients but fermentation degrades these elements.

Very few studies have examined the dynamic changes of nutrients and antioxidant components during moringa fermentation, and almost none have reported on the antioxidant activity of moringa leaf powder after fermentation.

FIGURE 12.2 Moringa leaves and trees.

World-renowned cosmetic enterprises have established vast moringa plantations in Africa claiming to help feed and medicinally treat poor people, but their main interest is in the food-grade moringa seed bright yellow oil. The oil can be exported as an added value raw material, to be manufactured elsewhere as skin moisturizing compounds, for cleansing the skin, anti-aging, and hair treatment.

12.2.5 Olives (*Olea europaea*)

Table olives (*O. europaea*) have been produced extensively in North African countries for centuries, with Morocco, Algeria, Tunisia, and Egypt being the main producers of olives and olive oil. This fermented vegetable has several interesting nutritional and bioactive compounds that have even been proposed as functional foods with known microbial stability and as a source and carrier of probiotic lactic acid bacteria (LAB).

Considered one of the largest agricultural business sectors it is highly industrialized and along with following traditional methods, new sophisticated processes have been introduced. Refineries and cooperatives have made several attempts at industrial upgrading.

The wild olives (*O. europaea* subsp. *cuspidata*) are native to the South African region but are inedible although it is considered a very valuable tree medicinally by indigenous people, used to boost the immune system, lower blood pressure, combat viral, fungal, and bacterial infections and improve circulation and brain function.

12.2.6 Tropical Wild Fruits

The traditional fermented products include beverages from indigenous fruits often mixed with functional herbal medicinal products. In Africa, blessed with huge agrobiodiversity, there are more than 50 wild indigenous tree species that bear highly valued edible fruits essential for food security, health, and nutrition. Africans live surrounded by vast readily accessible wild fruit and

FIGURE 12.3 *Uapaka kirkiana* or wild loquat (*mushuku/muzhanje*); *Ziziphus mauritiana* in an open informal market. (*See colour plate section for colour representation of this figure.*)

vegetables normally free and easy reaching. They are an important source of income for poor people, but their value chains are poorly developed.

Wild fruits such as *Adansonia digitata* (baobab), *Ziziphus mauritiana* (Indian jujube), and *Sclerocarya birrea* subsp. *caffra* (marula) provide micronutrients such as vitamins and minerals. These seasonal wild fruits are collected and consumed either fresh or made into various products such as jams, pickles, chutneys, and spontaneously fermented juices.

Just an example, *Uapaca kirkiana*, wild fruit (Figure 12.3) rich nutritionally and high in phytochemicals antioxidants, was reported growing in miombo areas in Angola, Burundi, Malawi, Mozambique, Tanzania, Zaire, Zambia, and Zimbabwe, but in need of being scientifically exploited and incorporated into the diets to help solve some nutrition-related concerns. The significance of the *U. kirkiana* fruit tree as a food resource as compared to other indigenous fruit trees has been researched and its fermentation resulted in the degradation of phytases leading to increased bioaccessibility and bioavailability of magnesium, iron, calcium, and zinc content.

In Zimbabwe, *Uapaca* pulp infused with cold water is left to ferment until the mixture turns cloudy. The solution is utilized to manufacture a thin porridge by adding milled maize. Sometimes the pulp is allowed to ferment into a candied alcoholic beverage called *mutandavira*. This seasonally abundant wild fruit fermented on a small scale (household-level craft industry) has the potential to turn into a large-scale production of *makumbi* (industrially processed fermented foods).

12.3 THE DIVERSITY OF GUT MICROBIOTA

The establishment and development of gut microbiota over the lifecycle moved from the previously accepted dogma that the mammalian healthy placenta and fetus were germ-free and considered to be sterile, to an understanding that the placenta and fetus are microbially colonized under conditions critical to the developing newborn's immune system. Our actual knowledge is that in-utero humans harbor unique prenatal microbiomes.

Many species of bacteria, specifically those found in the invisible universe of the human microbiota, are non-pathogenic commensals from the phyla Firmicutes, Bacteroidetes, Actinobacteria, Proteobacteria, and Verrucomicrobia. The gut microbiota form an interactive and complex micro-ecosystem composed of trillions of microorganisms. This vast community plays vital roles in mucosal immunity, nutrient metabolism, and intestinal epithelium differentiation. Its collective metabolic activities and host interactions influence both normal physiology and disease susceptibilities.

Having a stabilized gut bacterial community is one of the preconditions for a host to perform its normal physiological functions, metabolism, and immune functions. Temporal fluctuations in the composition of gut microbiota may result in multiple gastrointestinal diseases and even affect mental health.

Indeed, the digestive process is dependent on gut maturation and function of the mucus layer, motility, hormones, enzymes, and the activity of the microbiota, and its efficiency is affected by critical interactions between these highly diverse factors and microbial communities.

Our understanding regarding the activity and interactions among microorganisms in biological processes like fermentation has recently been revolutionized by advanced omics techniques and next-generation gene sequencing methods. Metaproteomics workflows also is a promising technique for advancing understanding of the functional roles of the microbiome in disease. However, research is required to develop databases of disease-associated microbial biomarkers since current models lack information regarding the interaction between different microorganisms.

Among the various molecular techniques used to study bacterial diversity, culture-independent methods (e.g. cloning of 16S rRNA and denaturing gradient gel electrophoresis) rely on the principle of amplifying rRNA or functional gene PCR products to study microorganisms within their original environments.

These molecular methods, in conjunction with additional procedures such as chemical analysis, imaging, and stable isotope labeling, can assist species identification of the immense diversity of the microbiota and the complex interactions among microbial community composition and kinetics.

The recent move from studying microbiology to microbiomes, with awareness of diverse microbial life and their symbiotic relationships, and how host and environmental factors affect gut microbial ecology, may in the future promote well-being, health, and lifespan by guiding therapeutic strategy development.

The discipline of microbiome and probiotic investigation has expanded rapidly, and despite the continuous development of new methods and equipment, these techniques well-used to characterize microbial pathogens in human patients are not yet implemented in detailed studies of the microbiome of fermented foods and beverages and depend on cellular-level activities and biochemical process-level phenomena.

In a recent clinical trial with 36 healthy adults, it was shown that while microbiota diversity was stable in a fiber-rich diet, with a high-fermented-food diet a relevant raise in the diversity of gut microbiota was evident, with a simultaneous decrease of molecular markers of inflammation, modulating the gut microbiome, which may favorably impact the immune system (Figure 12.4).

The countless blending of foods as substrates for microbial enzyme fermentation yields products with vastly different sensory and nutritional profiles. Fermentation improves the bioavailability of certain macro- and micronutrients, induces reductions in anti-nutritional compounds, and generates nutrients and bioactive compounds of microbial metabolism with specific nutritional and potential health functionalities.

LAB (e.g. *Lactobacillus*, *Streptococcus*, *Pediococcus*, and *Leuconostoc*), widely used in the fermentation of animal and vegetable products, also exhibiting probiotic properties, are known to produce several biogenic compounds, currently known as metabolic by-products or "postbiotics," including fatty acids, sugars, enzymes, bacteriocins, and bioactive peptides.

LAB may ferment dairy, meat, and vegetable products, whereas acetic acid bacteria (e.g. *Gluconobacter, Gluconacetobacter,* and *Komagataeibacter*), associated with the biotechnological

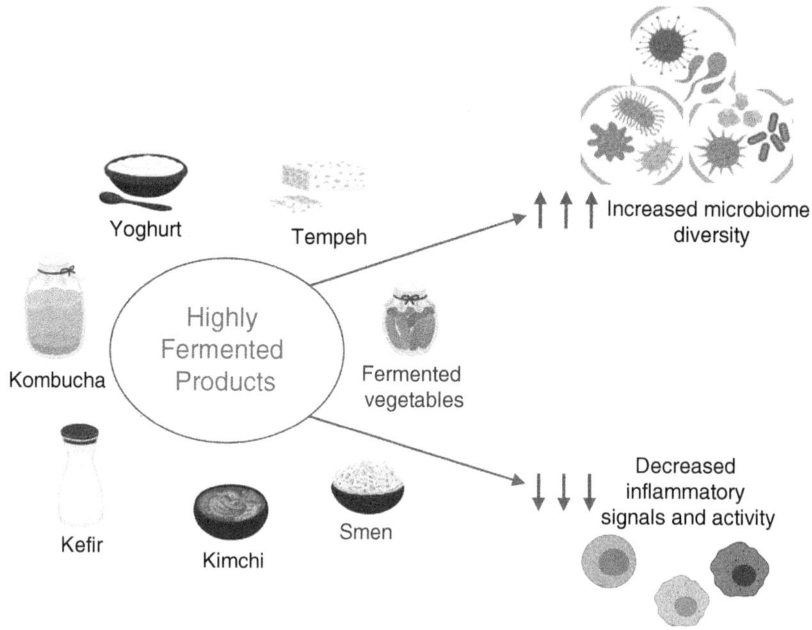

FIGURE 12.4 A diet high in fermented food or beverage decreases inflammatory proteins, reduces molecular signs of inflammation, and improves immune response. Therefore, improving digestive health with a high-quality fermented diet may improve immunity against infections and diseases. (*See colour plate section for colour representation of this figure.*)

process of vinegar and cellulose production, are responsible for the fermentation of cocoa bean, acidic beverages (e.g. kombucha, milk kefir, water kefir), and vinegars made from a wide range of base ingredients.

Alcoholic fermentations, where some 95% of sugars in the substrate are converted into ethyl alcohol and carbon dioxide, are mainly driven by yeasts (e.g. *Saccharomyces cerevisiae*) to produce beer, wine, and bread.

Alkaline fermentations make use of fungal (e.g. *Aspergillus, Penicillium, Rhizopus, Mucor, Humicola, Thermoascus, and Thermomyces*) enzymes (e.g. amylase, glucosidase, glucose oxidase, protease, pectinase, cellulase, invertase, laccase, ligninase, lípase) during the production and maturation of cheese, fermented meats, and fermented soybean products.

The presence of non-specific spontaneously fermenting microorganisms can hinder successful monitoring of the fermentation process, leading to finished products of variable quality. Some undesirable noxious contaminants and metabolites of the unintended microbiota will be released into the fermented food matrix. Commonly emerging toxic compounds in fermented foods and beverages include mycotoxins, biologically active amines (e.g. dopamine, histamine, norepinephrine, serotonin, and tyramine), and cyanogen glycosides (e.g. linamarin, dhurrin, prunasin, and amygdalin).

Research on these important topics is scarce while extension services and smallholders' family farmers do not have adequate knowledge of specific substrates and types of microbiota inocula. Therefore, the final fermented product quality at the rural African household level remains unknown and difficult to be standardized.

Further research is still needed to clarify some of our present understanding of the fermentation of foods and beverages regarding the role of microbiota in micronutrient bioavailability, bioaccessibility and metabolism, and the related biological mechanisms.

12.4 TRADITIONAL FERMENTATION STRATEGIES

In most African countries, traditional village-art practices and age-old systems are still used for agro-processing including fermentation. Fermented foods and beverages are produced by the controlled growth and enzymatic activities of microorganisms, through basically four main fermentation processes (lactic, acetic, alcoholic, and alkaline).

Frequently it implies tossing the mixture to be fermented with a previously fermented mash melange. Clay crocks and pots or repurposed polyethylene drums are used for traditional fermentation, covered with a cloth that creates a slightly anaerobic environment or completely closed to create stricter anaerobic conditions.

It is important to address the common misconception that fermented foods are the same thing as probiotics. Substantial confusion exists between fermented foods and beverages and the probiotic concept. Contrary to general expectations, fermented foods are generally not sources of probiotics. Nevertheless, some fermentation processes perpetuate the multiplication of pre-existent beneficial live microbes called probiotics. They are not probiotics, although they may contain them, as their live microbial content is undefined.

These live microbes beneficially affect the host by improving its gastrointestinal microbial equilibrium. LAB dictate the fermentation operation and lead to a low pH, eliminating the evolution of a pathogenic bacterial community, thereby preventing spoilage and increasing the shelf-life and safety of the food.

The current status of fermented products commercially available in Africa is less than ideal. Prevailing obstacles include failure to identify the constituent microorganisms and deficient feasibility of probiotic microorganisms which lead to inadequate shelf-life. Given the potential benefits of manipulating the human microbiota for beneficial effects, there is a clear need for improved regulation of fermented foods and beverages and carried probiotics.

There are various reasons why foods may be fermented. Perhaps the most important one is to produce a variety of flavors of existing foods. Others are to mask some undesirable flavors, to make some inedible foods edible, and to mainly preserve some foods and beverages. There is a wide geographical variation of numerous fermented foods and beverages across the continent thus impossible to mention all of them.

Considerable ways of rustic farm food processing and concoction are often used at rural dwellers and community levels to enhance the bioavailability of micronutrients including de-hulling, threshing, drying, grinding, soaking, cooking, wet, or dry milling, fermentation, and germination.

Most yeasts can ferment hexose sugars (e.g. glucose, fructose, galactose, or mannose) to ethanol anaerobically (Figure 12.5) and this has been well studied since Pasteur in the mid-nineteenth century, where the glycolytic pathway, a series of reactions used by aerobic organisms generate energy (ATP) through the oxidation of acetyl-CoA derived from carbohydrates, fats, and proteins into carbon dioxide and chemical energy.

Pentoses (e.g. xylose or arabinose), however, are not easily fermented by *S. cerevisiae*, the preferred microorganism in crop-based glucose fermentation.

Within the human microbiome, ethanol is not a truly endogenous substance of human cells, but enters them from the blood, originating from the microbial enzymes' fermentation of carbohydrates in the gastrointestinal tract. However, the problem of ethanol synthesis inside human cells remains unresolved since the presence of alcohol dehydrogenase and aldehyde dehydrogenase in all cell types including neurons also argues for an intracellular origin of ethanol.

Fermentation, via the action of microbial phytase enzymes, which hydrolyzes phytic acid, the primary source of phosphorus in plant seeds, lowers inositol phosphates, decoupling the complex links with calcium, zinc, and iron, enhancing the absorption of these minerals.

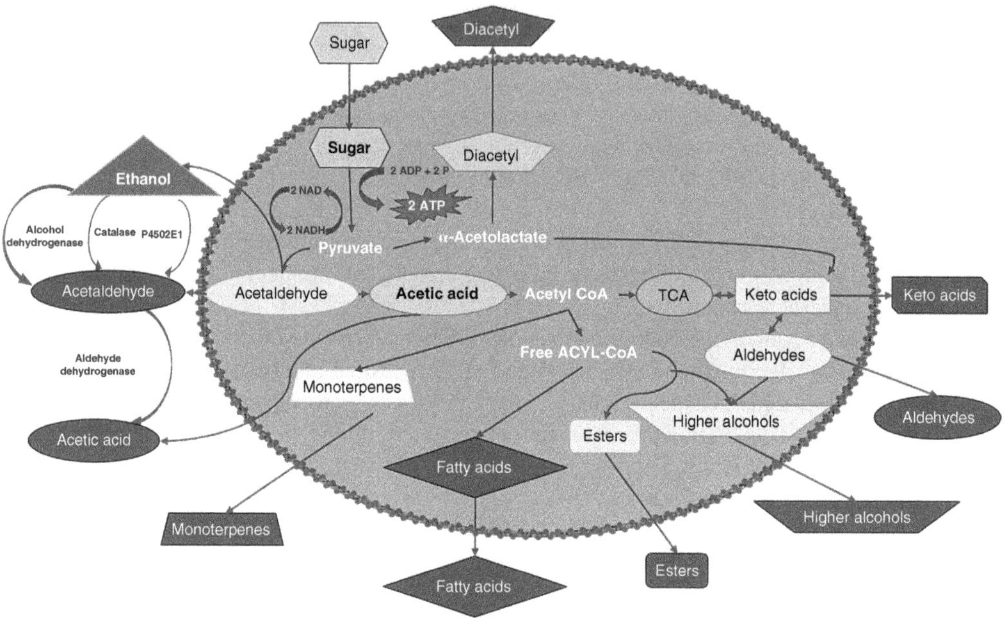

FIGURE 12.5 Mitochondrial yeast metabolism, maltose being the best sugar source. Secondary metabolites and metabolism intermediates are shown in ethanol formation. (*See colour plate section for colour representation of this figure.*)

Fermented beverages have been produced and consumed for ages, contributing to the nutrition of many communities and cultures worldwide. Traditional fermentation has evolved for millennia as a process of raw food conservation and a way to yield new foods with improved sensorial characteristics, such as appearance, texture, taste, and smell, and probably not to improve nutritional value.

Fermented foods constitute a significant component of African diets. Most African countries have a history of food processing passed down for generations and fermented products comprise some 5–40% of the diet in some populations. Fermented milk (yogurt, curds, and cheese) and, to a lesser extent, meat and fish as well as wine and beer, are the result of traditional methods of preservation, rooted in ancient tradition but with several innovations and technological adaptations introduced over time.

Northern African countries between the Sahara and the Mediterranean Sea (Morocco, Algeria, Mauritania, Egypt, Libya, Sudan, Tunisia, and Western Sahara) share many traditional fermented foods resulting from the blend of nutritional habits brought by the Greeks, Nubians, Phoenicians, Egyptians, Arabs, Roman civilizations, and their interactions with those of the original Berber inhabitants.

Olives, produced extensively in northern African countries (Morocco being the largest producer and Tunisia with the widest diversity of olive germplasm), are rarely consumed directly as a natural fruit due to their extreme bitterness but are used instead for the extraction of oil and, to a lesser degree, as table olives (after curing). Indeed, olives are inedible without fermentation that removes bitter phenolic compounds.

Coffee, tea, and cocoa (*Theobroma cacao*) are produced in Africa and consumed worldwide, but largely unperceived by consumers as "fermented" food products. Fermentation of coffee, tea, and cocoa occurs upstream in the food manufacturing process in contrast with milk fermented products, where the finished goods are subject to fermentation and are natural carriers of live microbiota.

The post-harvest processing stage occurs rudimentarily, and cocoa seeds are deliberately fermented for some 7 days, unprocessed green coffee grains for 10–25 days, and in the case of post-fermented teas, fresh tea leaves may be fermented from several months up to several years.

Kombucha, a sweetened fermented black or green tea that is non-alcoholic or low in alcohol content (<0.5%), may have originated from the Manchurian region of China where the drink is traditional but it is now widespread and popular in many African countries.

An alternative substrate to produce kombucha, is *Brassica tournefortii* (Brassicaceae), an annual invasive herb that grows wild in rocky and sandy environments in the Mediterranean regions of North Africa (Figure 12.6). Fermented for about 14 days with a symbiotic culture of several indigenous bacteria (LAB, *Acetobacter* and *Gluconobacter*) and yeasts (*Saccharomyces* spp. and non-*Saccharomyces* spp.), produces a tea beverage.

The production of coffee originates from Ethiopian highlands and, centuries ago, it was reported the production of an alcoholic beverage from fermented ripe coffee (*Coffea arabica*) fruit that also is termed coffee berries. Coffee beans, which are the seeds of the coffee fruit, undergo a fermentation stage that is performed by indigenous bacteria and yeasts present on the surface of coffee berries. Today over 80 bacterial genera have been identified in coffee fermentations, and the yeast population is represented mainly by *Saccharomyces* sp., followed by *Torulaspora delbrueckii*, *Pichia kluyveri*, *Hanseniaspora uvarum*, *Hanseniaspora vineae*, and *Meyerozyma caribbicca*.

Sudan is a particular country that still holds strong traditional values in a rapidly changing world. A special type of wafer-thin pancake or flatbread (*kissra*), which is made from naturally fermented durra sorghum or corn, constitutes the staple diet (Figure 12.7). This is eaten with

FIGURE 12.6 *Brassica tournefortii* on the left. Coffee beans in the middle. Kombucha on the right.

FIGURE 12.7 Typical Sudanese *kissra* making and a full meal with several fermented foods.

FIGURE 12.8 Ripe bananas being prepared for fermentation in Burundi.

various types of condiments for stews (*kawal*), sauces, and soups made from a mixture of dried meat, roots, okra, onions, spices, and peanut butter.

Several food waste products are fermented to produce edible food products in rural Sudan, comprised of bones, hides, and edible insects. Fresh bones are fermented simply by laying them on roofs made of straw or reeds, allowing slow fermentation for weeks or even months into a variety of products, or placing chopped, smaller pieces of bone into fermenting basins.

Several traditional fermented foods and beverages are produced at the household level in Burundi. These comprise cereal and banana-based beverages (*urwarwa, isongo, impeke,* and *kanyanga*) (Figure 12.8), milk products (*urubu, amateregua,* and *amavuta*), and cassava-based fermented foods (*ikivunde, inyange, imikembe,* and *ubswage*).

In most African countries, the unwholesome practice of itinerant or stationary street food vending and consumption represents a growing market, supplying daily ready-to-eat food and beverages and fast takeaway for most urban residents. Although solving major social and economic problems, these activities may result in risks to the health and safety of stakeholders along the value chain.

In few family settings, the eccentric manufacture of some indigenous beverages is as bizarre as to also include car battery fluid, marijuana, magic mushrooms, tobacco, and hallucinogenic herbs in the preparation.

12.5 FERMENTED NON-ALCOHOLIC STARCHY FOODS

Currently, "non-alcoholic" is a regulatory term and the laws regarding it vary across the globe, while in Africa none or little information exists on the extent to which countries are addressing alcohol consumption and alcohol-related harm with a great variation in the strength of alcohol control policies. Elsewhere, there is extensive literature on non-alcoholic fermented beverages from all cereals as they are rich in bioactive compounds and high nutritional values, although deficient in lysine, an essential amino acid. Their phenolic compounds have also strong antioxidant potential and scavenge harmful free radicals.

The market for viable, native-born, and non-alcoholic beverages is constantly growing all over the world. Traditional low-alcoholic fermented beverages and non-alcoholic fermented beverages are vastly consumed in Africa and represent a subgroup of fermented beverages that have received rather little attention from consumers and scientists alike. Their types and traditional uses in European societies are characterized as "functional" foods based on their probiotic culture content.

Fermentation commits to food security, especially in African agro-pastoralist communities. Nevertheless, there is not enough data regarding the dietary role, nutrient composition, health benefits, and other relevant aspects of diverse ethnic non-alcoholic fermented beverages consumed by Africans.

12.6 COMMON AFRICAN FERMENTED ALCOHOLIC BEVERAGES

The production of African alcoholic beverages, usually designated as wines, is handcraft-manufactured by the age-old method of chance inoculation and uncontrolled fermentation; therefore, it is expected there will be variations in the quality and stability of the product and resulting short shelf-life. The distinctive sensorial attributes identified with these beverages are due to mixed culture fermentations, the specific flavor from each fruit substrate, and the outcome secondary metabolites that are yet to be fully elucidated for each beverage. The alcohol content varies a lot, usually between 4% and 41% (*v*/*v*).

12.6.1 Palm Wine

Palm wine, known by numerous names throughout Africa, and also popular in India and South America, is an alcoholic beverage widely consumed in tropical regions, produced from the spontaneously fermented sap of numerous tropical plants of the *Palmae* family (*Cocos nucífera, Elaeis guineensis, Phoenix dactylifera, Nypa fruticans, Caryota urens, Borassus aethiopum,* and *Raphia hookeri*). The diversity of microorganisms in the natural unfermented palm sap yields the specificity of different palm wines influencing the flavors and aromas of the final product.

12.6.2 Banana Wine

The banana (*Musa acuminata*) alcoholic beverage is common in many countries with different designations (e.g. *mbege* in Tanzania, *tonto* in Uganda, *urwagwa* in Rwanda, *enika* in Mozambique). The process of fermentation sometimes involves adding germinated, dried, and ground sorghum to the diluted banana juice.

12.6.3 Honey Wine

Honey wine (e.g. *tej, kuri*) is home processed, but also commercially available where honey is supplemented with sugar before fermentation. The honey wine, common in Cameroon and Ethiopia, is more complex as its preparation involves the addition of herbs, tree barks, and bittering agents from *Rhamnus prinoides* (Gesho). African commercial honey wine (mead) is made using pure cultures of yeasts and they produce aromatic volatile compounds that include alcohols, organic acids, esters, volatile fatty acids, carbonyl compounds, and volatile phenols.

12.6.4 Pineapple Wine

Pineapples (*A. comosus*) are cultivated mainly in Angola, and they have a significant proportion of sugar and acids that make them ideal for wine production. The production of wine involves the fermentation of peeled, sliced, and pressed pineapple. Pineapple wine aroma complexity has been well identified, influencing specific flavors and tastes. Interesting to note is that pineapple peels can be fermented alone, adding sugar, to produce a weaker wine.

12.6.5 Fan Palm Wine

Fan palm (*H. petersiana*) is a common tree species in such areas as Botswana, Namibia, Zimbabwe, Zambia, Tanzania, Angola, Rwanda, Burundi, and many other Sub-Saharan countries. In Botswana, the drink is called *muchema* and in Namibia *omalunga,* but its production is formally forbidden. There is no scientific description of the microbiota, biochemical, and safety profile of fan palm wine, and further research is needed.

12.6.6 Wild Watermelon Wine

The alcoholic beverage traditional beer (*setopoti*) is made from the fresh material or wastes of wild forms of watermelons (*C. lanatus*) with seeds intact. Usually, these are then fermented using moomelo (malt) which metabolizes the sugar in the watermelon into alcohol. It is brewed in semi-arid regions of Botswana, Namibia, South Africa, and Zimbabwe, in a clay pot which is then tightly closed (Figure 12.9). The preparation is then allowed to spontaneously ferment for one to twodays undisturbed. After fermentation, the alcoholic beverage is sieved to remove seeds and solids and served as a smooth fermented beverage.

12.6.7 Cashew Apple Wine

While the external nut from the cashew (*A. occidentale*) fruit is traditionally used and even exported, the succulent part has a sweet–sour taste, contains lots of antioxidants, and has five times the vitamin C of an orange. Native to north-eastern Brazil, Portuguese missionaries took it to East Africa and India. An indigenous spirituous liquor (*cana*), known as *feni* in India, is also prepared from crushed cashew apples or coconut palm sap (Figure 12.10).

12.6.8 Marula

The marula plant (*S. birrea*) has three subspecies very common throughout Sub-Saharan Africa. The spontaneous three-day fermentation of their ripe fruits is known by several names as *omagongo* in Namibia, *buganu* in Eswatini, *bojalwa-jwa-morula* in Botswana, *mukumbi* in Zimbabwe, and *ocanhu* in Mozambique (Figure 12.11).

FIGURE 12.9 Preparation of wild water mellon wine.

FIGURE 12.10 Cashew apple fruit on the left with the unique external seed and women preparing cashew fruit wine in Guiné-Bissau.

FIGURE 12.11 Traditional preparation of amarula drink ("ocanhu") in Mozambique. (*See colour plate section for colour representation of this figure.*)

12.6.9 Other Mixed Brews

Several alcoholic beverages resulting from the fermentation of mixed cereals and fruits are well known. *Kachasu* (*nipa*, *lukutu,* or *tototo*) is like *waragi* of Uganda and *chang'aa* of Kenya. It is usually brewed using maize meal, but bulrush (*Cyperus papyrus*) or finger millet meal (*Eleusine coracana*), various fruits such baobab (*A. digitata*), masau/maçanica (*Z. mauritiana*), tambarind (*Tamarindus indica*) and banana peels may be used as alternative sources of soluble carbohydrates (Figure 12.12).

Fermented foods and beverages are more popular than ever before, while research into the health benefits of fermented foods is relatively new. Not all fermented foods contain live organisms; beer and wine, for example, sometimes undergo steps that remove the organisms, and other fermented foods like bread are heat-treated and the organisms are inactivated.

Traditional and modern dietary practices utilize fermented foods and beverages, contributing significantly to the food chain value and belonging to a category of foods called "functional foods" (e.g. probiotics, prebiotics, stanols, and sterols) by having an additional characteristic, i.e. health-promotion or disease prevention effect.

Traditional fermentation exhibits an exceptionally precious cultural patrimony in most African countries. Many fermented foods serve as main course meals, beverages, and popular food condiments. Those which serve as main meals and beverages are usually products of carbohydrate-rich raw materials. Some of the most essential ones in this group include "*gari*" from cassava, "*ogi*" and "*amahewu*" from maize, and "*kaffir*" beer from sorghum. *Amasi* is predominantly an outcome of spontaneous milk fermentation and *amahewu* is manufactured by spontaneous fermentation of cooked maize or sorghum meal over a one to three-day period, both non-alcoholic drinks.

12.7 SOME AFRICAN FERMENTED ANIMAL PRODUCTS

Milk is a major source of energy and nutrients, but can also be a vehicle for zoonotic foodborne diseases when raw milk is used with inadequate hygiene conditions. Pathogens (e.g. *Streptococcus infantarius* subsp. *infantarius*) may lead to spoilage and reduce food safety and security. Fermentation done carefully can help mitigate the impact of poor handling and storage conditions by enhancing shelf life and food safety.

Milk from cows, camels, and does is important in some pastoral civilizations of Africa. There is an extraordinary range of local products and by-products where tradition, culture, and environmental conditions and constraints play a crucial role in the diversity of end products. Generally, there is scarce hygiene in milk production and processing, giving rise to careless fermentations that shorten shelf life, leading to spoilage and hazards to human health.

In North Africa, the diet is limited to foods of animal origin as opposed to vegetable foods, mainly cereals, and a variety of centuries-old dairy products are known and still highly appreciated by consumers. The most popular of them are *jben*, *lben*, and *smen* (Figure 12.13), and they are being increasingly marketed throughout the region in informal markets.

In Algeria, *smen/dhan* is a cultural butter product made by empirical methods, used as taste and aroma enhancers on several dishes, which is a long-duration traditional and spontaneously fermented butter made from various sources of whole raw milk. Artisanal cheese from fermented milk is named *klila*.

In South Africa, small-scale local farmers, using just raw milk, produce well-matured hard cheese *karoo crumble,* while in Sudan, Somalia, Ethiopia, and Kenya camel milk has potential but has not been fully explored.

FIGURE 12.12 Images of masau (*Ziziphus mauritiana*), tambarind (*Tamarindus indica*), millet (*Cenchrus americanus*), bulrush (*Scirpoides holoschoenus*), monkey orange msala (*Strychnos spinosa*), and baobab (*Adansonia digitata*) fruit.

FIGURE 12.13 Milk and butter *smen*.

In Egypt where some 70% of milk production is used to manufacture cheese, its consumption is significantly higher than in the other countries of North Africa, while there are more diversified and elaborated cheese types, among which brined cheeses are the most dominant. White and salty *domiati* cheese is made by combining camel with cow and sometimes buffalo's milk with or without the addition of a starter culture.

In Cape Verde, distinct types of cheese are produced according to different islands of the archipelago. Local goat cheese from raw milk from the Planalto Norte region, produced in difficult harsh environmental conditions from grazing animals, is sold as fresh cheese or, less frequently, as ripened cheese.

In Nigeria, *wara* is an unripened cheese or a milk curd achieved by adding a coagulant to fresh milk with no aging or pressing process. It is not a product of fermentation and not being a cheese, is highly perishable, and does not last more than two to three days.

In Ghana and other West African countries, *nunu* produced from raw milk is a spontaneously fermented yogurt-like dairy product consumed as a basic food product.

In Senegal, fermented milk product *(lait caillé)*, is prepared in wooden bowls (*lahals*) with a bacterial biofilm to steer the fermentation process.

Fermentation has been used as a natural flavor-enhancing tool process, usually the result of microbial enzyme interactions (e.g. proteinase, amylase, mannase, cellulase, and catalase) with plant enzymes and their activities on vegetable proteins of legumes or oil seeds.

Traditional fermented proteins from oilseeds generate tasty food condiments rich in polypeptides. These include *iru/dawa/sumbala* from African locust bean (*P. biglobosa*); *ugba* from alkaline fermentation of African oil bean seeds (*P. macrophylla*) and *ogiri* from melon seeds (*Cucumeropsis mannii*) among others. All are known to be good sources of proteins and vitamins. These fermented condiments can assist consumers regarding the supply of antioxidants, polypeptides, organoleptic, and probiotic characteristics, and antibiotic activity.

Worldwide, the usage and design of fermentation processes have experienced notable improvement in recent years. Despite the increased scientific interest in African traditional fermented foods and beverages, the age-old techniques used for their production in Africa have not improved as a result.

The importance of fermented foods in the nutrition of Africans is now better appreciated, as a result of which efforts are now being made to industrialize some of the processing in some cases even through studying the microbiome of African fermented foods using high-throughput DNA sequencing techniques.

12.7.1 Non-dairy African Fermented Products

The fermentation of milk proteins has been well described for many years. However, only some 11 fermented meat products are made in Africa, mainly in Sudan. In *heirta*, for instance, all offal plus muscle meat is chopped, added to milk, and fermented twice. The product is used to make a sauce for *aceda* stiff porridge from Sudan.

Miriss is prepared by fermentation of visceral fat. The fat is kneaded with a quantity of an ash preparation, *combu*, and fermented for up to six days. The product is very white and extremely foul-smelling.

At least three fermented products are made from fresh bones and meat scrapings. *Beirta* is prepared from he-goat meat. *Dodery*, for example, is prepared from marrow-impregnated joint bone endings. These are chopped, sun-dried, and pounded into a paste which is mixed with *combu* and fermented for up to five days. *Um-tibey* is best prepared from gazelle's meat and rumen.

In Cape Town, South Africa, despite the recent (2022) success of a novel green innovative precision fermentation technology on cellular agriculture using microorganisms as a blueprint to create animal-free milk recombinant proteins, identical to those found in cow's milk, one cannot expect a replacement of indigenous fermented foods.

12.7.2 Fish Products

Africans have many different fermented fish products all of them raising safety questions. The Nile Valley probably has a greater variety of fermented fish products than other regions of Africa. Sudan has at least four distinct such products. *Kejeik* represents the common, dry, split fish of Africa. *Fesseilch* is fermented whole fish made of a particular species and is known also in Egypt. However, *mindeshi*, a fish paste, and *terkin*, a fish sauce paste, appear not to be known in other parts of Africa. However, these fermented fish products could also be a source of nematode parasites and heavy metals if poorly processed.

12.8 POPULAR AFRICAN TRADITIONAL FERMENTED FOODS AND BEVERAGES

Fermented products can play an important role by contributing to the livelihoods of rural and peri-urban dwellers alike, through enhanced food security, and income generation via a valuable small-scale enterprise option. The strong link between fermented foods and food shortages is revealed by the fact that when a family becomes rich several fermented foods are no longer prepared. Cereal grains account for more than 60% of food materials used in the preparation of indigenous fermented foods in Africa.

Traditional beverages are drinks that are indigenous to a particular area and have been developed by the natives of that area themselves using techniques that are ancient and from locally available and home-grown raw materials. A consortium of yeasts and bacteria is known to carry out fermentation and contribute to the unique sensorial properties of fermented beverages.

Fermented plant foods may be classified as (i) those made from cereal grains (e.g. maize, sorghum, millet, rice, wheat), such as *kenkey, mahewu, ogi*, and *injera*; (ii) those made from pulses, nuts, and other seeds, such as the African locust bean tree *dawadawa*; (Figure 12.14); (iii) those from tubers (cassava, aroids, potatoes), such as *gari* and *farinha mandioca;* (iv) those from fruits

FIGURE 12.14 The multipurpose dawadawa tree, known as the African locust bean tree (*Parkia biglobosa* and *Parkia filicoidea*).

and vegetables, such as *gundruk* (fermented leaf spinaches) and *chakalaka*; and (v) beverages derived from fruit wastes (pineapple peel wastes). Fruit vinegar and pickled fruit (lemon, lime, and mango) are also common.

The fermented locust bean seed is used to prepare condiments to flavor soups and stews, commonly called *netetu*, *afinti*, *iru*, *dawadawa*, *ogiri*, *farroba*, or *soumbala*. Although the seeds and fruit pulp are popular because of their culinary uses, the other parts of the tree such as the bark, roots, and stems all have significant medicinal roles in African traditional medicine. They are often used to treat diabetes, malaria, diarrhea, wounds, and dental caries.

Data on the profile of microorganisms involved in spontaneous fermentation of non-grain-based alcoholic beverages as compared to grain (maize, sorghum, millet, rice, wheat)-based alcoholic beverages in the Sub-Saharan Africa region is scarce.

In the Sub-Saharan region, non-grain-based alcoholic beverages are produced from inexpensive substrates such as wild or cultivated fruits, tree saps, roots, and tubers whereas grain-based alcoholic beverages are made from cassava, small grains, maize, sorghum, and millet.

Sorghum beers are considered both an energy-boosting food and an alcoholic drink, consumed extensively in various ceremonies. Commonly named opaque beers throughout Africa, sorghum malt beers are produced in most countries with different designations. In Cameroon, it's called *amgba*; in Burkina Faso, *dolo*; in Zimbabwe, *doro, chibuku,* and *chikokivana* are produced as an ardent spirit from a blend of coarse mealie meal and millet malt; in Rwanda, *ikagage*; in Tanzania, *mtama* and *togwa* are produced from sorghum, maize, and Panicum grass; in Nigeria and Ghana, *burukutu* and *pito*; in Ethiopia, *borde* is made from maize, barley, or wheat and their malts; in Sudan, *merissa;* in Chad, *bilibili;* in Ivory Coast, Togo, and Benin, *tchapalo*; in South Africa, *kaffir*. In Uganda, *obushera* refers to a group of traditionally fermented or unfermented alcoholic or non-alcoholic sorghum beverages and *bushera* from millet.

Nigeria produces over 900 million liters of sorghum beer annually. One of these cereal-based alcoholic beverages, *chibuku,* another sorghum-based beer, has made it to the commercial markets in Botswana, Zimbabwe, and South Africa. In Mozambique, a cassava-based beer lager is commercially produced in the north (Figure 12.15).

Kawal, a strong-smelling fermented product made of the leaves of a wild and reputedly toxic African legume (*Senna obtusifolia*), is believed to have helped people endure and survive famines in South Sudan and Chad.

In Sub-Saharan Africa, fermented alcoholic beverages are produced from various inexpensive local raw materials such as palm sap, *marula* fruit, and *Grewia flava* wild berries' *mogwana* and *khadi*.

FIGURE 12.15 Commercially cassava beer is produced in Mozambique, brand name *Impala*. *Chibuku* sorghum beer is made in several countries. *Burukutu* preparation in Nigeria.

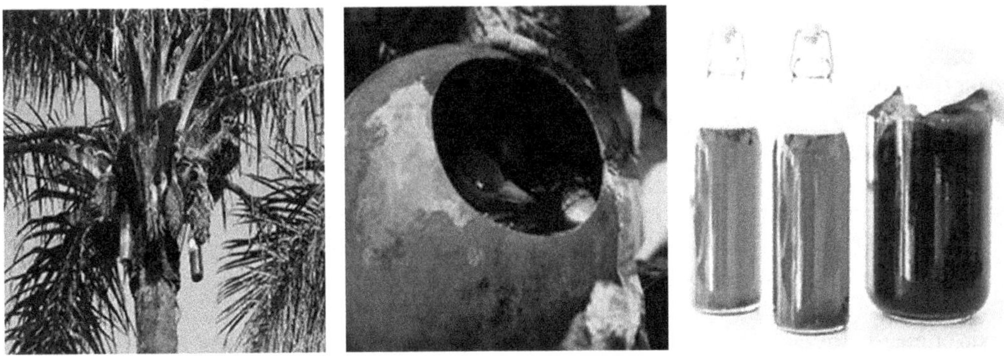

FIGURE 12.16 Palm wine created from the sap of various species of palm trees; hibiscus wine. (*See colour plate section for colour representation of this figure.*)

The non-cereal beverages vary across Africa, based on sugarcane, watermelons, banana, cassava, palm sap, pineapple, and others. In Ethiopia, an affordable wine (*tej*) is home-produced from a blend of honey and sugar as fermentable substrates. In Rwanda and The Democratic Republic of Congo, an artisanal banana beer (*urwaga*) is homemade or commercially produced from the fermentation of the *M. acuminata* variety. Palm wine, traditionally produced or commercially pasteurized, is known by several local names, with *malafu* being a common designation.

In Ghana, sorrel drink (*sobolo*) is an immune booster drink naturally high in antioxidants made from *Hibiscus sabdariffa* extract, while *asaana/elewonyo* is a refreshing tisane made from hibiscus, maize, and caramelized sugar (Figure 12.16). The seeds of this plant are widely used in alkaline fermentation to produce food condiments popularly known as *bikalga* (Burkina Faso), *dawadawa botso* (Niger), *datou* (Mali), *furundu* (Sudan), and *mbuja* (Cameroon).

12.9 EXAMPLES OF UNDESIRABLE EFFECTS OF FERMENTED PRODUCTS

Often, emphasis is made on the beneficial aspect of fermented food products. However, some of the microbiota used in the fermentation of foods and beverages may become harmful under certain undesirable conditions and pose a serious food safety challenge. Post-fermentation contamination of products may also affect the hygiene of the products thereby becoming disruptive to health and deleterious to life.

12.9.1 Mycotoxins in Indigenous Fermented Foods and Beverages

Most home-produced and processed grains and peanuts from which traditional beverages are made across the continent are frequently and excessively contaminated with different mycotoxins caused by poor farming, handling, post-harvesting, and storage practices that portray the region. Mycotoxins are non-toxic before metabolism, but after ingestion, they are metabolized to both toxic and non-toxic compounds, such as Cytochrome P450 enzymes.

There are many transcripts on the array and abundance of mycotoxins in the sources used in traditional beverage processing, nonetheless, few reports have analyzed mycotoxins in the final fermented products themselves.

The suggestion of making use of microorganisms to eliminate mycotoxins emerged long ago. Biological decontamination using microorganisms has revealed new opportunities. Mycotoxins are fungal metabolites commonly occurring in food, which pose a health risk to the consumer. The main fungi that infect cereals, roots, and tubers belong to *Aspergillus, Penicillium, Fusarium, Alternaria, or Claviceps* genera, they are associated with the production of mycotoxins (aflatoxins, fumonisins, zearalenone, and deoxynivalenol), aflatoxins, ochratoxin A, and patulin being the main risks in tropical crops.

Many countries have regulations specifying the maximum allowed concentration of mycotoxins, but there are no regulations requiring measures to reduce contamination. Only detoxification methods on feeds, but not in foods, have obtained a positive EFSA (European Food Safety Authority) scientific opinion and may be used. Sorghum, unlike other cereals, does not have yet legislation that regulates the maximum content of mycotoxins in grains for its commercialization.

Two main methods of microbial decontamination are identified in the literature: adsorption to the cell wall compounds (peptidoglycan, glucomannan, D-glucan) and biotransformation to less toxic or non-toxic compounds, thanks to the expression of appropriate enzymes. Atoxigenic strain-based biological control is a natural, nontoxic technology that uses the ability of native atoxigenic strains of *Aspergillus flavus* (one of the fungi that produce aflatoxin) to naturally outcompete their relatives, aflatoxin producers.

However, the levels of mycotoxin contamination in the raw materials should still not exceed the accepted levels established in international Regulations. For example, fumonisin esterase, which can degrade fumonisin mycotoxins and is produced by a genetically modified *Komagataella pastoris* yeast strain, has been approved by the EU but only for animal use. Therefore, in Africa, only good agricultural and manufacturing practices can help control the presence of mycotoxins as food contamination, and these are not common disciplines.

The risk of mycotoxin exposure continues in developing countries due to a lack of food security, poverty, and malnutrition. The occurrence of mycotoxin contamination in food is more prevalent in tropical and subtropical countries resulting in acute and chronic mycotoxicoses in humans and animals. Yeasts, especially *S. cerevisiae* and *Pichia kudriavzevii,* and LAB occur as part of the natural microbial population in spontaneous food fermentation and as starter cultures in the food and beverage industry.

Cooking and fermentation do little to reduce potential mycotoxin content, however, yeasts and LAB have immense potential as tools in tackling the problem of fungi and mycotoxins in cereal-based foods and beverages. The yeast *Saccharomyces boulardii, a* strain of *S. cerevisiae,* classified as a probiotic agent, has been used successfully in the treatment or prevention of mycotoxin activity. The question remains on the toxicity of products of enzymatic degradation and undesired effects of fermentation with non-native microorganisms on the quality of the final product.

The presence of uncertain microorganisms hinders the monitoring of the fermentation process, leading to finished products of variable quality. Some unwanted toxic contaminants and metabolites of the microbes are released into the food base during fermentation. Despite some progress, mycotoxins, bacterial toxins, biogenic amines, and cyanogen glycosides still frequently occur as toxic compounds in fermented foods and beverages.

Research is scarce while extension services and smallholders' family farmers do not have adequate knowledge of the specific substrates and type of the microbiota *inocula*. Therefore, the final fermented product quality at the household level remains unknown and difficult to be standardized.

Literature abounds on the microbiology of traditionally fermented foods and beverages, and mixed fermentation resulting in the production of diverse products. This microbiota has the enzyme complement to produce beneficial vitamins and amino acids for the host.

These foods have an important role in the African diet and can contribute to food security by increasing the availability of cheap, nutritious food and supporting livelihoods. However, the presence of foodborne pathogens and antibiotic-resistant bacteria in may constitute a health risk to consumers.

Furthermore, mixing different traditional and modern styles of eating and drinking, namely the use and abuse of diet sodas raises new hazards through the consumption of aspartame with an impact on several neurodegenerative diseases, including autism.

Methanol is found naturally in certain fruits and vegetables and the ingestion of as little as one to two teaspoons of this easily absorbed will cause death. Sugar will produce a very small amount of methanol, which is produced more from grains and their husks and organic fibers. It is also found naturally in trace amounts in fruit juice and distilled spirits such as whisky, wine, and beer. It may also be produced as an unintended by-product during the fermentation process.

Nevertheless, several hazards may occur through pathogenic contamination or the yield of methanol, which may impact health. The incidence of methanol contamination of traditionally fermented beverages is increasing globally resulting in the death of several people. Methanol is produced during fermentation by the hydrolysis of naturally occurring pectin in traditionally fermented beverages and can be linked to the activities of pectinase-producing yeast, fungi, and bacteria.

In addition to alcoholic drinks and accidental poisoning, another source of methanol is aspartame, which is used in over 6000 products, is a popular synthetic non-nutritive sweetener in diet sodas, and is nevertheless considered to be a potential carcinogenic. Acceptable Daily Intake (ADI) for aspartame of 40 mg/kg body weight was established in the early 80s.

Upon ingestion, aspartame is broken, converted, and oxidized into formaldehyde in various tissues. Aspartame (Figure 12.17) is fully broken down in the gut into aspartic acid and phenylalanine, which are absorbed and enter the body. In addition, the methyl group from the modified phenylalanine is released in the gut to form methanol.

Besides the exogenous methanol, there is endogenous metabolic methanol since in normal, healthy individuals, methanol and its short-lived oxidized metabolites, formaldehyde, and formic acid occur as a result of fermentation by gut bacteria and metabolic processes involving S-adenosyl-L-methionine.

These metabolites are associated with metabolic acidosis, visual dysfunction, and neurological symptoms. Regardless of the source, low levels of methanol in the body are maintained by physiological and metabolic clearance mechanisms, avoiding cell hypoxia.

During ripening, pectin in fruits is broken down by pectin methyl esterase resulting in the formation of methanol. Cases of methanol contamination have been reported in most traditional beverages at trace levels. Concentrations of 6–27 mg/l in beer and 10–220 mg/l in spirits are not harmful. There is however need for research to focus on other possible sources of methanol in locally traditionally fermented beverages.

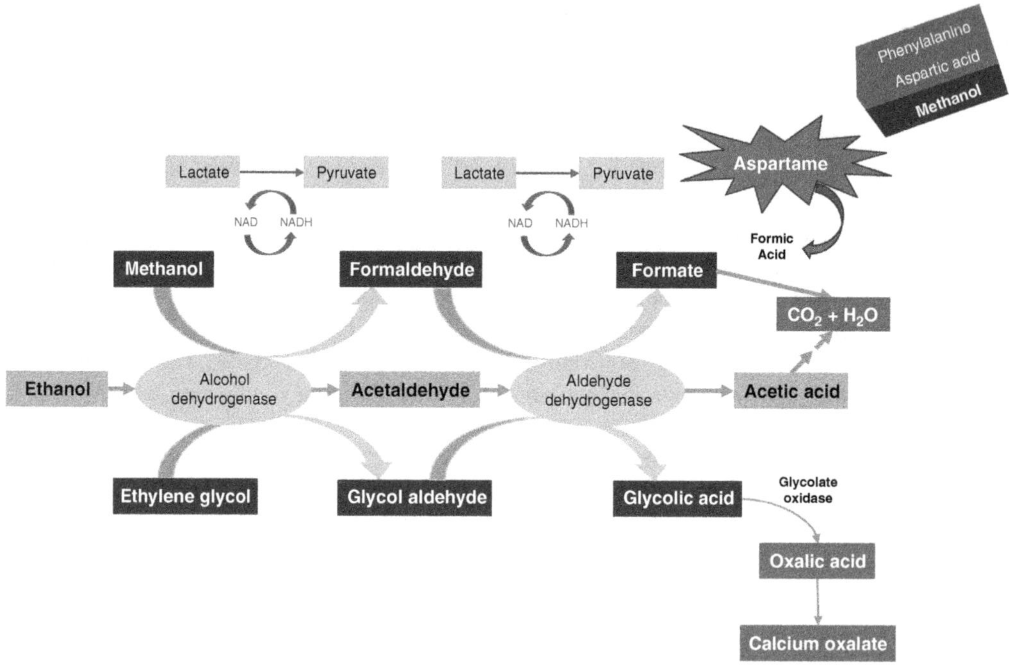

FIGURE 12.17 Aspartame, a well-known artificial sweetener of sodas, candy, chewing gum, and energy drinks, is a synthetic chemical composed of the amino acids phenylalanine and aspartic acid, with a methyl ester. The methanol produced by the metabolism of aspartame is absorbed and quickly converted into formaldehyde and then completely oxidized to formic acid. The acceptable daily intake (ADI) for aspartame is 40 mg/kg of body weight. Ethanol, drinking alcohol, is mainly produced from maize fermentation. Ethylene glycol, used as anti-freeze, if ingested is rapidly absorbed from the gastrointestinal tract and slowly absorbed through the skin or lungs. (*See colour plate section for colour representation of this figure.*)

In Mozambique, after a funeral, at least 75 people died and 235 suffered from severe intoxication after consuming traditional *pombe* (Swahili for "beer"), a type of traditional East African beverage made from malted sorghum, millet, maize, and sugar. The compound is boiled two or three times in huge pots, strained, when wanted clear, through a bag of matting, and allowed to ferment: after the third day, it becomes as sour as vinegar.

It was originally suggested that crocodile bile was used as the lethal toxin in the brew. However, the cause was identified as being contamination by the bacterium *Burkholderia gladioli* (previously named *Pseudomonas marginata*) which produces bongkrekic acid. This is implicated in outbreaks of food-borne illness involving also coconut- and corn-based products in other countries such as Indonesia and China. It is responsible to produce a lesser-known mitochondrial toxin and alter cellular apoptosis.

In Nigeria, recent findings have shown that Ochratoxin A has been detected in a non-alcoholic cereal-based fermented beverage (*kunun-zaki*), which also showed contamination by high counts of spoilage and pathogenic microorganisms (*Bacillus*) which may be responsible for its short shelf-life. In Nigeria, in 2015, a total of 89 persons died following the consumption of locally produced ethanol beverage, containing 16.3% methanol, called *kaikai/ogogoro/apeteshi* or illicit gin. *Kaikai* is produced mostly from the sap of raffia palm and oil palm and to a lesser extent from other palms such as date palm, and nipa palm. *S. cerevisiae* has been used as a catalyst for the production of ethanol for thousands of years. But recent studies have shown that there are different strains of *S. cerevisiae* involved in traditional ethanol fermentation in Africa.

In Kenya, alcohol-related deaths are common, with hundreds of people dying every year from poisoned liquor often sold in the brewer's home, others blinded, after drinking illegal alcohol laced with methanol. The drink is often known locally as *changaa* or *kumi kumi*.

In Cameroon, consumers who can no longer afford to buy good-quality whiskies now purchase fake, locally brewed alcoholic drinks which are dangerous to their health. Many people have developed liver problems and died from consuming fake whiskies, namely during the pandemic.

Despite the low prevalence of classical risk factors such as heavy drinking and tobacco smoking, there is evidence that significant levels of ethanol and acetaldehyde were produced during fermented milk *mursik* production in Kenya. This may justify the high incidence of esophageal squamous cell cancer in Africa even in young people (<30 years of age) due to their carcinogenic potential.

The microbiota in fermented products impacts the organoleptic profile, the physicochemical characteristics as well as human health. Microbiota can also release from the food matrix toxic compounds, the most notorious being decarboxylation products of amino acids (e.g. biogenic amines, histamine, tyramine, and putrescine, by decarboxylation of histidine, tyrosine, and ornithine, respectively) and mycotoxins.

12.9.2 Antimicrobial Resistance

The existence of antimicrobial resistance (AMR) in the environment, a growing public health concern, has been well documented although transmission routes of antimicrobial agents and genes across different ecosystems remain obscure. Multidrug-resistant bacteria is a natural phenomenon and has been identified in unexpected extreme weather sites such as the Sahara Desert and both Poles. In Africa, most countries have not established AMR surveillance systems.

During fermentation, bacteria multiply significantly, and various genes, including AMR genes, can be exchanged by bacteria during horizontal gene transfer processes. If any of those bacteria harbor AMR genes, the number of bacteria with resistance genes could be increased in the final products (e.g. fermented dairy products) and lead to critical health considerations.

Microorganisms present in each traditional fermented product, where the starting culture strains of fermented foods are not monitored, could act as resistance superbugs since several reports indicate that fermented produce could be considered as vehicles of antibiotic-resistant bacteria and, thus, AMR genes (e.g. *poxtA* gene) may be horizontally transferred to other bacteria including pathogens and commensals through the food.

12.10 CONCLUSIONS

Africa is well-known for its exotic traditionally fermented food and beverages produced using a wide range of raw materials, microorganisms, and fermentation processes. These artisanal practices need rejuvenation following traditional forms of art practice or craft making. But until these fermentation practices come under new improved safety and risk control they will remain beyond any official compliance with national regulatory standards. There also is a lack of appropriate surveillance programs that can supply information on consumption patterns and a lack of epidemiological data, which are necessary to properly evaluate the link between fermented products and health.

Indigenous crops and fruits constitute an important part of human diets in many Sub-Saharan African countries, particularly in rural areas and during droughts. To promote and expand the utilization of these products when fermented, knowledge of their nutritional composition is essential.

Despite the high potential of these heritage-fermented foods and beverages, only limited and outdated data on basic nutrient composition exist on raw indigenous foods. Traditional methods of manufacture should take advantage of technical progress to assure a reasonable margin of health benefit, and at the same time conserve the authenticity and origin(s) of these products, trying to transfer these ancestral technologies to small and medium industrial scales.

Notwithstanding scientific advances, our knowledge of the effects of fermented foods and beverages, and the precise associations of their microbiota on human health remains incompletely understood. Again, fermentation-dependent biomarkers have not yet been well documented.

While African indigenous fermented products offer a vast genetic potential of undiscovered microbiota strains that possess valuable technical characteristics, these fermented outcomes may also serve as vehicles of pathogenic and antibiotic-resistant bacteria and genetic determinants. Significant research and data gaps exist regarding the microbiological safety of these food products, which warrant urgent attention.

The restrictions and inconsistencies in the current portfolio of evidence imply that no absolute conclusions can yet be depicted on the potential health benefits of traditional fermented products although it may be expected that the microbiomes of fermented foods will acquire a leading role in future nutrition and therapeutics.

In summary, there is very scarce preclinical and clinical confirmation for the validity of most traditional fermented foods and beverages in gastrointestinal health and disease.

FURTHER READING

Abriouel, H., Knapp, C.W., Gálvez, A. et al. (2017). Chapter 29 – Antibiotic resistance profile of microbes from traditional fermented foods. In: *Fermented Foods in Health and Disease Prevention* (ed. J. Frias, C. Martinez-Villaluenga, and a.E. Peñas)), 675–704. Academic Press. https://doi.org/10.1016/B978-0-12-802309-9.00029-7.

Anyogu, A., Olukorede, A., Anumudu, C. et al. (2021). Microorganisms and food safety risks associated with indigenous fermented foods from Africa. *Food Control* 129: 108227. https://doi.org/10.1016/j.foodcont.2021.108227.

Bautista-Gallego, J., Arroyo-López, F.N., Bordons, A. et al. (2019). Editorial: New trends in table olive fermentation. *Frontiers in Microbiology* 10: 1880. https://doi.org/10.3389/fmicb.2019.01880.

Bell, V., Ferrão, J., and Fernandes, T. (2018). Fermented food guidelines for children. *Journal of Pediatrics and Pediatric Medicine* 2 (1): 1–4. http://www.pediatricsresearchjournal.com/articles/fermented-food-guidelines-for-children.pdf.

Bell, V., Ferrão, J., and Fernandes, T. (2018). Fermented products in nutrition. *Advances in Plants & Agriculture Research* 8 (2): 138–140. https://doi.org/10.15406/apar.2018.08.00303. https://medcraveonline.com/APAR/APAR-08-00303.pdf.

Bell, V., Ferrão, J., Pimentel, L. et al. (2018). One health, fermented foods, and gut microbiota. *Food* (7): 195. https://doi.org/10.3390/foods7120195.

Benkerroum, N. (2013). Traditional fermented foods of north African countries: technology and food safety challenges with regard to microbiological risks. *Comprehensive Reviews in Food Science and Food Safety* 12 (1): 54–89. 10.1111/j.1541-4337.2012.00215.x.

Bielik, V. and Kolisek, M. (2021). Bioaccessibility and bioavailability of minerals in relation to a healthy gut microbiome. *International Journal of Molecular Sciences* 22 (13): 6803. https://doi.org/10.3390/ijms22136803.

Dimidi, E., Cox, S.R., Rossi, M. et al. (2019). Fermented foods: definitions and characteristics, impact on the gut microbiota and effects on gastrointestinal health and disease. *Nutrients* 11 (8): https://doi.org/10.3390/nu11081806.

Dirar, H.A. (1996). The fermented foods of Africa: their potential. The 50th anniversary UNIDO. Vienna, Austria. https://open.unido.org/api/documents/4806319/download/the%20fermented%20foods%20of%20africa%20-%20their%20potential%20%2821156.en%29

Dorokhov, Y.L., Komarova, T.V., Petrunia, I.V. et al. (2012). Methanol may function as a cross-kingdom signal. *PLoS one* 7 (4): e36122. https://doi.org/10.1371/journal.pone.0036122.

FAO, ECA and AUC (2021). *Africa – Regional Overview of Food Security and Nutrition 2021: Statistics and Trends*. Accra, FAO https://doi.org/10.4060/cb7496en.

Flibert, G., Tankoano, A., and Savadogo, A. (2016). *African cassava* traditional fermented food: the microorganism's contribution to their nutritional and safety values – a review. *International Journal of Current Microbiology and Applied Sciences* 5 (10): 664–987. https://doi.org/10.20546/ijcmas.2016.510.074.

González-González, F., Delgado, S., Ruiz, L. et al. (2022). Functional bacterial cultures for dairy applications: towards improving safety, quality, nutritional and health benefit aspects. *Journal of Applied Microbiology* 133 ((1): 212–229. https://doi.org/10.1111/jam.15510.

Johansen, P.G., Owusu-Kwarteng, J., Parkouda, C. et al. (2019). Occurrence and importance of yeasts in indigenous fermented food and beverages produced in sub-Saharan Africa. *Frontiers in Microbiology* 10: 1789. https://doi.org/10.3389/fmicb.2019.01789.

Kapravelou, G., Martínez, R., Martino, J. et al. (2020). Natural fermentation of cowpea (*Vigna unguiculata*) flour improves the nutritive utilization of indispensable amino acids and phosphorus by growing rats. *Nutrients* 12 (8): 2186. https://doi.org/10.3390/nu12082186.

Kumar, M.R., Azizi, N.F., Yeap, S.K. et al. (2022). Clinical and preclinical studies of fermented foods and their effects on Alzheimer's disease. *Antioxidants* 11 (5): 883. https://doi.org/10.3390/antiox11050883.

Li, K.J., Brouwer-Brolsma, E.M., Burton-Pimentel, K.J. et al. (2021). A systematic review to identify biomarkers of intake for fermented food products. *Genes & Nutrition* 16: 5). https://doi.org/10.1186/s12263-021-00686-4.

Mattiello, S., Caroprese, M., Matteo, C.G. et al. (2018). Animal productions in development cooperation projects. Typical dairy products in Africa from local animal resources. *Italian Journal of Animal Science* 17 (3): 740–754. https://doi.org/10.1080/1828051X.2017.1401910.

Motlhanka, K., Zhou, N., and Lebani, K. (2018). Microbial and chemical diversity of traditional non-cereal based alcoholic beverages of sub-Saharan Africa. *Beverages* 4 (2): 36. https://doi.org/10.3390/beverages4020036.

Mudau, F.N., Chimonyo, V.G.P., Modi, A.T. et al. (2022). Neglected and underutilised crops: a systematic review of their potential as food and herbal medicinal crops in South Africa. *Frontiers in Pharmacology* 12: https://doi.org/10.3389/fphar.2021.809866.

National Research Council. (1992). Applications of biotechnology to traditional fermented foods. https://pdf.usaid.gov/pdf_docs/pnabz036.pdf

Nieminen, M.T., Novak-Frazer, L., Collins, R. et al. (2013). Alcohol and acetaldehyde in African fermented Milk *Mursik* – a possible etiologic factor for high incidence of esophageal cancer in Western Kenya. *Cancer Epidemiology, Biomarkers & Prevention* 22 (1): 69–75. 10.1158/1055-9965.EPI-12-0908.

Nzigamasabo, A. and Nimpagaritse, A. (2009). Traditional fermented foods and beverages in Burundi. *Food Research International* 42: 588–594. https://doi.org/10.1016/j.foodres.2009.02.021.

Obafemi, Y.D., Oranusi, S.U., Ajanaku, K.O. et al. (2022). African fermented foods: overview, emerging benefits, and novel approaches to microbiome profiling. *npj Science of Food* 6: (15). https://doi.org/10.1038/s41538-022-00130-w.

Ohimain, E.I. (2016). Methanol contamination in traditionally fermented alcoholic beverages: the microbial dimension. *SpringerPlus* 5: 1607). 10.1186/s40064-016-3303-1.

Olasupo, N.A. and Okorie, P.C. (2019). African fermented food condiments: microbiology impacts on their nutritional values. *In: Frontiers and New Trends in the Science of Fermented Food and Beverages (ed. R.L. Solís-Oviedo and Á. de la Cruz Pech-Canul). IntechOpen.* https://doi.org/10.5772/intechopen.83466.

Onyeaka, H., Ekwebelem, O.C., Eze, U.A. et al. (2021). Improving food safety culture in Nigeria: a review of practical issues. *Foods* 10 (8): 1878. https://doi.org/10.3390/foods10081878.

Shi, H., Yang, E., Li, Y. et al. (2021). Effect of solid-state fermentation on nutritional quality of leaf flour of the drumstick tree (*Moringa oleifera* Lam.). *Frontiers in Bioengineering and Biotechnology* 9: https://doi.org/10.3389/fbioe.2021.626628.

Staudacher, H.M. and Nevin, A.M. (2019). Fermented foods: fad or favourable addition to the diet? *Lancet Gastroenterology and Hepatology* 4 (1): 19. https://doi.org/10.1016/S2468-1253(18)30392-3.

Tóth, A.G., Csabai, I., Maróti, G. et al. (2020). A glimpse of antimicrobial resistance gene diversity in kefir and yoghurt. *Scientific Reports* 10: 22458. https://doi.org/10.1038/s41598-020-80444-5.

Wafula, E.N., Franz, C.M.A.P., Rohn, S. et al. (2016). Fermentation of African indigenous leafy vegetables to lower post-harvest losses, maintain quality and increase product safety. *African Journal of Horticultural Science* 9: 1–13. https://www.researchgate.net/publication/303696475.

Wastyk, H.C., Fragiadakis, G.K., Perelman, D. et al. (2021). Gut-microbiota-targeted diets modulate human immune status. *Cell* 184 (16): P4137–4153.E14. https://doi.org/10.1016/J.CELL.2021.06.019.

WFP. (2022). WFP at a glance. https://www.wfp.org/stories/wfp-glance Accessed July 20, 2022.

CHAPTER 13

FERMENTED FOODS OF SOUTH ASIA

GOMATHY M[1], SABARINATHAN K.G[1], RAJAKUMAR D[1], MANOJ M[2], BHIMANI H.D[3], PRIYA JOHN[4], THARANEE WIJAYARATNE[5], SAGARIKA EKANAYAKE[5], UMA MAHESWARI[1], ANU S. RAJAN[6], DILIP SAIKIA[7,8], DILIP TAMANG[9], SIMANA BORA[7,8], THERADIMANI M[1], AND RAMYA P[1]

[1]Department of Agricultural Microbiology, Tamil Nadu Agricultural University, Tamil Nadu, India

[2]Department of Microbiology, Navsari Agricultural University, Gujarat, India

[3]Division of Microbial & Environmental Biotechnology, Navsari Agricultural University, Gujarat, India

[4]Department of Plant Pathology, Navsari Agricultural University, Gujarat, India

[5]Department of Biochemistry, Faculty of Medical Sciences, University of Sri Jayewardenepura, Nugegoda, Sri Lanka

[6]Department of Agricultural Microbiology, Kerala Agricultural University, Kerala, India

[7]Department of Physics, Silapathar College, Silapathar, Assam, India

[8]Department of Botany, Bahona College, Assam, India

[9]Kakajan Govt. M.E. School, Assam, India

CONTENTS

Microbial Fermentations in Nature and as Designed Processes, First Edition. Edited by Christon J. Hurst.

13.1 INTRODUCTION

Since ancient times, fermented foods have been consumed as a part of our daily diet. Fermentation is considered one of the most common and oldest methods for the preservation of foods. Fermentation of foods is usually done to increase the nutritive value, taste, flavor, and pharmacological properties, and reduce antinutritional factors in addition to enhancing the shelf life of food by preserving excess raw material. It was also considered to be a good means for preserving food to use in lean seasons and during famine. It is a cost-effective food preservation method mostly being done at the household level or in cottage industries until now (Reddy and Joshi 2019).

Traditional fermented products are very diverse hence indigenous to each region. India has not only got rich, diversified traditions and culture but also numerous artisanal fermented food products. The local grapevine juices are used to make Soma juice or Soma rasam, which is the earliest documented fermented product in India and has been known since the early Vedic era. In addition, it is known to have medical properties and has been used in worship. Idli and Curd were mentioned as early as 700 BCE in ancient literature and signified the popularity of Indian fermented foods. The fermented products varied based on the nature of the raw material used, the microflora involved, and the cultural conditions prevalent in a particular region (Sekar and Mariappan 2007). The method of preparation depends on the traditional knowledge of regional native people which they acquired from the ethno-cultural knowledge of their lineages. It is produced as a basic staple diet (idli, dhokla, curd, koozh), pickling (vegetable pickle), confectionery (jalebi, athirasam), alcoholic beverages (wine, beer), etc. (Sathish Kumar et al. 2012).

In the modern era, one of the severe problems faced by most people is malnutrition which includes undernutrition, inadequate vitamins or minerals, obesity, overweight, and other diet-related non-communicable diseases (NCDs). It is the right time to switch attention to more easily prepared and cheaper fermented food substitutes to manage malnutrition. Moreover, the consumption of fermented foods will improve digestion and immunity.

Fermented foods are produced worldwide using various manufacturing techniques, raw materials, and microorganisms. Alcohol fermentation produces ethanol, and yeasts are the predominant organisms (e.g. wines and beers). Lactic acid fermentation (e.g. fermented milks and cereals) is mainly carried out by lactic acid bacteria (LAB). *Acetobacter* converts alcohol to acetic acid in the presence of excess oxygen. There are many native or traditional fermented dishes and drinks that can be made, and this is still considered a domestic skill.

One of the most significant sources of dietary fiber, vitamins, minerals, carbs, and proteins for people worldwide is cereal grains. Cereals can have lower nutritional value and inferior sensory qualities than milk and milk derivatives. The lower protein concentration, lack of lysine and other necessary amino acids, the limited availability of starch, the presence of phosphates bound as phytic acid, and specific antinutrients such as tannins, and polyphenols, and the coarse texture of the grains are the causes of this inferior quality (Chavan and Kadam 1989). To enhance the nutritional qualities of cereals, a variety of processing techniques, such as cooking, sprouting, milling, and fermentation, have been used. However, fermentation is likely the most effective (Mattila-Sandholm et al. 2002). The level of carbohydrates and some non-digestible poly and oligosaccharides generally decreases with the spontaneous fermentation of cereals. During fermentation, B group vitamins' accessibility may be increased, and some amino acids may be synthesized. Additionally, fermentation creates the ideal pH conditions for the enzymatic breakdown of phytate found in cereals which increases the amount of soluble iron, zinc, and calcium. Fermentation affects the nutritive value of foods and the effect is variable. Recent research has demonstrated the presence of probiotics, which are perceived as microbial food supplements, in these naturally fermented foods, which are helpful to consumers. These live microorganisms enhance gut health by reestablishing a healthy level of gut bacteria. This chapter focuses on fermented foods from different parts of south Asia and provides a basic discussion about food fermentation and mentions different types of food fermentation and the role of fermented foods in food safety and food preservation.

13.2 ORIGIN, HISTORY, AND ROLE OF FERMENTATION

Fermentation is a desirable process that biochemically modifies primary food products (carbohydrates) using microorganisms and their enzymes (Karovičová and Kohajdová 2005). It is a natural process, partly decaying when fruits and vegetables are considered. Fermentation improves the texture, taste, nutritional value, and aroma while extending the shelf life of the fermented food product (Fellows 2000). Consequently, fermented food products offer various health benefits to consumers.

13.2.1 Food Preservation

Preservation of food and beverages through fermentation has been used to extend the shelf-life of foods for millennia. Fermentation was not invented, but rather discovered. The concept of beer likely was discovered when a few barley grains were left in rain. In earlier times, when there was a need to protect excess food from spoilage and microbial poisoning, the fermentation technique of food preservation came into practice. Fermentation is a valuable food preservation technique that both preserves and creates nutritious foods. In modern days for biopreservation and large-scale production defined strain starter systems are used to extend food shelf-life and to improve food safety as they produce a wide range of antimicrobial and proteinaceous substances which competitively reduce undesirable microflora in food products and ultimately preserve food.

13.2.2 Fermented Foods

Since neolithic times, human beings are known to make fermented foods. About 10 000 BCE humans were naturally fermenting notoriously poor-holding dairy foods, particularly milk of camels, goats, sheep, and cattle. For earlier societies, this transformation of basic food into fermented food was a miracle as this fermentation was not done intensely as it happened spontaneously due to naturally existing microflora. Metchinoff noted that the x lifespan (87 years) of Bulgarians was due to greater consumption of fermented milk and the process of fermentation was due to *Bulgarian bacillus* (Clime 2022).

That microbe now has the official name *Lactobacillus delbrueckii* subsp. *bulgaricus*. The earliest types of fermented food were beer, wine, leavened bread, and cheese, followed by East Asian fermented foods such as yogurt, pickles, sauerkraut, vinegar, and traditional alcoholic beverages.

13.2.3 Fermentation Ecosystem

Fermentation is supported by a rich and dense collection of microbes. Fermentative microbes interact and support one another in a complex food web with the waste products of some species serving as nutrients for other species (Bowen 1999). A fermentation ecosystem embodies the succession of species, partitioning of resources, disturbance, and equilibrium found in larger ecosystems. Today, all human cultures rely on fermented food products and the microbes that produce them. Fermentation is not only a component of human ecology but also has an ecology of its own. Fungi and bacteria are the two kingdoms of life in fermentation ecosystems (Scott and Sullivan 2008). In simple words, raw food provides the ideal environment for a pioneer microbe to grow, use nutrients, alter chemical characteristics, especially pH, and produce metabolic wastes that create suitable conditions for another microbe to establish itself. Likewise, each microbe successively helps by creating favorable conditions for its successor and this ultimately forms a fermentation ecosystem.

13.2.4 Role of Microflora in Fermented Foods

The major role in fermented food is played by microorganisms which may be either spontaneous microflora or specific starter cultures. These microorganisms convert the raw material into organoleptically edible foodstuffs, valued for being nutritionally rich with increased palatability and safety. Indigenous fermented foods are largely dominated by LAB and also other bacteria, yeast, and fungi. The microorganisms used in fermentation are probiotic in nature, which means they will enhance the consumer's health and act as an integral component of the gut microflora thereby maintaining intestinal microbial balance and overall health. The predominant LAB is of two types viz., homo fermentative and heterofermentative by producing lactic acid alone or a combination of lactic acid, alcohol, and carbon dioxide (Angmo et al. 2016; Rawat et al. 2018).

Many LAB are considered to be beneficial as probiotic organisms with suggestions having been made that consumption of these organisms offers benefits that positively impact consumer health. LAB may help to control intestinal infections by competing against pathogenic bacteria for nutrients, improving the immunity of the consumer, and stabilizing the intestinal flora of prolonged antibiotic users (Ranasinghe and Perera 2016). *Lactobacillus acidophilus* and *Streptococcus lactis* produce lactic acid, acetic acid, and bacteriocins which may help to inhibit pathogenic infections (Ershidat and Mazahreh 2009). *Lactobacillus* strains additionally synthesize peroxide and pyroglutamate-like compounds that may inhibit pathogens (Ershidat and Mazahreh 2009). It has been speculated that LAB might contribute to reducing serum cholesterol levels and therefore aid in the prevention of hypercholesterolemia (Parvez et al. 2006). Furthermore, *Streptococcus thermophillus* and *Bifidobacterium* produce folic acid and act as good sources of vitamins (Crittenden et al. 2003).

13.2.5 History of Fermented Foods

Almost every culture and geographic area seem to have produced fermented foods for millennia. People have been teaming up with natural microbes for an even much longer time. Ancient Chinese used fermented grains for rice wine as early as 4000 years ago ("Fermentation: A History" 2017). The origins of fermentation are hard to track down, however, historians have traced signs of fermentation in food and beverage preparation dating as far back as 10 000 BCE (Table 13.1).

TABLE 13.1 History of fermented foods.

Dates	Development
10 000 BCE	Spontaneous dairy fermentation, first documented yogurt in North Africa by spontaneous fermentation of milk in searing temperature
7000 BCE	The world's oldest brewed beverage made up of fermented rice, honey, and fruit was discovered on Neolithic Chinese pottery
3500–300 BCE	Ancient Egyptians perfect the art of leavening bread, they first experimented dough fermentation with yeast
2000 BCE	Pickling of cucumbers in middle east, Pickled cucumbers originated in the Tigris Valley (modern day Iraq)
500 BCE	The first documented use of fermented food as medicine was use of moldy soybean curds as antibiotic for boils and carbuncles by Chinese
200 BCE	Tea leaf fermentation was first documented in ancient Chinese civilizations the name of fermented food product – kombucha
500–1000 CE	Cereal-legume fermentation of diverse cereals and legumes led to a variety of fermented products produced around the world
1856	Louis Pasteur discovered that fermentation requires live cells
1910–1935	Identification of bacterial strains, Elie Metchnikoff discovered the bacterial strain *B. bacillus* now known as *L. delbrueckii* subsp. *bulgaricus* in fermented milk
1970–2021	Production of probiotics, Metchnikoff called lactobacillus as probiotic

Source: Debret (2021).

13.3 FERMENTATION OF FOODS

13.3.1 Types of Fermentation

There are many different types of fermentation that each produce different fermented foods that humans consume. Different acids which are produced during fermentation processes extend the preservation period of food without losing its quality and desired characteristics.

Based on the end-product formed, different types of fermentation have been distinguished and some of these are depicted in Figure 13.1.

13.3.2 Lactic Acid Fermentation

It is the anaerobic fermentation in which LAB obtain energy by metabolizing sugars and converting them into lactic acid. Rabdi is a famous lactic acid-fermented milk product, prepared in the North-West part of India using pearl millet. Dhokla is an indigenous fermented food made through lactic acid fermentation mainly found in Gujarat and Maharashtra. Shrikhand is a popular western Indian lactic acid fermented food (Sathe and Mandel 2016). Dahi rich in LAB shows probiotic effects (Agarwal and Bhasin 2002).

13.3.3 Alcoholic Fermentation

In alcoholic fermentation, yeast or bacteria utilize and convert sugars, such as glucose, sucrose, and fructose, into energy molecules. During that anaerobic process, the microbes produce ethanol and carbon dioxide. Alcoholic fermentation involves glycolysis (glucose to pyruvate) and fermentation (pyruvate to ethanol and CO_2). It results in the production of various alcoholic beverages, like wines, beers, and distilled liquors. Specific microbial strains impart the characteristic flavors, associated with different alcoholic beverages (Ashwathi 2016). Soor, chhang, and kanji are important alcoholic beverages in Northern India, especially in Uttarakhand. Sur-madda feni is the ethnic alcoholic beverage in Goa.

13.3.4 Acetic Acid Fermentation

Acetic acid fermentation by microbial oxidation converts alcohol into acetic acid under aerobic conditions. Acetic acid bacteria (AAB) are used for the industrial production of vinegar. Vinegar is a fermented food typically containing 4 g of acetic acid per 100 ml. Acetic acid is produced

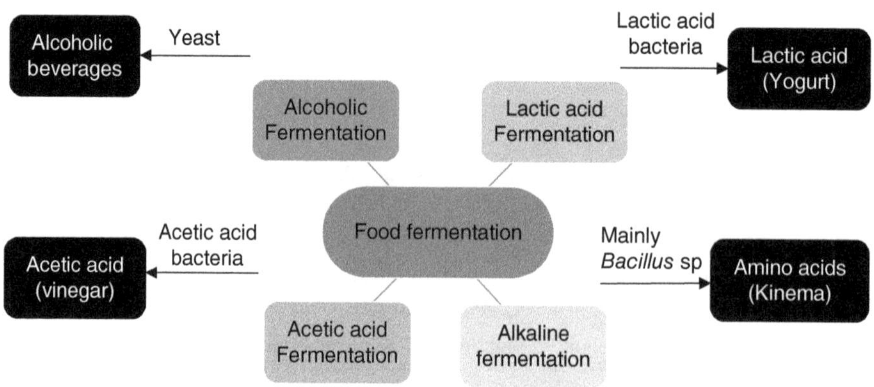

FIGURE 13.1 Types of food fermentation. (*See colour plate section for colour representation of this figure.*)

commercially by surface fermentation and submerged fermentation methods. In both methods, first yeast produces ethanol from sugar, then later acidic acid bacteria convert the ethanol to acetic acid (Dashmeet 2018). Kombucha is AAB fermented tea in North India especially popular in Ladakh and Himachal Pradesh.

13.3.5 Alkaline Fermentation

In alkaline fermented foods, during fermentation, the protein content of raw materials is broken down into amino acids and peptides releasing ammonia which raises the pH of final products and gives a strong ammonia smell to the food. Most of the alkaline fermentations are achieved by spontaneously using a mixed bacterial culture dominated by *Bacillus subtilis* (Wang and Fung 1996). Alkaline fermented foods are generally safe due to their high pH. Alkaline fermented foods mostly are consumed in North-East India compared to North-West India.

13.3.6 Leavened Bread Fermentation

In bread fermentation, the substrate is carbohydrates present in bread and the strain of yeast is *Saccharomyces cerevisiae*. The carbohydrates are broken down into simple sugars by the yeast through aerobic and anaerobic respiration producing alcohol and carbon dioxide. Leavening is a process that causes dough to rise through the formation of carbon dioxide bubbles. Sourdough breads are fermented with LAB and yeast. Loochi is the leavened fried bread found indigenous to Northern India, found especially in Jammu and Kashmir.

13.4 SOUTH ASIAN COUNTRIES AND INDIGENOUS FERMENTED FOODS – INDIA

Fermentation is among the techniques and procedures discovered to preserve food items for a long periods while maintaining flavor, taste, and, most importantly, nutrients. It is considered one of the oldest forms of food technology discovered on the Indian subcontinent (Deka 2012). Regional fermented foods reflect the traditions, culture, and legacy of the ethnic people who reside there. Recent reports have shown that foods and alcoholic beverages contain different types of microbial flora thereby enhancing the nutrient quality and durability.

13.4.1 Indigenous South Indian Fermented Foods

Tamil Nadu India is divided into multiple regions and, as a result, people's eating habits and the methods used to ferment certain products vary according to location. Based on this fact, the indigenous fermented foods can be divided into the following types depending upon the substrate used: (1) fermented cereal foods, cereal-legume/legume-based products; (2) fermented milk products; (3) fermented vegetables; and (4) fermented beverages.

Fermented Cereal Foods In their respective regions, fermented dishes made from cereal are regarded as staple foods. The majority of dishes, including idli, dosa, dhokla, adai dosa, kallappam, koozhu, ambali, jalebi, athirasam, pazhaiya soru (fermented sour rice), etc., are regularly consumed by local inhabitants. Cereals and millets such as rice, wheat, maize, ragi, cumbu, and legumes like black gram, green gram, fenugreek, Bengal gram, and red gram are mainly used in indigenous fermented food preparations.

Idli and Dosa, which involve natural/spontaneous fermentation, are prepared by fermenting rice, black gram dhal, fenugreek seeds, and table salt. Adai dosa is prepared by grinding and fermenting boiled rice, Bengal gram, red gram, black gram, and green gram with spices. In the dish of Kallappam, boiled or raw rice is fermented after the addition of coconut water. The excessive leftover cooked rice is fermented overnight with the addition of salt and water and is popularly termed and commonly consumed as Pazhaiya soru for breakfast. Appalam/Papadam is prepared by fermenting black gram flour for four to six hours and then making it into flat, spicy, circular dried wafers which are eaten after frying in oil. Koozh/Kali is yet another of the oldest types of fermented food, and it is prepared by fermenting ragi flour or cumbu flour, boiled rice, curd, and salt. Ambali is native to Karnataka and Tamil Nadu which is like that of koozh. Due to their high energy level and health benefits, koozh, palaya soru, and ambali was once consumed regularly for breakfast with buttermilk, green chili, onions, and pickles to improve the flavor. Such indigenous cereal-/legume-based fermented foods are dominated by a few typical microorganisms viz. *Lactiplantibacillus plantarum, Leuconostoc mesenteroides, Limosilactobacillus fermentum, Levilactobacillus brevis, Enterococcus faecalis,* and *Saccharomyces* sp. Athirasam is a dessert similarly prepared by fermenting rice and jaggery (Figure 13.2).

Idli with sambar and Chutney

Athirasam

Pazhaiya soru

FIGURE 13.2 Fermented foods of Tamil Nadu.

Fermented Milk Products Products made from fermented milk have nutritional and therapeutic value. South India is a tropical region, thus long ago people began making sour milk because it is more stable and contains higher-quality nutrients than fresh milk.

Vedas and Upanishads mention curd, the oldest fermented milk product, as having originated between 400 and 6000 years ago (Indian holy books). Curd preparation and consumption have been documented in India since 2000 BCE (Yegna Narayan Aiyar 1953). Curd is revered as a sacred food and is consumed during numerous religious rituals and festivals among Hindus and Buddhists. It is an ethnic Indian fermented milk product that originated from a combination of ancestor knowledge, artisanal skills, and the particular consortia of bacteria that were abundant in the area. Over time, it has become an essential part of their culinary heritage. The well-studied microbial cultures in curd are *Lactococcus lactis* subsp. *lactis, Lactococcus cremoris, L. lactis* subsp. *lactis bv. diacetylactis, Streptococcus thermophilus, L. delbrueckii* subsp. *bulgaricus, L. acidophilus,* and *Lactobacillus helveticus*. Curd is used for prophylaxis in Ayurveda and is also known to have antimicrobial activity against various pathogens. Buttermilk is yet another popular fermented milk platter across India and is relished as a refreshing drink. Buttermilk is traditionally made as a byproduct of butter by churning curd while simultaneously adding water until butter granules are created and separated. Additionally, it can be made by combining curd with liquid, sugar, spices, etc. Since curd serves as the foundation for buttermilk, its nutritional value, bio-functional characteristics, and microbial profile are strikingly similar and loaded with phospholipids. South India has got its own basket of traditional fermented milk products. Traditional South Indian raita called Pachadi is a product made from curd that is mildly spiced with vegetables, coconut, ginger, curry leaves, and mustards, and served as a side dish with rice, Idli, and dosa.

Fermented Vegetables One of the earliest techniques for food preservation is pickling. Fresh, fully developed fruits and vegetables are washed, cut into the necessary sizes, and then the necessary amounts of salt, mirchi (green chili peppers), oil, and other ingredients are added. Mango pickle is among the most well-known pickles and they are all made similarly. To achieve a variety of flavors, tastes, and qualities, pickles are typically made from immature or mature unripe mangoes, with or without peeling or destemming. Pickle making is best suited for acidic and fibrous kinds of fruits. Salting the pickles limits the growth of gram-negative bacteria and promotes the growth of LAB. *L. mesenteroides, Leuconostoc fallax*, and *L. plantarum* are the predominant microorganisms in all varieties of pickles.

Fermented Beverages The fermented sap of some types of palm trees is known as palm wine. In South India, palm wine has much popularity (Shamala and Sreekantiah 1988). Palm wine is prepared by collecting and fermenting fresh palm sap (10–12% sugar) from palmyra and date palm. The sugar in the sap is partially fermented to alcohol, which aids in the product's preservation. Soon after the sap is collected, fermentation begins and within an hour the alcohol content reaches up to 4%. Toddy is an alcoholic beverage created by fermenting coconut palm flower sap. The sugar content ranges from 12–15% in fresh coconut sap. After fermentation, it is white, sweet, and flavorful with an alcohol content ranging between 4 and 6%. Yeasts found in the sap and those added from a previous batch of toddy cause a natural fermentation. Overall, lactic and acetic acids make up 1–1.5% of palm wine, which also contains 5% ethanol, B12, and ascorbic acid.

Kerala Traditional fermented dishes made from rice are popular in Kerala. Many of these products have complex and unknown microbiology. Most of these foods undergo a natural fermentation, which includes a blend of yeast, bacterial, and fungal cultures. While some microbes may interact simultaneously others may operate sequentially, changing the dominant flora as the fermentation progresses. After cleaning, the cereal grains are soaked in water which slowly increases the growth of natural microbial flora and is subsequently dominated by LAB. The fermentable sugars produced by the activity of endogenous amylases serve as a source of energy for the LAB.

The common fermenting bacteria are species of *Leuconostoc, Lactobacillus, Streptococcus, Pediococcus, Micrococcus, and Bacillus*. The fungi genera *Aspergillus, Paecilomyces, Cladosporium, Fusarium, Penicillium, and Trichothecium* are predominant in certain products. The common fermenting yeasts are species of *Saccharomyces*, which result in alcoholic fermentation. Some of the traditional fermented foods of Kerala are the Appam, sour rice or Pazhankanji, Unniappam, Neyyappam, etc.

Appam The origin of appam is attributed to St. Thomas Christians who are Malayali Syriac Christians of Kerala (also called Syrian Christians, Malankara Nasrani, or Nasrani Mappila). Their origin is traced to Thomas the Apostle whose evangelistic work in Kerala lasted 20 years (from 52 to 72 AD). The word "appam" may have derived from the Syriac verb "apya" ("apo" in West Syriac). Interestingly, all traditional bread used by the Syrian Christians (Nasranis) of Kerala ends with the word appam (Kallappam, Paalappam, Achappam, Cheeppappam, Neyyappam, Unniyappam, etc.).

Appam is a type of pancake made from fermented rice batter and coconut milk. It is a traditional breakfast in Kerala. Appam is thick in the center but thin on the edges. They can be made using yeast, or toddy for fermentation, and the taste of each variant changes with the ingredients. Appams have crisp and brown edges when the batter is left to rise using yeast. Kallappams are small and thick pancakes flavored with spices and are sweet. They are not swirled as it spreads itself. Both sugar and salt are mixed with the rice batter to enrich the taste. A plate of Appam is relished with vegetable or chicken or mutton stews. It makes a healthy breakfast for both vegans and non-vegans. Naresh Kumar et al. (2018) narrated the efficacy of bacteriocin produced by *Pediococcus pentosaceus* isolated from appam batter against *Acinetobacter baumannii*. This was the first report of the antimicrobial activity of LAB isolates against Gram-negative bacteria in the fermented food appam. Pal et al. (2005) isolated bacteriocinogenic LAB from Appam and 10 isolates showed good antimicrobial activity against Gram-positive bacteria such as *Bacillus cereus, Staphylococcus aureus, Listeria monocytogenes* as well as Gram-negative bacteria such as *Pseudomonas aeruginosa, Vibrio parahaemolyticus*, and *Aeromonas hydrophila* (Figure 13.3).

Neyyappam Neyyappam is a popular traditional sweet dish of Kerala and Tamil Nadu. It is a fried sweet, brown in color made from rice flour, jaggery, and coconut milk. The traditional method takes almost 16–18 hours for preparation. The rice is soaked overnight (10–12 hours), ground to paste with coconut milk, and mixed with jaggery and cardamom powder. The batter is left for fermentation for four to six hours and the batter then is deep-fried in a hemispherical shape pan. As its name implies, neyyappam is fried in ghee. It has a very low shelf-life of about two to three days at room temperature. The fermentation process is responsible for the soft fluffy nature of the dish (Figure 13.3).

Unniyappam Unniyappam is a popular teatime snack in Kerala. It is a sweet dish made of rice, jaggery, and bananas with a crispy exterior and a spongy interior. It is a ritual food, made during festivals and given in temples as prasadam. Rice powder, banana, jaggery, and cardamom are used for preparing unniyappam. The term unniyappam is derived from the words "unni" meaning small kid and "appam" meaning rice cake. The batter is fermented for four to five hours and is deep-fried with coconut oil or ghee. There is a specific pot called unniyappachatti for making this snack. Adding ingredients like sesame seeds, cumin seeds, and dry ginger powder gives it a richer, more complex flavor profile (Figure 13.3).

Appam

Neyyappam

Unniappam

FIGURE 13.3 Fermented foods of Kerala.

Sour Rice/Pazhankanji Sour rice, Pazhankanji, poita bhaat, pazhaya sadam, and pakhala are different versions of fermented rice. It is usually consumed as breakfast or lunch. The principal raw ingredients of this dish are rice and water. The cooked rice is cooled down to room temperature and adequate water is added to it. It is left overnight for fermentation at room temperature. Sometimes *dahi* (Indian yogurt), salt, spices, and leafy vegetables are also added to increase the taste. LAB like *L. delbrueckii* subsp. *bulgaricus*, *Lacticaseibacillus casei*, *Pediococcus acidilactici*, *E. faecalis*, *S. thermophilus*, *Microbacterium flavum*, and *Saccharomyces* sp. are responsible for this fermentation. Vitamin B and Vitamin K are increased due to this fermentation process and is a rich source of minerals like iron, sodium, potassium, and calcium. Probiotics are added to fermented foods to enhance the nutrients and preserve intestinal health. Probiotics are healthy food-borne microorganisms which include LAB. They enhance the food by adding a wide range of micronutrients including vitamins, minerals, and vital amino acids. LAB and starter cultures change many of the biochemical and organoleptic features of their fermentation substrate into metabolites (Giri et al. 2018). Kowsalya et al. (2022) assessed the probiotic properties of LAB isolated from naturally fermented rice gruel of South India and reported that the *P. pentosaceus* strain PRK1 (MT019527), *L. plantarum* strain PRK7 (MT022517) and *L. plantarum* strain PRK11 (MT022576) were identified by 16s rRNA gene sequencing (Kowsalya et al. 2022). This fermented rice restores healthy intestinal flora and can prevent gastrointestinal ailments.

13.4.2 Indigenous Fermented Foods of North-West India

In India using indigenous knowledge of food fermentation more than 350 types of major and region-specific fermented foods and alcoholic beverages are produced either naturally or by adding mixed starter cultures. Some of the indigenous fermented foods of North-West Indian states are listed in the tables below (Table 13.2).

Dhokla is a cereal and pulse-based fermented food prepared by the fermentation of Bengal gram and rice, found mainly in Gujarat. Khaman is similar to dhokla, but made with only Bengal gram dhal and prepared by steaming the leavened and acidified batter. Raabdi is mostly prepared in the north-west part of India, especially in the western region of Rajasthan by mixing and fermenting the flour of pearl millet, barley, wheat, sorghum, or maize with homemade buttermilk. Handvo is eaten as a snack by Gujarati people, it is prepared by using a fermented batter of rice and a mixture of dals such as urad dal, chana dal, and moong dal followed by baking. Shrikhand is prepared by partially removing whey from fresh curd through muslin cloth to obtain chakka which is further mixed with sugar and flavorings. Raita serves as a side dish to meals in Gujarat and Rajasthan, it is prepared instantly using various seasonal vegetables and fruits mixed into plain dahi. Khandvi is the traditional Gujarati snack prepared with gram flour and buttermilk and seasoned with fried sesame seeds and other spices. Some additional fermented foods are dahi, chhash, kadhi, and chakka (Sreeja and Prajapati 2020). These ingredients like Bengal gram, rice, urad dal, chana dal, moong dal, and sesame seeds are from the crops mostly cultivated in Gujarat. The ingredients like pearl millet, barley, wheat, sorghum, and maize are mostly cultivated in Rajasthan (Table 13.3).

Doli ki roti is consumed by the ethnic Punjabi community who migrated from Pakistan to India and it is prepared by fermenting wheat flour and spices. Keem is used as a starter for preparing soor, an alcoholic beverage. Jalebi is made by mixing yogurt and maida and is consumed mostly in North India, especially served in Uttarakhand, Uttar Pradesh, and Haryana. Soor is an alcoholic beverage made by fermentation and distillation of dough made up of different fruits, traditional herbs, and barley flour mostly cultivated in Uttarakhand. Sez is a semi-fermented food prepared from rice mostly consumed by the tribal community Bhotiya. Bhallae are prepared from black

TABLE 13.2 Indigenous fermented foods of Gujarat and Rajasthan.

Fermented Food	Substrate	Description
Dhokla	Cereal, spices	Rich in lactic acid bacteria, steamed pancake like consistency, slightly salty-sour taste
Raabadi	Cereal	Sour in taste, soft, and slippery texture
Handvo	Cereal	Spicy with cake-like texture
Shrikhand	Cow/buffalo milk	Dewatered dahi, paste consistency, sweet–sour taste
Khandvi	Cereal	Used as beverage, haspleasant taste
Khaman	Cereal	Soft, spongy texture, made of Bengal gram dhal

Source: Sreeja and Prajapati (2020).

TABLE 13.3 Indigenous fermented foods of Uttarakhand, Uttar Pradesh, Haryana, and Punjab.

Fermented Food	Substrate	Description
Doli Ki Roti	Whole wheat flour, spices	Like puri, golden-brown color
Keem	Barley flour, roots, leaves	Flavored curry and sticky
Jalebi	Maida, dahi, water	Deep-fried and crispy
Soor beverage	Cereals flour, fruits	Blood purifier and relieves urinary troubles
Sez	Rice	Used by bhotiyas, commonly used as snacks
Bhallae	Black gram	Deep-fried patties

Source: Beniwal et al. (2020).

gram and are consumed as snacks mostly in Punjab and Haryana. Some additional fermented foods are Balam, chhang, jamma, arijia, jann, chartayshya, and wari (Beniwal et al. 2020). These ingredients like wheat and black gram are mostly cultivated in Punjab and Haryana whereas rice and barley are mostly cultivated in Uttarakhand (Table 13.4).

Aska is prepared by cooking rice flour dough. Gulgule is consumed as a sweet snack in rural and urban areas, it is prepared from a batter of wheat flour with sugar. Chulli is indigenous to the tribal district of Kinnaur and it is prepared from the dried wild apricots of Himachal Pradesh. Sura is a fermented beverage prepared from kodra millet (*Eleusine coracana*), mostly produced in the Lug valley of the Kullu district. Angoori/kinnauri is the traditional alcoholic beverage prepared from red and green grapes which are cultivated in Kinnaur. Some additional fermented foods are Babru, Bagpinni, Bhaturu, Siddu, and Thuktal (Kanwar and Bhushan 2020; Table 13.5).

TABLE 13.4 Indigenous fermented foods of Himachal Pradesh.

Fermented Food	Substrate	Description
Aska	Rice flour	Special dish prepared during festivals
Gulgule	Wheat flour	Common snacks, cooked as flat pancake
Pakk	Barley, butter, lassi	Eaten as staple food
Khoru	Buttermilk and spices	Used as food in normal diet
Angoori/kinnauri	Grapes	Used as distilled drink

Source: Kanwar and Bhushan (2020).

TABLE 13.5 Indigenous fermented foods of Madhya Pradesh and Goa.

Fermented Food	Substrate	Description
Haria	Rice	Provides protein and carbohydrate source
Handia	Rice and ranu	Used as alcoholic beverage
Rabadi	Wheat, barley, pearl	Slurry-like food and mild acidic
Sur	Coconut palm	Used directly as beverage, acidic taste
Sonnas	Goan red rice	Rice cake leavened with sur
Kazara-bollo	Wheat or rye grains	Wheat or rye cake leavened with sur

Source: Sharma et al. (2020); Furtado and Velho-Pereira (2020).

Haria is a fermented beverage made from rice, mostly prepared by tribal people of Central India. Handia is a traditional drink prepared indigenously from rice and locally available plant parts. Rabadi is a tribal dish of Central India prepared by fermentation of crushed maize with buttermilk. Some additional fermented foods of Madhya Pradesh are Peja, Pendum, Ranu goti, Paneer, and Mahua (Sharma et al. 2020). Sur is tapped from the tender part of coconut inflorescence or spadix by incising and it is consumed as a beverage or used as a source of leaven in making bread or jaggery. Caju feni is an alcoholic beverage made through the distillation of cashew apple juice and awarded as a geographical indicator of Goa. Hittli is prepared by fermentation of urad dal added with coconut kernel and later it is steamed. Ambil and amboli are the other urad dal fermented preparations in Goa (Furtado and Velho-Pereira 2020). These ingredients like rice and maize are mostly cultivated in Madhya Pradesh and coconut palm, cashew apple, and urad dal are mostly cultivated in Goa.

13.4.3 Indigenous Fermented Foods of North East – Indo-Nepal Himalayan Region

The Himalayan mountain range in India stretches from Jammu and Kashmir to the easternmost state of Arunachal Pradesh. It also covers a major geographical portion of our neighboring country Nepal. This Indo-Nepal Himalayan region is one of the major sites for biological diversity and as well a diverse group of ethnic people dwells in this belt. Though this mountain range spreads between two different countries there are huge similarities in terms of food habits, culture, and rituals among different ethnic groups along the Himalayan belt ending at mainly Arunachal Pradesh. The eastern Himalayas touch a slight portion of Assam which also shares food habit similarities. Being connected with age-old mountains from early periods, the Himalayan people are engaged with nature and natural resources for their sustainable uses. There are hundreds of ethnic communities spreading throughout the Himalayan range with hundreds of fermented food items and this chapter highlights some of the very popular foods and alcoholic beverages which are commonly prepared by the ethnic people along the Indo-Nepal Himalayan regions, mainly covering Uttarakhand, Sikkim, Darjeeling, Nepal, and North Eastern India such as Assam, Arunachal Pradesh, and Meghalaya. Some of the common and popular fermented foods and alcoholic beverages are briefly described as follows (Figure 13.4).

Gundruk Throughout Himalayan regions of India and Nepal, gundruk is a popular ethnic fermented vegetable with an annual production of 3.2 kg/house and daily per capita consumption of gundruk is 1.4 g in the Indian state of Sikkim (Tamang et al. 2007). As the raw material, the local vegetable rayo-sag (*Brassica rapa* subsp. *oleifera*), mustard, and cauliflower are used for the gundruk preparation. The raw material is allowed to wilt under mild sunlight shredded, crushed mildly, and pressed in an airtight earthen jar or container. It takes 10 days to complete the fermentation process where the container is kept in warm places. When Gundruk is taken out from the container, again sun dried for five to seven days. The ethnic people also use dried gundruk as pickles and soup for more than two years. The different types of microorganisms isolated from gundruk are *L. fermentum, L. plantarum, L. casei, Lacticaseibacillus paracasei* subsp. *paracasei*, and *P. pentosaceus* (Tamang et al. 2005). Several different types of organic acids such as lactic, acetic, citric, malic, and acetic acids are present in gundruk (Karki 1986). It is also rich in other chemicals such as cyanides and isothiocyanates components, alcohols, esters, phenylacetaldehyde, etc. (Karki et al. 1983).

Sinki Sinki is another popular traditional ethnic fermented food prepared from radish taproot. This is locally prepared in the Himalayan regions of India and Nepal. The process called pit fermentation is used to prepare sinki. Radish taproots are wilted for two to three days, crushed,

FIGURE 13.4 Fermented foods of North East–Indo-Nepal Himalayan region.

squeezed, and pressed tightly into the pit (average depth 3 ft) with the same diameter. The pit is plastered using mud and then warmed by constructing a fire in the pit. The pit subsequently is lined with paddy straw and, after putting the crushed radish taproots into the pit, the pit is covered with dry leaves. Heavy weights such as stones are placed atop the leaves to restrict air passage into or out of the pit to avoid contamination. Fermentation takes place for 22–30 days. On completion of fermentation, the freshly fermented sinki is cut into small pieces and for three to five days it is sundried. Sinki is then stored at room temperature for more than two years in airtight containers to avoid contamination. Sinki is normally taken as a pickle mixed with salt, chili, onion, etc., and is also added to curry for enhancing the taste. Some of the common microbes isolated from sinki are *L. plantarum, L. brevis, Lactobacillus casei,* and *L. fallax* (Tamang and Sarkar 1993; Tamang et al. 2005).

Mesu/Kharisa Bamboo shoot is another popular fermented food from the Himalayan regions of India and Nepal which is locally called Mesu in Sikkim, Darjeeling, Nepal, Khorisa, Baanh-Gaaj in Assam, etc. Himalayan regions are found to be favorable locations for the growth of Bamboo and the ethnic people utilize it as a vegetable for making pickles. For Mesu, young bamboo shoots of different Bamboo species such as *Bambusa tulda, Dendrocalamus sikkimensis,* and *Dendrocalamus hamiltonii* are used. During mesu preparation, the defoliated young shoots are chopped and pressed in either a glass container or a hollow bamboo stem. The container or stem is covered airtight and the whole fermentation process occurs in anaerobic conditions for approximately 15–20 days. In Assam, during Khorisa preparation a little amount of turmeric powder is mixed with the finely chopped shoots as a coloring agent. The fermented bamboo shoot contains microbes like *L. plantarum, L. brevis, Latilactobacillus curvatus, Leuconostoc citreum,* and *P. pentosaceus* (Tamang and Sarkar 1996; Tamang et al. 2008).

Khalpi Another popular fermented food is Khalpi which is prepared from ripened cucumber and has been cut into pieces and sun-dried for three to four days. Normally a bamboo vessel is used for the fermentation process. The processing time to complete this fermentation takes five to seven days after which the khalpi is preserved as a pickle. The isolated microbes are *L. fallax, P. pentosaceus, L. brevis,* and *L. plantarum* from khalpi (Tamang et al. 2005).

Selroti One of the most popular and common fermented food is selroti from the Indo-Nepal Himalayan regions which is mainly prepared during rituals and festivals. Selroti's raw materials are rice flour, wheat flour, sugar, butter, milk, curd, ripened banana, and spices such as cardamom, cloves, fennel, nutmeg, and cinnamon. All the raw materials are mixed thoroughly and water is added so that the mixture becomes a free-flowing batter. Fermentation takes place for five to six hours at room temperature. Oil is then heated up in a cast iron frying pan and the batter is placed in the oil by squeezing it to create a ring-like shape. When the fried batter becomes a brownish-golden color, the selroti is drained out from the pan and readily served.

Selroti is a common fermented food eaten in Nepalese communities residing in the Himalayan regions of India, including Sikkim, Darjeeling, Assam, Arunchal Pradesh, Meghalaya, etc. Like other fermented food items, it is enriched with various microbes such as *L. mesenteroides, Enterococcus faecium, P. pentosaceus, L. curvatus, S. cerevisiae, Pichia kluyveri, Debaryomyces hansenii, Hyphopichia burtonii,* and *Zygosaccharomyces rouxii* (Yonzan and Tamang 2010).

Jaanr/Kodo Ko Jaanr Among fermented alcoholic beverages, jaanr, kodo ko jaanr, and chyang are very common and popular around the Himalayan regions of India and Nepal where they are mainly prepared by the ethnic people of this region. The production procedure is very lengthy and the same for both Jaanr and Kodo ko jaanr although the substrate used is different. Kodo ko jaanr is prepared from finger millet (*E. coracana*) and following fermentation has an alcoholic strength of 4.8% and a sweet taste (Tamang 2010). Jaanr instead is prepared from rice (*Oryza sativa*). The rice or millet is cooked and spread on a mat for cooling and yeast powder (*S. cerevisiae*) then is mixed with the grain and kept for five days at room temperature either in an aluminum vessel or bamboo basket. After the fermentation period, lukewarm water is poured over the mass and left for 15–20 minutes. The milky white extract obtained after fermentation is called Jaanr and the residues of Jaanr are used as animal feeds mainly for pigs. Jaanr is rich in minerals such as calcium, magnesium, manganese, iron, potassium, phosphorus, riboflavin, niacin, and folic acid. Jaanr or Chyang are very important beverages to the people of this region as they contain a lot of calories (Basappa 2002).

Raksi Raksi is another type of alcoholic beverage of the ethnic people of this region with an alcoholic content of 22–27% v/v (Tamang 2005). It is produced by a distillation process from traditional fermented cereal beverages and enriched with a characteristic aroma (Kozaki et al. 2000). The fermented mass contains buckwheat, molasses, potato, etc. The distillation unit consists of a large cylindrical vessel, a perforated container, and a collector for the raksi. The condenser which is a metallic vessel with cold water is kept at the top of the distillation unit. The entire setup is heated with firewood burning either on a hearth or in an earthen oven. Once the distillate is fully collected, it is stored in glass bottles. The ethnic people consume Raksi as a treatment for body aches and tiredness due to a long day of hard labor.

13.5 INDIGENOUS FERMENTED FOODS OF SRI LANKA

Sri Lanka is a tropical island in the Indian Ocean situated at the southern tip of the Indian subcontinent. It has a conspicuous ethnic diversity that resulted from occupations by the Portuguese, Dutch, Malaysians, Arabians, and South Indians at different eras of its history. Hence, Sri Lankan cuisine is influenced by the tastebuds of those ethnic groups who have left their legacies diversifying the local culinary features.

The array of fermented food products available in Sri Lanka is limited. Their fermented food is consumed either as a raw meal or an appetizer by locals. The choice of which fermented food products will be prepared and served typically depends on the occasion and is in concurrence with ethnic and cultural values. However, some foods that previously have been associated with a particular ethnic community are now popular among all Sri Lankans irrespective of their cultural origin. That commonality concurs with the fact that Sri Lankans are widely known for their generosity when it comes to sharing food.

Those fermented food products of Sri Lanka discussed in this chapter are categorized into milk-based foods, cereal and legume-based foods, fruit- and vegetable-based foods, and either meat- or fish-based foods (Figure 13.5).

13.5.1 Milk-based Fermented Food Products

Curd and yogurt are the most popular and commonly consumed dairy-based fermented food products in Sri Lanka. They are consumed by locals as a dessert after a main meal and are well-known for their health benefits for the growth and overall development of people of all ages (Ekanayake 2016). Both of these dairy products are the results of lactic acid fermentation by LAB, rich in dairy probiotics.

FIGURE 13.5 Fermented foods of Sri Lanka. (*See colour plate section for colour representation of this figure.*)

Curd, locally known as "Mee kiri", is a traditional fermented food product that has been served as part of royal cuisine for ancient Sri Lankan kings, hence it is considered a timeless delicacy (Ekanayake 2016; Priyashantha et al. 2021). Mee kiri is the only commercially available buffalo milk-derived fermented product in Sri Lanka, although numerous buffalo milk dairy products are available worldwide (Priyashantha et al. 2021). The preparation of mee kiri involves several steps. Within three hours of milking, buffalo milk is boiled for 20–30 minutes and cooled to 35–40 °C. A small amount of starter culture consisting of LAB such as *L. acidophilus*, *Bifidobacterium bifidum*, and *L. delbrueckii* subsp. *bulgaricus* is inoculated to initiate fermentation (Ekanayake 2016). The culture is obtained as either freeze-dried pellets, a portion of domestically maintained mother culture, or a back-slopping inoculum of previously fermented mee kiri. At the household level of mee kiri production, one-day-old mee kiri, also known as "muhun", is added before the boiled milk is cooled down to room temperature, stirred firmly, and poured into clay pots (Priyashantha et al. 2021). During fermentation, the starter culture converts lactose in the milk to lactic acid, reducing the pH of the medium. Once the casein isoelectric point of pH 4.6 is reached, casein starts coagulating and forming curd. After inoculation, to attain acceptable curdling 3–5 hours is required which in some instances can be extended to 36–72 hours. The filled clay pots are finally sealed and wrapped with either oil paper or natural banana leaf. Mee kiri is self-stable for seven days if the storing temperature is 4 °C and 1–3 days if stored at room temperature (Da Cruz et al. 2007).

Whey, also referred to as "mohoru" and "moru", is the liquid portion that separates by syneresis after the formation of curd. It often is consumed as a protein drink to boost muscle mass (Priyashantha et al. 2021). Mee kiri is generally served with palm treacle, "kithul pani", and traditionally offered at Buddhist alms-giving ceremonies demonstrating its importance in Sri Lankan culture. Some favor coconut syrup or table sugar over kitul pani in accordance with their preference.

"Dee kiri" is the curd prepared from cows' milk. Buffalo milk has a higher total solid content and less water compared to cow milk therefore mee kiri is thicker (Abd El-Salam and El-Shibiny 2011). Mee kiri is firmer in texture since buffalo milk has a higher casein-to-protein ratio (Priyashantha et al. 2021). Due to the lack of β-carotene in buffalo milk which is the pigment responsible for giving a yellow color, mee kiri is whiter compared to dee kiri (Abesinghe et al. 2020). Peroxidase activity of buffalo milk is higher in buffalo milk, hence the preservative ability of mee kiri is higher (Kumar et al. 1999). Compared to dee kiri, mee kiri is nutritionally rich owing to its high protein, fat, and mineral (calcium and phosphate) content (Priyashantha et al. 2021). Curd promotes digestion, lowers cholesterol, reduces hypertension, and boosts immunity (Priyashantha et al. 2021; Sankaranarayanan and Amaresan 2019). Reduction of lactose due to fermentation and the presence of many β-galactosidase-containing LAB allows lactose-intolerant individuals to consume mee kiri even though it is a dairy product (Shiby and Mishra 2013). Mee kiri is also used to treat diarrhea due to rotavirus infection and extended antibiotic use (Priyashantha et al. 2021).

Yogurt is a fermented dairy product produced by bacterial fermentation of milk. In Sri Lanka, conventionally cow milk is commonly used to prepare yogurt, using *S. thermophilus* and *L. delbrueckii* subsp. *bulgaricus* as starter cultures (Lourens-Hattingh and Viljoen 2001). In recent years, *L. acidophilus* and species of *Bifidobacterium* have been used for the formulation of yogurt in addition to the conventional microorganisms known as probiotics that produce lactic acid by fermenting milk lactose (Ranasinghe and Perera 2016).

13.5.2 Cereal and Legume-based Fermented Food Products

"Appa" or hoppers is a soft-textured fermented rice pancake with a spongy middle part surrounded by a crispy outer fringe consumed mostly as breakfast or dinner or occasionally as a snack by Sri Lankans (Mihiranie et al. 2020). Initially, during hopper preparation, rice grains are washed and soaked overnight whereupon naturally occurring LAB would grow in the material. The well-soaked rice grains are ground and mixed with coconut water and water to get a smooth rice batter. As a raising agent, the locally available traditional alcoholic drink, palm toddy, which is a source of LAB (*Weissella paramesenteroides, L. plantarum*, and *Lactobacillus fermentum*), is added to give it a unique taste (Ekanayake 2016; Satish Kumar et al. 2013). Nowadays, due to the unavailability of toddy, a piece of bread or a yeast is added as an aid for fermentation (Ekanayake 2016). The mixture is left at room temperature under aerobic conditions for eight hours to rise, consequently allowing the endogenous amylase to accelerate the fermentation process. The hopper mixture finally attains an acidic pH (4.0 or less) which helps to repress the growth of pathogenic bacteria. Coconut milk is added to the fermented mixture after which hoppers are cooked on a hot griddle. Hoppers are either savory or sweet with a high glycemic index.

Savory hoppers (egg hoppers and milk hoppers) are eaten together with a mixture of red onions and red chili known as "lunu miris" and with the addition of jaggery, sweet hoppers, or "pani appa" (Ekanayake 2016). "Kee roti" also known as "gedi appa" is a type of flat appa that does not have a crispy outer fringe but only the middle spongy part. Soon after cooking, it is soaked in sugar syrup prepared of sugar, water, coconut milk, appa batter, salt, and condiments like cloves and cardamon (Mihiranie et al. 2020).

"Diyabath" in which mild fermentation takes place is considered a Sri Lankan household remedy for gastric acidity (Mihiranie et al. 2020). Diyabath is prepared by soaking leftover cooked rice in water overnight at room temperature, often mixed with coconut milk and garnished with shallot, green chili, and salt, and consumed as a whole breakfast food. Diyabath has been around in Sri Lankan culture for a very long time. It was the method followed by ancestors to preserve the previous days' leftover cooked rice without the aid and availability of refrigeration. Diyabath is a functional food and in addition to its remedial action against gastric acidity, it can delay hunger and therefore is highly favored as a diet food. LAB responsible for the fermenting process abundantly produce vitamin B complex, B_{12}, and participate in removing antinutrients in cooked rice (Jayawardena and Wansapala 2015). Diyabath prepared with long-grain rice varieties is recommended for diabetic patients. Cooked long-grain rice varieties show slow digestibility owing to their high amylose content and act as a low glycemic index food leading to reduced blood glucose levels upon consumption (Mihiranie et al. 2020; Tao et al. 2019).

Dosa and idli have an Indian origin but are popular in Sri Lanka due to inhabitation by South Indians. Dosa is a flat, thin, crispy pancake prepared from a fermented batter of rice and black gram or lentils and generally served with a chutney or sambar. Dosa is a low-calorie food and is generally consumed at breakfast and suggested to younger children (under 10 years) and elders due to its digestibility and high nutritional density (Latha et al. 2019). Due to its gluten-free profile and low glycemic index, it is recommended for people suffering from celiac disease and diabetic conditions (Latha et al. 2019). *L. mesenteroides, E. faecalis, Bacillus amyloliquefaciens, Candida glabrata, L. fermentum, L. lactis, L. delbrueckii*, and *L. plantarum* are the LAB participating in the fermentation process of dosa (Latha et al. 2019).

Idly is a small, round, spongy, white-colored breakfast food served with sambar or chutney. Black gram and rice, soaked for four to six hours, are mashed and mixed in a ratio of 1 : 2, left overnight for fermentation, combined with water, salt, and fenugreek seeds to prepare the idly batter, and then steamed (Latha et al. 2019; Wikramanayake 1996). Black gram consists of naturally fermenting LAB like *L. mesenteroides, Lactobacillus delbrueckii, L. fermentum, Loigolactobacillus coryniformis, L. plantarum, L. lactis,* and *E. faecalis* that are responsible for the fermentation and black gram can be substituted with soybean, lentils, green gram, and chickpea according to preference (Latha et al. 2019; Satish Kumar et al. 2013; Sridevi et al. 2010). Idly is considered a weight loss diet component that can reduce the risk of high blood pressure, stroke, and cardiovascular diseases while boosting the probiotic action benefitting human health (Latha et al. 2019).

Soya tempeh (Figure 13.5), locally referred to as "soya-karawala", is a fermented soybean product that originated in Indonesia and was introduced to Sri Lanka in 1981 with the "The Home Level Training and Cookery Demonstration" (Sri Lanka: Soya Pioneer in the Third World). It is prepared naturally by the addition of a live starter culture mixture and a subsequently controlled fermentation process (Astuti et al. 2000; Selvi and Das 2019). Lactic acid fermentation initiates during the soaking of the soybeans, and acidifies the soybeans, preventing the growth of spoilage microorganisms (Ashenafi and Busse 1991; Nout et al. 1987). It is then followed by fungal fermentation performed by filamentous fungi *Rhizopus* spp. and *Fusarium* spp., the former of which inhibits the production of aflatoxin and biosynthesizes vitamin B (Murata et al. 1968). Fermentation results in a tempeh cake after one to two days that has various health benefits. Soya originally shows limited digestibility due to the presence of anti-nutrients like trypsin inhibitors, lectins, and oligosaccharides that cause flatulence (Nout 2015). Fermentation significantly enhances digestibility and improves the accessibility of soya nutrients by breaking down complex antinutrients to simple compounds (Kiers et al. 2000). Niacin in tempeh lowers cholesterol, provides high calcium and copper boosting bone health, may possess anticancer and anti-inflammatory properties, and seems to aid in improving the function of the immune system and maintaining blood sugar levels (Selvi and Das 2019).

13.5.3 Vegetable-based Fermented Products

Fruits and vegetables are naturally highly perishable in nature therefore difficult to store. Since ancient times the possibilities of preserving fruits and vegetables have been explored to finally discover the preservation method of fermentation. Fruits are rich in antioxidants, phytochemicals, vitamins, minerals, dietary fibers, carotenoids, and flavonoids making them more suitable for probiotics (Yoon et al. 2004). Like fruits, vegetables are dense with minerals, vitamins, carbohydrates, and phytochemicals (Yadav et al. 2014). Most fruits and vegetables are classified as low-acid foods due to their low acidic levels making them prone to microbial deterioration (Ekanayake 2016). By fermentation, fruit and vegetable-based probiotic products can be prepared to reduce the related post-harvest losses. Dairy allergens like lactose and cholesterol limit the consumption of dairy probiotics for some people and there can be comparative health benefits since fruits and vegetables are free of such allergens (Luckow and Delahunty 2004).

In Asia, a classic fruit and vegetable-based fermented food product that has been widespread for centuries is pickles. One of the most common pickles found in Sri Lanka is "lunu dehi," a pickled lime usually eaten with a rice meal. Mature, ripened, yellow-colored limes are cut into quarters, and salt is added as alternative layers with a 4 : 1 ratio in a fermenting container until the container is ¾ full, tightly covered by a cloth, and kept aside. During the first 24 hours, brine is formed initiating the fermentation process. To avoid the raw taste of lime and for effective marination, limes used for the pickle are sundried for a prolonged period until the lime cores acquire a white color (Ekanayake 2016). Depending on the temperature, fermentation takes about one to four weeks and the final sour lime pickle produced can be stored for up to five years. Depending on the final taste preference vinegar and spices like chili powder, or flakes, cloves, and cardamon are added. "Bilin" (*Averrhoa bilimbi*) pickle is also prepared by locals as a condiment following the same procedure as lunu dehi to enhance the overall flavor of main meals.

Vegetable-based fermented products, Sinhala "achcharu" and "wambatu moju", are also regarded as integral elements in traditional Sri Lankan cuisine (Liyanage 2020). Sinhala achcharu is one of the main dishes prepared and served during the April Sinhala and Tamil New Year season. In an earthen clay pot, red onions, green chili, and sliced carrots (raw papaya and baby jackfruit are optional) are separately simmered in vinegar and mixed. Over the vegetable mix, a combination of crushed ginger, garlic, mustard seed, pepper, turmeric, and salt boiled in vinegar is poured, mixed, and the whole is packed in glass bottles. Typically, Sinhala achcharu is left to mature for two days before consumption during which fermentation takes place developing the characteristic taste and aroma (Liyanage 2020).

Wambatu moju is a requisite dish served during almost every special occasion or celebration in Sri Lankan culture. It is an eggplant-based dish similar to Sinhala achcharu where fried sliced eggplant, sautéed red onions, and green chili are mixed with a sauce prepared from ground mustard, cloves, cardamom, cinnamon, chili powder, salt, sugar, and vinegar. Fermentation is allowed to take place by storing "moju" in a glass container for a few days before consumption.

"Polos achcharu", or pickled baby jackfruit (*Artocarpus heterophyllus*), is a fruit-based fermented food consumed by locals in Sri Lanka (Liyanage 2020). Thinly sliced baby jackfruit, onions, and green chili are mixed with ground mustard seed, chili powder, salt, and vinegar to prepare the pickle, stored in a clay pot, and kept to mature for two to three days before consumption. Jackfruit is a rich source of dietary fibers, vitamins, minerals, and calories. The presence of these compounds makes jackfruit a good laxative and an antioxidant important for human vision function (Mondal et al. 2013). Globally, depending on the maturation stage of jackfruit used, there are two types of pickled jackfruit: green and sweet. As these products are prepared without using any chemical preservatives, they can be stored for about 12 months with unchanged quality parameters like color, texture, taste, and flavor (Mondal et al. 2013).

For the production of fermented capsicum *L. mesenteroides, L. plantarum, L. fermentum, P. pentosaceus, Saccharomyces* species, heterotrophic species *Enterobacter cloacae, Citrobacter freundii, Micrococcus* species, and *Pseudomonas* species are involved (Mahinda 2005). Wet and dry salt methods produce a salt concentration of 5% and 5.5% and result in pH 3–4 following 12 days of capsicum fermentation.

13.5.4 Fish- and Meat-based Fermented Food Products

In Sri Lanka, all fermented fish products are created following the osmotic dehydration principle based on initial salting and drying, which is one of the oldest preservation methods. During salting, salt (NaCl) is diffused through the permeable fish tissues releasing water from the tissues by osmosis giving a characteristic aroma and a flavor to the final product (Jamas et al. 2019).

Maldive fish (Figure 13.5), famously known as "umbalakada", are presently produced locally and imported from India but used to be imported from the Maldives. Widely used as a flavoring agent in almost all local curries, umbalakada is a slightly salted, smoked, dried loin of skipjack tuna. Initially, skipjack tuna or other tuna varieties are beheaded, degutted, and horizontal cuts are made. Salt is added onto the surface and insides of the cuts and kept for 30 minutes, after which the fish is cooked and smoked for 10–12 hours or overnight. Smoking is carried out for another two to four hours after separating the cut pieces from the central bone and removing all the bone scales. The smoked pieces are rolled in ash and finely sun-dried for 7–10 days. Inadequate drying would lead to toxigenic fungal growth by such organisms as *Aspergillus flavus, Aspergillus tamarii, Aspergillus niger, Aspergillus ochraceus,* and *Penicillium citrinum*, spoiling the fish and risking consumer health (Mohamed 2013).

"Jaadi" is a country-specific, highly salted fermented fish product commonly prepared using Indian mackerels, seer, and sardines (Ekanayake 2016; Waisundara et al. 2016). Jaadi preparation is a low-cost procedure where the curing process involves stacking the slightly opened or sliced fish on top of a paste of salt and "goraka" (*Garcinia gummi-gutta*) or alternative layers of salt and pods of goraka in a wide-mouth bottle. Goraka is responsible for maintaining a pH of 4 in the curing medium, partially hydrolyzing the flesh of fish (Jayasinghe et al. 2000). The bottle is sealed to avoid contamination and stored in a dry place for several months during which the fermentation takes place. The reduction in pH and water activity favors the growth of LAB which is the predominant microorganism in fermented fish products and further contributes to the pH reduction by fermenting available carbohydrates (Paludan-Muller et al. 2002; el Sheikha et al. 2014). These days, beheaded, deskinned tilapia (*Oreochromis niloticus*) is used to produce jaadi as a cottage industry along the coastal areas. Given the high yield of tilapia, as a way to reduce the post-harvest losses, 1 kg of tilapia is cured with 500 g of goraka and 100 g of salt to develop acceptable sensory properties (Ekanayake 2016; Lakshmi et al. 2011). Cholesterol, which is the main sterol in fish is susceptible to oxidation-producing cholesterol oxidation products (COPs) responsible for dysfunctions in cholesterol metabolism, cell membrane damage, cytotoxicity, and mutagenicity

(Hovenkamp et al. 2008; Otaegui-Arrazola et al. 2010). Traditionally preserved Jaadi uses fatty fish shown to contain high amounts of COPs known for their toxicity such as cholesteroltriol and 25-hydroxycholesterol (Ubhayasekera et al. 2012). Therefore, regular consumption of jaadi may contribute to the progression of atherosclerosis in certain populations (Ubhayasekera et al. 2012).

Meat-based fermented products are less common in Sri Lanka. Salami and pepperoni, produced using a *P. pentosaceus* starter culture are not ethnic meat products of Sri Lanka but gaining popularity owing to the tourist industry, especially among urban areas. Meat together with the starter culture is frozen and kept at room temperature for three days. Then, for three to four hours the cultured meat is kept in a 35–40 °C steam chamber and transferred to a controlled atmospheric room for three weeks while gradually decreasing the temperature and relative humidity (Ekanayake 2016). Artificial flavors are added to the produced salami or pepperoni product in addition to the flavor naturally acquired by five weeks of bacterial fermentation.

Thus, traditional as well as newly introduced fermented foods have become a permanent part of Sri Lankan cuisine irrespective of diverse ethnicities, cultures, and religions and have contributed to enhancing the food variety available in the country.

13.6 INDIGENOUS FERMENTED FOODS OF BHUTAN

The most unique feature of Bhutanese food is its spiciness. Chilies are one of the essential parts of almost every meal. Most of the Bhutanese would not enjoy a meal that was not spicy. Rice is a staple food in Bhutanese food culture. Bhutanese use both red rice and white rice during their food preparation. Most Bhutanese meals consist of rice as a major portion along with side dishes consisting of meat or vegetables. The Ema Datshi is the national dish of Bhutan, which is a spicy mixture of chilies and the delicious local cheese known as Datshi. Momos, which are steamed flour dumplings stuffed with pork, beef, or cabbages and cheese are another traditional food of Bhutan.

13.7 FERMENTATION – AN ECONOMIC APPROACH

The use of fermented foods is the best economical approach to preserve foods and enhance their nutritional level. But the market success of fermentation processes depends upon the fermented product and recovering all the preparation costs along with desired profits. The fermentation process and its product must compete in a way that is economically sound compared to similar products in the market. A fermented product's economic position depends on its cost of production and distribution which include media components, labor cost, losses from contamination, sterilization, yield, product recovery, product purity, waste disposal, and research costs. For a fermentation process to be economically viable the costs should be managed, raw material should be cheap, necessarily employing high yielding strains for fermentation, simple product recovery and purification process, effective use of heat and power, saving labor, successful automation, etc. Strain improvement for better fermentation can also be an economic approach.

13.8 FOOD SAFETY AND FOOD PRESERVATION

Although fermented food is less likely to be contaminated and cause food infection or in-toxification there are cases of microbial food-borne infection reported in fresh cheese, sausages, and fermented cereals due to the usage of contaminated raw materials for fermentation, lack of pasteurization, and sub-optimal fermentation starters that reduce the safety of fermented foods. Optimization of starter culture by conventional, mutational, or DNA recombinant technology may increase the safety of fermented foods (Nout 1994). For the preservation of food, LAB play a key role as they produce a variety of primary and secondary metabolites like organic acids and antibiotics. They also secrete a wide range of bacteriocins that act against food pathogens. Utilization of such naturally produced antagonists helps to improve the safety of fermented foods and extend the shelf life of foods (Ross et al. 2002).

13.9 FOOD SAFETY RISKS IN TRADITIONAL FOODS

Traditional foods are those naturally raised and consumed in their original form or processed by natural fermentation, drying, or cooking. Many fermented foods are prepared from traditional foods and vice-versa (Prakash 2016). Many traditional foods possess immense nutritional value but unfortunately, food safety risks are to be considered which can involve microbial food in-toxification due to the synthesis of bacterial toxin or mycotoxin during fermentation by bacterial or fungal inoculants, utilization of toxin-contaminated raw materials, and production of toxic by-products like ethyl carbamate and biogenic amines. The combination of poorly controlled naturally fermented foods, and consumption without proper cooking, may reduce the food safety of traditional foods (Nout 1994). The use of high-quality raw ingredients, proper sanitation, and ingredient preparation, appropriate NaCl concentration, temperature, and adequate fermentation time are the factors to be considered for reducing food safety risks (Nielson-Stowell 2021).

13.10 SUMMARY AND FUTURE PROSPECTS

In history, indigenous fermented foods played a significant role in every civilization and nation. The knowledge of food fermentation is passed over from generation to generation from indigenous people. Consumption of fermented food will benefit humans and the whole fermentation ecosystem. Fermentation of food will preserve food and improve the nutritional quality of food due to the presence of a wide range of microbial metabolites. There are several indigenous foods in South Asia specific to each regional part which is not yet being investigated and those offer a potential field for future research. Documentation of the diverse fermented foods of South Asia needs to be addressed completely. The goal of understanding different types of fermentation paves our way forward and provides future thrust to investigate humanity's diverse relationships with fermented foods. At present local indigenous people are untapped sources of knowledge regarding traditional fermented foods. Utilizing their knowledge about various fermentation processes would lead to the creation of new tradition-based foods with good nutritional value and taste. Proper hygienic preparation of fermented foods would reduce food safety risks. Proper awareness and marketing of fermented foods are required to reduce malnutrition among people.

13.11 CONCLUSION

The traditional fermented meals and beverages in a majority of South Asian countries have been thoroughly researched and documented. Although, most of the traditional fermented foods and beverages in India have not yet been studied thoroughly. Only a small number of popular fermented dishes, such as idli, dosa, and curd, have been investigated so far. There is a need to research the microbial diversity linked to fermented foods since it may provide a sizeable gene pool that must be preserved for the next generation. This is the appropriate time to market well-accepted traditional fermented foods, some of which may have efficient beneficial probiotic bacteria that are yet unknown. Our scientific interest is overflowing with anticipation of colossal human-beneficial qualities that are yet to be understood about the combinations of food and microbe that create fermentations. Hence, it is perhaps advisable to include as many fermented foods as anyone can in their diet. Fermentation increases the shelf life of food, preserving it by using the microbial ends products such as lactic, alcohol, and acetic acid. It is known to add beneficial microorganisms to the gut which modulate the digestive system and enhance proper digestion. It increases the levels of vitamins, especially B vitamins in food. Most importantly, South Indian traditional food is mainly prepared as the fermentation makes the food more digestible and reduces antinutritional factors in some foods. In the case of milk, some people having lactose intolerance problems can safely consume fermented milk products. Most importantly, fermentation can make food that can be consumed more pleasantly with improved flavor.

REFERENCES

Abd El-Salam, M.H. and El-Shibiny, S. (2011). A comprehensive review on the composition and properties of buffalo milk. *Dairy Science & Technology* 91 (6): 663–699. https://doi.org/10.1007/s13594-011-0029-2.

Abesinghe, A.M.N.L., Priyashantha, H., Prasanna, P.H.P. et al. (2020). Inclusion of probiotics into fermented buffalo (*Bubalus bubalis*) milk: an overview of challenges and opportunities. *Fermentation* 6 (4): 121. https://doi.org/10.3390/fermentation6040121.

Agarwal, K.N. and Bhasin, S.K. (2002). Feasibility studies to control acute diarrhoea in children by feeding fermented milk preparations Actimel and Indian Dahi. *European Journal of Clinical Nutrition* 56: 56–59.

Angmo, K., Kumari, A., and Bhalla, T.C. (2016). Probiotic characterization of lactic acid bacteria isolated from fermented foods and beverage of Ladakh. *LWT – Food Science and Technology* 66: 428–435.

Ashenafi, M. and Busse, M. (1991). The microflora of soak water during tempeh production from various beans. *Journal of Applied Bacteriology* 70 (4): 334–338. https://doi.org/10.1111/j.1365-2672.1991.tb02945.x.

Ashwathi, P. (2016). 5 main types of fermentations. *Biology Discussion*. https://www.biologydiscussion.com/organism/metabolism-organism/5-main-types-of-fermentations/50854.

Astuti, M., Meliala, A., Dalais, F.S. et al. (2000). Tempe, a nutritious and healthy food from Indonesia. *Asia Pacific Journal of Clinical Nutrition* 9 (4): 322–325. https://doi.org/10.1046/j.1440-6047.2000.00176.x.

Basappa, S.C. (2002). Investigations on Chhang from finger millet (*Eleucine coracana* Gaertn.) and its commercial prospects. *Indian Food Industries* 21 (1): 46–51.

Beniwal, A., Ghosh, T., Bhardwaj, K.N. et al. (2020). Ethnic fermented foods and beverages of Uttarakhand, Uttar Pradesh, Haryana, and Punjab. In: *Ethnic Fermented Foods and Beverages of India: Science History and Culture* (ed. J.P. Tamang), 621–645. Singapore: Springer.

Bowen, R. (1999). Fermentation microbiology and ecology. *Vivo Pathophysiology*. https://rabowen.org/hbooks/pathphys/digestion/herbivores/microbes.html.

Chavan, J.K. and Kadam, S.S. (1989). Nutritional improvement of cereals by fermentation. *Critical Reviews in Food Science and Nutrition* 28 (5): 349–400.

Clime, K. (2022). Beyond sauerkraut: A brief history of fermented foods. *Living History Farms*. https://www.lhf.org/2014/03/beyond-sauerkraut-a-brief-history-of-fermented-foods/#:~:text=Evidence%20of%20a%20fermented%20alcoholic,in%20Babylon%20around%203000%20BCE.

Crittenden, R.G., Martinez, N.R., and Playne, M.J. (2003). Synthesis and utilisation of folate by yoghurt starter cultures and probiotic bacteria. *International Journal of Food Microbiology* 80 (3): 217–222. https://doi.org/10.1016/S0168-1605(02)00170-8.

da Cruz, A.G., Faria, J.d.A.F., and van Dender, A.G.F. (2007). Packaging system and probiotic dairy foods. *Food Research International* 40 (8): 951–956. https://doi.org/10.1016/j.foodres.2007.05.003.

Dashmeet. (2018). Acetic acid: biosynthesis and fermentation process. Bio Technology. https://www.biotechnologynotes.com/industrial-biotechnology/acetic-acid/acetic-acid-biosynthesis-and-fermentation-process-industries-biotechnology/13778 (accessed 19 Apirl 2023).

Debret, C. (2021). History of fermentation around the world. *Stacker* https://stacker.com/stories/23704/history-fermentation-around-world#:~:text=Human%2Dcaused%20fermentation%20dates%20back,making%20the%20first%20documented%20yogurt.

Deka, S.C. (2012). Mini review on fermented foods and beverages of the north-East India. *International Food Research Journal* 19 ((2): 377–392.

Eat Cultured. (2017). Fermentation: A history. https://eatcultured.com/blogs/our-awesome-blog/fermentation-a-history (accessed 19 April 2023).

Ekanayake, S. (2016). Ethnic fermented foods and beverages of Sri Lanka. In: *Ethnic Fermented Foods and Alcoholic Beverages of Asia* (ed. J.P. Tamang), 139–150. India: Springer https://doi.org/10.1007/978-81-322-2800-4_6.

Ershidat, O.T.M. and Mazahreh, A.S. (2009). Probiotics bacteria in fermented dairy products. *Pakistan Journal of Nutrition* 8 (7): 1107–1113. https://doi.org/10.3923/pjn.2009.1107.1113.

Fellows, P. (2000). *Food Processing Technology*. CRC Press https://doi.org/10.1201/NOE0849308871.

Furtado, I. and Velho-Pereira, S. (2020). Ethnic fermented foods and beverages of Goa. In: *Ethnic Fermented Foods and Beverages of India: Science History and Culture* (ed. J.P. Tamang), 139–155. Singapore: Springer.

Giri, S.S., Sen, S.S., Saha, S. et al. (2018). Use of a potential probiotic, *Lactobacillus plantarum* L7, for the preparation of a rice-based fermented beverage. *Frontiers in Microbiology* 9: 473.

Hovenkamp, E., Demonty, I., Plat, J. et al. (2008). Biological effects of oxidized phytosterols: a review of the current knowledge. *Progress in Lipid Research* 47 (1): 37–49. https://doi.org/10.1016/j.plipres.2007.10.001.

Jamas, A.L.A., Pinto, B.V.V., Estanech, A.F.C. et al. (2019). Fermented fish products. In: *Fermented Food Products* (ed. A. Sankaranarayanan, N. Amaresan, and D. Dhanasekaran), 239–252. Boca Raton, FL: CRC Press https://doi.org/10.1201/9780429274787-14.

Jayasinghe, P.S., Bamunuarachchi, A., and Fonseka, T.S.G. (2000). Survey on the quality of Jaadi available Sri Lankan market. *Journal of the National Aquatic Resources Research and Development Agency* 36: 26–34.

Jayawardena, J. and Wansapala, M. (2015). Study on developing micronutrients enriched spread using fermented cooked rice. *International Journal of Innovative Researh in Technology* 1 (12): https://www.researchgate.net/publication/283676557.

Kanwar, S.S. and Bhushan, K. (2020). Ethnic fermented foods and beverages of Himachal Pradesh. In: *Ethnic Fermented Foods and Beverages of India: Science History and Culture* (ed. J.P. Tamang), 189–208. Singapore: Springer.

Karki, T. (1986). Some Nepalese fermented foods and beverages. In: *Traditional Foods: Some Products and Technologies* (ed. J.P. Tamang), 84–96. Mysore: Central Food Technological Research Institute.

Karki, T., Itoh, H., and Kozaki, M. (1983). Chemical changes occurring during gundruk fermentation Part II-2 Flavor components. *Lebensmittel-Wissenschaft und-Technologie* 16: 203–208.

Karovičová, J. and Kohajdová, Z. (2005). Lactic acid-fermented vegetable juices-palatable and wholesome foods. *Chemical Papers* 59 (2): 143–148.

Kiers, J.L., van Laeken, A.E.A., Rombouts, F.M. et al. (2000). In vitro digestibility of Bacillus fermented soya bean. *International Journal of Food Microbiology* 60 (2–3): 163–169. https://doi.org/10.1016/S0168-1605(00)00308-1.

Kowsalya, M., Sudha, K.G., Ali, S. et al. (2022). In-vitro assessment of probiotic properties of lactic acid bacteria isolated from naturally fermented rice gruel of South India: lactic acid bacteria from fermented rice gruel of South India. *Journal of Microbiology, Biotechnology and Food Sciences* 12 (1): 1–9, e4908-e4908.

Kozaki, M., Tamang, J.P., Kataoka, J. et al. (2000). Cereal wine (jaanr) and distilled wine (raksi) in Sikkim. *Journal of the Brewing Society of Japan* 95 (2): 115–122.

Kumar, R., Bhatia, K.L., Kumar, R. et al. (1999). Standardization of method for lactoperoxidase assay in milk. *Lait* 79: 269–274.

Kumar, R.S., Kanmani, P., Yuvaraj, N. et al. (2012). Traditional Indian fermented foods: a rich source of lactic acid bacteria. *International Journal of Food Sciences and Nutrition* Early Online: 1–14. https://doi.org/10.3109/09637486.2012.746288.

Kumar, K.N., Devadas, S.M., Murugan, S. et al. (2018). Production and characterization of bacteriocin by lactic acid bacterium-*Pediococcus pentosaceus* NKSM1 isolated from fermented 'appam' batter. *Journal of Pure and Applied Microbiology* 12 (3): 1315–1330.

Lakshmi, W., Prassanna, P., and Edirisinghe, U. (2011). Production of Jaadi using Tilapia (*Oreochromis niloticus*) and determination of its physcio-chemical and sensory properties. *Sabaragamuwa University Journal* 9 (1): 57. https://doi.org/10.4038/suslj.v9i1.3734.

Latha, S., Hemamalini, A., Kumar, S.D. et al. (2019). Ethnic probiotic foods of South India and their health benefits. In: *Fermented Food Products* (ed. A. Sankaranarayanan, N. Amaresan, and D. Dhanasekaran), 77–92. CRC Press https://doi.org/10.1201/9780429274787-6.

Liyanage, C.J. (2020). Traditional and ethnic foods of Sri Lanka—safety aspects. *Nutritional and Health Aspects of Food in South Asian Countries* 127–141. https://doi.org/10.1016/B978-0-12-820011-7.00019-8.

Lourens-Hattingh, A. and Viljoen, B.C. (2001). Yogurt as probiotic carrier food. *International Dairy Journal* 11 (1–2): 1–17.

Luckow, T. and Delahunty, C. (2004). Which juice is 'healthier'? A consumer study of probiotic non-dairy juice drinks. *Food Quality and Preference* 15 (7–8): 751–759. https://doi.org/10.1016/J.FOODQUAL.2003.12.007.

Mahinda, T.D.C.I. (2005). Studies on fermentation of capsicum. Master thesis. University of Kelaniya, Kelaniya.

Mattila-Sandholm, M.P., Crittenden, R., Mogensen, G. et al. (2002). Technological challenges for future Probiotic foods. *International Dairy Journal* 12 (2-3): 173–182.

Mihiranie, S., Jayasinghe, J.K., Jayasinghe, C.V.L. et al. (2020). Indigenous and traditional foods of Sri Lanka. *Journal of Ethnic Foods* 7 (1): BioMed central ltd. https://doi.org/10.1186/s42779-020-00075-z.

Mohamed, S. (2013). toxigenic fungi and mycotoxin production in maldive fish (smoked dried tuna fish). PhD thesis (Food Technology) Massey University of Palmerston North, New Zealand.

Mondal, C., Remme, R.N., Mamun, A.A. et al. (2013). Product development from jackfruit (*Artocarpus heterophyllus*) and analysis of nutritional quality of the processed products. *Journal of Agriculture and Veterinary Science* 4 (1): 76–84.

Murata, K., Miyamoto, T., and Taguchi, F. (1968). Biosynthesis of B vitamins with Rhizopus oligosporus. *The Journal of Vitaminology* 14 (3): 191–197. https://doi.org/10.5925/jnsv1954.14.191.

Nielson-Stowell, A. (2021). The safety of fermented food. The Fermentation Association. https://fermentationassociation.org/the-safety-of-fermented-food (accessed 19 April 2023).

Nout, M.J.R. (1994). Fermented foods and food safety. *Food Research International* 27 (3): 291–298.

Nout, R. (2015). Quality, safety, biofunctionality and fermentation control in soya. In: *Advances in Fermented Foods and Beverages: Improving Quality, Technologies and Health Benefits* (ed. W. Holzapfel), 409–434. Elsevier Inc. https://doi.org/10.1016/B978-1-78242-015-6.00018-9.

Nout, M.J.R., de Dreu, M.A., Zuurbier, A.M. et al. (1987). Ecology of controlled soyabean acidification for tempe manufacture. *Food Microbiology* 4 (2): 165–172. https://doi.org/10.1016/0740-0020(87)90032-3.

Otaegui-Arrazola, A., Menéndez-Carreño, M., Ansorena, D. et al. (2010). Oxysterols: a world to explore. *Food and Chemical Toxicology* 48 (12): 3289–3303. https://doi.org/10.1016/j.fct.2010.09.023.

Pal, V., Jamuna, M., and Jeevaratnam, K. (2005). Isolation and characterization of bacteriocin producing lactic acid bacteria from a south Indian special dosa (Appam) batter. *Journal of Culture Collections* 4 (1): 53–60.

Paludan-Muller, C., Valyasevi, R., Huss, H.H. et al. (2002). Genotypic and phenotypic characterization of garlic-fermenting lactic acid bacteria isolated from som-fak, a Thai low-salt fermented fish product. *Journal of Applied Microbiology* 92 (2): 307–314. https://doi.org/10.1046/j.1365-2672.2002.01544.x.

Parvez, S., Malik, K.A., Ah Kang, S. et al. (2006). Probiotics and their fermented food products are beneficial for health. *Journal of Applied Microbiology* 100 (6): 1171–1185. https://doi.org/10.1111/j.1365-2672.2006.02963.x.

Prakash, V. (2016). Introduction: the importance of traditional and ethnic food in the context of food safety, harmonization, and regulations. In: *Regulating Safety of Traditional and Ethnic Foods* (ed. V. Prakash, O. Martin-Belloso, L. Keener, et al.), 1–6. Academic Press.

Priyashantha, H., Ranadheera, C.S., Rasika, D.M.D. et al. (2021). Traditional Sri Lankan fermented buffalo (Bubalus bubalis) milk gel (Meekiri): technology, microbiology and quality characteristics. *Journal of Ethnic Foods* 8 (1): BioMed Central Ltd. https://doi.org/10.1186/s42779-021-00105-4.

Ranasinghe, J. and Perera, W. (2016). Prevalence of *Lactobacillus bulgaricus* and *Streptococcus thermophilus* stability in commercially available yogurts in Sri lanka. *Asian Journal of Medical Sciences* 7 (5): 97–101. https://doi.org/10.3126/ajms.v7i5.14326.

Rawat, K., Kumari, A., Kumar, S. et al. (2018). Traditional fermented products of India. *International Journal of Current Microbiology and Applied Sciences* 7 (04): 1873–1883.

Reddy, L.V. and Joshi, V.K. (2019). Indigenous fermented foods of India with special reference to Andhra Pradesh and adjoining areas. *International Journal of Food Fermentation Technology* 9 (2): 73–87.

Ross, R.P., Morgan, S., and Hill, C. (2002). Preservation and fermentation: past, present and future. *International Journal of Food Microbiology* 79 (1–2): 3–16.

Sankaranarayanan, A. and Amaresan, N. (2019). Diversity of global fermented food products: an overview. In: *Fermented Food Products* (ed. A. Sankaranarayanan, N. Amaresan, and D. Dhanasekaran), 3–24. Boca Raton, FL: CRC Press https://doi.org/10.1201/9780429274787-1.

Sathe, G.B. and Mandal, S. (2016). Fermented products of India and its implication: a review. *Asian Journal of Dairy & Food Research* 35 (1): 1–9.

Satish Kumar, R., Kanmani, P., Yuvaraj, N. et al. (2013). Traditional Indian fermented foods: a rich source of lactic acid bacteria. *International Journal of Food Sciences and Nutrition* 64 (4): 415–428. https://doi.org/10.3109/09637486.2012.746288.

Scott, R. and Sullivan, W.C. (2008). Ecology of fermented foods. *Human Ecology Review* 15 (1): 25–31.

Sekar, S. and Mariappan, S. (2007). Usage of traditional fermented products by Indian rural folks and IPR. *Indian Journal of Traditional Knowledge* 6: 111–120.

Selvi, A. and Das, N. (2019). Fermented soybean food products as sources of protein-rich diet: an overview. In: *Fermented Food Products* (ed. A. Sankaranarayanan, N. Amaresan, and D. Dhanasekaran), 167–180. CRC Press https://doi.org/10.1201/9780429274787-11.

Shamala, T.R. and Sreekantiah, K.R. (1988). Microbiological and biochemical studies on traditional Indian palm wine fermentation. *Food Microbiology* 5 (3): 157–162.

Sharma, I., Kapale, R., and Kango, N. (2020). Ethnic fermented foods and beverages of Madhya Pradesh. In: *Ethnic Fermented Foods and Beverages of India: Science History and Culture* (ed. J.P. Tamang), 287–303. Singapore: Springer.

el Sheikha, A., Ray, R.C., and Panda, S.H. (2014). African fermented fish products in scope of risks effect of post-harvest quality parameters on ultra-sonication treatment of Khoonphal (Haematocarpus validus) of Meghalaya, North-East India View project microbial enzymes in food applications view project. *International Food Research Journal* 21 (2): 425–432. http://www.ifrj.upm.edu.my.

Shiby, V.K. and Mishra, H.N. (2013). Fermented milks and milk products as functional foods—A review. *Critical Reviews in Food Science and Nutrition* 53 (5): 482–496. https://doi.org/10.1080/10408398.2010.547398.

Sreeja, V. and Prajapati, J.B. (2020). Ethnic fermented foods and beverages of Gujarat and Rajasthan. In: *Ethnic Fermented Foods and Beverages of India: Science History and Culture* (ed. J.P. Tamang), 157–187. Singapore: Springer.

Sridevi, J., Halami, P.M., and Vijayendra, S.V.N. (2010). Selection of starter cultures for idli batter fermentation and their effect on quality of idlis. *Journal of Food Science and Technology* 47 (5): 557–563. https://doi.org/10.1007/s13197-010-0101-6.

Tamang, J.P. (2005). *Food Culture of Sikkim, Sikkim Study Series*, vol. 4. Gangtok: Information and Public Relations Department. Government of Sikkim.

Tamang, J.P. (2010). Diversity of fermented beverages. In: *Fermented Foods and Beverages of the World* (ed. J.P. Tamang and K. Kailasapathy), 85–125. New York: CRC Press, Taylor & Francis Group.

Tamang, J.P.a. and Sarkar, P.,.K. (1993). Sinki—a traditional lactic acid fermented radish tap root product. *The Journal of General and Applied Microbiology* 39: 395–408.

Tamang, J.P. and Sarkar, P.K. (1996). Microbiology of mesu, a traditional fermented bamboo shoot product. *International Journal of Food Microbiology* 29: 49–58.

Tamang, J.P., Tamang, B., Schillinger, U. et al. (2005). Identification of predominant lactic acid bacteria isolated from traditional fermented vegetable products of the eastern Himalayas. *International Journal of Food Microbiology* 105 (3): 347–356.

Tamang, J.P., Thapa, N., Rai, B. et al. (2007). Food consumption in Sikkim with special reference to traditional fermented foods and beverages: a micro-level survey. *Journal of Hill Research* Supplementary Issue 20 (1): 1–37.

Tamang, B., Tamang, J.P., Schillinger, U. et al. (2008). Phenotypic and genotypic identification of lactic acid bacteria isolated from ethnic fermented tender bamboo shoots of north East India. *International Journal of Food Microbiology* 121: 35–40.

Tao, K., Yu, W., Prakash, S. et al. (2019). High-amylose rice: starch molecular structural features controlling cooked rice texture and preference. *Carbohydrate Polymers* 219: 251–260. https://doi.org/10.1016/j.carbpol.2019.05.031.

Ubhayasekera, S.J.K.A., Jayasinghe, P., Ekanayake, S. et al. (2012). High cholesterol oxidation in pickled mackerel (*Rastrelliger kanagurta*) from Sri Lanka. *European Journal of Lipid Science and Technology* 114 (6): 695–700. https://doi.org/10.1002/ejlt.201200017.

Waisundara, V., Jayawardena, N., and Watawana, M. (2016). Safety of fermented fish products. *Regulating Safety of Traditional and Ethnic Foods* 149–168: https://doi.org/10.1016/B978-0-12-800605-4.00008-6.

Wang, J. and Fung, D.Y. (1996). Alkaline-fermented foods: a review with emphasis on pidan fermentation. *Critical Reviews in Microbiology* 22 (2): 101–138.

Wikramanayake, T.W. (1996). *Food and Nutrition*, vol. 3. Hector Kobbekaduwa Agrarian Research and Training Institute.

Yadav, D.N., Sharma, M., Chikara, N. et al. (2014). Quality characteristics of vegetable-blended wheat–pearl millet composite pasta. *Agricultural Research* 3 (3): 263–270. https://doi.org/10.1007/s40003-014-0117-7.

Yegna Narayan Aiyar, A.K. (1953). Dairying in ancient India. *Indian Dairyman* 5: 77–83.

Yonzan, H. and Tamang, J.P. (2010). Microbiology and nutritional value of selroti, an ethnic fermented cereal food of the Himalayas. *Food Biotechnology* 24 (3): 227–247.

Yoon, K.Y., Woodams, E.E., and Hang, Y.D. (2004). Probiotication of tomato juice by lactic acid bacteria. *Journal of Microbiology (Seoul, Korea)* 42 (4): 315–318.

CHAPTER 14

THE ENSILING FERMENTATION OF FORAGE CROPS

ZWI G. WEINBERG
Forage Preservation and By-Products Research Unit, Agricultural Research Organization, The Volcani Institute, Rishon Le Zion, Israel (Retired)

CONTENTS

14.1 INTRODUCTION

Rations for ruminants include concentrates such as cereal grains and soybean meal which supply energy and protein; roughages in the rations supply mainly fibers which are necessary for the adequate function of the rumen. In addition, there are vitamin and mineral supplements. Such balanced rations are referred to as total mixed rations (TMR) which are designed according to the needs of each sort of animal and its phase of life (e.g. lactating cows, dry cows, heifers, bull calves, beef cattle, small ruminants, etc.). In ruminants, the feed is mainly digested and metabolized in the rumen by huge populations of various bacteria and protozoa which inhabit it. The rumen is actually a large (about 150 l in mature cattle) anaerobic isothermic fermentation vessel functioning at a constant pH of 6–7 (Van Soest 1994, pp. 152–177).

Microbial Fermentations in Nature and as Designed Processes, First Edition. Edited by Christon J. Hurst.

Roughages comprise whole forage crops, the major worldwide sources include whole forage corn, alfalfa, and other legumes, various grasses, sorghum, and small grain cereals such as whole crop wheat, barley, and oats (Figure 14.1). Ruminants may obtain roughages either by grazing forage grasses on pastures or being fed at the barn. In many areas of the world forage crops (and pastures) are seasonal which implies that during the harvest season there are abundant amounts of crops that cannot be used within a short time. On the other hand, during out-of-season periods such as cold winters or dry seasons, there may be no fresh crops available. Therefore, on modern large ruminant farms, forage crops are harvested during the season at peak yields and nutritional quality (at maximal content of digestible nutrients) and are preserved to insure year-round supply for the animals. The alternative preservation methods of forage crops are hay-making and ensiling. Hay-making is based on drying forage crops in the field to such a degree that they remain stable, usually <15% moisture; however, not all plants are suitable for drying because of their morphology (e.g. thick stems like in corn and sorghum). Even during the drying of legumes with thin stems, the protein-rich leaves dry long before the stems, become brittle and drop off and much of the nutritional value is lost as leaf shedding. Furthermore, in some areas, the dry season begins after the plants become over-mature and their nutritional value declines. If rain occurs during drying, it may slow down the process and the crops may spoil. Ensiling of forage crops is less weather dependent than hay-making but it requires initial significant capital investments in storage structures (silos) and machinery. Nevertheless, ensiling is widespread in modern agricultural systems, especially in developed countries.

Ensiling had been practiced in ancient Egypt and Greece and there is mention of ensiled fodder in the Old Testament (Isaiah, Chapter 30, verse 24: "and the oxen eat sour mixed fodder"). In the

FIGURE 14.1 Some forage crops for ensiling. (*See colour plate section for colour representation of this figure.*)

modern era, ensiling experiments began by farmers in Europe and the USA in the nineteenth century. More on the history of silage can be found in Wilkinson et al. (2003). The amount of annual silage production in the 1990s in Europe and North America was estimated at over 500 million tons of dry matter (Wilkinson et al. 2003). The price of corn silage in the silo in the USA is $50–80 per ton wet (Farm Progress, website).

Ensiling requires background knowledge in many different areas: ensiling is an agricultural operation that comprises the interphase between growing forage crops and ruminant nutrition and therefore one has to have some knowledge in these subjects; chemistry, biochemistry, and microbiology are crucial to understanding the basics of the ensiling fermentation; engineering is important for the appreciation of silo structures and the machinery involved throughout the process; plastic properties are important for adequate sealing materials and scientific skills and statistics are useful for adequate testing of various relevant issues. Many of these topics are covered in the special monograph Silage Science and Technology edited by Buxton et al. (2003).

The objectives of the current chapter are as follows:

1. Briefly summarize the basic biochemical and microbiological processes of ensiling fermentation of forage crops.
2. To describe the practical steps involved in silage making on commercial farms.
3. To describe and discuss the recent results pertaining to the microbiome composition of the silage ecosystem obtained with the state-of-the-art techniques of molecular biology.

14.2 THE ENSILING FERMENTATION

In general, microorganisms derive their energy (in form of ATP), reducing balance (in form of NADH or NADPH) and building blocks necessary for their sustainability by either aerobic respiration or anaerobic fermentation. There are also marine purple bacteria that derive their energy from anaerobic photosynthesis using bacterial chlorophyll and various electron donors (e.g. Ehrenreich and Widdel 1994). Fermentation is a process in which mono and di-saccharides are metabolized under anaerobic conditions, resulting in smaller organic substances such as ethanol, short-chain organic acids, and carbon dioxide and this process is coupled with ATP formation and adequate redox potential.

Ensiling is a preservation method of moist crops that is based on anaerobic solid-state fermentation by lactic acid bacteria (LAB). These bacteria convert water-soluble carbohydrates (WSC) into organic acids, mainly lactic acid, and as a result, the pH of the ensiled biomass decreases, and the ensiled forage is preserved as long as it is not exposed to air. In a way, ensiling is similar to sauerkraut made from cabbage, although the scales are different. Spontaneous ensiling begins when epiphytic LAB develop in the silage as soon as air is depleted and the environment in the silage becomes anaerobic. Lactic acid and volatile fatty acids (VFAs) (such as acetic, propionic, and butyric) which are formed inhibit spoilage and pathogenic microorganisms such as yeasts, molds, enterobacteria, and clostridium.

To understand how the ensiling fermentation takes place, it is important to have in mind the abiotic and biotic conditions which lead to successful ensiling. In general, microorganisms are affected by four abiotic factors: temperature, available moisture (water-activity, Aw), pH of their habitat, and availability or lack of oxygen. Microorganisms are referred to as psychrotrophics, mesophilic, and thermophiles according to the optimal temperature at which they develop best (cool, 0–20; medium, 20–40, and warm, >50 °C, respectively). The terms strict aerobes, strict anaerobes, facultative anaerobes, and microaerophilic describe the requirements and tolerance of microorganisms to air. The microorganisms for which the combination of these factors are most suitable become the predominant microbial group of a specific ecosystem whereas other groups are inhibited. Yeasts and molds, for example, tolerate lower pH and Aw values (more acidic and drier environment), however, molds require air to thrive. That is why, for example, yeasts and certain molds develop on acidic citrus peels and in stale bakery products which are aerated. Yeasts and molds develop in aerated parts of silages which were not compacted or sealed properly. Another example pertains to bacteria of the genus *Clostridium* which are spore-formers, the spores of which are very heat resistant; some *clostridium* species are saccharolytic, some are proteolytic, and some are both. Some species are highly toxigenic. *Clostridium* are strict anaerobes and also require high moisture conditions (Aw > 0.93). That is why clostrisdial growth may develop in anaerobic cans of moist food products of neutral pH values (e.g. cans of sweet corn, peas, uncured meat) which were not sterilized properly, with dire consequences. They may also develop in moist soil-contaminated silages in which the pH values remained relatively high. However, *Clostridium* cannot develop at pH values below 4.2–4.5, and they are also inhibited by nitrates which are found in some forage crops naturally or evolve from fertilizers (Weissbach and Ohff 1979).

In addition to the abiotic conditions, microorganisms usually depend also on the nutritional constituents of the medium on which they develop which supply energy, amino acids, vitamins, and minerals.

The LAB are the crucial bacteria for the ensiling fermentation which is basically a lactic acid fermentation. Similar lactic acid fermentation-based products include the manufacture of pickles, sauerkraut, salami sausage, and yogurt. The LAB are micro-aerophylic (which means that they thrive best in the presence of little oxygen), mesophilic (function best at 20–40 °C), and quite acid tolerant (minimal pH values for certain genera are as low as 3.7–4.0). The minimal water-activity requirements of LAB are around 0.87–0.88. LAB depend on carbohydrate and amino-acid supply from the media in which they develop. With this in mind, it is possible to appreciate the composition of the ensiled forage which leads to successful ensiling: its water content should be 600–700 g/kg (or 300–400 g/kg dry matter, DM), somewhat lower for legumes; it should contain at least 30–50 g/kg in DM of fermentable sugars (WSC) and the silage should become anaerobic as soon as possible. If the crop is too moist (>700 g/kg moisture), clostridial species (*Clostridium butyricum* and *Clostridium tyrobutyricum*) take over and they convert lactic to butyric acid resulting in higher pH values and spoilage: heat resistant clostridial spores originating from silage may penetrate milk, survive pasteurization and may spoil hard cheeses (the "thousand eyes" phenomena caused by sudden high fermentation gas pressure). Nitrates evolving from fertilization of the ensiled crop or occurring naturally inhibit *clostridium*. On the other hand, if the crop is too dry the ensiling fermentation is too slow resulting in considerable losses.

An additional important factor is the buffering capacity of the ensiled crop, which is a measure of its resistance to pH decrease even if acids are formed. If the crop is high in proteins and minerals, its buffering capacity is high and the pH remains high (>5.0). Examples of crops with low buffering capacity are corn and other cereals (200–250 mE/kg DM) whereas those with high buffering capacity are alfalfa and other legumes (500–550 mE/kg DM) (McDonald et al. 1991). The latter should be ensiled at lower water content (ca. 500 g/kg) to overcome the high buffering capacity.

Table 14.1 presents typical concentrations of fermentation products, from data compiled from various sources. The results in this table clearly indicate the effects of DM content and buffering capacity (in the case of alfalfa) on the final silage fermentation characteristics. The acids present and their concentration affect the final pH of the silage; acetic acid usually evolves from the

TABLE 14.1 Typical fermentation end-products of various silage.

Silage Type	DM (g/kg)	pH	Lactic Acid	Acetic Acid	Butyric Acid	Ethanol
Alfalfa	290	5.0	25	20		
	400	4.6	44	10		
	660	5.4	4	2		
Grass (high moisture)	180	4.8	20		55	
Corn	350	3.8	50	7		
Corn-commercial	325	3.7	38	39		2
Wheat (milk stage)	310	3.7	34	21	8	
Wheat (dough stage)	395	4.1	17	17	5	
	445	4.7	21	1		13
Wheat-commercial	390	3.9	35	22		
	315	4.2	20	15		

Results are in g/kg DM.

activity of *Lentilactobacillus buchneri*, butyric acid from the activity of *clostridium,* and ethanol may indicate yeast activity.

The LAB include many genera which differ in their abiotic requirements and mode of carbohydrate fermentation. It is common to distinguish between homo and hetero-fermentative LAB: homo LAB convert hexoses solely (>85%) to lactic acid under anaerobic conditions, as summarized in the following equation:

$$C_6H_{12}O_6 \rightarrow 2CH_3CHOHCOOH$$

Lactiplantibacillus plantarum (*Lactobacillus plantarum*) is an important homo-fermentative LAB species in the ensiling fermentation.

Hetero-fermentative LAB also convert hexoses to lactic acid but, in addition, other acids and fermentation products are formed, such as acetic acid and carbon dioxide.

$$C_6H_{12}O_6 \rightarrow CH_3CHOHCOOH + CH_3COOH + CO_2$$

L. buchneri (*Lactobacillus buchneri*) is an important hetero-fermentative LAB species in the ensiling fermentation. It can produce acetic acid from both WSC and lactic acid, as follows:

$$CH_3CHOHCOOH \rightarrow CH_3COOH + CO_2$$

Pahlow et al. (2003) also distinguish between obligate homo-fermentative LAB which homo-ferment hexoses but cannot ferment pentoses because they lack phosphoketulase; facultative hetero-fermentative LAB that can also ferment pentoses; and obligate hetero-fermentative which ferment hexoses to other products in addition to lactic acid.

Since lactic acid is a stronger acid than VFAs (its pK is 3.8 as compared with 4.8 of acetic acid) the resulting silage pH of hetero-fermentation is higher than that of homo-fermentation; the CO_2 produced represents a loss of nutritive material in the form of gas. Therefore, the ensiling fermentation by homo-LAB is usually preferred. However, VFAs inhibit detrimental yeasts and molds upon aerobic exposure (Moon 1983) and therefore, it is beneficial to have some hetero-fermentative activity during the ensiling fermentation to enhance the silage aerobic stability during feeding.

The most abundant LAB genera in forage and silage include *Lactobacillus* (some species of which are homo and some hetero-fermentative), *Pediococcus* (homo-fermentative), *Leuconostoc* (hetero-fermentative), *Enterococcus*, and *Lactococcus* (both homofermentative).

14.3 THE FOUR PHASES OF ENSILING

The initial phase: an aerobic phase when air is still present between the ensiled forage particles and the pH is still close to neutral. During that phase, aerobic microorganisms which are acid sensitive are still active until the air is depleted. These include yeasts and molds, enterobacteria, and others. Enzymes such as proteolytic enzymes are also active, and responsible for much of the amino acid nitrogen converted to ammonia. This phase usually lasts between 24 and 48 hours, depending on the moisture content of the forage, its buffering capacity, and the degree of silage compaction.

The fermentation phase: epiphytic LAB comprise only a tiny fraction of the microorganism population of the standing crop (Daeschel et al. 1987). Pahlow et al. (1995) found that after the crop has been chopped (as part of the silage-making protocol) the LAB population increases to 3 log CFU/g. It was explained that chopping releases manganese from the plant cells which helps to detoxify peroxides produced by the LAB themselves under the previous aerobic conditions. Once the air is depleted, LAB proliferate and become the predominant population and lactic acid fermentation sets on. The fermentation stage of ensiling lasts between two and seven days, depending on crop composition (moisture, WSC, buffering capacity) and the quality of silage preparation. During this period the temperature of the silage rises to above 40 °C, due to the heat generated by the microbial activity. The atmosphere within the silage is enriched with CO_2 and N_2 (Weinberg and Ashbell 1994). At the beginning of the lactic acid fermentation the genera which are acid sensitive, *Enterococcus* and *Leuconostoc*, are active. When the pH decreases to below 5.0 *Lactobacillus* and *Pediococcus* take over until the pH values decrease to 3.7–4.0, and then the fermentation terminates (Woolford 1984). At the peak of fermentation, the numbers of LAB may reach 9–10 log CFU/g and then they decrease to 5–6 log CFU/g.

The biochemistry of the ensiling fermentation depends on the enzymatic "toolbox" of the microorganisms which are prevalent at different stages of the process. Hence, various fermentation products might be detected in different silages which affect their odor and flavor, either attractive to the animals or deterring them. Pyruvate is the metabolic junction of anaerobic and aerobic sugar metabolism by microorganisms. Metabolic anaerobic pathways lead to the production of lactic acid (by the Embdem-Meyerhof glycolysis), acetyl-P (phosphogluconate/phospho-ketolase pathway), pyruvate and CO_2 (pentose phosphate pathway), ethanol (lactate and pyruvate dehydrogenases, mainly by yeasts), acetate (pyruvate format lyase), acetoin, 2,3 butanediol, acetone, butyrate, and butanol (mainly by *Clostridium*).

The stable phase: after the silage has stabilized, the general consensus was that no major changes take place, as long as air does not penetrate. However, about 60 days after ensiling the pH rises by 0.3–0.5 units, because the heterofermentative species *L. buchneri* becomes active and it produces acetic acid from lactic acid. This process increases pH and decrease in lactic acid concentration and losses due to CO_2 production; however, the bright side of *L. buchneri* activity is the enhanced aerobic stability of the silage due to the acetic acid which inhibits aerobic yeasts and molds (Weinberg and Chen 2013).

Feed-out phase: when the silage is fed to the ruminants, it is inevitably exposed to air. Air may penetrate silages in bunker silos 1–2 m from the unloading surface (Weinberg and Ashbell 1994). This may result in the development of yeasts and molds which spoil the silage. Yeasts produce relatively high concentrations of ethanol in the silage which does not contribute to its preservation and CO_2 which represent losses of nutritive matter. The major yeast genera found in silages include *Candida, Pichia,* and *Saccharomyces*. Molds reduce the nutrient contents of the silage, change the color and texture to gray-black and sticky and some molds produce mycotoxins which are hazardous to both humans and animals. The major mold genera found in deteriorating silages include *Aspergillus*, *Fusarium,* and *Penicillium,* some species of which may produce mycotoxins. Yeast and mold activity results in heating of the silage which in turn becomes brown due to caramelization or Millard-type reactions, reducing further its nutritional value. However, if the silage contains enough VFA (such as acetic acid produced by *L. buchneri*) its aerobic stability is enhanced because yeasts and molds are suppressed by them (Moon 1983; Weinberg et al. 1993).

14.4 STEPS OF ON-FARM ENSILING

Ensiling is a large-scale on-farm multi-step operation that involves hundreds to thousands of tons of forage. To ensure successful ensiling the steps should be carried out carefully and should be synchronized precisely.

The following steps are taken during the entire ensiling operation and use: harvesting, wilting, chopping, application of additives, transportation to the silo, compaction, sealing, and finally, unloading for feeding. Figure 14.2 shows the loading of chopped crop onto a transportation wagon (left upper panel), unloading of the chopped crop at a bunker silo (center upper panel), compaction by tractors (right upper panel), sealed silage (bottom left), and a scrapper for smooth unloading (bottom right).

Harvesting: should consider the maturation stage of the ensiled forage which meets the highest possible yields, the best nutritional composition for ruminants, and the best ensiling properties. Since silage should supply mainly roughage, the requirements are that the plant cell wall should

FIGURE 14.2 The major steps of silage making. (*See colour plate section for colour representation of this figure.*)

be highly digestible. In other words, the NDF digestibility should be as high as possible (NDF, neutral detergent fiber which comprises cellulose, hemicellulose, pectin, and lignin, Van Soest 1994, pp. 75–94). As the plant matures, a secondary cell wall is formed which is rich in lignin which comprises condensed phenylpropanoid polymers of high molecular weight. Often a lignocellulose complex is formed. These "woody" structures are nutritionally unavailable for ruminants and, as a result, the NDF digestibility decreases considerably during plant maturation. These considerations dictate the maturity stage which is recommended for harvesting crops for silage. The maturity of forage crops is determined according to the growth stage of the plants which includes germination, vegetative growth, blooming (in legumes)/flowering, filling of the grains, and drying of grains by the accumulation of starch in cereals. Forage crops for ensiling are usually harvested at the early reproductive stages, from blooming or flowering through the soft grain stages. Factors that may affect crop development and quality include fertilization (nitrogen supplementation might increase the crude protein [CP] content of the crop) and climate conditions: late freezing may kill the crop, droughts may inhibit its development, and high temperatures may increase the stem/leaf ratio. In corn, prolonged droughts enhance mycotoxin production by molds in the standing crop (Mahanna and Chase 2003).

Irrigation of corn fields for silage is stopped a week before harvesting. Maize is harvested at the 1/3 to ½ milk line when starch starts to accumulate in the kernels (Figure 14.3, Thurlow 2013). For more information on corn growth stages see Agronomic Spotlight (2018). Legumes (such as alfalfa, clover) are harvested at 5–10% bloom. Wheat for silage is the major winter forage crop in Israel because it can grow in winter rains. It is usually harvested at the milk stage of maturity (which means that when the soft grains are squeezed a milky liquid exudates). If wheat is harvested at the earlier flowering stage, the NDF digestibility is higher, but yields are lower and the dry matter content is well below the minimal 300 g/kg which is recommended for ensiling. If harvested later, at the dough stage, the crop is already too dry and its digestibility decreases (Ashbell et al. 1997). In Israel, in years when Jewish Passover coincides with the milk stage of wheat, harvest is performed earlier at the flowering stage; this is because grains at the

(a) (b)

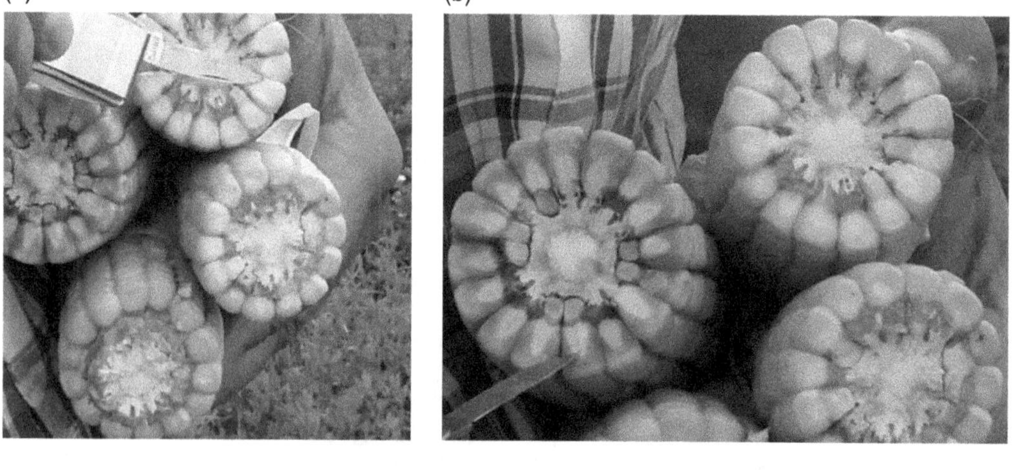

FIGURE 14.3 Assessing corn grain maturity by milk line. The Milk Line (the border between the deep yellow and the milky parts) proceeds from the external perimeter toward the center of the cobs as the plant matures (A and B, respectively). Source: Courtesy of Kable Thurlow, Grazing Educator & Beef, Michigan State University Extension. (*See colour plate section for colour representation of this figure.*)

TABLE 14.2 The effect of maturity stage on yields and wheat plants composition (g/kg DM).

Maturity	Yields t/ha	DM	WSC	Ash	CP	NDF	ADL
Shooting	6.4–8.4	170–190	—	108	133	562	57
Flowering	8.2–9.3	200–245	62–110	80–110	96–132	585–640	60–90
Milk	11.3–13.0	280–390	51–136	62–110	77–104	510–600	50–100
Dough	15.2–16.4	355–466	30–32	57–91	81–90	480–510	57–67

DM, dry matter; WSC, water-soluble carbohydrates; CP, crude protein; NDF, neutral detergent fiber; ADL, acid detergent lignin. Source: Adapted from Kennelly and Weinberg (2003).

milk stage of specific Biblical cereals including wheat are considered as Khametz (of leavening and souring potential) and cannot be handled or fed during the holiday. Table 14.2 summarizes the compositional changes in wheat plants during maturation. At the dough, the cell wall fraction decreases due to the grain filling with starch, but the NDF digestibility decreases as well.

Wilting: this means, field drying of the harvested crop for a few hours to a couple of days to reach the required DM content for ensiling. Some regions in the world, such as northern Europe, are cool and wet and the crops which are harvested are too moist for adequate ensiling. Therefore, it is necessary to plan the harvest just before some sunny and dry days to enable wilting to at least 300 g/kg DM. In Israel, it is more convenient to separate the mowing of the wheat from picking it up for chopping and transportation to the ensiling site. Inevitably, the wheat is wilted for a few hours. During wilting, plant respiration continues and the plants are susceptible to aerobic microorganisms and the proteolytic activity of the plant's enzymes. These result in some losses of nutrients. Legumes are necessary to wilt to higher DM content (400–500 g/kg) to overcome their high buffering capacity and the resulting dry silage is referred to as "haylage". The thin leaves of legumes, which are high in protein, dry faster than the stems and become brittle and fall off (leaf shedding) which adds to the field losses. Therefore, it is recommended to shorten the wilting period as much as possible. Sometimes legumes are co-ensiled with a cereal forage crop high in WSC such as whole-crop corn which results in improved ensiling characteristics and high CP content (Wang et al. 2021).

Chopping: before ensiling the crop is chopped to a theoretical particle length of 10–20 mm to enable its consolidation (compaction) in the silo. Chopping increases the surface area of the ensiled crops resulting in more efficient microbial fermentation activity. However, shorter particles than 10 mm lose their effectiveness as efficient fibers necessary for adequate rumen function. Pahlow et al. (1995) found that immediately after chopping the LAB population increases threefold. Chopping is followed by immediate pick up to the transportation vehicle by a blower. At this step, silage additives may be applied.

Silage additives: silage additives may be applied at ensiling to enhance the ensiling preservation and minimize ensiling losses. Biological additives comprise various species and strains of LAB. Homo-fermentative silage inoculants which comprise such species as *L. plantarum* and *Pediococcus pentosaceus* are applied at 10^5 CFU/g crop to produce large amounts of lactic acid in the silage within a short period. As a result, the pH value decreases faster to around 4.0 and the activity of proteolytic enzymes and detrimental microorganisms is inhibited and losses are minimized. Figure 14.4 shows the effect of homo lactic LAB inoculants on the rate of pH decrease in wheat silage (adapted from Keshri et al. 2019). In cereal silages, homo-fermentatitve LAB additives may reduce the aerobic stability because of the lack of sufficient VFAs which inhibit aerobic yeasts and molds (Weinberg et al. 1993). Some inoculant strains are believed to exert probiotic effects on the animals and improve their performance (increased daily live weight gain, feed intake, and milk yield)

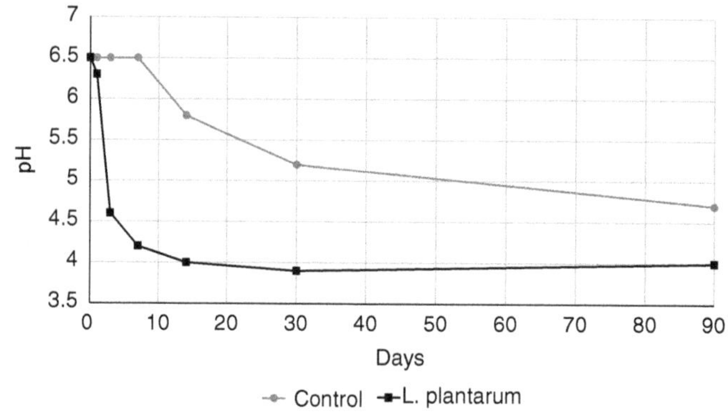

FIGURE 14.4 The effect of applying *L. plantarum* at ensiling on rate of pH decrease in wheat silage. Source: Adapted from Keshri et al. (2019).

which makes such inoculants even more cost-efficient (Weinberg and Muck 1996). If the silage is susceptible to aerobic exposure, *L. buchneri* may be added alone or in combination with homolactic inoculants. The acetic acid it produces inhibits aerobic yeasts and molds. Publications that summarize inoculant effects on silage usually describe laboratory experiments with a single crop with specific DM content treated with specific species or strains. Therefore, the results of such studies may be limited to the conditions of the described experiments. To explore general trends Oliveira et al. (2017) performed a meta analysis on the effects of homo-fermentative and facultative hetero-fermentative Lactobacillus inoculants. This analysis revealed that such inoculants enhanced the ensiling fermentation and DM recovery of grasses and legumes but did not affect the ensiling of corn, sorghum, and sugarcane silages, maybe because the latter forage crops ensile easily. Surprisingly there was no inoculant effect on the aerobic stability of silages. These inoculants had a positive effect on milk yields.

Chemical additives such as Ca-propionate and sulfur-based additives are sometimes used as silage preservatives or aerobic stability enhancers but acid additives are corrosive to machinery and less convenient to use. More details on silage additives can be found in Kung Jr. et al. (2003).

Transportation to the silo: the chopped forage is loaded into trucks by a blower which is set up adjacent to the chopper. Transportation distances should be as short as possible to avoid excessive aerobic activity and heating and to save fuel costs.

Consolidation (compaction): the ensiled particles must be compacted in the silo in order to expel as much air as possible from within the silage. Silages vary in their degree of compaction but 230 kg DM/m^3 is considered satisfactory. The method of compaction depends on the type of silo: in vertical tower silos the ensiled particles are compacted by their own weight; in bunker silos, tractors are moving slowly back and forth after each wagon unloads the chopped crop; baled silage are compacted by a press just before wrapping with plastic layers. Bag silages which are stored in heavy-duty long plastic sleeves are compacted during filling by the filling machine. It should be mentioned that the compaction of moist crop particles (<250 g/kg DM) results in viscous dark seepage which represents considerable losses and heavy environmental pollution. Compaction of particles that are too dry (>400 g/kg DM) is incomplete because dry plant particles are resilient and difficult to press, leaving more air within the silage which results in aerobic activity followed by excessive heating and browning reactions such as caramelization and Maillard type of reactions (condensation products of carbohydrates and proteins which cannot be utilized by the animals) (Muck et al. 2003).

Ideal sealing Poor sealing

FIGURE 14.5 Ideal (left) vs. poor sealing (right). (*See colour plate section for colour representation of this figure.*)

Sealing: The ensiled forage must be sealed off the air because air is detrimental to silage and enables the activity of aerobic spoilage microorganisms, especially yeasts and molds (Woolford 1990). In steel or concrete tower silos no sealing action takes place; however, bunker silos must be covered with polyethylene UV-resistant sheeting of 200 μm thickness to prevent air penetration. Special care should be given to the "shoulders" of the bunker silo, where the walls meet the upper layers of the silage, these are the areas that are most susceptible to air penetration and spoil the fastest. The plastic sheeting is anchored to the top layer of the silage by placing used tires, cotton pulp, or special plastic nets which are secured with sandbags at the shoulders and minimize tearing by birds and rodents (Figure 14.5).

Transportation, compaction, and sealing should be done as quickly as possible in order to minimize the exposure to air of the chopped forage; it may take several days to complete the ensiling operation in some very large bunker silos.

Unloading for feeding: when the silage is unloaded for feeding it is inevitably exposed to air. Air might penetrate the silage from the unloading front 1–2 m (Weinberg and Ashbell 1994). Some silages are susceptible to spoilage upon aerobic exposure (e.g. corn silage) if there are not enough VFAs to inhibit yeasts and molds. It is recommended to remove the plastic sheeting as little as possible and plan to unload the silage at a fast rate, to prevent the development of yeasts and molds. Silages are unloaded either with a front load bucket tractor, with a special scrapper which leaves a smooth undisturbed front, or with a block cutter.

Silo types: in some countries tower silos made of concrete or metal are used. Forceful blowers or conveyor belts are used to load the forage from the top of the tower and the ripe silage exits either from an opening at the bottom or scrapped from the top. Bunker silos are made of concrete slanted walls (to withstand the pressure) and concrete floors. These should be treated periodically to prevent corrosion of the concrete by coating it with tar. In Israel, silages are prepared and stored in very large bunker silos of several thousand tons at regional feeding centers which replaced small individual bunker silos. This arrangement is more cost-efficient. Silages are also stored in 500–700 kg bales wrapped with 6–8 layers of polyethylene. Baled silages enable flexibility in the transport of silages to remote farms or pastures as supplement feed. The outer surface of such silages is susceptible to molding. Bag silos are long sleeves made of heavy-duty plastic equipped with valves to release internal gas pressure. A special machine evacuates the air from the bags by vacuum and presses the forage inside. Figure 14.6 shows the various types of silos for silage.

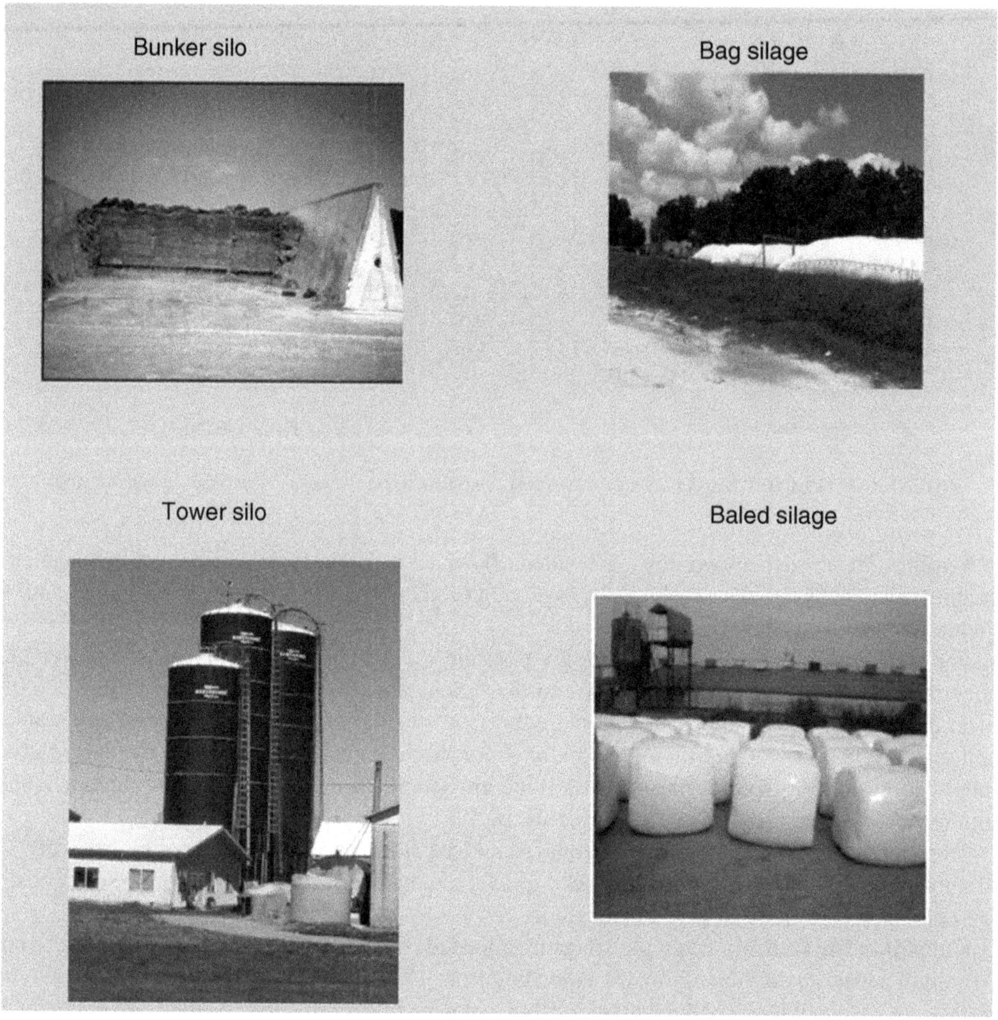

FIGURE 14.6 Silage silo types. Source: The tower silo image, courtesy of Hanson Silo Company. (*See colour plate section for colour representation of this figure.*)

Sources of ensiling losses: Ensiling is an expensive operation and therefore, losses should be minimal. Ten percent loss of nutrients during ensiling is considered normal. Silage price covers such losses. Losses evolve from hetero-fermentative microbial activity in which nutritive materials are lost as CO_2 and likewise will be lost from activity from aerobic microorganisms (especially yeasts and molds) during the aerobic or feed-out phases. True protein is lost by the conversion of amino acids to ammonia nitrogen by the proteolytic activity of the plant's enzymes in the initial aerobic stage. In moist silages, clostridial activity and seepage result in losses. Butyric acid produced by *Clostridium* results in an obnoxious smell and reduced animal intake (feed consumption). Seepage from moist silages is high in biological oxygen demand (BOD) and therefore, it is detrimental to the environment and may pollute freshwater sources. In silage parts that are exposed to air (in bunker silos at the corners between the top layer and the walls) heavy molding results in sticky dark silage which is nutritionally poor and might contain hazardous mycotoxins and therefore, must be discarded. Losses should be minimized by fast and proper silage-making procedures and by the use of additives as described earlier.

FIGURE 14.7 Anaerobic mini-silos (Weck™, Wehr-Offlingen, Germany). The springs which secure the lids enable gas release from inside only but not air penetration. (*See colour plate section for colour representation of this figure.*)

On-farm assessment of silage quality: the following sensorial parameters can be used: color, should be green; brownish to dark colors indicate excessive heating and losses; texture: non-sticky free-flowing particles; smell: should be pleasantly acidic; temperature (by hand feeling): not warm, close to ambient; dots of yeasts or visible molds indicate potential spoilage. The top layers and the shoulders of the silage should be examined carefully. Sticky dark moldy stuff should not be fed.

Laboratory silage analysis: preservation quality of the silage is determined by measuring pH values, the concentration of lactic acid and VFAs, crude protein and ammonia nitrogen, and ash content. It is possible to assess the aerobic stability of silages, usually by measuring temperature increases during open storage. It is expressed as the time it takes for a silage sample to heat 2 °C above the ambient temperature.

The nutritional value of the silage is determined by its *in vitro* DM and NDF digestibility (Tilley and Terry 1963), by anaerobic incubation of dried and ground samples at 39 °C in buffered rumen fluid taken from ruminants. The digestibility is expressed as a sample weight loss fraction during this incubation. The effect of feeding silage on animal performance can be measured *in vivo*. Some commercial laboratories determine the nutritional constituents of silages making use of calibrated NIRS (near-infrared spectroscopy) which is simple, quick, and relatively inexpensive, given that the analysis is calibrated for specific forage types.

Ensiling experiments can be performed in the laboratory in mini-silos of anaerobic jars (Figure 14.7) or in vacuumed air-tight plastic bags. In such experiments, enough replicates are prepared for each tested treatment and it is also possible to sample replicates at different time points throughout the storage period.

14.5 ENVIRONMENTAL ASPECTS OF ENSILING

With increasing concern about global warming due to greenhouse gas emissions and earth pollution it is paramount to recognize that silage contributes to these aspects. The following is a very brief account of this broad issue. As mentioned above, ensiling results in CO_2 emission due to inevitable ensiling fermentation losses. Another greenhouse gas under research pertains to methane emission by ruminants. The question of whether certain silages or silage additives could mitigate methane emission by ruminants is being studied extensively (e.g. Van Gastelen et al. 2019; Ellis et al. 2016).

In addition, moist silages release effluent with a high pollution potential to freshwater sources. Moist silages which underwent butyric acid fermentation also result in unpleasant odors. With this regard, it is worth mentioning the emission of volatile organic compounds (VOC) resulting from silage fermentation into the atmosphere. These include various volatile acids, alcohols, aldehydes, and ketones which might affect air quality and ozone formation in the atmosphere. The magnitude and impact of such emissions are not as yet clear (Hafner et al. 2013).

Another nuisance evolves from huge amounts of remaining used plastic sheeting (usually polyethylene) which is practically un-degradable. Over the years researchers tried to develop biodegradable or edible organic plastic films (e.g. polymers of lactate) in forms of spray which would solidify upon contact with air; however, such polymers are mechanically weak they are not yet widely used and have not yet proven to protect silages under farm conditions.

14.6 UNCONVENTIONAL ENSILING

Ensiling can be applied for the preservation of different plant materials, as follows.

TMR: Comprise the entire components necessary for the nutritional requirements of the animals. They include roughage, concentrate (cereal grains and soybean meal), by-products, minerals, and vitamins. Such rations can be specially designed for different types of animals, e.g. lactating cows, dry cows (before calving when milking is stopped), heifers, calve bulls, beef, small ruminants, etc. Weinberg et al. (2011) showed that commercial-baled TMR for lactating cows ensiled at DM content of 500 and 650 g/kg had final pH of 4.3 with minimal losses. Preservation of TMR by ensiling would enable flexibility in use and shipment.

High moisture corn: In North America, silage of high-moisture corn grains is common as a concentrate for both dairy and beef cattle. This technology involves ensiling of corn grains (or ears) at a moisture content of around 250–300 g/kg, which saves the necessity to fully dry them. This is much different from silages of whole crop forages which are ensiled at 600–700 g/kg moisture. The grains should be ground or rolled before packing into a silo to facilitate consolidation and exclusion of air. The lowest pH achieved in such silages is around 5.0 and the lactic acid is less than 20 g/kg DM. Such silage is susceptible to aerobic deterioration by yeasts and molds and there is some risk of mycotoxins. Treatment with acetic or propionic acid can be used to inhibit the aerobic yeasts and molds (Buchanan-Smith et al. 2003; Wardynski et al. 1993).

Wet by-products (BPs): Agricultural BPs are a result of food processing and surplus. The estimated annual global production of wet BP is a billion tonnes (Orosz and Davies 2015). The major wet BP worldwide includes BP from the sugar industry (sugar beet pulp and sugar cane tops and bagasse), citrus pulp and peels from the juice industry, BP from the processing of other fruits and vegetables, wet brewers' grains, and more. The BP contains energy, fiber, or protein and could be used for ruminant feeding. If not used, large quantities of rotting agricultural BP become an environmental nuisance: they attract insects, birds, and rodents and emit obnoxious odors. Moreover, seepage released from such large piles is high in BOD and COD values and may contaminate freshwater sources. Some BP are seasonal and during the season large quantities are produced that cannot be used immediately. Therefore it is necessary to preserve them. Sun drying is usually not an option because of their thickness and high moisture content (>800 g/kg); fuel drying is usually not cost-efficient. Hence ensiling is an option. However, the high moisture and carbohydrate content dictates special solutions. For example, ensiling of citrus pulp and peels alone resulted in large volumes of seepage and considerable losses due to intensive yeast activity (Ashbell et al. 1987). The proposed solution is to ensile such materials in mixtures with some dry feedstuff such as chopped straw or soybean and almond hulls (Ashbell et al. 1995).

The ENLAC process (enzyme-assisted lactic acid fermentation): Plants contain a large variety of biologically active materials that could be used as medicines, cosmetics, and food ingredients. It has been suggested to ensile compatible plants along with cell-wall hydrolyzing enzymes and LAB inoculants to facilitate the extraction of valuable target substances. The pH of such silages 4.5–5.0 is optimal for the enzyme activity, and ensiling enables flexibility in the timing of the extraction operation. Softening of the cell walls by the enzyme activity might facilitate and improve extraction efficiency. It has been found that ENLAC improved the extraction of protein and chlorophyll from alfalfa and polyphenols from sage and rosemary (Weinberg et al. 1990, 1999). However, when ENLAC was tried on lemon grass and lemon eucalyptus, the acidic conditions prevailing in the ensiled biomass resulted in changes in the essential oils which were not desirable from the aspect of their odor profiles. This might have been caused by acid-catalyzed Prins-type condensation in which the carbonyl and carbon=carbon double bonds of the terpenes making up the essential oils are protonated which results in changes in their chemical structure and consequently their odor (Dudai et al. 2001). The challenge is to find plants that can be ensiled easily and contain worthy substances which still maintain their biological activity under ensiling conditions. A mini-review of ENLAC trials so far can be found in Weinberg and Szakacs (2018).

14.7 DEVELOPMENT OF SILAGE RESEARCH TOOLS

Silage research has been going on intensively since the 1950s first in western and northern Europe and since the 1980s also in North America. Now, much silage research is performed also in Brazil, Japan, China, and South Africa. Surprisingly, not much silage research output is coming out of areas such as Russia, India, Latin (Spanish) America, most of Africa, and Australia. The reasons could be that these areas rely more on pastures and silage is not common, or that their research facilities are not adapted for silage research or they publish in local journals in local languages. Since the 1970s an International Silage Conference has been taking place about every three years, each time in a different country with subsequent publication of the proceedings.

The early silage research employed classical microbiology and biochemistry tools. Many laboratories and on-farm experiments have been performed over the years pertaining to ensiling of newly introduced crops or cultivars, silage additive testing, aerobic stability of silages, silage rumen digestibility at different crop stages of maturity, animal performance as a response to feeding different silages, etc. Microbial dynamics during ensiling was studied until quite recently by culturing, isolation, and identification of microorganisms using classical microbiological techniques. Such methods are still widely used for the simple assessment of microbial populations in silages, using selective media and incubation conditions. This approach provides limited and general data as it is not specific even to the genera level. Furthermore, some microorganisms which are involved in ensiling may not be culturable. However, such techniques combined with complementary chemical analysis such as determination of pH, fermentation products (by HPLC or GC), and aerobic stability tests enable researchers to assess the quality of the ensiling fermentation and to compare between experimental treatments.

Culture-independent molecular techniques, such as denaturing gradient gel electrophoresis (DGGE), PCR-amplified DNA fragments and cloning and sequencing of 16S rRNA, and high-resolution melt analysis of the bacterial genome have provided more detailed information regarding microbial diversity and microbial succession during ensiling (e.g. Parvin and Nishino 2010; Stevenson et al. 2006; McGarvey et al. 2013; Ogiy et al. 2016). The recent introduction of next-generation sequencing (NGS) technology has resulted in fast inexpensive and accurate information on environmental/ecological microbial community composition. NGS has been employed to study changes in the microbiome during ensiling of various crops such as whole-crop corn, various grasses, small-grain cereals, and alfalfa (Eikmeyer et al. 2013; Zhao et al. 2016; Duniere et al. 2017; Ni et al. 2017; Kraut-Cohen et al. 2016; Keshri et al. 2018, 2019). Experiments that studied the effect of applying LAB inoculants on the microbial composition showed that inoculants comprised of *L. plantarum* indeed took over at the early stages of ensiling and became the dominant species (e.g. Ogiy et al. 2016; Keshri et al. 2019). A review of the molecular techniques used in silage microbiome research can be found in McAllister et al. (2018).

NGS is the state-of-the-art technology that enables detailed taxonomic profiling of the microorganisms involved in ensiling fermentation down to the genus and species levels. Some studies which employed NGS tried to correlate microbial species with metabolome results (Keshri et al. 2019; Xu et al. 2020) but the quantity of information is still overwhelming and warrants more research. Applying meta-analysis of broad data pertaining to the microbiome and metabolome could shed light on the most important species that play a role in most ensiling fermentations.

Table 14.3 is a humble trial to compare the silage microbiome data obtained from the relevant literature. The results of the wheat and corn silages are quite consistent while the legume silage differs. Kraut-Cohen et al. (2016) determined the microbiome of several commercial corn, mixed corn-sorghum, and wheat silages at feed centers in Israel. Two out of the three wheat silages had close to 100% *Lactobacillus*. However, one wheat silage had only 8% *Lactobacillus*, 20% *Bacillus*, 2% *Bifidobacterium*, and 2% Clostridium_senso_stricto_R. Indeed, the latter silage had a higher pH value (4.2) and lower lactic acid content (21 g/kg DM) as compared with the other wheat silages (3.9–4.0 and 37 g/kg DM, respectively). It also contained a high concentration of acetic acid (36 g/kg DM).

NGS provides information down to the genus and species level and has identified many species the role of which in the ensiling fermentation is yet unclear. Studies have detected a myriad of species, some of which may seem surprising in that context. For example, Keshri et al. (2019) detected during the course of fermentation many species among which are *Levilactobacillus parabrevi*, *Companilactobacillus paralimentarius*, *Levilactobacillus brevis*, *Companilactobacillus*

TABLE 14.3 Results of microbiome analysis of 90–100 days laboratory silages compiled from several references.

Genus	Wheat[a]	Corn[b]	Corn[c]	Corn[d]	Alfalfa[e]
Lactobacillus	60 (90)	95 (98)	99	100 (99)	15 (62, 48)
Weissella	30	1			30 (4, 9)
Pediococcus	9				30 (0, 20)
Pantoea					4 (8, 5)

Results are expressed as relative abundance (%) at the genera level. Numbers in parenthesis give the abundance of silages treated with *L. plantarum* and *L. buchneri*, respectively.

[a] Keshri et al. (2018).

[b] Keshri et al. (2019).

[c] Xu et al. (2019).

[d] Xu et al. (2020).

[e] Ogunade et al. (2017).

crustorum, and *Acinetobacter junii*. In this study, after one day of ensiling, *Weissella* was the predominant genera while after seven days this genera decreased to 7% and Lactobacillus increased to 40–60%, without major differences between the control and silages treated with *L. plantarum*. Xu et al. (2019) detected *Lactobacillus acetotolerance* (45%, 58%, and 30%), *L. buchneri* supsp. *Silagei* (45%, 12%, and 45%), *L. buchneri* (3%, 5%, and 2%), and *Companilactobacillus farciminis* at (2%, 12%, and 10%) in control, versus *L. plantarum* and *L. buchneri* for inoculated silages. Xu et al. (2020) mentioned over 40 species in silages many of which seem rare, never heard of before in this context, and in minor abundance. (Since the results are given as bar charts in which each species is presented by a different color and hue, some hues are too similar and therefore, it is very difficult to read these charts.)

In some studies, fungal genera were also determined. Fungi which are mainly active when air is present, are detrimental to silages as they result in nutritive losses and some molds produce hazardous mycotoxins in it. Keshri et al. (2018) showed that in laboratory corn silages *Meyerozyma, Acremonium,* and *Cladosporium* were the major genera in un-ensiled corn replaced by *Candida* and *Monascus* at later stages of ensiling. Saccharomycetales were the major order found in small grain silages after 90 days and upon aerobic exposure (70%) (Duniere et al. 2017).

Recent results obtained with NGS confirm results obtained with classical methods of microbiology and chemical analysis: during the ensiling fermentation the diversity of bacterial communities declines with time and in well-preserved silages, *Lactobacillus* becomes the predominant genus; silages that were treated with inoculants comprised of Lactobacillus, this genus became dominated by Lactobacillus faster as compared with un-inoculated silages. However, the myriad of other genera and species detected at various stages of ensiling points out the complexity of the silage ecosystem which surely warrants more research. Again, the question that comes to mind is **what is the role of the rare different genera and species in the silage fermentation or the silage ecosystem?** The state-of-the-art molecular and metabolome analytical tools may help to shed light on these issues in the future.

14.8 CONCLUSIONS

Ensiling is a large-scale on-farm operation that is performed outdoors where conditions may be much different from those in a closed, clean, and isolated laboratory. If principles are followed closely silage making is usually successful and inevitable losses are under control. Rarely, a whole silo molds for some failure in compaction and sealing and then losses are significant. Silage spoilage may be associated with health risks for both animals and humans.

Since the beginning of modern ensiling experimental research supported the farmers of this industry. Based on research and development, the quality and preservation of silage ever improve with an increasing understanding of the factors which affect its quality. The development of recent molecular tools such as NGS of microbial genomes boosts the research significantly and opens venues for more sophisticated research which would enable a better understanding of the silage ecosystem under various conditions.

ACKNOWLEDGMENT

I thank my friends and colleagues Dr. Gili Ashbell and Mrs. Yaira Chen for long years of excellent research partnership from which I have learned a lot.

REFERENCES

Agronomic Spotlight (2018). Corn growth stages and GDU requirements. https://www.corn-states.com/app/uploads/2018/07/corn-growth-stages-and-gdu-requirements.pdf (accessed May 31, 2022).

Ashbell, G., Pahlow, G., Dinter, B., and Weinberg, Z.G. (1987). Dynamics of orange peel fermentation during ensiling. *Journal of Applied Bacteriology* 63: 275–279.

Ashbell, G., Weinberg, Z.G., and Hen, Y. (1995). Studies of quality parameters of variety ensiled broiler litter. *Animal Feed Science and Technology* 52: 271–278.

Ashbell, G., Weinberg, Z.G., Bruckental, I. et al. (1997). Wheat silage: the effect of cultivar and stage of maturity on yield and degradability *in situ*. *Journal of Agriculture and Food Chemistry* 45: 709–712.

Buchanan-Smith, J., Morris, J.R., and Smith, T.K. (2003). High moisture grain and by-products. In: *Silage Science and Technology* (ed. D.R. Buxton, R.E. Muck, and J.H. Harrison), 825–854. Madison, WI, USA: Agronomy Monograph 42, American Society of Agronomy.

Buxton, D.R., Muck, R.E., and Harrison, J.H. (ed.) (2003). *Silage Science and Technology*. Madison, WI, USA: Agronomy Monograph 42, American Society of Agronomy.

Daeschel, M.A., Anderson, R.E., and Fleming, H.P. (1987). Microbial ecology of fermenting plant materials. *FEMS Microbiology Reviews* 46: 357–367.

Dudai, N., Weinberg, Z.G., Larkov, O. et al. (2001). Changes in essential oil during enzyme-assisted ensiling of lemongrass (*Cymbopogon citratus* Stapf.) and lemon eucalyptus (*Eucalyptus citroidora* Hook). *Journal of Agricultural and Food Chemistry* 49: 2262–2266.

Duniere, L., Xu, S., Long, J. et al. (2017). Bacterial and fungal core microbiomes associated with small grain silages during ensiling and aerobic spoilage. *BMC Microbiology* 17: Article 50.

Ehrenreich, A. and Widdel, F. (1994). Anaerobic oxidation of ferrous iron by purple bacteria, a new type of photosynthetic metabolism. *Applied and Environmental Microbiology* 60: 4517–4526. https://doi.org/10.1128/AEM.60.12.4517-4526.1994.

Eikmeyer, F.G., Kofinger, P., Poschenel, A. et al. (2013). Metagenome analysis reveal the influence of the inoculant Lactobacillus buchneri CD034 on the microbial community involved in grass ensiling. *Journal of Biotechnology* 167: 334–343. https://doi.org/10.1016/j.jbiotec.2013.07.021.

Ellis, J.L., Hindrichsen, I.K., Klop, G. et al. (2016). Effect of lacticacid bacteria inoculation on methane emission and productivity of Holstein Friesian dairy cattle. *Journal of Dairy Science* 99: 7159–7174. https://doi.org/10.3168/jds.2015-10754.

Hafner, S.D., Howard, C., Muck, R.E. et al. (2013). Emission of volatile organic compounds from silage: compounds, sources and implications. *Atmospheric Environment* 77: 827–839. https://doi.org/10.1016/j.atmoseny.2013.04.076.

Kennelly, J.J. and Weinberg, Z.G. (2003). Small grain silage. In: *Silage Science and Technology* (ed. D.R. Buxton, R.E. Muck, and J.H. Harrison), 749–779. Madison, WI, USA: Agronomy Monograph 42, American Society of Agronomy.

Keshri, J., Chen, Y., Pinto, R. et al. (2018). Microbiome dynamics during ensiling of corn with and without *Lactobacillus plantarum* inoculant. *Applied Microbiology and Biotechnology* 102: 4025–4037. https://doi.org/10.1007/s00253-018-8903-y.

Keshri, J., Chen, Y., Pinto, R. et al. (2019). Bacterial dynamics of wheat silages. *Frontiers in Microbiology* 10: 1532–1548. https://doi.org/10.3389/fmicb.2019.01532.

Kraut-Cohen, J., Tripathi, V., Volchinski, V. et al. (2016). Temporal and spatial assessment of microbial communities in commercial communities in bunker silos. *Applied Microbiology and Biotechnology* 100: 6827–6835. https://doi.org/10.1007/s00253-016-7512.

Kung, L. Jr., Stokes, M.R., and Lin, C.J. (2003). Silage additives. In: *Silage Science and Technology* (ed. D.R. Buxton, R.E. Muck, and J.H. Harrison), 305–360. Madison, WI: Agronomy Monograph 42, American Society of Agronomy.

Mahanna, B. and Chase, L.E. (2003). Practical applications and solutions to silage problems. In: *Silage Science and Technology* (ed. D.R. Buxton, R.E. Muck, and J.H. Harrison), 855–895. Madison, WI: Agronomy Monograph 42, American Society of Agronomy.

McAllister, T.A., Duniere, L., Drouin, P. et al. (2018). Silage review: using molecular approaches to define the microbial ecology of silage. *Journal of Dairy Science* 101: 4060–4074.

McDonald, P., Henderson, A.R., and Heron, S.J.E. (ed.) (1991). *The Biochemistry of Silage*, 2e. Marlow, UK: Chalcombe Publications.

McGarvey, J.A., Franco, R.B., Palumbo, J.D. et al. (2013). Bacterial population dynamics during the ensiling of *Medicago sativa* (alfalfa) and subsequent exposure to air. *Journal of Applied Microbiology* 114: 1661–1670.

Moon, N.J. (1983). Inhibition of the growth of acid tolerant yeasts by acetate, lactate and propionate and their synergistic mixtures. *Journal of Applied Bacteriology* 55: 454–460.

Muck, R.E., Moser, L.E., and Pitt, R.E. (2003). Postharvest factors affecting ensiling. In: *Silage Science and Technology* (ed. D.R. Buxton, R.E. Muck, and J.H. Harrison), 251–304. Madison, WI, USA: Agronomy Monograph 42, American Society of Agronomy.

Ni, K., Minh, T.T., Tu, T.T. et al. (2017). Comparative microbiota assessment of wilted Italian ryegrass, whole crop corn, and wilted alfalfa silage using denaturing gradient gel electrophoresis and next-generation sequencing. *Applied Microbiology and Biotechnology* 101: 1385–1394.

Ogiy, S., Chen, Y., Pasvolski, R. et al. (2016). High resolution melt analysis to confirm the establishment of *Lactobacillus plantarum* and *Enterococcus faecium* from silage inoculants during ensiling of wheat. *Grassland Science* 62: 29–36.

Ogunade, I.M., Jiang, Y., Pech Cervantes, A.A. et al. (2017). Bacterial diversity and composition of alfalfa silages as analyzed by Illumina MiSeq sequencing: effects of *Escherichia coli* O157:H7 and silage additives. *Journal of Dairy Science* 101: 2048–2059. https://doi.org/10.3168/jds.2017-12876.

Oliveira, A.S., Weinberg, Z.G., Ogunade, I.M. et al. (2017). Meta-analysis of effects of homo-fermentative and facultative hetero-fermentative lactic acid bacteria inoculation on silage fermentation and aerobic stability and the performance of dairy cows. *Journal of Dairy Science* 100: 4587–4603. https://doi.org/10.3168/jds.2016-11815.

Orosz, S. and Davies, D.R. (2015). Short and long-term storage of wet by-products fed by ruminants. In: *Proceedings of the 17th International Silage Conference* (ed. J.L.P. Daniel, G. Morais, D. Jurge, and L.G. Nussio), 200–243. S.P., Brazil: Piracicaba.

Pahlow, G., Muller, T., and Lier, D. (1995). Einfluss der Ernteverfahrensauf die Nachweisbarkeit epiphytischer Laktobakterien von Futterpflanzen. *Wirtschfliches Futter* 41: 306–326. (in German).

Pahlow, G., Muck, R.E., Driehuis, F. et al. (2003). Microbiology of ensiling. Chapter 2. In: *Silage Science and Technology* (ed. D.R. Buxton, R.E. Muck, and J.H. Harrison), 31–93. Madison, WI, USA: Agronomy Monograph 42, American Society of Agronomy.

Parvin, S. and Nishino, N. (2010). Succession of lactic acid bacteria in wilted rhodes grass silage assessed by plate culture and denaturing gradient gel electrophoresis. *Grassland Science* 56: 51–55.

Stevenson, D.M., Muck, R.E., Shinners, K. et al. (2006). Use of real time PCR to determine population profiles of individual species of lactic acid bacteria in alfalfa silage and stored corn stover. *Applied Microbiology and Biotechnology* 71: 329–338.

Thurlow, K. (2013). Assessing Corn Maturity, using milk line. https://www.youtube.com/watch?v=X8qu3ID4Dtx0 (accessed June 7, 2022).

Tilley, J.M.A. and Terry, R.A. (1963). A two stage technique for the *in vitro* digestion of forage crops. *Journal of British Grassland Society* 18: 104–111.

Van Gastelen, S., Dilkstra, J., and Bannink, A. (2019). Are dietary strategies to mitigate enteric methane emission equally effective across dairy cattle, beef cattle and sheep? *Journal of Dairy Science* 102: 6109–6130. https://doi.org/10.3168/jds.2018-15785.

Van Soest, P. (1994). *Nutritional Ecology of the Ruminant*. Ithaca, NY: Cornell University Press.

Wang, M., Gao, R., Franco, M. et al. (2021). Effect of mixing alfalfa with whole-plant corn in different proportions on fermentation characteristics and bacterial community of silage. *Agriculture* 11: 174–185. https://doi.org/10.3390/Agricultre11020174.

Wardynski, F.A., Rust, S.R., and Yokoyama, A.T. (1993). The effect of microbial inoculation of high-moisture corn on fermentation characteristics, aerobic stability and cattle performance. *Journal of Animal Science* 71: 2246–2252.

Weinberg, Z.G. and Ashbell, G. (1994). Changes in gas composition in corn silages in bunker silos during storage and feedout. *Canadian Agricultural Engineering* 36: 155–158.

Weinberg, Z.G. and Chen, Y. (2013). Effects of storage period on the composition of whole crop wheat and corn silages. *Animal Feed Science and Technology* 185: 196–200.

Weinberg, Z.G. and Muck, R.E. (1996). New trends and opportunities in the development and use of inoculants for silage. *FEMS Microbiological Reviews* 19: 53–68.

Weinberg, Z.G. and Szakacs, G. (2018). Extraction of bio-active plant materials by simultaneous enzyme treatment and lactic acid fermentation – a mini review. *Israel Journal of Plant Sciences* 65: 31–37.

Weinberg, Z.G., Szakacs, G., Linden, J.C., and Tengerdy, R.P. (1990). Recovery of protein and chlorophyll from alfalfa by simultaneous lactic acid fermentation and enzyme hydrolysis (ENLAC). *Enzyme and Microbial Technology* 12: 921–925.

Weinberg, Z.G., Ashbell, G., Hen, Y. et al. (1993). The effect of applying lactic acid bacteria at ensiling on the aerobic stability of silages. *Journal of Applied Bacteriology* 75: 512–518.

Weinberg, Z.G., Akiri, B., Putievski, E. et al. (1999). Enhancement of polyphenol recovery from Rosemary (*Rosmarinus officinalis*) and Sage (*Salvia officinalis*) by enzyme-assisted ensiling. *Journal of Agricultural and Food Chemistry* 47: 2959–2962.

Weinberg, Z.G., Chen, Y., Miron, D. et al. (2011). Short communication: preservation of total mixed rations for dairy cows in bales wrapped with polyethylene stretch film – a commercial scale experiment. *Animal Feed Science and Technology* 164: 125–129.

Weissbach, F. and Ohff, R. (1979). Studies on nitrate breakdown in fodder conservation. *Archiv fur Tiernehrung* 29: 50. (in German).

Wilkinson, J.M., Bolsen, K.K., and Lin, C.J. (2003). History of silage. Chapter 1. In: *Silage Science and Technology* (ed. D.R. Buxton, R.E. Muck, and J.H. Harrison), 1–30. Madison, WI: Agronomy Monograph 42, American Society of Agronomy.

Woolford, M.K. (1984). *The Silage Fermentation*, Microbiological Series 14. New York, NY: Marcel Dekker.

Woolford, M.K. (1990). The detrimental effects of air on silage. *Journal of Applied Bacteriology* 68: 101–116.

Xu, D., Ding, N., Ke, W. et al. (2019). Modulation of metabolome and bacterial community in whole crop corn silage by inoculating with homofermentative *Lactobacillus plantarum* and heterofermentative *L. buchneri*. *Frontiers in Microbiology* 23: 3299–3214. https://doi.org/10.3389/fmicb.2018.03299.

Xu, D., Wang, N., Rinne, M. et al. (2020). The bacterial community and metabolome dynamics and their interactions modulate fermentation process of whole crop corn silage prepared with or without inoculants. *Microbial Biotechnology* Special Issue Article, Open access. 16p. https://doi.org/10.1111/1751-7915.13623.

Zhao, Y., Yu, J., Lin, J. et al. (2016). Material and microbial changes during corn silage and their effects on methane fermentation. *Bioresource Technology* 222: 89–99. https://doi.org/10.1016/j.biortech.2016.09.113.

SECTION V

A CLOSING PERSPECTIVE OF MICROBIAL FERMENTATION

CHAPTER 15

THE INTERSECTION OF MICROBIAL FERMENTATION AND EVOLUTIONARY ECOLOGY: WE PROVIDE THE HABITAT, YOU PROVIDE THE FERMENTATION

CHRISTON J. HURST[1,2]
[1]Cincinnati, OH, USA
[2]Universidad del Valle, Santiago de Cali, Valle del Cauca, Colombia

CONTENTS

15.1 INTRODUCTION TO THE COMPETITIVE ECOLOGY OF MICROBIAL FERMENTATION

Evolution is a game played by biological groups. The evolutionary progression that began with prokaryotic microbes and then continued onward to the creation of unicellular eukaryotic microbes, vascular plants, and vascular animals, has been a sequence of competitive efforts. The goal of each participating biological group has been to increase both the amount of physical territory and the number of ecological functions to which that group can establish a claim (Hurst 2021a). In keeping with the proposal by Charles Elton (1927), that physical territory is defined as a habitat. Within its habitat, the functions performed by a species collectively define the species' niche. This understanding of niche has been described as the Eltonian Niche. Each individual species must find a suitable habitat and successfully control some of the functions which are present within its habitat

Microbial Fermentations in Nature and as Designed Processes, First Edition. Edited by Christon J. Hurst.

because, otherwise, the species would lose its existence. When considering fermentation, the microbial habitats include rhizosphere soil, sections in an animal's gastrointestinal tract such as the rumen, wort for making beer, dough that is destined to become bread, and dairy curds that are to become cheese. The niches occupied by fermenting microbes within those habitats include solubilizing the phosphates in soil, decomposing lignocellulosic materials in rumen contents, creating ethanol as well as liberating carbon dioxide in either the wort or bread dough and producing lactic acid in the dairy curds.

A niche can be visualized as a hypothetical volume, or niche space, which has complex surface geometry (Hurst 2022). The relative volume of a species' niche space is directly related to the ability of that species to control ecosystem functions within its habitat. Controlling a greater amount of ecological functions can be visualized as controlling a larger volume of niche space. The surface geometry of a species niche space represents those variables and respective parameter ranges which define the symbiotic interactions of that species (Hurst 2022).

Ideally, each biological group would like to control all of the ecological functions within its habitat. As each biological group struggles to gain control over an ever greater amount of ecological functions, the group's selfish and possessive goals will help to drive its speciation whenever that action becomes either allowed or necessary. Speciation is the division of a biological species into daughter groups, each of which has greater specialization and is better fit to control specific niche activities within the available habitats. That perspective of biological goals accounts for the commonalities and variations between phylogenetically related microbes that perform similar functions. That perspective of evolutionary ecology also can be envisioned as representing the competitive efforts by each biological group to control for itself the greatest possible volume of niche space (Hurst 2022). A biological genus lays claim to the sum of all ecological functions and thus all the niche space controlled by its constituent species. At a larger taxonomic scale, a biological family lays claim to the sum of all ecological functions and thus all the niche space controlled by its genera.

Cooperative symbiotic interactions are established by partnering species (Fisher et al. 2017) and it has been suggested that a host microbiome is an ecosystem on a leash (Foster et al. 2017). Those symbiotic interactions have evolved and are maintained to enable the partners to have a better collective chance at competitively controlling niche activity. A host animal, plant, or microbe encourages its cooperative microbial symbionts by providing for those symbionts a habitat that is physically and chemically suitable. Assuring adequacy of the habitat can involve the host supplying necessities such as nutrients, maintaining an appropriate level of moisture, and assuring a pH that is appropriate for the desired niche activities of its intended microbial symbionts. The early history of scientific research on microbial symbiosis has been summarized by Hurst (2021b). Classic examples are the symbioses that create lichens (Frank 1877) and root nodules (Schindler 1884).

Each fermentative microbial function will have been honed by evolution to optimize energetic opportunities. We also must understand that a balance exists between beneficial versus detrimental perspectives for many of the microbial fermentation activities. For example, fermentative recycling processes, which have important roles in soil, sediments, and the gastrointestinal tract of animals, can seem detrimental when they contribute to disease in animals and plants.

It seems possible that using microbial fermentation to produce and preserve food items may represent a collaborative genetic selection process. We provide a habitat that is suitable for those microbes which are prevalent in spontaneous fermentations of food. Those microbes have become our symbiotic partners and would have selected for a human population that could survive ingesting the microbes. Similarly, those microbes which are spontaneously present in the silage materials that we produce as fodder for our ruminant livestock would have exerted selection pressure on our livestock. In cooperative response, we have learned to design food and silage fermentation processes in ways that would favor the growth and eventual ingestion of the most easily survivable microbial populations. That collaborative selection process has not been perfect as evidenced by the opportunistic pathogenicity associated with some of the microbes that exist in fermented food products and fermented fodder (Chapter 10 by Bourdichon et al.). Careful design of microbial starter cultures can help to assure the quality and safety of the produced fermentation products (Chapter 11 by Laranjo).

15.2 EXAMPLES OF SYMBIOTIC NICHE COOPERATIONS BETWEEN MICROBES AND THEIR ANIMAL HOSTS THAT DEPEND UPON MICROBIAL FERMENTATION

Lignocellulosic degradation is a microbial activity that exists widely in nature. The cooperative activities of lignocellulose degrading symbiont microbes housed in the stomachs of ruminant mammals, and the similarly functioning symbiont microbes housed in the cellulosic comb material of termites, represent a means by which those animals rely upon a fermentation service function performed by the participating microbes. By evolving to use fermentative microbial partners in these ways, the ruminants and termites have gained niche space that is associated with utilizing otherwise recalcitrant lignocellulosic materials as a valuable source of energy (Chapter 3 by McKenney, Chapter 4 by Ungerfeld et al., Chapter 5 by Goux et al.). Simultaneously, the symbiont microbes which achieve these fermentation processes gain a share in the niches occupied by their ruminant mammal and termite partners. We have engineered those natural processes and applied them for our composting of plant materials (Chapter 1, Section 1.8 pages by Hurst) and also for our use in anaerobic digestion (Chapter 1 Section 1.8 pages 91 by Hurst, Chapter 5 by Goux et al.).

Organic polyphosphates are valuably incorporated as phytates into plant seeds as a means of provisioning the new generation of plants with an initial source of phosphate. Ruminant mammals which ingest those seeds gain biological access to the phosphates by depending upon symbiont gastrointestinal microbes that produce phytases (Chapter 1, Section 1.2.3 pages 174 by Hurst). By evolving to use fermentative microbial partners in this way, the ruminants have gained a share in niche space that is associated with utilizing those recalcitrant organic polyphosphate materials. Simultaneously, the symbiont microbes gain a share in the niches occupied by those ruminant mammals. Humans use engineered microbial processes to produce phytases, and those phytases then are used as a food supplement for nonruminant livestock animals that do not have partnerships with phytase-producing symbiont microbes. Supplying phytase as a dietary supplement to those nonruminant livestock animals is a cooperative interaction by which humans expand the niches of those nonruminant livestock by allowing the nonruminants to utilize the otherwise inaccessible phytates contained in seed materials (Chapter 1, Section 1.2.3 pages 174 by Hurst). In that instance, humans are cooperative symbionts of the livestock.

15.3 EXAMPLES OF USING SYMBIOTIC NICHE COMPETITIONS THAT RELY UPON MICROBIAL FERMENTATION TO PRODUCE FOOD FOR HUMANS AND LIVESTOCK

Spontaneous fermentations begin with a mixture of naturally present microbial species that engage in niche competitions between themselves. Processes by which we use such microbial fermentations to create food and fodder must be designed to preferentially facilitate the growth and metabolic activity of those microbial constituents that will achieve our desired outcome. The oldest way of achieving those preferred outcomes is by establishing within the fermenting materials a set of habitat conditions that will allow our preferred microorganisms to have competitive superiority.

At the end of a competitive fermentation process, the initial variety of microbes may have been reduced to the point that only one or a very few microbial species will successfully demonstrate niche activity. That reduction in microbial species composition and activity often results from the eventually dominant microbes having become winners by modifying the material in ways that made an increasingly unsuitable habitat for the losers. Spontaneous fermentations still are used to prepare food for humans and are most common at the level of individual households. Commercial food fermentations increasingly are relying upon designed starter cultures which can offer greater food safety and greater constancy of product characteristics (Chapter 9 by Takeshita and Oda, Chapter 10 by Bourdichon et al., Chapter 11 by Laranjo, Chapter 13 by Gomathy et al.).

Production of alcoholic beverages (Chapter 6 by Liu) occurs in stages and largely begins as a competition among microbes whose niche activity allows them to produce low levels of ethanol by fermenting simple sugars present in the liquid (Chapter 7 by Thomas, Chapter 8 by Suther). The increasing concentration of microbial metabolites, particularly ethanol, which are present in the fermenting material changes that habitat until most of the initially active microbes no longer can continue their niche activities (Chapter 1, Section 1.5.4 pages 84–85 by Hurst). *Saccharomyces cerevisiae*, which has a greater tolerance for ethanol, can continue performing niche activities in the fermenting product until eventually imposing habitat changes that end its own niche activities (Chapter 1, Section 1.5.4 pages 84–85 by Hurst, Chapter 9 by Takeshita and Oda).

Creating cheese from milk relies upon beginning with a habitat that allows production of lactic acid from lactose during rapid fermentative microbial niche activity. The produced lactic acid quickly changes that habitat in ways that prevent spoilage microbes from gaining control. The niche activity of those lactic acid-producing microbes then is slowed by adding salt which changes the habitat. Further production of lactic acid can occur but happens only more slowly within that changed habitat. The changed habitat instead facilitates niche activity by other fermentative microbes which complete the development of the product (Chapter 1, Section 1.5.6 pages 85–87 by Hurst).

Fermentation pickling of fruits and vegetables depends upon those plant materials being submerged in a salt brine. The salt importantly creates a habit that desirably favors microbes that will produce lactic acid by fermenting sugars present in the plant materials. Without that salt, the submerged plant materials would be in a habitat that instead favors niche activity by microbes that cause spoilage (Chapter 1, Section 1.5.2 pages 82–83 by Hurst).

Silage is a fermentation pickling of plant materials that depends upon those materials being rapidly acidified by microbes which produce lactic acid by fermenting naturally present sugars. Without rapid acidification, other groups of microbes could achieve gains from niche competition within the fermenting material. Most undesirable among those other microbes are producers of butyric acid and fungal toxins. Successful rapid production of lactic acid changes the habitat in ways that suppress unwanted niche activity by those other microbes (Chapter 1, Section 1.6.2 pages 90 by Hurst, Chapter 14 by Weinberg).

15.4 EXAMPLES OF USING MICROBIAL FERMENTATION TO PRODUCE INDUSTRIAL CHEMICALS

We use fermenting microbes in a cooperative manner to produce a wide range of chemicals (Chapter 2 by Viñals et al.). These applications historically have been termed Industrial Fermentations, but that term easily can be misunderstood because the food products including beverages that I mentioned earlier in this chapter often are fermented on industrial levels. Fodder for our livestock animals tends to be fermented on agricultural lands where it will be used rather than sold. However, if those fermented animal feed materials are sold then their production certainly is commercial.

Often, the microbes used to produce chemicals have been genetically engineered for their specific purpose. That is a difference from the microbes used in food production which frequently have been genetically selected but not genetically engineered. The microbial activity that produces these desired chemical products often is not fermentative metabolism, but rather is either aerobic respiration or anaerobic respiration. By tradition, the processes will be described as a fermentation (Chapter 1, Section 1.9 pages 91–96 by Hurst).

Ethanol and hydrogen destined to be burned as fuels are two examples of the substances that begin as microbial fermentation products. For marketing purposes, those two chemical products often are termed bioethanol and biohydrogen. An extensive catalog of microbially generated enzymes useful as detergent ingredients is represented in the marketed consumer products of wealthier countries.

The food industry has developed numerous applications for microbial products. Those include using compounds that are relied upon in food processing, compounds used as food additives and preservatives, and microbially produced materials that get used in food packaging.

Numerous microbial products are used as biochemical intermediates and chemical intermediates for non-food related industrial applications. These variously have applications in the leather, metal, paint, perfume, plastic, textile, polymer production, and woodworking industries.

Many of the dietary supplements that we use for ourselves and for our domesticated animals are made by microbial fermentations. There also is a broad range of pharmaceuticals whose manufacturing has been based upon microbial fermentation. Their uses include treatments for angina, cataract prevention, diabetes, hypercholesterolemia, hypertension, immunosuppressants, anticancer compounds, and antimicrobials.

Last in the listing, but not least in importance, are the biological insecticides which we produce using industrial microbial fermentations.

15.5 SUMMARY

Microbial fermentation processes are ecological tools. The wide range of symbiotic associations that have been constructed based upon use of those tools very notably includes gastrointestinal fermentation by ruminant mammals. Many groups of termites accomplish a similar process by using symbiont microbes housed in their cellulosic comb material to perform degradation of lignocellulosic materials. The termites subsequently consume those fermented comb materials. Humans use the same general concept when relying upon symbiont microbes to fermentatively produce food items that subsequently will be consumed either by us or by our livestock animals. We have extended that concept even further by using microbial fermentations to produce industrial chemicals and biochemicals, including pharmaceuticals. The microbes used to produce those industrial chemicals also can be perceived as our symbionts. When considering all of these symbiotic interactions, we could describe them by stating to the participant microbes "We provide the habitat, you provide the fermentation."

REFERENCES

Elton, C.S. (1927). *Animal Ecology*. New York.: Macmillan.

Fisher, R.M., Henry, L.M., Cornwallis, C.K. et al. (2017). The evolution of host-symbiont dependence. *Nature Communications* 8: 15973. https://doi.org/10.1038/ncomms15973.

Foster, K., Schluter, J., Coyte, K. et al. (2017). The evolution of the host microbiome as an ecosystem on a leash. *Nature* 548: 43–51. https://doi.org/10.1038/nature23292.

Frank, A.B. (1877). Über die biologischen Verkältnisse des Thallus einiger Krustflechten. *Beiträge zur Biologie der Pflanzen* 2: 123–200.

Hurst, C.J. (2021a). Discovering the symbiotic nature of microbial life: summarizing milestone publications from 1866 through 1947. In: *Microbes: The Foundation Stone of the Biosphere, Advances in Environmental Microbiology*, vol. 8 (ed. C.J. Hurst), 213–241. Basel, Switzerland: Springer International Publishing AG.

Hurst, C.J. (2021b). The game of evolution is won by competitive cheating. In: *Microbes: The Foundation Stone of the Biosphere, Advances in Environmental Microbiology*, vol. 8 (ed. C.J. Hurst), 545–593. Basel, Switzerland: Springer International Publishing AG.

Hurst, C.J. (2022). Ecosystems function like interlocking puzzles: visually interpreting the concept of niche space plus a brief tour through genetic hyperspace. In: *Assessing the Microbiological Health of Ecosystems* (ed. C.J. Hurst), 1–29. Chichester, United Kingdom: Wiley.

Schindler, F. (1884). Zur Kenntniss der Wurzelknöllchen der Papilionaceen. *Botanisches Centralblatt* 5 (18): 84–88. https://babel.hathitrust.org/cgi/pt?id=uc1.c079535293&view=1up&seq=110.

INDEX

Microbial Fermentations in Nature and as Designed Processes, First Edition. Edited by Christon J. Hurst.

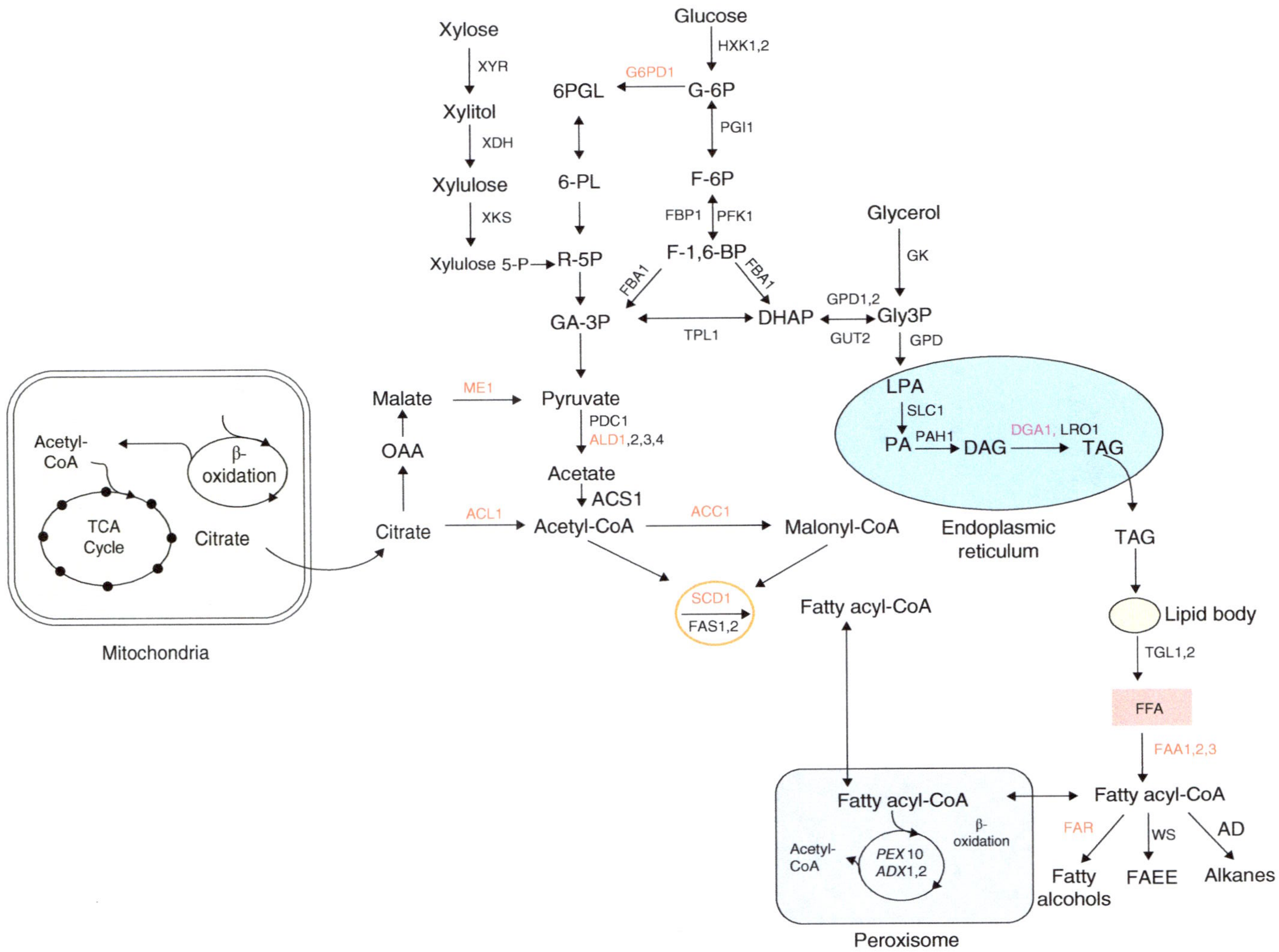

FIGURE 2.1 Metabolic pathways of *Rhodosporidium toruloides*. The metabolites connected by arrows (a single arrow represents a step) and involved enzymes (named in the center of the arrow) are represented. Abbreviations are shown for the commonly used names of metabolites and enzymes. The endoplasmic reticulum, mitochondria, lipid body, and peroxisome are also present as cellular compartments. The enzymes with names in green coincide with genes that have been previously defined. FA is an abbreviation for fatty acid; FAME is an abbreviation for fatty acid methyl ester; FFA is an abbreviation for free fatty acid; FAEE is an abbreviation for fatty acid ethyl ester; TAG is an abbreviation for triacylglycerol, and VLFA is an abbreviation for very long chain fatty acid.

Microbial Fermentations in Nature and as Designed Processes, First Edition. Edited by Christon J. Hurst.

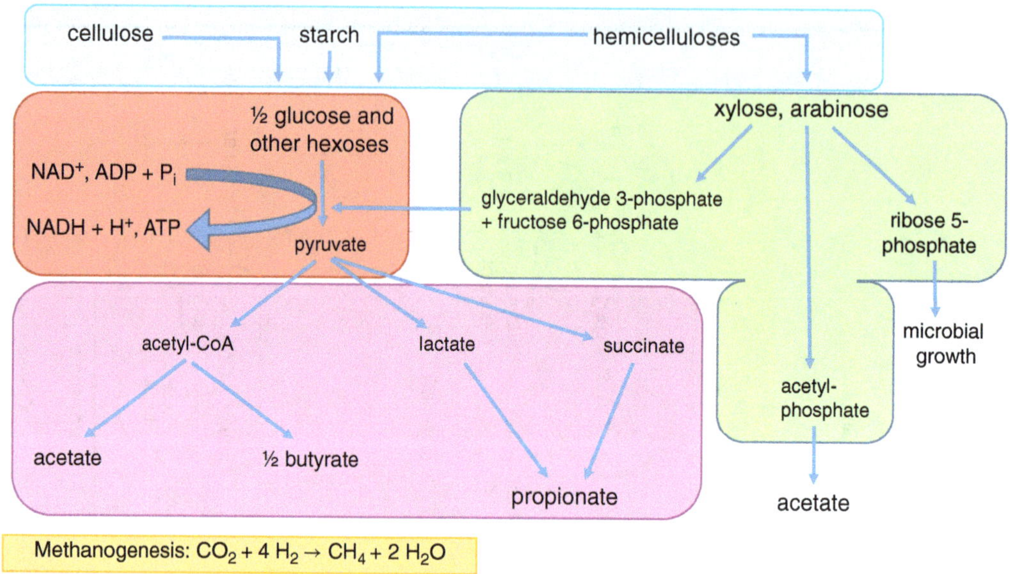

FIGURE 4.1 Overall scheme of rumen digestion of polysaccharides (inside of sky blue squircle), fermentation of hexoses and pentoses to pyruvate via glycolysis (inside of red squircle), pyruvate metabolism to volatile fatty acids (inside of purple squircle), pentose cycle and phosphoketolase cleavage (inside of green squircle), and hydrogenotrophic methanogenesis (inside of orange squircle).

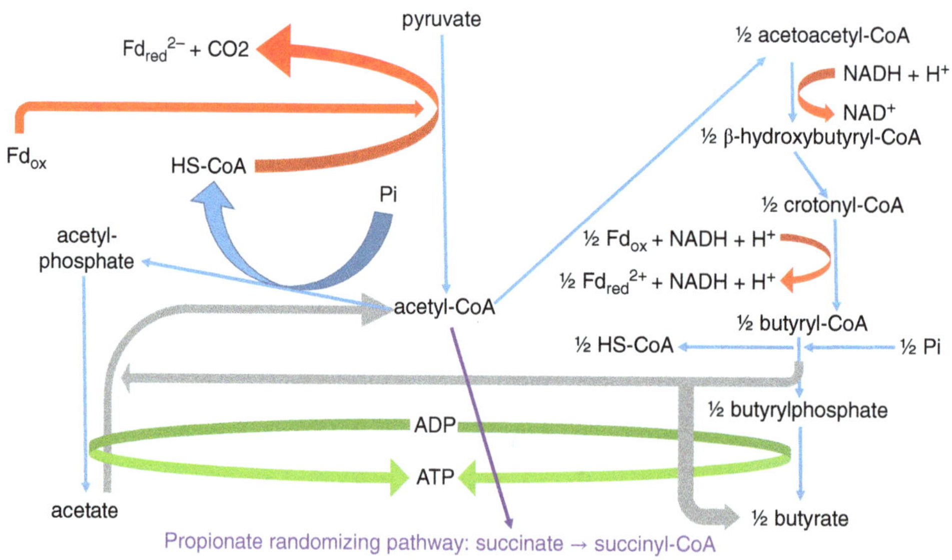

FIGURE 4.2 Metabolism of pyruvate to acetate and butyrate. Main routes of carbon compounds are marked with sky blue arrows. Alternative pathway involving butyryl-CoA: acetate CoA transferase is shown in gray arrows. Link with propionate randomizing pathway involving succinyl-CoA: acetate CoA transferase is shown in purple. Generation of ATP by substrate-level phosphorylation is shown with light green arrows. Reactions of cofactor reduction are shown in red arrows. Abbreviations: ADP, adenosine diphosphate; ATP, adenosine triphosphate; CO_2, carbon dioxide; Fe_{ox}, oxidized ferredoxin; Fe_{red}^{2-}, reduced ferredoxin; H^+, proton; HS-CoA, coenzyme A; NAD^+, oxidized nicotinamide adenine dinucleotide; NADH, reduced nicotinamide adenine dinucleotide; P_i, phosphate group.

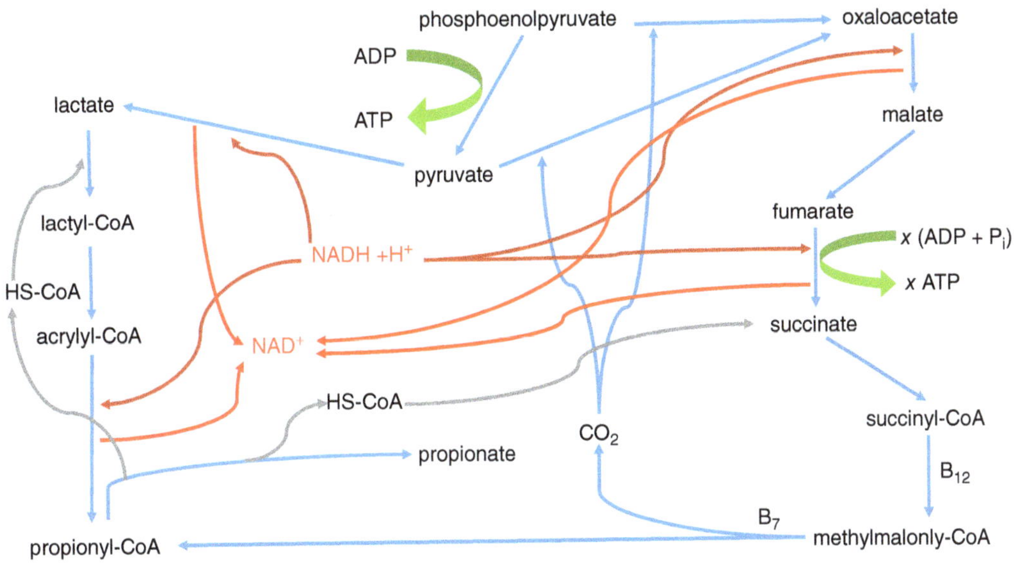

FIGURE 4.3 Propionate formation via randomizing (succinate) and non-randomizing (acrylate) pathways. Main routes of carbon compounds are marked with sky blue arrows. Reactions involving transfer of coenzyme A are shown with gray arrows. Generation of ATP by substrate level and electron transport-linked phosphorylation is shown with light green arrows. Reactions of cofactor re-oxidation are shown in red arrows. Abbreviations: ADP, adenosine diphosphate; ATP, adenosine triphosphate; B_7, biotin; B_{12}, cobalamin; CO_2, carbon dioxide; H^+, proton; HS-CoA, coenzyme A; NAD^+, oxidized nicotinamide adenine dinucleotide; NADH, reduced nicotinamide adenine dinucleotide; P_i, phosphate group.

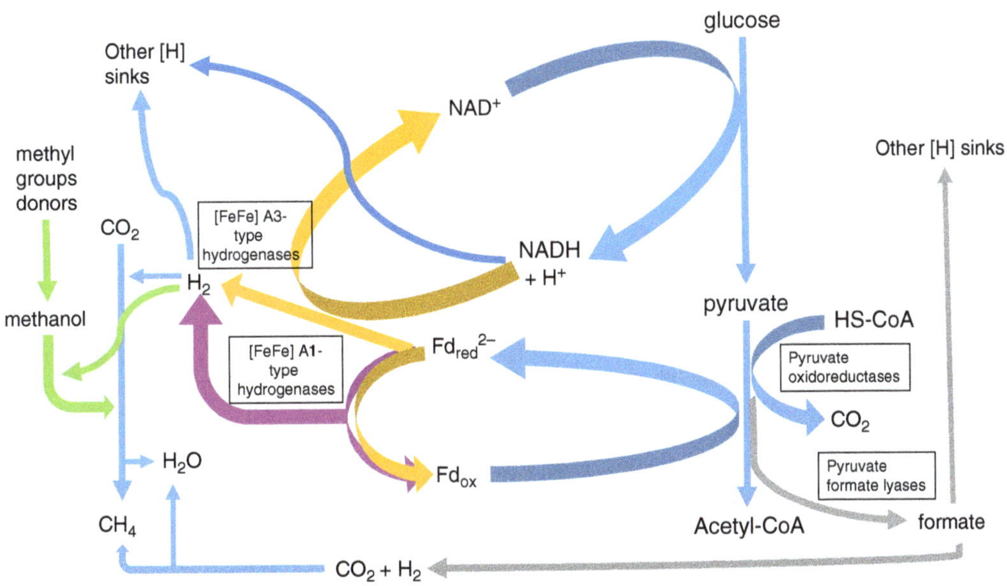

FIGURE 4.4 Main pathways of metabolic hydrogen transactions in rumen fermentation. Reduction of NAD^+ to NADH in glycolysis and of ferredoxin in pyruvate oxidative decarboxylation is depicted in sky blue arrows. Formate formation in pyruvate oxidative decarboxylation and dissociation into carbon dioxide (CO_2) and dihydrogen (H_2) previously to its incorporation into methane (CH_4) and other electron sinks is shown in gray arrows. Under low dihydrogen concentration NADH is re-oxidized by [FeFe] A3 type hydrogenases in confurcation with reduced ferredoxin (Fd_{red}^{2-}; yellow arrows). Under relatively high dihydrogen concentration NADH donates electrons in intracellular reactions and reduced ferredoxin is re-oxidized to its oxidized species (Fd_{ox}) by fermentative, dihydrogen-evolving [FeFe] A1 type hydrogenases (purple arrows). Hydrogenotrophic methanogenesis incorporates dihydrogen and carbon dioxide (sky blue arrows). Incorporation of metabolic hydrogen into methyl-reducing and fermentative methylotrophic methanogenesis is depicted in light green arrows.

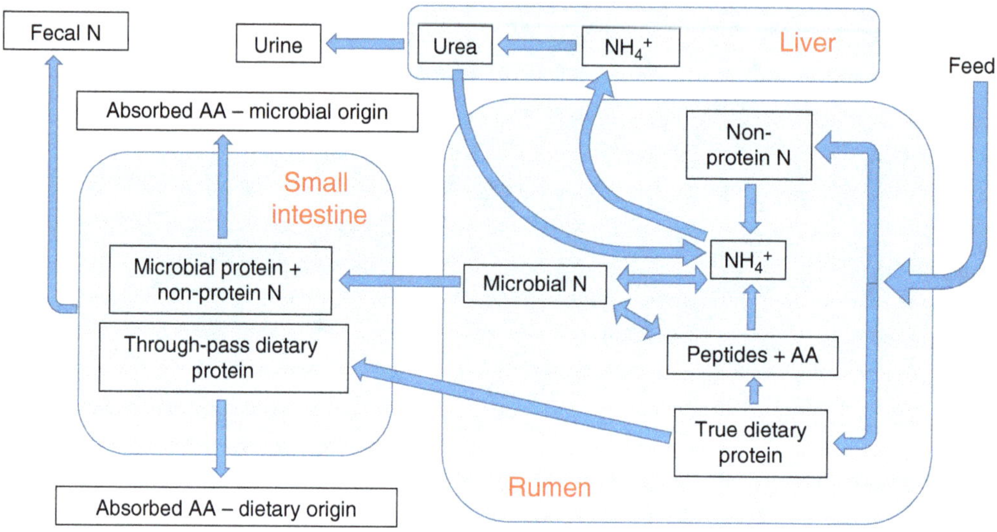

FIGURE 4.5 Simplified scheme of rumen, small intestine, and liver nitrogen metabolism. Endogenous protein is not depicted for simplicity. Liver synthesized urea is recycled to the rumen through the rumen wall and in saliva (not depicted for simplicity). Abbreviations: AA, amino acids; N, nitrogen; NH_4^+, ammonium.

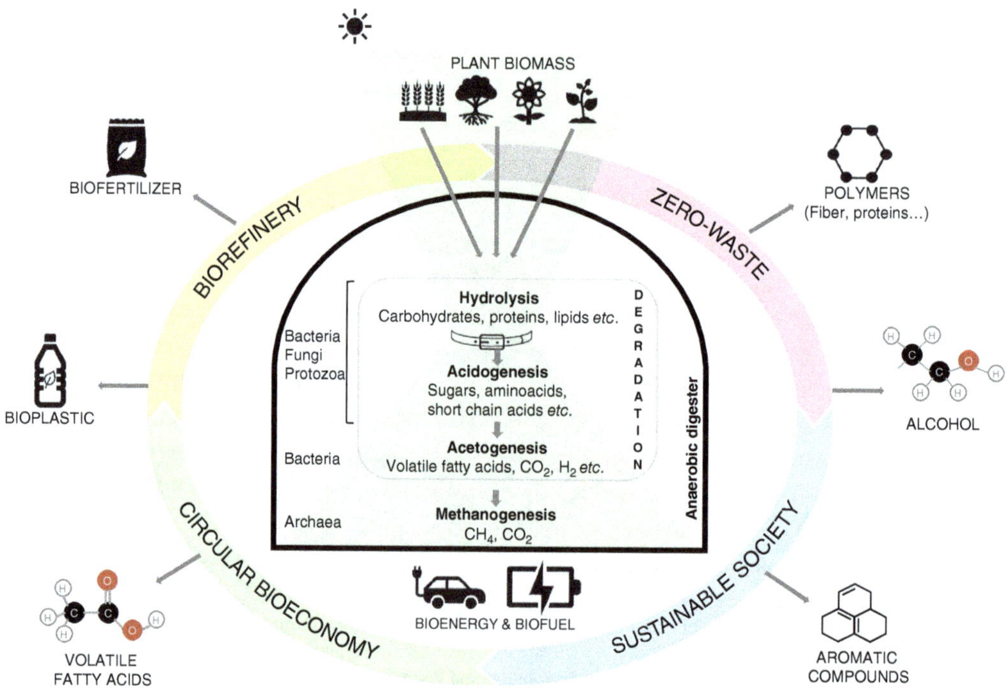

FIGURE 5.1 Schematic overview of the anaerobic digestion of plant biomass and its position within the lignocellulose biorefinery concept.

(a)

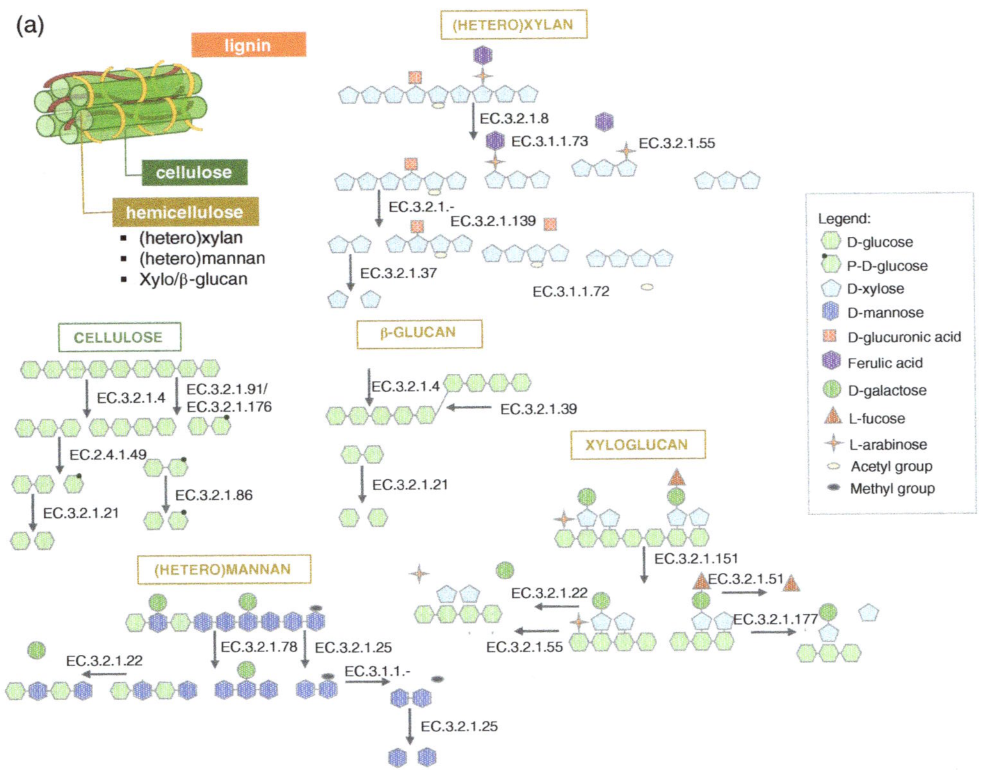

(b)

EC number	Function	Main CAZyme families
3.2.1.4	Endo-β-1,4-glucanase	GH5, 6, 7, 8, 9, 10, 12, 44, 45, 48, 51, 74, 124
3.2.1.91 3.2.1.176	Exo-β-1,4-glucanase	GH5, 6, 7, 9, 48
3.2.1.21 3.2.1.86	(phosphor)β-glucosidase	GH1, 3, 4, 5, 9, 30, 116
2.4.1.49	Cellodextrin phosphorylase	GH94
3.2.1.8	Endo-β-1,4-xylanase	GH5, 8, 10, 11, 30, 43, 51, 98, 141
3.2.1.37	β-xylosidase	GH1, 2, 3, 30, 39, 43, 51, 52, 54, 116, 120
3.1.1.72	Acetylxylan esterase	CE1, 2, 3, 4, 5, 6, 7, 12
3.1.1.73	Feruloyl esterase	CE 1
3.2.1.139	α-glucuronidase	GH 4, 67
3.2.1.55	α-L-arabinofuranosidase	GH 2, 3, 43, 51, 54, 62
3.2.1.78	endo-β-1,4-mannanase	GH 5, 26, 45, 113, 134
3.2.1.25	β-mannosidase	GH 1, 2, 5, 164
3.2.1.22	α-galactosidase	GH 4, 27, 31, 36, 57, 97, 110

FIGURE 5.2 Simplified structure of the main lignocellulosic polysaccharides and their enzymatic degradation mechanisms (a). Lignin degradation pathway is omitted here. Classification of main enzymes involved in lignocellulosic polysaccharides degradation (b). EC, enzyme commission number; CAZyme, carbohydrate-active enzyme; GH, glycoside hydrolase; CE, carbohydrate esterase.

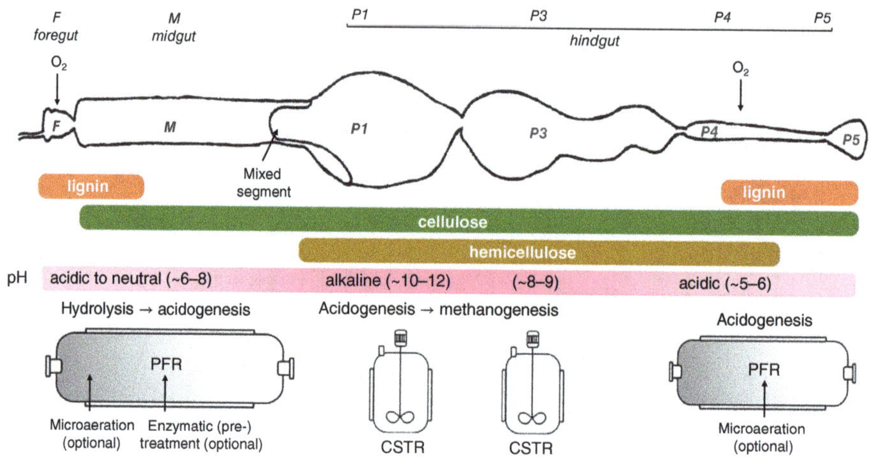

FIGURE 5.3 Characterization of the termite gut system and proposed exploitation of its potential toward improved lignocellulose utilization in engineered systems. Figure inspired by the work from Godon et al. (2013) and further modified. The highly compartmented gut model is characteristic of soil feeding termites, e.g. *Labiotermes* sp. The gut compartments involved in the degradation of the main lignocellulose fractions were identified after Marynowska et al. (2022). PFR, plug flow reactor; CSTR, continuously stirred tank reactor.

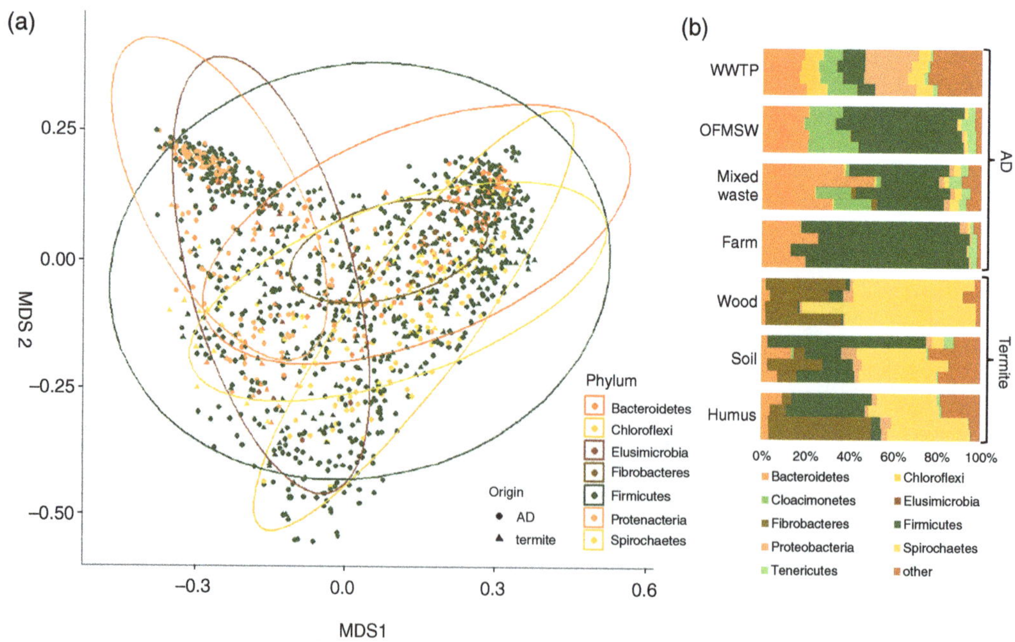

FIGURE 5.4 Comparison of microbial communities and their polysaccharide degrading potential in the anaerobic digestion (AD) reactors and the termite gut (Termite) system. Multidimensional scaling (MDS) of the calculated Jaccard similarity coefficient comparing the genomic content of the carbohydrate active enzymes (CAZymes) encoded in microbial genomes from the two systems (a). This subfigure was compiled based on microbial genomes generated in previous studies (Campanaro et al. 2020; Hervé et al. 2020). Main bacterial phyla present in the different types of AD reactors (i.e. alimented with different feedstocks), and higher termites feeding on distinct substrates (b). This subfigure was compiled based on (Calusinska et al. 2018b; Marynowska et al. 2020). WWTP, wastewater treatment plant, OFMSW, organic fraction of municipal solid waste.

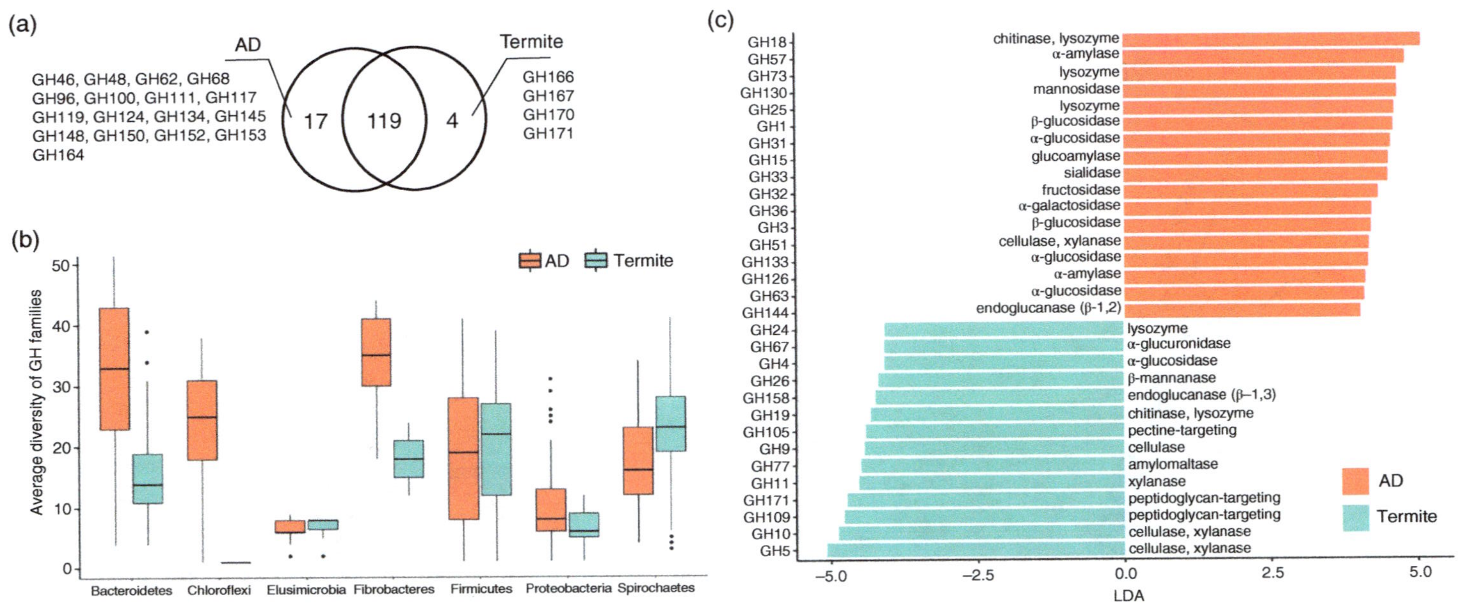

FIGURE 5.5 Glucoside hydrolases in microbial genomes from the anaerobic digestion reactors (AD) and the termite gut system (Termite). (a) Shared diversity of glucoside hydrolase (GH) families in bacterial genomes from AD and the termite gut systems. (b) Average GH family diversity per microbial genome (only genomes with >70% completeness and <10% contamination were used), displayed for the main bacterial phyla. All pair-wise comparisons are statistically significant (t.test, $p<0.005$). (c) Linear discriminant analysis (LDA) of GH gene families differentially enriched in microbial genomes (community-wise) in the two systems. This figure was compiled based on microbial genomes generated in previously published studies (Campanaro et al. 2020; Hervé et al. 2020).

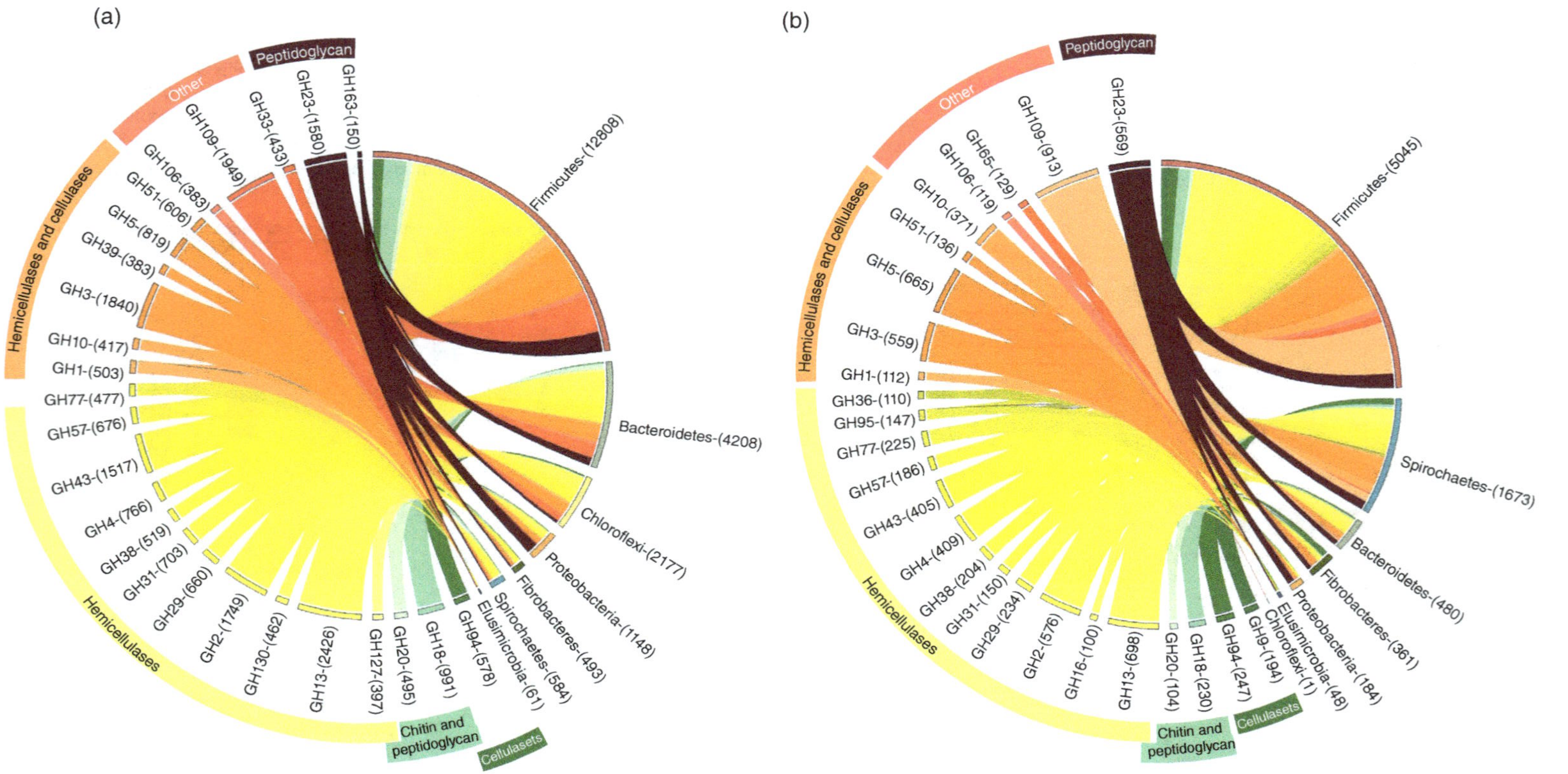

FIGURE 5.6 Comparison of the main glycoside hydrolase (GH) families targeting the different lignocellulose fractions and other non-lignocellulosic polysaccharides and their phylogenetic distribution among major bacterial phyla, for the anaerobic digestion reactors (a) and the termite gut system (b). This figure was compiled based on microbial genomes generated in previously published studies (Campanaro et al. 2020; Hervé et al. 2020).

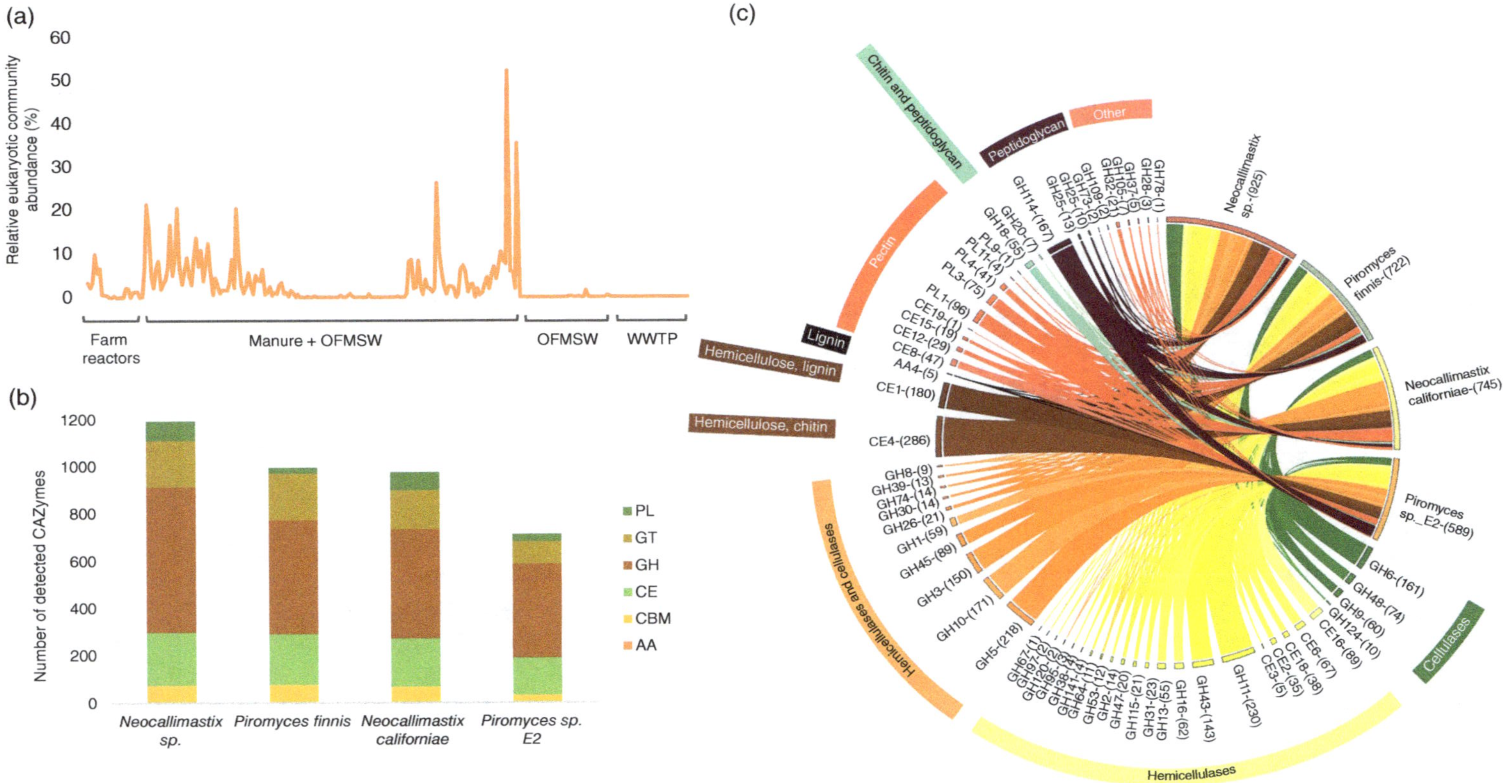

FIGURE 5.7 Anaerobic fungi in full-scale anaerobic digestion systems (a) and their lignocellulolytic potential (b and c). OFMSW, organic fraction of the municipal solid waste; WWTP, wastewater treatment plant; PL, polysaccharide lyase; GT, glycosyl transferase; GH, glycoside hydrolase; CE, carbohydrate esterase; CBM, carbohydrate-binding module; AA, auxiliary enzyme. This figure was compiled based on own, unpublished data (a), and relying on the analysis of Neocallimastigomycota genomes available in public databases (b and c).

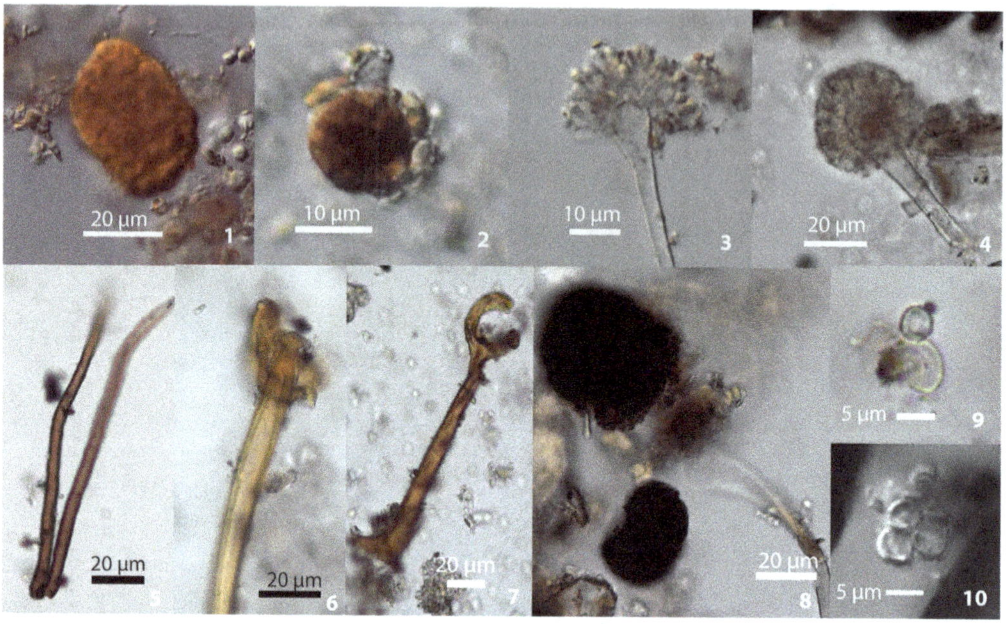

FIGURE 6.1 Examples of ancient fungal elements found on Neolithic pottery vessels in China. 1, 2: *Monascus* cleistothecia; 3, 4: *Aspergillus* vesicles; 5: cf. *Rhizopus* hyphae; 6: *Rhizopus* sporangium, showing structure of collarette and columella; 7: *Mucor* sporangium connecting to hypha; 8: *Rhizopus* or *Mucor* sporangia; 9: yeast cells in budding position; 10: a cluster of yeast cells. Source: These images appear courtesy of the author.

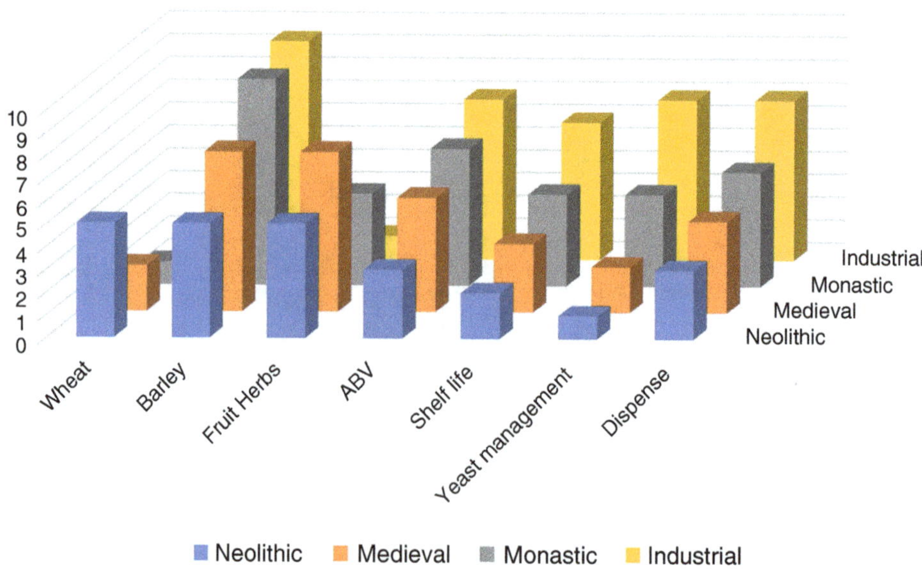

FIGURE 7.2 Profiles of major features of brewing through periods of development. Wheat – proportion used as a major ingredient: barley — proportion used as a major ingredient; fruit & herbs — prevalence in recipes, particularly as gruit; ABV — general level of alcohol achieved: shelf life — likely drinkable period (days to months); yeast management — awareness of value of yeast repatching; dispense — sophistication of draught and packaged product.

FIGURE 8.2 Allagash Brewing Company aging barrels. "CS" refers to "Coolship." Apple refers to the fresh pomades added while lavender refers to the lavender-smoked malt used. Beer was poured into barrels to begin aging on March 27th, 2019, November 20th, 2020, or January 4th, 2022. Source: These images appear courtesy of the author.

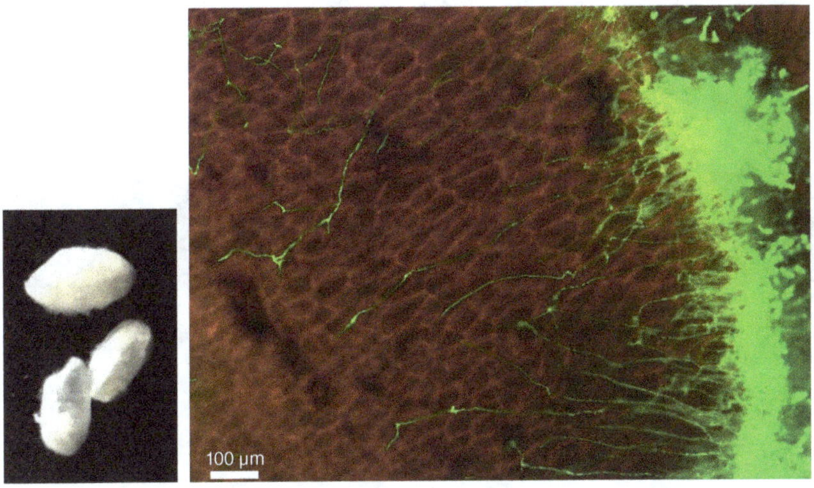

FIGURE 9.2 Haze komi and invasive growth of *A. oryzae* (green) into steamed rice (red).

FIGURE 9.3 A tank to create the fermentation mash and fresh sale.

FIGURE 12.3 *Uapaka kirkiana* or wild loquat (*mushuku/muzhanje*); *Ziziphus mauritiana* in an open informal market.

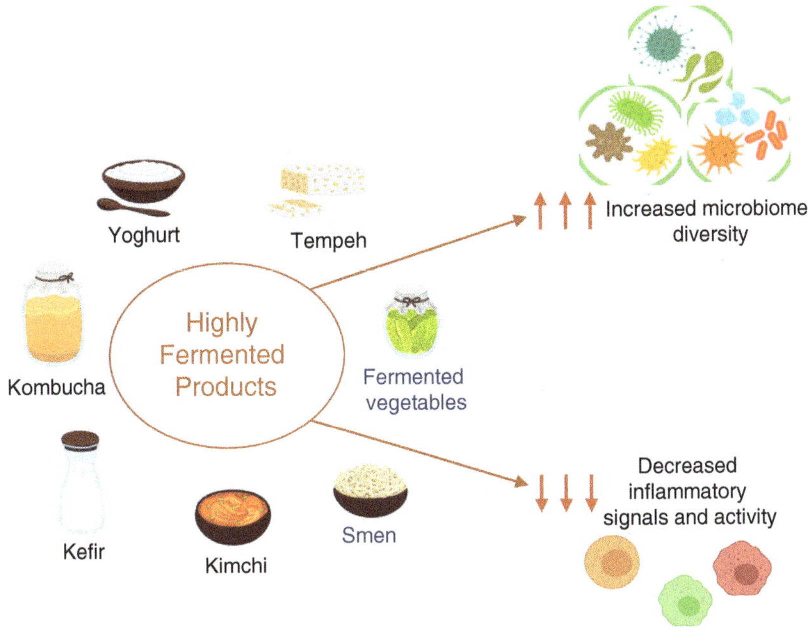

FIGURE 12.4 A diet high in fermented food or beverage decreases inflammatory proteins, reduces molecular signs of inflammation, and improves immune response. Therefore, improving digestive health with a high-quality fermented diet may improve immunity against infections and diseases.

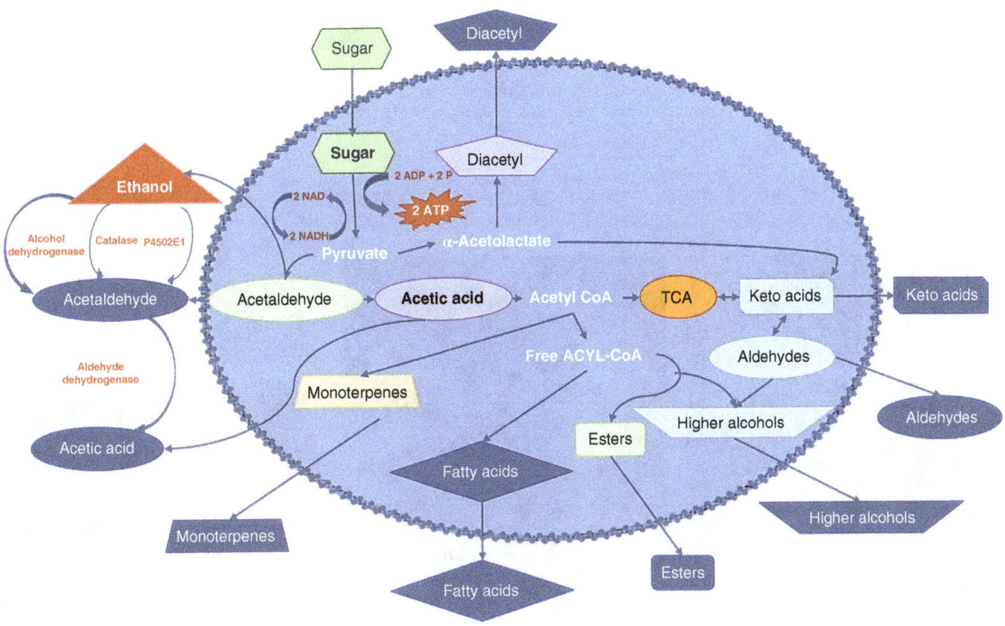

FIGURE 12.5 Mitochondrial yeast metabolism, maltose being the best sugar source. Secondary metabolites and metabolism intermediates are shown in ethanol formation.

FIGURE 12.11 Traditional preparation of amarula drink ("ocanhu") in Mozambique.

FIGURE 12.16 Palm wine created from the sap of various species of palm trees; hibiscus wine.

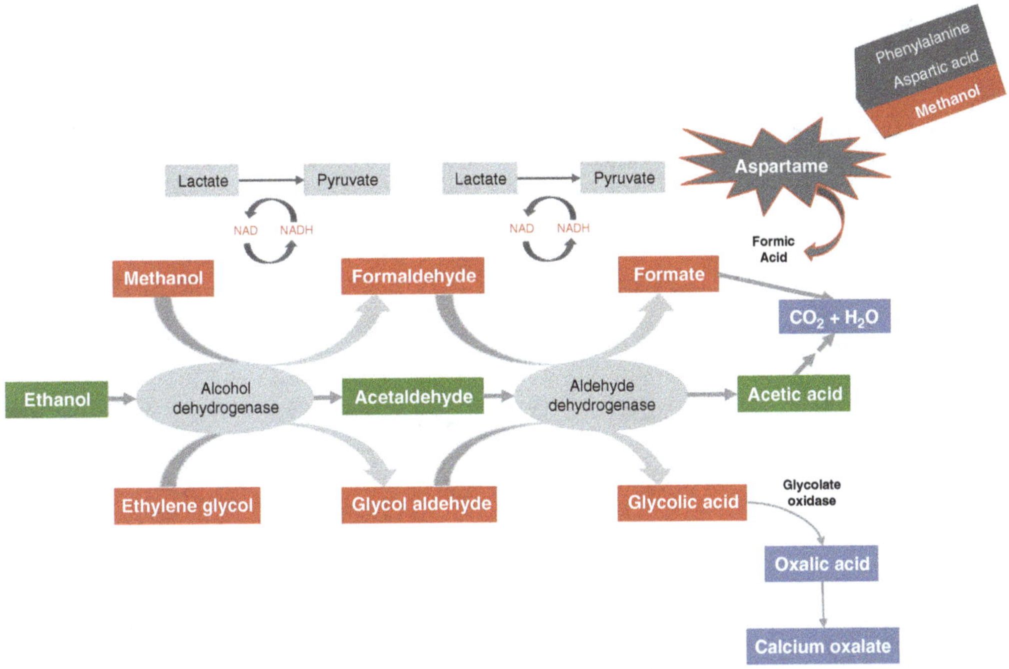

FIGURE 12.17 Aspartame, a well-known artificial sweetener of sodas, candy, chewing gum, and energy drinks, is a synthetic chemical composed of the amino acids phenylalanine and aspartic acid, with a methyl ester. The methanol produced by the metabolism of aspartame is absorbed and quickly converted into formaldehyde and then completely oxidized to formic acid. The acceptable daily intake (ADI) for aspartame is 40 mg/kg of body weight. Ethanol, drinking alcohol, is mainly produced from maize fermentation. Ethylene glycol, used as antifreeze, if ingested is rapidly absorbed from the gastrointestinal tract and slowly absorbed through the skin or lungs.

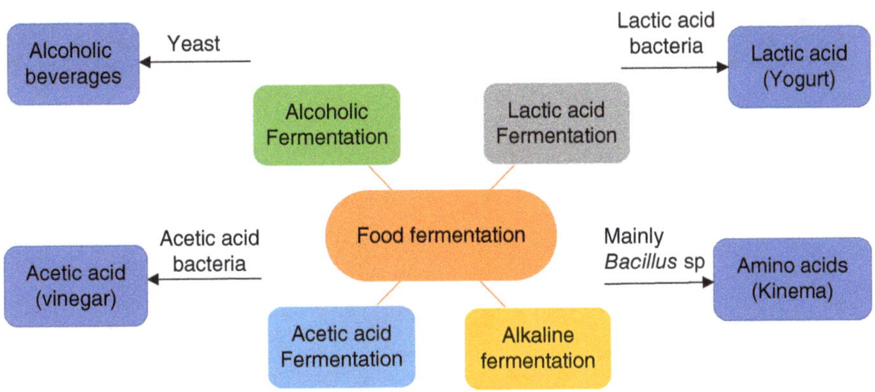

FIGURE 13.1 Types of food fermentation.

FIGURE 13.5 Fermented foods of Sri Lanka.

FIGURE 14.1 Some forage crops for ensiling.

FIGURE 14.2 The major steps of silage making.

FIGURE 14.3 Assessing corn grain maturity by milk line. The Milk Line (the border between the deep yellow and the milky parts) proceeds from the external perimeter toward the center of the cobs as the plant matures (A and B, respectively). Source: Courtesy of Kable Thurlow, Grazing Educator & Beef, Michigan State University Extension.

FIGURE 14.5 Ideal (left) vs. poor sealing (right).

FIGURE 14.6 Silage silo types. Source: The tower silo image, courtesy of Hanson Silo Company.

FIGURE 14.7 Anaerobic mini-silos (Weck™, Wehr-Offlingen, Germany). The springs which secure the lids enable gas release from inside only but not air penetration.

www.ingramcontent.com/pod-product-compliance
Lightning Source LLC
LaVergne TN
LVHW062214200926
848484LV00006B/1390
9781119849971